A Guide To Transformer Maintenance

A Guide To Transformer Maintenance

By

S. D. Myers

J. J. Kelly

R. H. Parrish

TMI®

Transformer Maintenance Institute,
Division, S. D. Myers, Inc.
Akron, Ohio

DIVISION OF SD Myers TRANSFORMER MAINTENANCE INSTITUTE

Acts 4:12

Library of Congress Catalog Card Number: 81-50169.
ISBN 0-939320-00-2

Printed in the United States of America

FOREWORD

Transformer Maintenance Institute's new book **A Guide to Transformer Maintenance** does a superior job on filling a long existing need for the transformer owner and operator. This book is of **major** importance when one considers such things as the present and future cost of replacing apparatus, the rising cost of transformer oil, and the complex problems associated with existing askarel (PCB) transformers.

To my knowledge only one other book, F.M. Clark's **Insulating Materials for Design and Engineering Practice**, published in 1962, has covered part of this area of concern. Much that was written then is now out of date.

As a Consultant to TMI, I have spent much time over a period of months on this project in reviewing the materials, editing and offering suggestions. The material reads easily; the book is written in language that can be both understood by non-electrical oriented personnel and yet technical enough to interest the electrical engineer in charge of transformers.

The bottom line today is saving money and preventing power outages, thus, preventing loss of service or production. **A Guide to Transformer Maintenance** is a **must** for the transformer owner/operator.

Edward L. Raab

Credentials of E.L. Raab

Currently, E.L. Raab is a consultant on liquid and solid insulations for power transformers. He is a fellow of American Society for Testing and Materials; a member of D-27 Committee on Electrical Insulation Liquids and Gases and its past chairman; member D-9 Committee on Solid Insulating Materials; senior member of American Chemical Society and Institute of Electrical and Electronics Engineers; past chairman of the Insulating Fluids Subcommittee of the Transformers Committee; past chief U.S. delegate to International Electrotechnical Commission Committee 10 on liquids and gases; past chairman of National Electrical Manufacturers Association Technical Oil Committee and American National Standards Institute C-107 on askarels, and so on.

Mr. Raab worked for the General Electric Company, Pittsfield, Mass., for over 40 years on research, development and application of liquid and solid insulations for transformers; first, as a development chemist and eventually as manager of the materials laboratory. Ed Raab, a graduate of Wesleyan University with a BA and MA in chemistry, has thus become a recognized authority, as he has authored more than 50 technical articles concerning insulating materials, especially liquid insulations.

PREFACE

The authors have written this book for the simple reason that in their long, long search for such a book as this, there was none to be found. But what they did find was a wealth of information from here and there, buried in technical presentations or dissertations on various and related subjects. As early as 1950, Stanley D. Myers, a graduate Electrical Engineer from the Class of 1944, University of Akron (Ohio), began accumulating this information in his files. Because the field of preserving and maintaining existing electrical equipment was of special interest to Mr. Myers, he became very active in soliciting information from all sources. Many of the articles that are in his possession came from unknown sources, with this simple notation—"Stan, I think you will be interested in this article."

Once Mr. Myers started his own business, the only clear vision he had was to clean transformer oil which had badly deteriorated, and to do this work while the transformer was energized.

But as time progressed it became apparent that while cleaning transformer oil was a good thing to do, there were better things to do. What were the better things? Why, cleaning from the transformer the oil decay products that saturated the paper insulation, as well as the sludges that were deposited throughout the transformer. A dirty transformer with clean oil (by either the oil reclamation process or oil replacement) remains an economic liability.

Mr. Myers worked out the procedures required to desludge energized transformers and over a period of 12 years has had the opportunity of witnessing the effectiveness of this new concept.

The one thing that continues to amaze the authors is the unending revelation of **new** information coming to their attention as they continue their study into this subject. Just about the time they feel there are no new horizons, another appears.

Mr. Myers, recognizing this fact, established the TMI (Transformer Maintenance Institute) division of S. D. Myers, Inc., to continue the exploration of this subject. TMI® is advertised as "The world's only learning and training center for in-service transformers." TMI has been holding seminars since 1975, both in its Akron, Ohio facilities and in various parts of the U.S.A. The comment of one of those attending a seminar was: "You owe it to industry to get this information out to them as fast as possible."

Ergo, The Book.

Individual contributions to this endeavor by the three stated authors should be recognized.

Stanley D. Myers, E.E.

Mr. Myers is an entrepreneur in the classic definition of the term. Although he has a degree in electrical engineering, he is really a salesman at heart. Any good salesman has in him a "teacher complex"—that is, he loves to research a subject, learn the obvious, and seek out the unknown factors required to master the subject. He recognized the difference between a good teacher and a good salesman is that the salesman asks for an order. He also has recognized that while an *academic* grasp of a subject is desirable, the *practical* application of the subject remains the ultimate goal. Where a chemist would recognize a concept, Stan visualized a service or a business opportunity. Therefore, throughout this book you will find the basic concepts of transformer maintenance as established by Mr. Myers in his own company. They appear as engineering and chemical concepts, as referenced in the selected bibliography. This in turn forms only a small part of TMI's reference library.

Some of the "firsts" established by S. D. Myers, Inc., most of which did not meet with ready acceptance, but have now become the standards of the industry, are:

Establishment of the first service in the USA of testing of transformer oils at the job site, with a MTL®—Mobile Test

Laboratory—(see Chapter 3, Part 1).

Proper classification of transformer oil using ACID and IFT test results and M.I.N. number (Myers Index Number ™ or oil quality index number—see Chapter 3, Part 2).

Establishment of the concept of transformer core and coil cleaning using SLUDGPURG® procedures (see Chapter 7, Part 3).

In conjunction with the MTL program, establishment of a rigorous 3 months' training program to put qualified men in the customer's plant to not only conduct the tests properly, but to spend whatever time is required to explain all the test procedures. In addition, and this was a big plus, to interpret the test results so that the customer could understand what all of those "funny numbers" mean.

The establishment of a training and resource institute to speak specifically to the maintenance of "in-service" transformers. To understand the boldness and courage of this move you need to understand the philosophy that existed in industry and utilities at the time this step was taken. The prevailing concept, heard hundreds—no thousands of times, as spoken by knowledgeable and influential people, was—"We haven't even bought a barrel of transformer oil in the past 20 years—why should we be concerned about our transformers? They just keep working away and I'm just going to let them alone." To undertake and underwrite the expense of setting up a training institute, to research thousands of articles, to find authoritative information, and to educate industry and utilities required a vision.

Mr. Myers has been extremely fortunate in pulling together a team of men who have a technical background to understand the concepts—and enough native curiosity to say, "I'm not willing to accept all of these ideas just because Mr. Myers says it; I want to know for myself." Using the resource materials already gathered by Mr. Myers, reading "The Consultors" that Stan had already written to promote the concept, the search began. These men exhausted all of the resources for technical information, employed a library research team to find additional materials, joined and became active in such societies as IEEE, ASTM, Doble Engineering Client Committees, ANSI, AAAS and others, called outside consultants from different parts of the country, and thus left no stone unturned.

Joseph J. Kelly

Mr. Kelly, Director of TMI, is a graduate of John Carroll University, Cleveland, Ohio. He is an active member of the Institute of Electrical and Electronics Engineers (IEEE), Power Engineering (PES), Industrial Applications (IAS), and Electrical Insulation Societies (EIS). He is also an active member of the American Society of Testing and Materials (ASTM).

Joe joined S.D. Myers, Inc., in 1972 after 13 years' experience in the chemical research center of a major rubber firm in Akron, Ohio.

Training in chemistry has given Mr. Kelly a different perspective of a transformer than the engineer. His conviction that a transformer can be defined as a chemical reactor that just happens to transform electricity is very evident throughout the book.

After having spent a couple of years in an intensive literature search he had the opportunity of studying in detail the engineering reports of over 100,000 transformers. Diagnosing the problem areas, making corrective recommendations, and reviewing the results after the corrective action was completed has made Mr. Kelly an expert in his field.

Mr. Kelly has both written and presented papers for *Plant Engineering®*, *Transactions of IEEE Industrial Applications Society*, Doble Engineering Client Conferences, as well as biennial IEEE/NEMA Electrical/Electronics Insulation Conferences (E/EIC).

Robert H. Parrish, E.E.

Mr. Parrish, Senior Technical Writer, TMI, holds a BSEE from the University of Akron, 1962. He is an active member of IEEE Power Engineering and Professional Communication Societies, American Association for the Advancement of Science (AAAS), American Scientific Affilation, as well as a charter member of the Association of Energy Engineers (in the company's effort concerning oil conservation).

Bob, with TMI since 1976, has been a writer for 17 years, with 11 years spent in the electrical/electronics field. He was employed as an aerospace circuit design engineer following graduation from Akron University.

Mr. Parrish has authored or co-authored many technical articles for *Plant Engineering®*, *Electric Light and Power*, as well as presentations before IEEE/IAS and E/EIC. His 1978 article entitled "Warning: TSCA May Be Hazardous To Your Professional Health" was voted as the best article that year by *Plant Engineering®*. International Correspondence School (ICS), Scranton, Penn., retained him in 1980 to update its home study text—"Transformers."

Though an engineer by training, Parrish is a research scientist and environmentalist by temperament. Throughout the writing of this book, his overriding goal has been to make an accurate, authoritative, and objective presentation, leaving to you—the transformer owner/operator—the final decision as to which way to go on a given problem.

Mr. Myers thus has taken the advice given to him following one of the TMI 3-day seminars conducted in Chicago, Illinois, "You have an obligation to get this material out to industry so they can take care of their transformers." Yet when all the major publishers of engineering books reviewed an outline of this book, they did not express any interest in publishing it. They did not see a ready market. With this scenario, Mr. Myers said, "If they don't want to print the book, we will publish it ourselves." After four years of research and writing, rewriting, reviewing, editing, and proofreading, the book has gone to press. Only time will tell who made the wrong decision!

The book's contents generally follow TMI's basic 4-day seminar outline. Portions of a number of chapters originally appeared as articles in trade magazines or as presentations given before technical societies. These articles have been edited to bring them up to date. For this reason, periodic supplements will be published to keep the reader aware of the latest transformer state-of-the-art.

Even when discussion centers in Chapter 1, Part 2, on transformer design, it is with equipment maintenance in mind, and how to relate design factors to practical construction and life expectancy. The primary theme throughout the book focuses on maintenance of the oil-filled transformer's "insulation system." For numerous authorities have stated that the **life** of the transformer is the **life** of the insulation system.

S.D. Myers
J.J. Kelly
R.H. Parrish

DISCLAIMER

A Guide to Transformer Maintenance is designed to provide accurate and authoritative information in regard to the subject matter covered. By using this maintenance guide the user agrees that Transformer Maintenance Institute, Div. S.D. Myers, Inc., and its authors will not be held liable for any damages resulting directly from the use of information contained herein or from any errors or omissions.

ACKNOWLEDGEMENTS

The authors gratefully acknowledge the assistance and cooperation given to them in the writing of this book.

First of all, they express their sincere thanks to Edward L. Raab (retired from the General Electric Co., Pittsfield, Mass.), for his work as a consultant in the capacity of reviewing and editing the full manuscript text; and to Ron Cook, Ross Gunther, Dr. Silas Kasambira, and Karl Stutz for input from their respective areas of expertise.

Ron Cook, an active member of the National Association of Corrosion Engineers (NACE) and ASTM, along with 10 years' experience in preventive maintenance of electrical apparatus, worked on various drafts of Chapter 8, Parts 2 and 3; Ross Gunther, a former educator and currently a TMI Instructor, wrote much of Chapter 10, as well as editing several other chapters; Karl Stutz, an active member of IEEE and ASTM and a field service engineer for seven years, wrote the original drafts for Chapters 5 and 8 (Parts 1 and 4); Dr. Silas Kasambira, also an educator, focused on the historical aspects of Chapters 1 and 2, subsequent drafts in Chapter 8, and pulled together the Chapter Selected Bibliographies.

In addition, the authors offer special thanks to those manufacturers and others who provided data or information, or supplied photographs or illustrations. They include:

Academic Press, Inc.
AGA Corp.
Allied Chemical Corp.
Barnes Engineering Co.
BES, Inc.
James G. Biddle Co.
Bonneville Power Administration
Chemical Week Magazine (McGraw-Hill)
Doble Engineering Co.
Dow Corning Corp.
Edison Electric Institute
Englehard Minerals and Chemical Corp.
Factory Mutual Engineering Corp.
Ferranti-Packard Transformers, Ltd., Toronto, Canada
Foster Transformers, Ltd., London, England
GEC Transformers, Ltd., Stafford, England
General Electric Co., (Pittsfield, Mass; Rome, Ga.)
GTE Sylvania, Hampton, Va.
H. K. Porter Co., Inc.
Hayden Books

Hartford Steam Boiler Inspection and Insurance Co.
Hughes Aircraft Corp.
Hipotronics, Inc.
Kemper Insurance Companies
McGraw-Edison Co., Power Systems Div.
McGraw-Hill Publishers
Micanite & Insulators Co., Ltd., Manchester, England
Mikron Instrument Co.
Moloney Electric Co.
Monsanto Chemical Co.
Ohio Brass Co.
Ohio Edison Co.
Panametrics, Inc.
Photovolt Corp.
Plant Engineering® Magazine (Technical Publishing Co.)
Power Magazine (McGraw-Hill)
RTE Corp.
Science Magazine
The Smithsonian Institution
Square D Co., Clearwater, Fla.
Thermotest Infra-Red Surveys
Trafo-Union, Stuttgart, West Germany
Transmission & Distribution Magazine (Cleworth Publishing)
Westinghouse Electric Corp., (Muncie, In; Sharon, Penn.)
John Wiley and Sons, Inc., Publishers
Xedar Corp.

Edward L. Raab

Ronald L. Cook

Ross O. Gunther

Dr. Silas T. Kasambira, Jr.

Karl E. Stutz

Finally, last but by no means the least, the authors express heartfelt thanks to those who transformed handwritten manuscripts into this published book.

Mr. Joe Stanley, Production Manager
Miss Margaret Griffith, Artist/Layout
Miss Berit Lane, Artist/Layout
Mr. Lorin Alexander, Artist
Mrs. Sheila Stropoli, Typesetting
Mrs. Debbie Lister, Typesetting
Miss Loretta Johnson, Proofreading
Mrs. Mary Jo Curley, Typing
Mrs. Mary Frank, Typing
Mrs. Jean Bell, Typing
Mrs. Anne Dargie, Typing

CONTENTS

Chapter 5 continued on next page

Chapter 6

Basic Protective Maintenance Practice for Oil-Insulated Transformers.................433

Chapter 7

Corrective Maintenance Procedures for Oil-Insulated Transformers 471

Chapter 8

Preventing Unscheduled Substation Outages........609

Chapter 9

What To Do About Askarel (PCB) Transformers (Past, Present, and Future) 691

Chapter 1

The Transformer—The Heart of Transmission and Distribution

Part 1—Historical Survey of Transformer Development and Application

Introduction

The engrossing history of the transformer dates back to the year of our Lord 1455, and the invention of the printing press. Printing fostered the birth of the modern electrical age through *written* communication of basic knowledge. Thus, Luigi Galvani published in 1791 results of an experiment with dead frogs' legs which he labeled, "animal electricity."

Only 12 copies of Galvani's book were printed because "he couldn't think of any more people who would be interested in this strange experiment." One copy was sent to Allesandro Volta, a physicist who followed up on the experiments, concluding:

> ...That two dissimilar metals in contact with the frog's nerve were necessary to make it twitch and that the nerve was only acting as a sensitive detector of the phenomenon.

This unusual event marked the discovery of primitive electromagnetism and led to Volta's invention of the voltage cell ("or voltaic pile") about 1800. Man had thus found how to create a new kind of electricity—an electric current which could be directed wherever wires would lead it in contrast to random accumulation of static electricity (Figure 1.1).

Following the discovery of electromagnetism and the development of the first battery, scientists from various academic disciplines invented many electrical devices, one of which—the transformer—has become the cornerstone of the electrical power industry.

Today, transformer technology continues to respond to societies' changing needs. Among other things, a reliable energy supply demands more than ever trouble-free transformer operation that is environmentally acceptable.

Figure 1.1 - Dr. Bern Dibner (industrialist, inventor, and science historian) at the handle of an electrostatic generator, the "oldest electrical device in the western hemisphere." The only practical use of "static" electricity was the lightning rod developed by Benjamin Franklin. (Courtesy of ***Science Magazine****).*

Much can be done to insure maximum transformer life. Protective maintenance (consisting of preventive and predictive, as well as corrective) remains the key. Nevertheless, an understanding of much background information is essential. As progressive steps in the history of transformers unfold, this technological dialogue will, undoubtedly, arouse your interest.

The Role of Transformers in Today's Economy

The commercial introduction of the transformer in 1885 overcame many technical and economic problems associated with the electrical industry. Before the invention of the transformer, electrical power was distributed as direct current (dc) at low voltage. This limited its use to the few urban areas where a large number of consumers could be supplied by relatively short distribution circuits. With dc power, costs of changing from one voltage to another became so prohibitive that such power became an obvious handicap to the budding electrical industry.

About the same time that the discovery of alternating current (ac) opened the way for a great future in the electrical power industry, successful demonstration using the first transformer took place at Great Barrington, Massachusetts in 1886.

At last it became feasible to generate electrical power at locations best suited for generator stations and economically transmit it to many load demands in rural areas hundreds of miles away.

Today, approximately 35 million transformers are operating in the United States. Their never ending job of stepping up and down the voltage used in every household and industrial concern has won them a unique role in our modern world. In a very real sense, the transformer has become the heart of an electrical substation. If it were not for this relatively quiet, sphinx-like device performing its timeless sleight of hand function, modern society today would not be experiencing an era of electrical revolution.

Essentials to Generate Electrical Voltage

Fundamental Laws of Electromagnetics

The evolution of the transformer dates back to the early 1800's when many scientists were performing experiments with electromagnetism. For example, in 1819, Hans Christian Oersted, a Danish physicist, was conducting an experiment using one of Volta's new batteries. By passing electricity through a wire, he discovered that a compass needle was deflected to the north by a nearby magnet (Figure 1.2). Oersted's wire experiment (Figure 1.3) demonstrated the relationship between electricity and magnetism.

Shortly thereafter, it was discovered that a looped wire produces a more intense magnetic field, as illustrated in Figure 1.4. The coiled wire resembles a "bar magnet" in its magnetic action on a compass needle and in the pattern of lines of force generated. Both of the foregoing progressive steps illustrate the *first law of magnetics*—a current-carrying conductor produces a magnetic field.

In 1820, Dominique Arago, a French physicist/astronomer, was conducting experiments similar to Oersted's. By passing an electric current through copper, he was able to reveal the effects of magnetism in soft iron. The same year, A.M. Ampere (a French mathematician) performed experiments that led to the development of the galvanometer by J.S. Schweigger (a German professor of physics and chemistry). The year 1825 saw the invention of the electromagnet (Figure 1.5) by William Sturgeon (an English self-taught physics teacher), an invention which was later perfected by Joseph Henry (an American physicist) between 1829 and 1832.

When a coil is wrapped around a bar of iron, the magnetic field becomes still more intense. Thus, a simple electromagnet is a current-carrying coil of wire with an iron core (Figure 1.6).

The electromagnet illustrates the *second law of magnetics*—a current-carrying conductor surrounding a stationary magnetic field produces an alternating voltage.

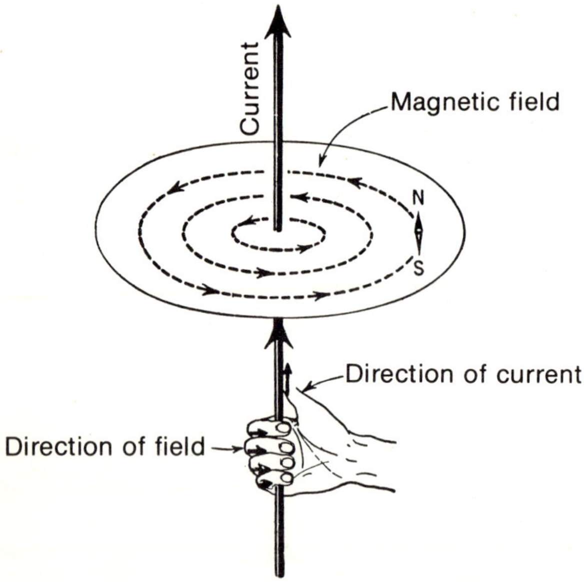

Figure 1.2 - A current produces a magnetic field; the right-hand thumb rule shows the direction of the current and the field.

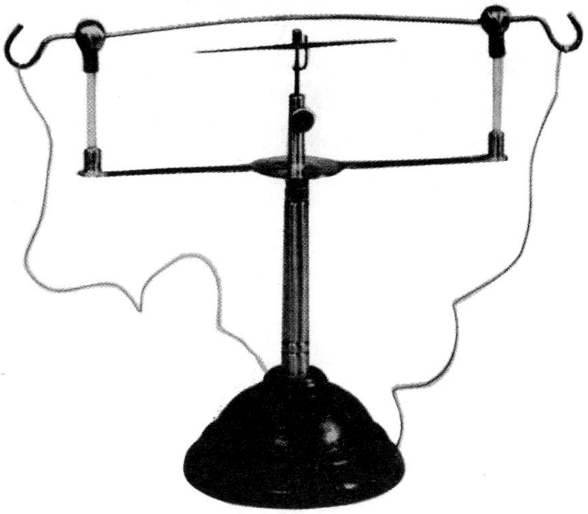

Figure 1.3 - A model of Oersted's wire experiment. (Courtesy of the Smithsonian Institution. USNM 47106-A; Smithsonian photo 31405-A).

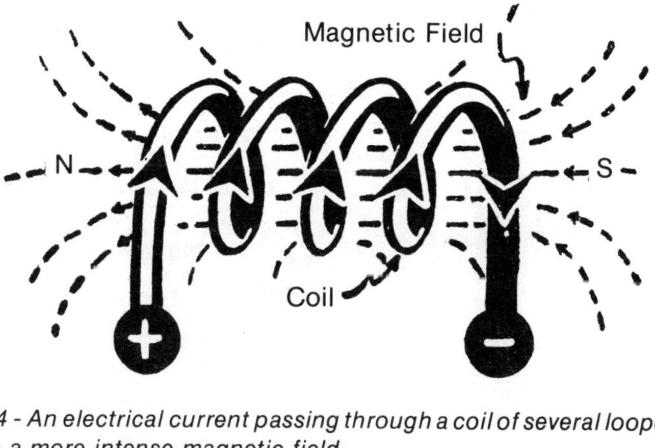

Figure 1.4 - An electrical current passing through a coil of several looped wires results in a more intense magnetic field.

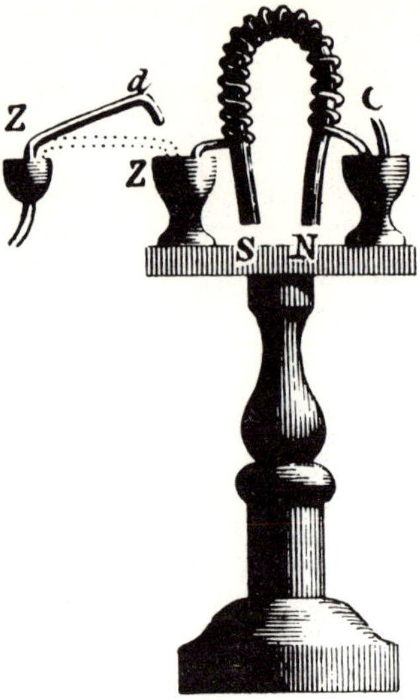

Figure 1.5 - Sturgeon's first electromagnet. (Courtesy of the Smithsonian Institution. From ***Transactions of the Society for the Encouragement of the Arts, Manufactures and Commerce****, 1824, vol. 43, plate 3).*

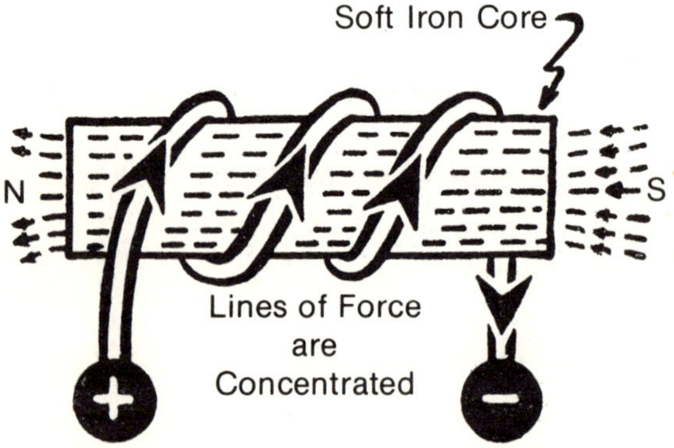

Figure 1.6 - An elementary electromagnet. Magnetic lines of force are dotted.

Primary Elements of a Transformer

Such was the state of the art concerning electricity when, between 1831 and 1832, two physicists—Sir Michael Faraday, at the Royal Institution in London, England, and Joseph Henry, at the Albany (New York) Academy in the United States—independently discovered the principle of electromagnetic induction by combining the two basic laws of magnetics. In so doing, they uncovered the primitive transformer concept—an iron core upon which are wound *two windings* (primary and secondary) suitably located with respect to each other and coupled together by mutual inductance (Figure 1.7).

Electrical energy with one voltage-amperage ratio can be converted or transformed into electrical energy of a different voltage-amperage ratio. Changing the magnetic field occurs when you deliberately vary the intensity of the current.

Since the flow of current in a looped wire produces a magnetic field, the ac produces a field which repeatedly expands and collapses, reversing polarity with each reversal of direction of the current. When this field cuts a conductor, a voltage is induced in it.

Faraday discovered that a current of electricity flowing in a coil of wire around a piece of soft iron would convert the iron into a magnet and that if this magnet were put into another coil of wire, a galvanometer connected to the terminals of the second coil would be deflected (Figure 1.8). Faraday's device (a primitive pulse transformer) demonstrated the principle on which transformers operate.

"As a fellow-laborer in the cause of American science,"* Charles G. Page developed primitive motors, an induction coil, and the first autotransformer, when he discovered that sparks could be produced between any tappings of a coil if its circuit is interrupted at any point (Figure 1.9).

Development of Modern Power Distribution Systems

For nearly half a century after the discovery of electromagnetic induction, the idea had "no practical value." No thought was given to the use of a "transformer" as a means for economical distribution of power.

At last, the concept did develop. In 1876, there were several important events that gave birth to the electrical industry. Among the highlights of the U.S. Centennial Exposition held in Philadelphia was Alexander

**As addressed by Joseph Henry.*

Graham Bell's exhibition of the telephone and two dc generators. In the same year at the Paris Exposition, Paul N. Yablochkov, a Russian electrical engineer, described a method of "breaking down" electrical energy by means of induction coils.

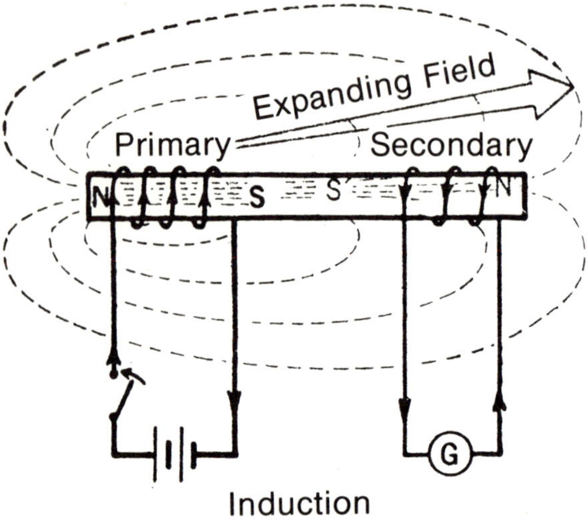

Figure 1.7 - The principle of electromagnetic induction applied. As the switch is closed, current builds up on the primary. This increasing current produces an expanding magnetic field which "sweeps" across the turns of the secondary, inducing a current in the secondary.

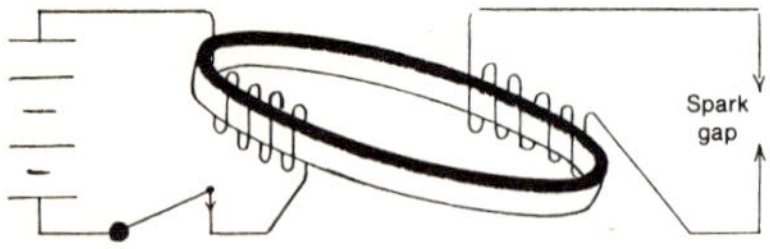

Figure 1.8 - M. Faraday had no real concept that this iron toroid ring device (1831) was a primitive transformer.

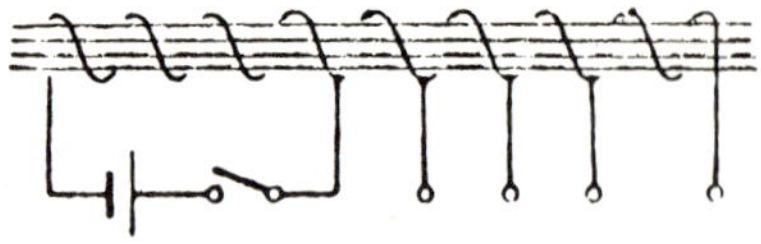

Figure 1.9 - C.G. Page's primitive autotransformer (1836).

In 1878, the Edison Electric Light Company was formed to support Edison's incandescent lamp research. This led to Edison's invention of the first practical incandescent lamp, and the development of his first dynamo. In 1879, Edison perfected the electric light.

In 1882, a Frenchman, Lucien Gaulard, and an Englishman, John D. Gibbs, were granted an English patent for an ac distribution system that used devices called "secondary generators" (transformers) to step up and step down voltages (Figure 1.10). The Gaulard-Gibbs electrical system, which had an open core, was first demonstrated in 1882 at the small Electrical Exhibition in the Westminster Aquarium in London, England, and at an exhibition in Turin, Italy, the following year.

Among the visitors at the Turin Exhibition of 1884 were three young Hungarian electrical engineers, Max Deri, Otto Blathy and Karl Zipernowsky. They recognized the limitations of the open iron core of the Gaulard-Gibbs system. Returning home, they modified the system and obtained two Austrian patents in 1885. That same year, the three engineers obtained the "privilege" of employing the term *transformer* for the first time (Figure 1.11).*

The Budapest Exhibition of 1885 afforded Messrs. Deri, Blathy, and Zipernowsky the first real opportunity to demonstrate the practical use of their transformer. The exhibition had far reaching effects, for it caught the curiosity of one important American, George Westinghouse, inventor of the railway air brake.

Westinghouse's genius immediately recognized the potentiality of the new device and system and acquired an option. He had several of the "secondary generators" shipped to his Union Switch and Signal Company where he commenced to redesign these units mechanically and electrically. In January of 1886, the Westinghouse Electric Company was formed to manufacture and promote the ac equipment. The following month, Westinghouse acquired the American rights to the Gaulard-Gibbs patents, and subsequently commissioned his chief electical engineer, William Stanley, to further develop the transformer (Figure 1.12).

Stanley worked relentlessly on the system before he moved to Great Barrington, Massachusetts for health reasons. There, he set up his own laboratory in an abandoned rubber mill and was ready to demonstrate the ac system on a practical scale. Thus:

**Figure 1.11 is located in the special Color Section, following Chapter 10.*

Figure 1.10 - Gaulard and Gibbs' "secondary generator," forerunner to the transformer. The original British patent is dated 1882. (Courtesy of the Smithsonian Institution).

Figure 1.12 - Stanley and his original "converter," now known as a shell-type transformer. (Courtesy of General Electric Co., ***Men and Volts****).*

> With some of the transformers he had built in Pittsburgh and others made at Great Barrington, and using the elm trees lining the quiet streets to carry his wires, he put together a system that served 13 stores, two hotels, two doctors' offices, a barbershop and the telephone and post offices. With testing details carried out while construction progressed, all was quickly ready for the grand public display. On the evening of March 20, 1886, the "big" 150-cp lamps were turned on to the admiration of the townsfolk crowding the streets and stores.*

The success of this demonstration led Westinghouse and his associates to establish the ac system on a commmercial scale. By September of 1886, the ac system had been tried in Lawrenceville, Pennsylvania, four miles from the Westinghouse plant, and had also been installed in Buffalo, New York, in November of the same year.

At this stage of the game, ac systems seemed to offer economical advantages until they were challenged by proponents of the dc system. Among the ac opponents were Thomas Edison and his associates who sought legislation to prohibit use of "the deadly alternating current."

The War of the Currents

Although the ac system demonstrated many economic advantages over the dc system, it still had two limitations—no practical motor, limiting its use to lighting, and no meter for measuring alternating current consumption.

The rapid electrical advances achieved by the Westinghouse Electric Company and William Stanley were not without controversy as to whether ac or dc current systems should be selected to meet U.S. electrical demands.†

AC Power Systems Gain the Upper Hand

The debate over the pros and cons of ac and dc systems which had started in 1886 came to a head in 1888. The Edison forces launched a direct attack on Westinghouse's ac system. They focused their efforts

**The first (dry type) transformers were rated as "20 lamp units," "40 lamp units," and so forth, in contrast to 20 kilovoltampere (KVA) units today.*

†William Stanley established his own company, SKF Transformer Co., in 1890. It was purchased by the General Electric Co. in 1903.

on a single issue—safety. The final shot at the ac system came when the Edisonians cited the first execution in New York's ac electric chair (1890) as an example of the dangers of ac. Throughout these attacks, the Westinghouse camp chose to ignore the challenge, and concentrated on perfecting the two shortcomings of the ac system.

In searching for solutions to ac's limitations, Elihu Thomson, an American professor of chemistry and mechanics, carried out extensive experiments with ac magnetic repulsion which he employed to produce a pioneering induction motor.* It was one of Thomson's demonstration lectures that challenged a young Yugoslav scientist, Nikola Tesla.

Tesla, upon being hired by Westinghouse, commenced on experiments to improve the ac system, including use of electricity by the utilization of two or more alternating currents—what later came to be known as the polyphase system.

O.B. Shallenberger, another Westinghouse engineer, accidentally discovered the principle that would later serve as a basis for measurement of ac energy consumption.

By the end of the 1880's, it became apparent that the status of the alternating system was gaining the upper hand over dc. By the summer of 1890, the advantage in the "battle of the currents" was in the ac court. In 1892, after overcoming a series of developmental problems, the Westinghouse Electric Company finally succeeded in marketing the Tesla motor.

A number of victories followed, in Westinghouse's favor. The decision in 1894 to employ the ac system to transmit energy from Niagara Falls to Buffalo, New York (22 miles), represents the high water mark of nineteenth century research and development and essentially sentenced dc transmission to the status of a dormant technology.

DC Technology Stages a Comeback

Early dc transmission operated at low voltages, so the system could not be employed for long distance transmission because of prohibitive copper costs. In addition, numerous power central stations had to be located very near consumers, resulting in the high cost of real estate and securing fuel and water supplies.

**One of Thomson's 632 patents, which also includes the use of mineral oil as a transformer cooling/insulating medium, obtained in 1887. Only Thomas Edison had more patents.*

Despite these limitations, revealed in the early years of the electrical power industry, high voltage direct current (HVDC) is emerging today in some cases as an economical way to transmit electricity. Edison may not have been wrong after all! Solid state equipment and laser technology make it possible to send dc current efficiently over great distances. The first U.S. HVDC system went on-line in 1977. It transmits power 456 miles from Center, North Dakota, to Duluth, Minnesota. Then, a series of silicone controlled rectifiers (SCR), called thyristor valves, reconvert the power to ac for distribution to homes and factories.

Numerous other examples the world over indicate the increasing role now being played by dc transmission systems (Figure 1.13).

HVDC transmission projects now in operation and under construction range in operating voltages from 500-800 KV (U.S. and Soviet Union) and transmission distance extended up to 1050 miles (1700 km) in Zaire, Africa. Less insulation, reduced right-of-way miles, and less energy loss are also distinct advantages of dc transmission.

Although dc power systems are proving to be advantageous, primarily for transmission purposes, ac systems have nevertheless continued to show flexibility in their many uses, especially for residential distribution.

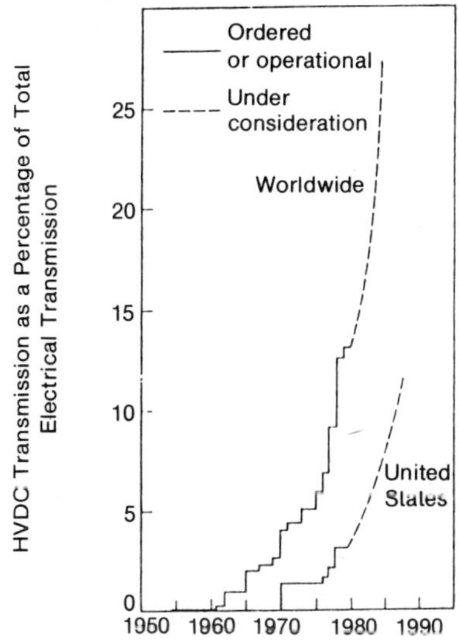

Figure 1.13 - Increasing role of dc transmission in the U.S. and worldwide.

With alternating current, it is possible to raise and lower the voltage economically by means of transformers. Because transformers have very high efficiencies (98-99.5 percent), they are very economical in stepping up ac power to high voltages and down to low voltages at which the power is used. On the contrary, it is not easy to convert high dc voltages to low ac voltages, at which the power can be used, nor is direct current readily converted from low to high voltages for transmission.

Flexibility of AC Systems

The practical flexibility of using alternating current is demonstrated by its wide usage in industrial concerns. At the present time, over 95 percent of the electric energy used commercially is generated as ac. Nevertheless, this does not mean that ac applicability in industrial and domestic settings is superior to that of dc. In fact, its major advantage is that it is flexible enough to be converted into dc energy and then used in the immediate vicinity of its place of generation. Such use has been demonstrated at Niagara Falls where the energy of the Falls is employed to generate ac energy, and then convert it to dc energy for use in surrounding electrochemical industries.

Because ac power can be transmitted economically over long distances, it is feasible to generate electrical energy in large quantities in a single central power station and then distribute it over a comparatively large area.

On the whole, ac power systems can be employed economically with the help of various types of transformers. The ANSI/IEEE Standard C57.12.80-1978 classifies transformers by size, insulation, and location.* Table 1.1 gives a typical summary of such classification.

**ACRONYM IDENTIFICATION*

ANSI — *American National Standards Institute*
IEEE — *Institute of Electrical and Electronics Engineers*
ASTM — *American Society for Testing and Materials*
IEC — *International Electrotechnical Commission*
BSI — *British Standards Institute (United Kingdom)*
NEMA — *National (U.S.) Electrical Manufacturers Association*

TABLE 1.1

CLASSIFICATION OF TRANSFORMERS[1]										
Size or End Use	Mechanical Form of Construction		Insulation-Coolant					Location		
	Core (Figure 1.19)[4]	Shell (Figure 1.21)[4]	Dry (air or gas)	Liquid Filled: Oil[3]	Liquid Filled: PCB	Compound Filled	Indoor	Outdoor Weather-proof	Pole-Type Mounted	Underground
Power (> 500 KVA)	X[2]	X	X	X[2]	X	X	X	X[2]	—	X
Distribution (< 500 KVA)	X	—	X	X[2]	X	X[2]	X	—	X[2]	X
SPECIAL PURPOSE										
Instrument	X	—	X	X[2]	—	X	X[2]	X	—	—
Arc Furnace	X[2]	X	—	X	—	—	X	—	—	X
Rectifier	X	—	X	X	—	—	X[2]	X	—	X
Railway Locomotive	X	—	—	—	X	—	—	X	—	—
Coal Mining	X	—	X[2]	—	X	—	—	—	—	X
Underground Distribution (Network, subway, etc.)	X	—	X	—	X[2]	—	—	—	—	X

[1]ANSI/IEEE Standard C57.12.80-1978, p. 8. Each X in Table 1.1 means that the particular item is applicable to the size or end use of a transformer.

[2]The majority of units fall into this category.

[3]The method of oil preservation is another type of specific classification.

[4]Figures 1.19 and 1.21 are located in the special Color Section, following Chapter 10.

Part 2—Transformer Design and Construction

Introduction

The transformer is defined as a static electric device, consisting of a winding, or two or more coupled windings, with or without a magnetic core. It transfers power by electromagnetic induction between circuits at the same frequency, but usually at changed values of voltage and current (either serving a step-up or step-down function). Unfortunately, as a static device, the transformer has been traditionally considered as having no moving parts. This is not true. On the contrary, the transformer is a unique piece of electrical equipment in that it has no *functional* moving parts, yet is moving continuously while energized.* The magnitude of the movement under stress can be as high as three inches for "stationary windings."

The generation of electricity requires motion between a magnetic field and conductors. In power plants, this motion is supplied by moving the wires through a fixed magnetic field. A transformer, however, depends upon a constantly changing magnetic field to transmit power. The wires are fixed and the magnetic field moves. This *continual* movement of oscillating magnetism in the core steel is set up by 60 Hertz ac power. This results in a 120 cycle mechanical vibration. Normal cyclical loading, as well as heat, load, and voltage/current power surges play a major rule in this mechanical vibration of the entire structure (expansion and contraction of materials). As an example, during a test of an operating transformer, three inches of distortion developed in the unit under a controlled short circuit.†

Fortunately, the day is now past when a transformer is looked upon as a piece of apparatus which requires no attention because its parts are *stationary*.

Major Parts of a Transformer

Various authorities differ on what they consider to be the primary components of a transformer. Nonetheless, all do agree that the basic power transformer consists of three essential parts (Figure 1.14).

Tap changers, load ratio controls, and switches are normally specified as* ***auxiliary equipment* *for the transformer.*

†*The transformer was specially built with a glass observation hole.*

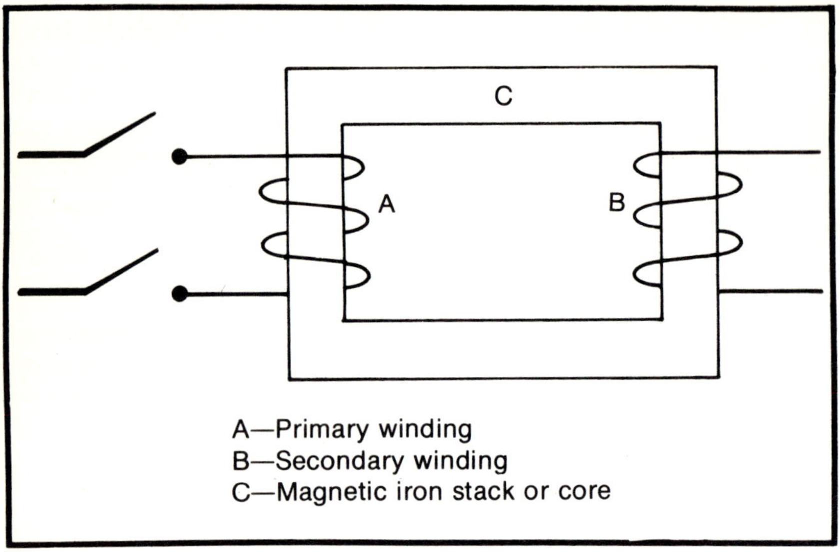

Figure 1.14 - Basic elements of a transformer.

The core, a magnetic circuit with a clamping structure, is the part of the transformer in which a magnetic field oscillates. The metallic composition of the core is a special high grade silicon sheet steel. A typical sheet of steel is 0.014 inches (0.3 mm) thick. These sheets are laminated into sections that are several inches wide. These core laminations help reduce eddy currents or currents induced in the iron parts of the unit. Each lamination in turn is coated with insulating material. This coating helps prevent magnetic losses and reduces heating losses.

The primary and secondary windings include clamping arrangements. The coils for larger units are usually made of copper, which is an extremely good conductor, and as such has a low resistance. Smaller units may use aluminum sheet wound secondaries. Thus, any heat generated in the windings due to current flow is held to a minimum. The two separated windings are insulated from each other and wound on a common core of magnetically suitable material, such as iron or steel.

Other major components include:

- Tank enclosure—It contains the transformer together with cooling surface and coolant. It serves as a surface to radiate heat to the surrounding air.

- Bushings—Leads and tappings from windings, together with their supporting arrangements. They are usually made of porcelain. The newest insulator material, made of polymerized resin and silica, is called Polysil™.* The inside of the bushing may be oil, paper, epoxy, or fiber glass. Bushings serve to insulate the primary and secondary windings from the tank (ground).
- Coolant (insulant) or cooling arrangement—Air, gas, oil, or synthetic liquid may be used.

Figure 1.15 illustrates these six major components of a power transformer. In addition, the principal parts of a transformer should include:

- Solid insulating materials of windings and core
- Taps and tap changing arrangement
- Protective gear, circuit breakers, protective relays
- Other *auxiliary* equipment (Figure 1.58)

Finally, we hold as the major premise for transformer maintenance that the liquid insulant and solid insulating materials make up "an insulation system." Furthermore, on the basis of our field experience and various authoritative opinions, the life of the transformer depends primarily on the life of this insulation system. This view has not received the emphasis it rightfully deserves, even though the principle holds for both liquid and dry type systems.

In subsequent sections, we will consider some of these diverse building materials that comprise a typical power transformer:

- Copper
- Aluminum
- Iron
- Brass
- Wood (Maple)
- Kraft paper
- Pressboard
- Mineral oil
- Vulcanized fiber
- Resins
- Porcelain

**Polysil is a registered trademark of the Electric Power Research Institute (EPRI).*

- Cements
- Polymer coatings
- Tapes
- Glass cloth
- Cord
- Gasket materials
- Internal paints

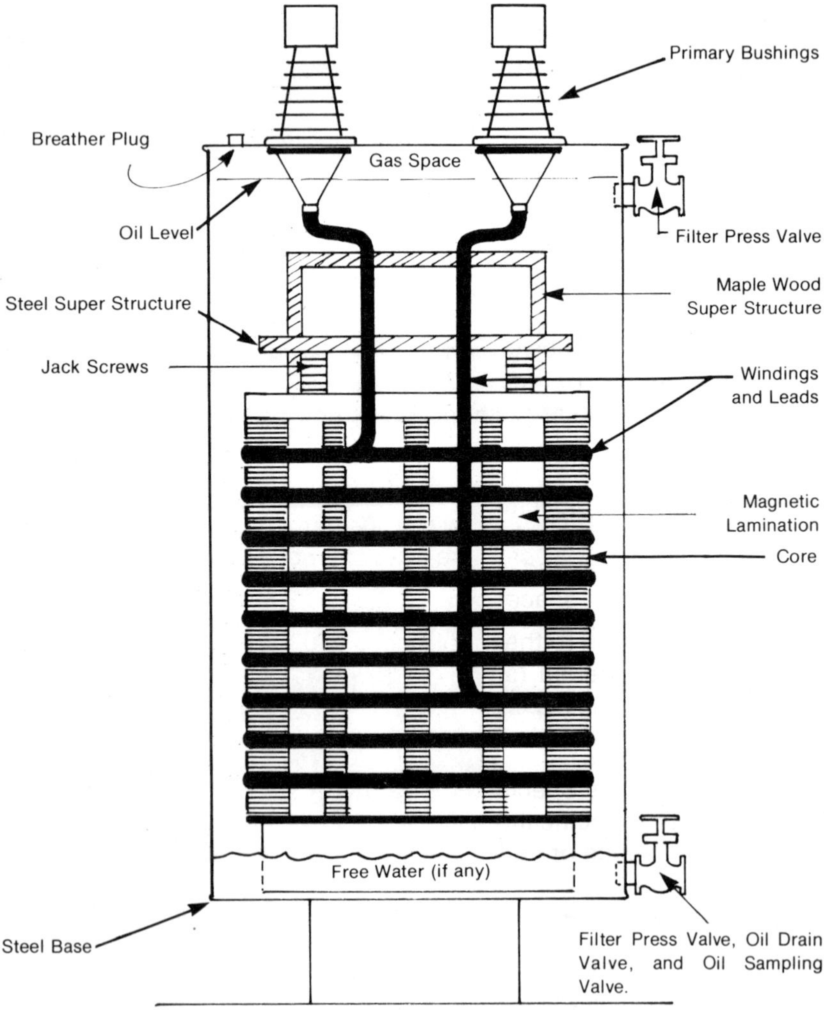

Figure 1.15 - Pictorial diagram of major power transformer parts.

Basic Design Procedures

Several recent authoritative books consider transformer design in great detail.* Therefore, minimum emphasis will be placed on this subject. Several remarks are, nevertheless, still in order.

The purpose and particular application that a transformer is intended to serve usually determines how it is constructed. Transformers are now designed both with and without the aid of a digital computer. Modern units are computer designed around customer specifications. Yet, the transformer still maintains some "old-fashioned flavor" because much of it is handmade, particularly larger units. In fact, there has been very little standardization for power transformers. Most units are still custom-designed to meet each purchaser's specifications. This factor adds to the cost, but more importantly, adds to the backlog of one to two years once an order is placed. Therefore, this time delay alone should behoove the transformer owner/operator to maintain units now in-service. While rebuilding time is less, rebuilding costs remain a high percentage of new equipment. Most owners can no longer afford to have standby units, even for critical areas. But even with a computer, modern design involves a number of factors, beginning with an understanding of the characteristic properties of the materials used in transformer construction.

The basic information the designer must initially have consists of:

1. Customer Specifications
   Significant factors include the primary supply system, the secondary load application, protective devices needed, environment, space, cost, and efficiency considerations.

   In addition to the traditional environmental aspects of altitude, wind, external (ambient) temperature, and atmospheric conditions, the maximum noise level (audio, radio, and TV) is an increasingly critical factor for many applications.†

2. Designer's Specifications
   These specifications are governed by the manufacturer's code of practice or national/international standards for a particular design. Among the key bits of information the designer must consider are:

*See Stigant, **J & P Transformer Book** (1973) and Feinberg, **Modern Power Transformer Practice** (1979).*

†*Chapter 1, Part 2, Transformer Testing.*

- General arrangement of coils with regard to core
- Type of component design
- Type of component material chosen
- Maximum allowable transformer losses
- Insulation material clearances and thickness
- Short circuit, impulse and switching surge withstand

Once an invoice is received, certain data such as heat rise, impedance, no-load losses, and short circuit withstand become guaranteed values.*

3. Transformer Selection and/or Construction Requirements
Significant standards that the designer must adhere to include:

a. U.S. (ANSI/IEEE)

1) C57.12.00-1980
General Requirements for Distribution, Power, and Regulating Transformers

2) C57.12.01-1979
General Requirements for Dry-Type Distribution and Power Transformers

3) C57.12.10-1977
Requirements for Transformers 230 KV and Below

4) C57.97-1971
Preparation of Specifications for Large Power Transformers with or without Load-Tap Changing

5) C57.17-1971
Requirements for Arc Furnace Transformers

6) ANSI C2
National Electrical (Safety) Code, 1981 Edition

7) C57.12.20-1971
Requirements for Load-Tap Changing Transformers 138 KV and Below

8) ASTM D-3487
Standard Specification for Mineral Insulating Oil Used in Electrical Apparatus

**Chapter 1, Part 2, Understanding Transformer Nameplates (Figure1.67).*

b. International (IEC)

1) IEC 76, Power Transformers (1967)

2) United Kingdom (B.S.)
Same as (IEC 76)

3) IEC 296, Specification for New Insulating Oils for Transformers and Switchgear (1969)

4. Goal of Transformer Design
The electrical designer seeks a satisfactory compromise of cost-benefit considerations. But at times this decision is difficult. For example, computer standardized design is the primary reason for the relative low cost of today's distribution transformer. In so doing, the manufacturer establishes a standard core loss, impedance, heat rise, and short circuit withstand suitable for the majority of his customers.

Variations, however, in the cost of energy and loading practices in different localities of the U.S. (West vs. Midwest) are dictating different core loss levels for a single transformer rating; obviously, additional tooling and machine set-up costs result from this benefit of lower energy costs. Clearly, this dilemma will most likely confront both the transformer designer and owner/operator in the years to come. With both goals in mind (cost and benefit), we will later contrast current and former transformer design practices.*

Reviewing Basic Transformer Principles

Basic Design Requirements of a Transformer

Figure 1.16 (A) illustrates these basic requirements in a transformer:

N = Number of turns of a non-magnetic metal, such as copper or aluminum
I = Current in amperes
E = Voltage in volts
ϕ = Flux or magnetic lines of force produced by ampere turns (NI)
M = Motion (magnetic field)

**Chapter 1, Part 3.*

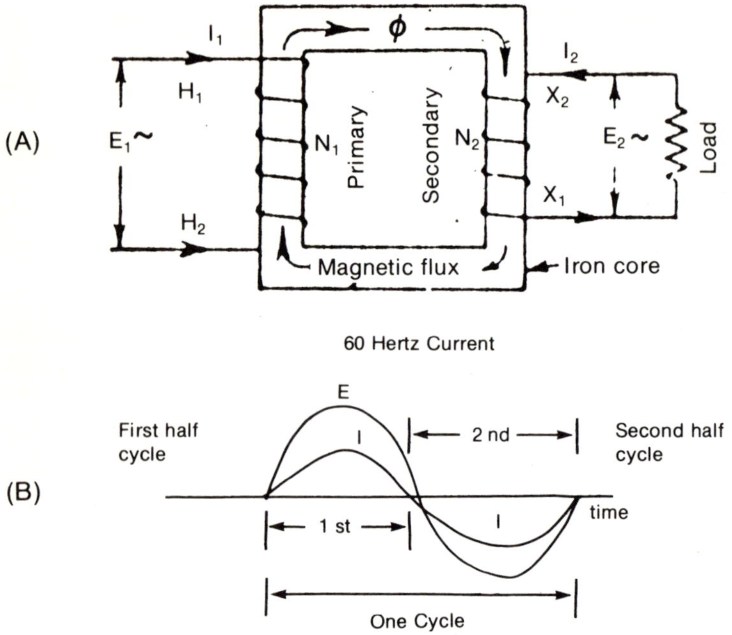

Figure 1.16 - (A) Essential requirements for a simple transformer. (B) In a 60 Hertz system each cycle or two alternations takes place in ***one*** *second. Result: generated voltage. The constantly changing magnetic field replaces the motion required in the primitive electromagnet (Figure 1.6).*

Basic Transformer Relationships

When an alternating voltage E_1, is applied to the primary winding (Figure 1.16 A), a current I_1, called the exciting current, will flow through and produce an alternating magnetic field with a flux of ϕ magnetic lines of force.

Magnetic flux ϕ will always flow over the path of least resistance. The flux would prefer passing down through the air space between the windings, then go through the secondary winding and do work. The flux that does pass between the two windings is called the "leakage flux." This leakage will also cause some heating of the core which in effect is a power loss.

This construction (Figure 1.16A) would thus not normally be used in a regular transformer because of the high leakage flux. In order to decrease this amount and force the flux to do work, the windings are assembled concentrically; or one inside of the other (Figure 1.17).

Figure 1.17 - A concentric winding assembly. The high voltage coil is being lowered over the low voltage coil. (Courtesy of the Moloney Electric Company). R. Bean ***Transformers for the Electric Power Industry*** *(1959).*

The lines of force cut across the secondary winding and induce voltage in each turn. This voltage is equal to $\frac{E_1}{N_1}$ because the voltage E is distributed uniformly across the turns of the primary coil. Likewise, the secondary winding will induce $\frac{E_2}{N_2}$ across the primary:

1. $$\frac{E_1}{N_1} = \frac{E_2}{N_2} \quad \text{or} \quad E_1 N_2 = N_1 E_2$$

2. $$\frac{E_1}{E_2} = \frac{N_1}{N_2}$$

Hence, this relationship—the *fundamental transformer equation*—states that the ratio of the primary and secondary voltage equals the ratio of the primary and secondary turns. The ratio $\frac{N_1}{N_2}$ is called the *turns ratio* of the transformer. Equation (2) may be written as:

3. $$\frac{I_1}{I_2} = \frac{E_2}{E_1}$$

which means the current ratio is inversely proportional to the voltage ratio.

Transformer Efficiency

The efficiency of a transformer is the ratio of output power to input power, expressed as a percentage, that is,

$$\text{Efficiency} = \frac{\text{output}}{\text{input}} \times 100$$

since $$\text{Input} = \text{output plus losses}$$

therefore, $$\text{Efficiency} = \frac{\text{output}}{\text{output plus losses}} \times 100$$

These values are expressed in watts or kilowatts. Equation (3) may be expressed as $I_1E_1 = I_2E_2$.

The product of primary voltage and current (primary power) equals the secondary voltage times current (secondary power). Power P is usually expressed on a transformer's nameplate as kilovolt-amperes (KVA) or megavolt-amperes (MVA).

Under full load operating conditions modern transformer efficiency ranges between 98 percent for a small unit in a distribution system and 99.5 percent for a bulk power unit in a transmission system.

Although all the foregoing equations disregard losses introduced in the transformer, they are used in practical applications with acceptable accuracy.

Transformer Losses

Though newer transformers operate at very high efficiencies, nevertheless, there is no such thing as an ideal transformer. In terms of design and operation, the losses in a transformer consist of two kinds: no-load and load losses.

1. No-Load (Iron) Losses
   Certain losses occur in a transformer regardless of the load, when the unit is connected to a source of voltage.

These losses include core losses, copper losses in the primary winding due to the flow of no-load current (exciting current) and dielectric losses in the insulation. In practice, while the last two losses are very small, the iron or steel losses are significant. Thus, one of the no-load losses is commonly termed "iron losses." The first major loss, known as hysteresis, is the power required to magnetize the core first in one direction and then in the other. As it is magnetized, the core actually expands and contracts ever so slightly, producing the characteristic hum associated with transformers. The second major contributor to the no-load loss is the eddy current loss or power loss due to the circulating currents in the core iron.

2. Load (Copper) Losses
   Load losses are those which occur in a transformer incident to the carrying of load. These losses are called copper losses, or I^2R losses. The greatest part of the load loss is due to power lost when the load current flows through the resistance of both the windings.

   The actual path of magnetic flux is somewhat less than ideal. There is a loss produced by the flow of small eddy currents in the copper conductors. The undesirable currents are set up because some of the lines of flux do not follow the regular steel magnetic path of the transformer but leak ("cut corners") through the air or other available path. Also, stray losses result from the effect of leakage flux and consist of circulating currents when two or more wires are parallel in each turn and any noncurrent-carrying parts, such as core clamps or core bolts.

3. Total Losses
   The total losses consist of the sum of the load losses and the no-load losses. These losses in transformers are converted to heat which is a power loss: therefore, proper design of the insulation system is required to dissipate this heat. Manufacturers usually specify the no-load loss, as well as the total losses of a transformer.

Transformer Cooling

As the core and coil assembly heats up from both load and losses it *releases* its heat to the surrounding insulating coolant, such as oil.

The wasted energy in the form of heat generated in transformers due to the foregoing iron and copper losses must be carried away to prevent excessive rise of temperature and injury to the insulation about the conductors.

The cooling method used must be capable of maintaining a sufficiently low average temperature. It must also be capable of preventing an excessive temperature-rise in any portion of the transformer, and the formation of "hot spots." This is accomplished, for example, by submerging the core and coils of the transformer in oil and allowing free circulation for the oil through oil ducts suitably located in the coil structure. The heated oil will tend to rise to the top, causing the cooler oil to drop toward the bottom of the tank, thus setting up a natural circulating current of oil within the tank or enclosure. The oil in contact with the tank, in turn, gives up its heat to the tank walls and to the surrounding air. In this manner, the temperature of the coil and core assembly is held to safe limits.

Primary Transformer Circuits

In the design of a transformer, three distinct circuits are considered.

- The *magnetic* core
- *Electric* conductors or winding
- The *dielectric* insulating circuit

Transformer losses may also be subdivided in terms of these three circuits. Discussion in the subsequent three sections will proceed from the circuit of the most losses in the core (at no-load); then to the windings (under load): and finally, to the *minor* losses of the insulating system (under the no-load condition). The dielectric losses in the insulation usually only become significant when subjected to high temperature, for example, excessive overload.* When this happens, you can get a thermal runaway condition leading to transformer failure.

Magnetic Cores

The Designer's Starting Point

Once a customer decides what a given transformer should be capable

**Chapter 7, Part 4.*

of doing, the electrical designer as his starting point, selects the type of core and core size. It is a fact that the core carries the magnetism that is needed.

The magnetic circuit determines the type of transformer—either core or shell mechanical form. In the core-form the magnetic circuit forms the center of the transformer. This design is generally used on power transformers of lower current and KVA ratings (Figures 1.18 and 1.19).*

In the shell-form the magnetic circuit surrounds and is a shell around the windings. The interleaved stack of rectangular coils extends above and below the magnetic circuit (Figures 1.20 and 1.21).† This design is generally used on transformers of large KVA ratings—50,000 KVA and above.

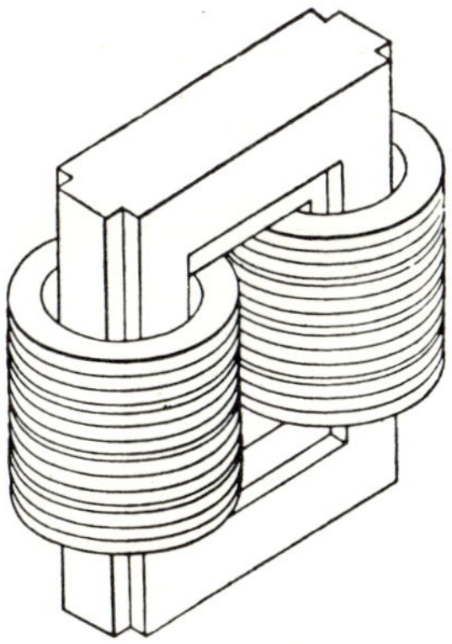

Figure 1.18 - An arrangement of winding used with core-form transformers.

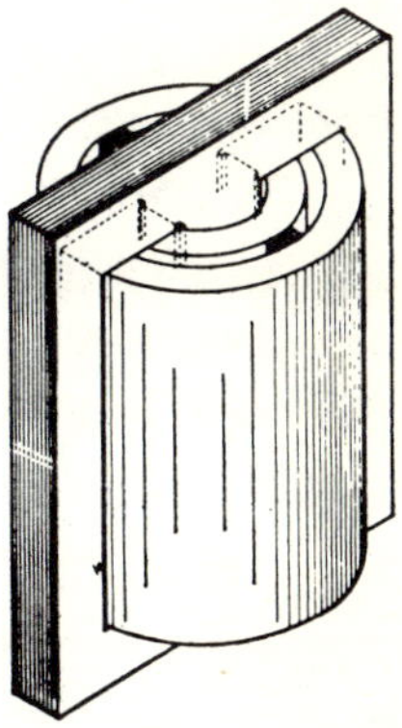

Figure 1.20 - An arrangement of winding used with shell-form transformers.

**Figure 1.19 is located in the Color Section, following Chapter 10. While the General Electric Co. has been wedded to core-form for large power transformers for years, Westinghouse Electric and McGraw-Edison use shell-form.*

†Figure 1.21 is located in the Color Section, following Chapter 10.

Core Design Calculations

Recall that magnetism is described in terms of flux or the number of lines of magnetic force per unit area (flux density). The total flux ϕ is the number of lines of force per unit of area (B) times the net cross-sectional area of the core (A).

1. $\phi = A \times B.$

The common formula for induced voltage (a sine wave voltage assumed, Figure 1.16B) is given by:

2. $V = 4.44\,\phi\,N\,f \times 10^{-8}$

where
V = effective value of induced voltage
N = number of turns in winding
ϕ = maximum total value of flux, in lines of force
f = frequency, in cycles per second

Recalling that when Michael Faraday discovered the phenomenon of electro-magnetic induction (Figure 1.7) the derivation of the common formula begins with the instantaneous relationship given by this expression, using a sine wave:

3. $V = \dfrac{d\phi}{dt}$

Assuming N turns in a magnetic field, each of the turns is linked with the same magnetic flux ϕ, which varies with time.

4. $V_a = \dfrac{N \times \text{change in flux per second}}{10^8}$

The key value is "the change in flux per second." If the ac with a frequency of f cycles per second produces an alternating magnetic field with a maximum flux value of ϕ lines of force, the lines of force threading each turn of the winding must change from $+\phi$ to $-\phi$ during the time of one-half cycle, or $\frac{1}{2f}$ sec (seconds). Therefore, it follows that the average change of flux in one second is $20 \div \frac{1}{2f} = 4\phi f$. Substituting the value of flux change per second into equation (4),

5. $V_a = \dfrac{N \times 4\phi f}{10^8} = \dfrac{4\,\phi\,Nf}{10^8}$

where V_a = average induced voltage. Knowing that $V = 1.11 V_a$, then the basic equation for induced voltage becomes:

6. $V = 1.11 \times \dfrac{4\phi\,Nf}{10^8} = \dfrac{4.44\,\phi Nf}{10^8}$

The actual design calculation is rearranged as:

7. $$\frac{V}{N} = \frac{4.44\,\phi\max f}{10^8}$$

A modern core has a *maximum* flux density B of 100,000 lines of force per square inch (Figure 1.22), in order to avoid excessive exciting current. Sufficient room for NI (ampere-turns) needed must be provided to set up the required flux. In the NI relationship only the "N" (number of turns) can be varied.

8. Ideally, $$\frac{V}{N} = 4.44\ (100{,}000)\ (60) \times 10^{-8}$$

$$= 4.44\ (10^5)\ (6)\ (10) \times 10^{-8}$$

$$= 4.44\ (6) \times 10^{-2}$$

$$= 0.266 \text{ volts/turn}$$

Again, under ideal conditions to generate one volt, a conductor of N turns moves from position A to B (½ cycle) and B to A (½ cycle) in one second at 60 Hertz per second through a field of 375,000 lines of force (Figures 1.23 and 1.24).

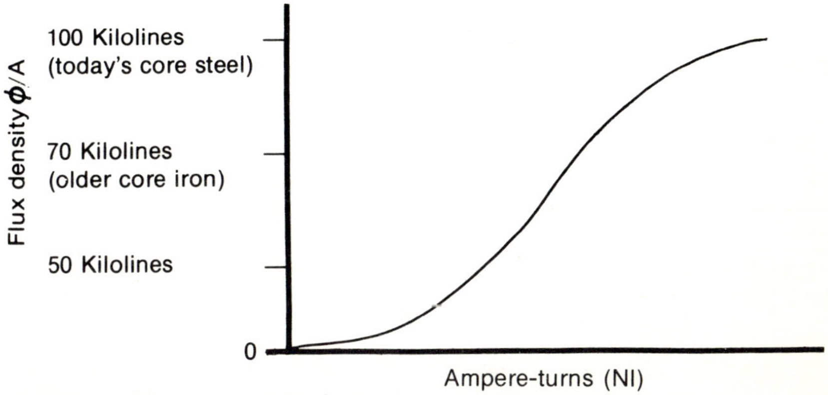

Figure 1.22 - A typical core saturation curve. Saturation refers to the maximum permissible flux density in silicon iron steels. Today's transformers are designed to operate at the knee of the curve.

9. $\frac{V}{N} = \frac{4.44 \, f\phi \, max}{10^8}$

$\frac{1}{1} = \frac{4.44 \, (60) \, \phi \, max}{10^8}$

$1 = 4.44 \, (60) \, \phi \times 10^{-8}$

$\phi \, max = \frac{1}{4.44 \, (60) \times 10^{-8}}$

$= \frac{1}{2.66 \times 10^{-6}}$

$= 1 \times \frac{10^6}{2.66}$

$= .375 \times 10^6$

ϕ max = 375,000 lines of force (or 375 kgauss/375 kOersted)

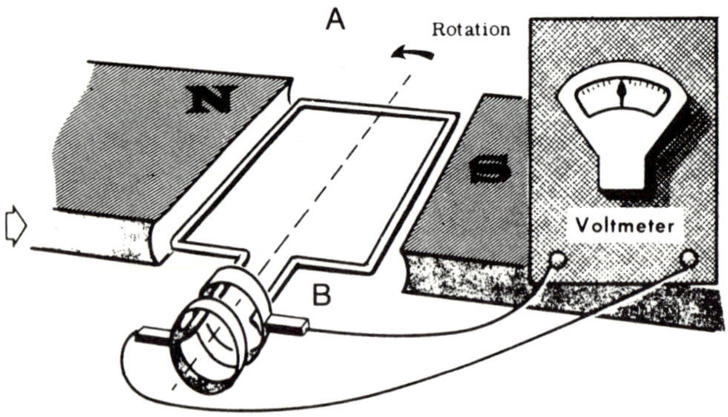

Figure 1.23 - A simple alternating current generator, a rectangular loop of wire rotating in a magnetic field between north and south poles of a magnet.

Next let's apply equation (1):

$\phi = A \times B$

to a specific example:

To generate 110 volts with one conductor (turn) would require how much core steel at 60 Hertz:

Plugging in typical values of ϕ max and B (Figure 1.22) we find the value of net area of the transformer-core cross section as:

$\phi = A \times B$

375,000 lines of force = A x 100,000 lines of force/in^2

$$A = \frac{375{,}000}{100.000}$$

= 3.75 in^2 core steel *per volt*

or when V = 110 volts, then

A = 110 (3.75) = 412.5 in^2 core steel

As a check, use the basic core design equation (2) in slightly different form.

$$V = \frac{4.44\ A\ B\ N\ f}{10^8}$$

$$A = \frac{(110)\ 10^8}{4.44\ (100{,}000)\ (1)\ (60)}$$

$$A = \frac{110 \times 10^2}{26.64} = 412.5 \text{ in}^2 \text{ core steel}$$

= 20.3 "x 20.3" piece of core

The designer must weigh the benefit of increased flux density B against corresponding cost of additional core materials.

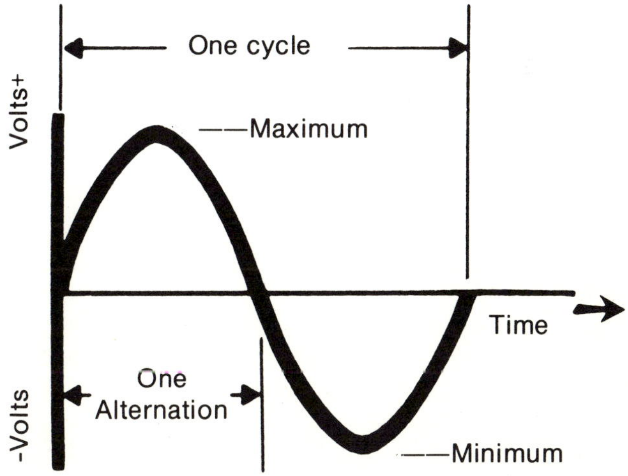

Figure 1.24 - The voltage induced by the alternating current generator (Figure 1.23).

Core Materials

1. Basic Magnetic Materials
   The cores of modern transformers still retain the basic features developed over 80 years ago—going back to thin, flat laminations of soft iron (Figure 1.6). The core has been made of steel because this material carries and confines the magnetic flow so much better than air; thus, the permeability of the steel is a measure of the number of times it is a better conductor of flux than is air. Ongoing technology has brought about improved quality in terms of core steels, core design, and manufacturing techniques. Table 1.2 summarizes these core steel improvements.

   The ideal core materials (minimizing hysteresis loss) would produce no friction between magnetic molecular particles as the ac magnetic field continually reverses its direction. To minimize eddy currents the lamination materials would be as thin as possible. However, the thickness can be not reduced beyond a certain point because the laminations would become too weak mechanically.

   In the early days of transformer manufacture, soft iron was used as the core material. Heat developed through friction took place during the early life of the transformer. It was found that small quantities of silicon (Si) alloyed with low carbon content minimized the frictional loss known as hysteresis. Today, only crystal-oriented iron silicon alloys of high purity are used in transformers.

   Two other major manufacturing changes have taken place. Until the mid-1930's iron-silicon alloys (4½ percent maximum Si) were produced from slabs by successive rollings at a red heat (1370°C, followed by annealing at approximately 900°C). This process produced sheets in which the constituent crystals were almost all random-oriented.

In 1935, N.P. Goss discovered that cold-rolling and high temperature annealing reduce the total core losses. The most efficient magnetization occurs when the magnetic lines of force flow parallel to the steel grain (high permeability and low hysteresis loss). Today, this oriented grade containing about 3 percent silicon is used almost exclusively on large power and distribution transformers where low core losses are a significant factor.

TABLE 1.2

IMPROVEMENTS IN TRANSFORMER CORE MATERIALS

Year	Type of Material	Thickness (mm)	Total Losses (Watts per kg)
1890-1905	Laminated soft iron (unalloyed steel)	0.35	5.90
1900	Swedish charcoal iron	0.35	3.50
1905	Enameled sheet steel lamination	0.35	—
1910	3.25% hot-spilled silicon steel[1]	0.35	1.75
1925	4% silicon steel	0.35	1.40
1935	3.2% cold-rolled grain-oriented silicon steel[2]	0.35	—
1945	3.2% cold-rolled grain-oriented silicon steel	0.33	0.57
1968	3% grain-oriented silicon steel (Nippon Steel)[3]	0.33	—
1970	3% grain-oriented silicon	0.30	0.40
1974	3% grain-oriented silicon steel (Kawaski)	0.30	—
1977	3% grain-oriented silicon steel (Allegheny Ludlow)	0.30	—
1980	Amorphous soft magnetic alloys[4]	—	0.25

[1]Hot-rolled non-oriented processing used until 1940.

[2]Cold rolled grain oriented processing used since the early 1940's.

[3]Japanese steel developed during the 1970's.

[4]Metallic glass will not be in production until a five-year evaluation is completed.

The ease of magnetization combined with desired hardness has made metallic glasses or amorphous metal a likely candidate to reduce core losses even more than conventional silicon steels (Figure 1.25).

Unique molecular structure is the basis for this desirable combination. Atoms are arranged irregularly (amorphous) in contrast to conventional metal's crystalline structures.

The first electric power application was made in 1980. A five-year evaluation will precede commercial availability. Table 1.3 and Figure 1.26 also illustrate possible advantages of using metallic glass cores in power transformers.

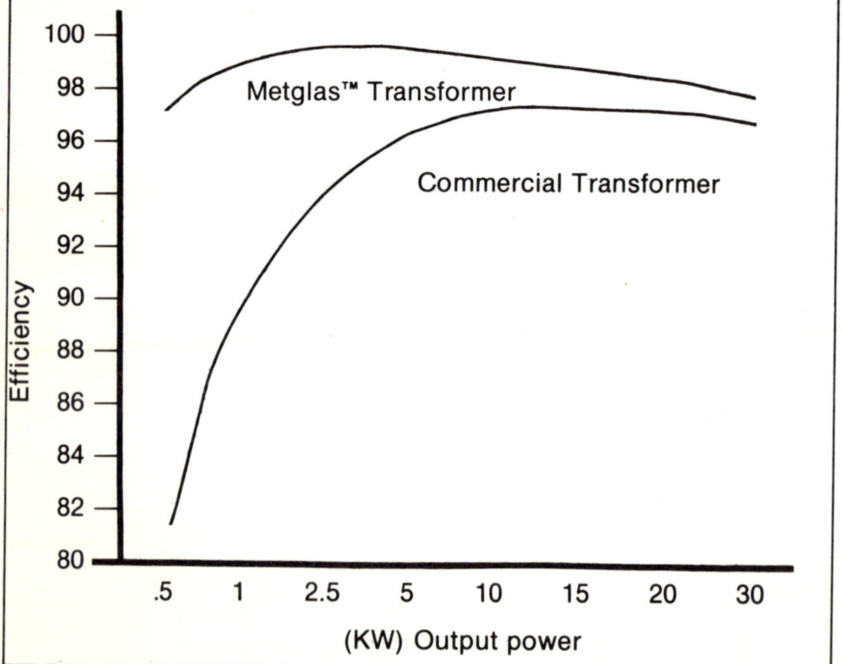

Note: the efficiency of the Metglas™ transformer at 30 KW (twice the rated output) is comparable to the commercial unit efficiency at 15 KW.

Figure 1.25 - Comparison of efficiency of Metglas™ and conventional 15 KVA transformers. (Courtesy of Allied Chemical Corp.). Metglas is a registered trademark of Allied Chemical Corp.

TABLE 1.3

PERFORMANCE COMPARISON OF TWO 15 KVA TRANSFORMERS[1]		
Operating Parameter	Core Material	
	Silicon Steel	Metallic Glass
Exciting current, amperes	2.5	0.12
Core loss, watts	112.0	14.00
Copper loss (15 KVA), watts	210.0	166.00
Total loss (15 KVA), watts	322.0	180.00
Energy saving, kilowatt-hours per year	0.0	1250.00
Temperature, °C	100.0	70.00

[1]The total loss for the conventional transformer is 322 watts, compared with 180 watts for the metallic glass core transformer.

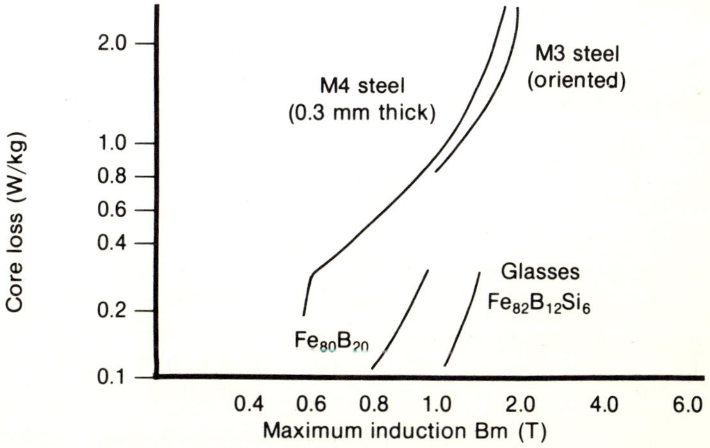

Figure 1.26 - Transformer core losses at 60 Hertz for two glass compositions compared with conventional magnetic steel of 0.06 mm thickness. J. J. Gilman, ***Science*** *(1980).*

2. Core Insulating Materials
   Eddy current losses on the core materials are also minimized by insulating adjacent laminations or thin sheets with an oil-proof coating. The laminating and insulating coating presents a high resistance path to the eddy currents, thereby reducing these losses.

   Traditionally, for most transformers this insulating coating has been varnish; core steel materials have inorganic coatings as delivered to the transformer manufacturer. The current trend is toward use of some form of phosphate and/or organic resin treatment. The phosphate treatment made prior to the final annealing stage of manufacture gives a more heat resistant coating than varnish, as well as preventing rusting during storage.

   On large power transformers (>30 MVA) magnesia applied during an early stage of manufacturing provides additional surface insulation. Traditionally, paper and mixtures of china clay and flour have been used, but their increased insulation resistance is negated by a relatively thick coating (.0125 inch vs. the newer .003 inch minimum for the newer treatments).

Types of Core Construction

Since transformer windings are normally cylindrical in shape, the ideal cross-section core form should be circular. It is possible, but would be uneconomical. In practice, by making a stepped pattern of individual steel lamination plates, a realistic approximation can be achieved (Figure 1.27).

The three basic types of core construction are the traditional butt and lap, mitered, and strip wound (Figures 1.28 through 1.30).

Most power transformers are built up from laminations laid flat, according to any one of 25 patterns, depending on the number of legs (or limbs) required. For larger transformers two different types of joints have been generally used. In older designs the butt and lap joint have been universally used with the hot-rolled non-oriented core steel (Figure 1.28).

The use of cold-rolled oriented material justified the design of a mitered joint (Figure 1.31 B). A butt lap core with joints made at approximately 45 degrees takes advantage of the magnetic direction properties of the cold-rolled steel (Figure 1.32).

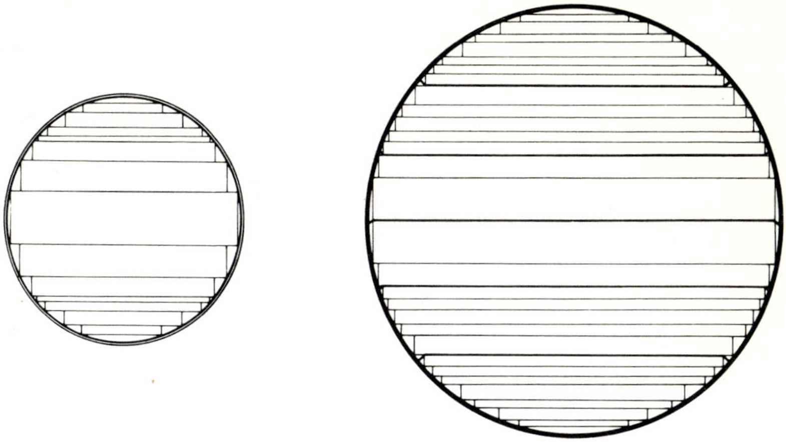

*Figure 1.27 - **Seven** step and **fourteen** step core sections. (Courtesy of **The J & P Transformer Book**).*

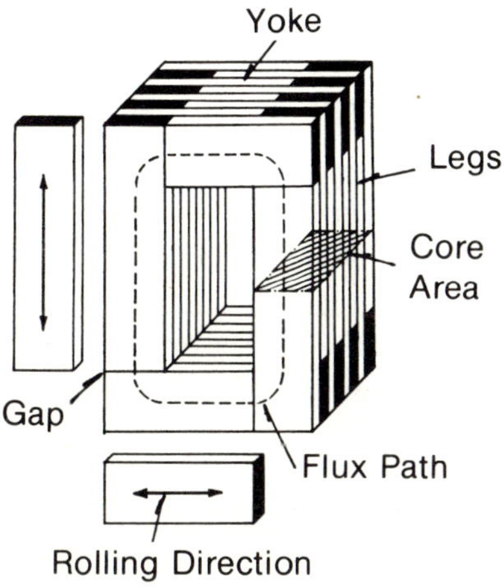

Figure 1.28 - Butt and lap core.

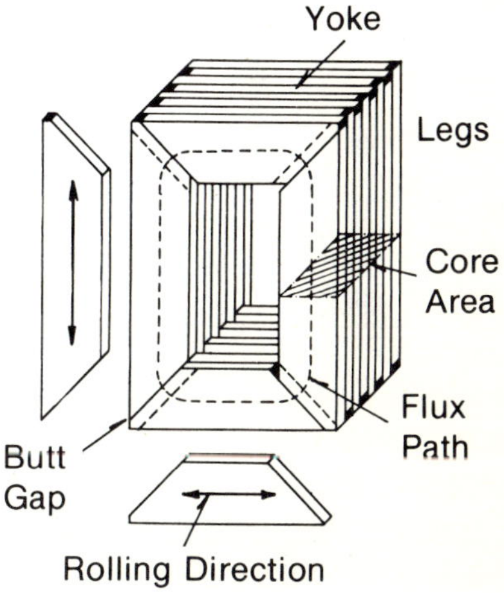

Figure 1.29 - Mitered core.

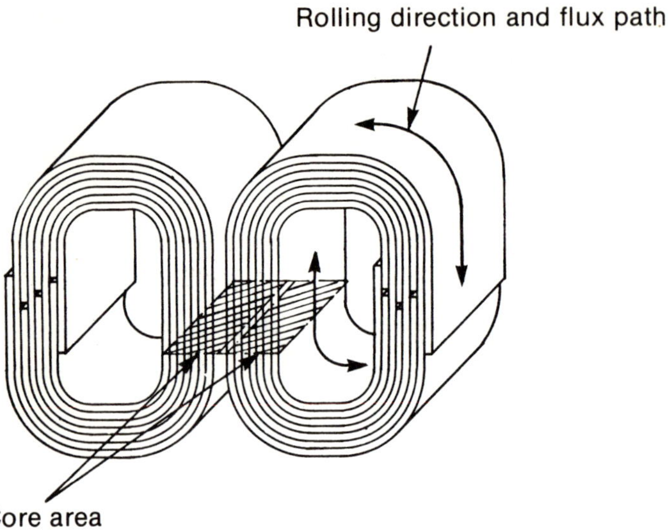

Figure 1.30 - Wound core.

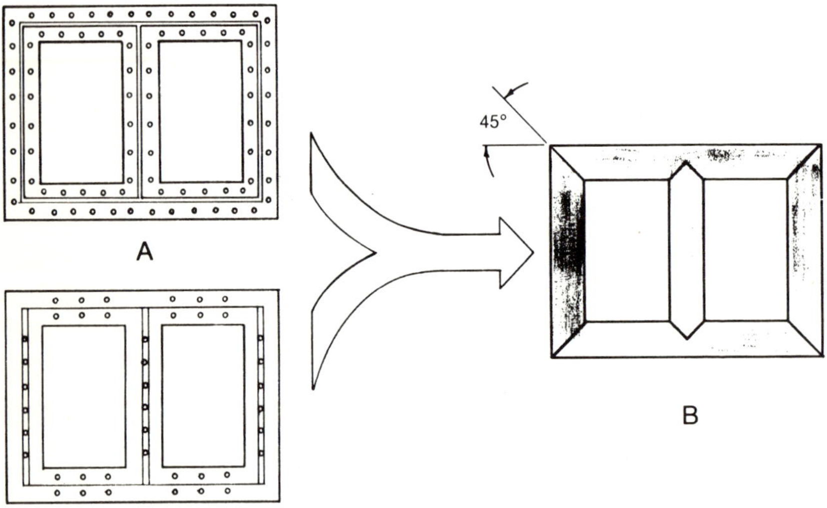

Figure 1.31 - New core steel necessitated changeover from the traditional core (A) to 45° mitered core (B). (Courtesy of Trafo-Union, Stuttgart, W. Germany).

Figure 1.32 - Three-phase mitered, taped core for a 160 MVA power transformer. (Courtesy of GEC Transformers Ltd., Stafford, England).

In both types the layers of laminations are placed so that the air gaps between lamination ends of one layer are overlapped by the lamination in the next layer. For any interleaved joint it is important to minimize the gap (and thus eddy currents) between abutting plates.

If the permeability of the core steel is 10,000 Oersteds (gausses), it means it is 10,000 times a better conductor of flux than air. If the core steel were removed, the magnetizing current to produce the same flux would increase 10,000 times. Therefore, it is extremely important in building the core that the sheets or plates are joined together precisely and with a minimum air gap. As an example, if the gap in a joint were only 1/1000 of an inch, the magnetizing current to push the flux across each gap would be equivalent to 10,000 times 1/1000 inch or 10 inches of steel for each gap. This gap would materially increase the exciting current.

This is why the test for the no-load excitation current is valuable. For example, if a unit is moved from one location to another, shipped over land by rail or truck, or is rewound, the no-load excitation current will verify a "good" transformer or one damaged in shipment.*

For small distribution transformers, a third type of design—cores built from strip wound loops—has proven feasible (Figure 1.30). The wound

**Chapter 1 (Transformer Factory Testing) and Chapter 5 (Field Testing).*

core is spirally constructed from a continuous strip of cold rolled steel and is cut at every other turn to permit assembly. Nevertheless, its practical use has been generally limited to smaller core construction.

For smaller transformers (distribution size) where *wound* cores are used, the manufacturer does stress annealing after cutting the core. Usually no second coating is added. For larger transformers, however, using stacked cores the steel laminations are frequently given an additional (thin) organic polymer coating.

Core Clamping Construction

The final step of core construction is proper clamping. Obviously, if laminations are not properly secure, vibration is the result, contributing further to the hum of the transformer, and unit failure may be inevitable.

Traditionally, power transformer core legs and yokes have been clamped together by means of insulated steel bolts passing through holes punched in the lamination (Figure 1.33).

This arrangement, though giving sufficient clamping, was not without its disadvantages. The number and size of clamping bolts were first reduced by use of high tensile strength steel. With the use of cold-rolled materials, the disadvantages became more serious. Thus, this traditional method has been fully superseded by a method that makes cores without any bolts whatever passing through the active iron section (Figure 1.34).

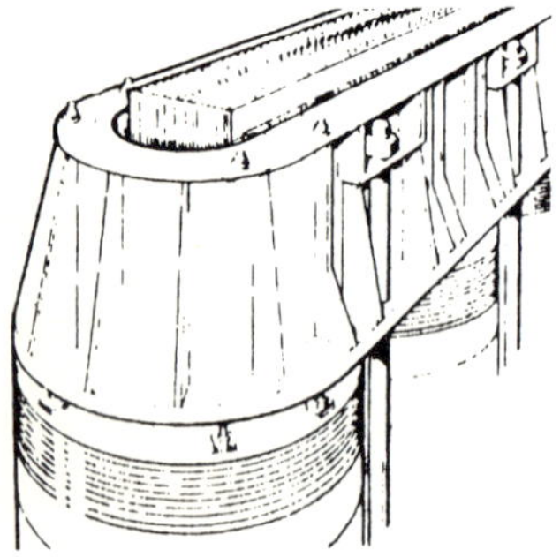

*Figure 1.33 - Typical bolted clamping structure used in older transformers.**

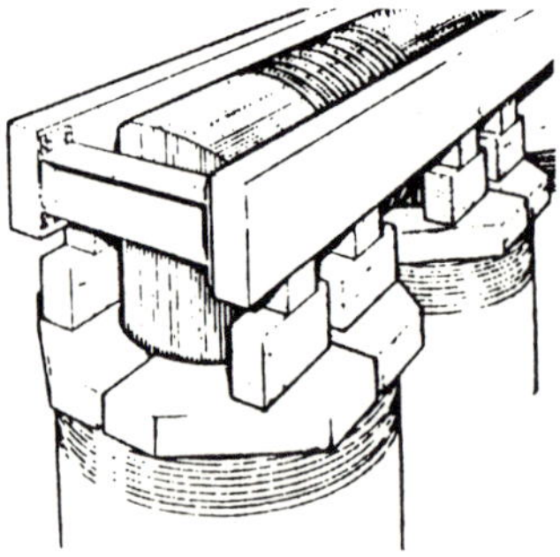

*Figure 1.34 - Typical simplified boltless clamping structure used in large modern transformers.**

**Courtesy of Trafo-Union.*

In the newest method, the leg laminations are held tightly together by strong tape applied on smaller cores, or by suitably spaced high strength resin-glass bands applied to the periphery of larger cores. The yoke laminations are secured by fabricated steel clamps (or in very large units by glass fiber bands). These clamps are lined with resilient packing to obtain uniform pressures and minimize the transmission of sound and vibration.

Completing the clamping arrangement, vertical tie bolts hold together the steel frames clamping the top and bottom yokes. Consequently, electrical and mechanical stresses are minimized and core bolt failures are eliminated.

Windings

Design Criteria

Coil design is the next major step. Transformer coils are designed to get the required number of turns into a minimum of space. At the same time the cross-section of the conductor must be large enough to carry the current without overheating and sufficient space must be provided for the insulator and for cooling ducts, if any.

Windings can be classified by two basic materials, three particular forms, and two basic types of construction. Although there is an overlap between these ranges, no one design meets all the requirements. Each form or type of winding has particular advantages for a given range of voltage and current.

Naturally, the ideal goal is to produce windings of minimum cost which have all the required characteristics to make them efficient yet provide long life.

In practice, the designer must choose the best combination of possible coil arrangements to achieve:

- Sufficient dielectric strength against various voltage stresses, such as lightning or switching surges*
- Adequate winding ventilation†
- Adequate mechanical strength‡ (for example, to withstand short circuit forces)

**Chapter 2, Part 4.*

†Chapter 7, Part 4.

‡Chapter 2, Part 4.

- Minimum cost
- Specified loss maximums

Winding Materials

1. Basic Electrical Materials

   The first material consideration depends essentially on economics where copper is generally used as a conductor (in spite of the development of some synthetic materials). Even though copper has a low specific resistance, its cost has brought about increasing use of aluminum (first used in 1952), especially in distribution and small power dry and liquid insulated transformers. Units as large as 12 MVA have given an excellent service record. During World War II when copper was unavailable, silver conductors were used in a number of transformers for U.S. government wartime plants. However, after the war, the silver windings were salvaged.

   The designer must consider various characteristics in choosing the particular conductor material. Table 1.4 summarizes these properties.

   The primary advantages of copper coils are:

   - Mechanical strength
   - Electrical conductivity (smaller coils)

   The primary advantages of aluminum coils are:

   - Cost of supply stability
   - Efficient heat dissipation (smaller capacities) for sheet wound only (not wire wound)
   - Reduction in weight

   Although it was thought that transformer oil would be more stable in the presence of aluminum than copper, this is not true. Copper is still normally present in the transformer in one form or another (tap changer contacts, and so forth), so that it acts as an accelerator (only) as oxidation takes place.*

2. Basic Material Forms

   Rectangular strips of copper, in parallel, either singly or in groups are generally employed for power transformers. Wire is

Oxygen is normally present in one form or another (Chapter 2, Part 3).

normally found only in the high voltage coils of distribution units, where the current requirement is fairly low. The more amperes (current) to be carried, the larger the wire must be. Eddy currents can be reduced by the use of rectangular wire as it has less surface area than a round wire of the same cross-sectional area.

To keep the resistance loss to a minimum, the cross-sectional area of the turns is adjusted; so as the current is increased, the cross section area is increased. As the area is increased it is necessary to divide it into a number of smaller areas. This also helps reduce the eddy currents losses. The ratio of the primary turns to the secondary turns will establish the voltage ratio between the two sections. Uniform current distribution is desired for maximum efficiency. Transposition of the coils (Figure 1.35) aids in making sure that each wire in the turn encloses the same amount of flux leakage and therefore produces the same volts per turn (or ampere-turns).

The use of sheets and foil has increased in recent years, but they are normally limited to distribution and small transformers, where again the currents are relatively low. Aluminum sheet winding provides more uniform conduction, greater short circuit withstand voltage, and better heat conduction.*

3. Coil Insulating Materials

   a. Insulation Within a Winding
   Insulation materials, providing a barrier for the voltage, must always be placed between individual turns and layers of the windings (Figure 1.36A). Kraft paper and pressboard paper are the two principal materials employed. Paper or synthetic enamel having sufficient dielectric strength is suitable for strip or wire. For low voltages, turn insulation enamel on wire is sufficient, but paper is still needed for larger insulation at higher voltages. In some cases, enamel and paper are both used.

   Regardless of the type of winding, paper spacers or wood are also provided to form an oil duct to allow circulation for

**Aluminum sheet windings are under development for possible use in very large power transformers.*

cooling (minimizing heat from copper losses), as well as for insulation (Figure 1.36B). The ducts have to be constructed to permit circulation of the liquid or air up through the coils.

b. Insulation Between Windings
Insulation is also required between windings or from coils to ground. Selected pressboard or synthetic resin-bonded cylinders and end insulation, such as paper washers, collars, or spacing blocks, are the major items.*

TABLE 1.4

COMPARISON OF PHYSICAL PROPERTIES OF ALUMINUM AND COPPER

Property	Aluminum	Copper
Electrical Conductivity at 20° C Annealed	62%	100%
Weight lbs. per Cubic Inch at 20° C	0.0975 (1.56 oz.)	0.322 (4.8 oz.)
Specific Heat	0.21	0.003
Melting Point °C	660	1083
Thermal Conductivity, 20° C Cal./Sq. Cm. Cm./Sec./° C	0.57	0.941
Mechanical Strength, Annealed 1000 psi	13	32
2500 KVA Transformer with 44 KV-HV winding	13,900 lbs.[1]	14,700 lbs.[1]

[1]Total weight of the transformer.

Details are illustrated in Figure 1.55. For certain very large transformers a so-called "soft" cylinder is used. This is a specially bonded laminated cylinder. It is said that it can be impregnated with oil, thus increasing impulse, hi-pot, and creepage strength.

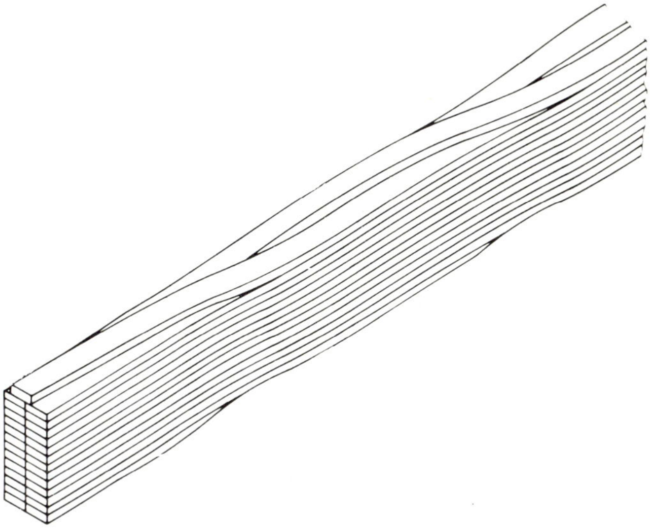

Figure 1.35 - Continuously transposed multi-strip conductor. (Courtesy of Feinberg, ***Modern Power Transformer Practice****).*

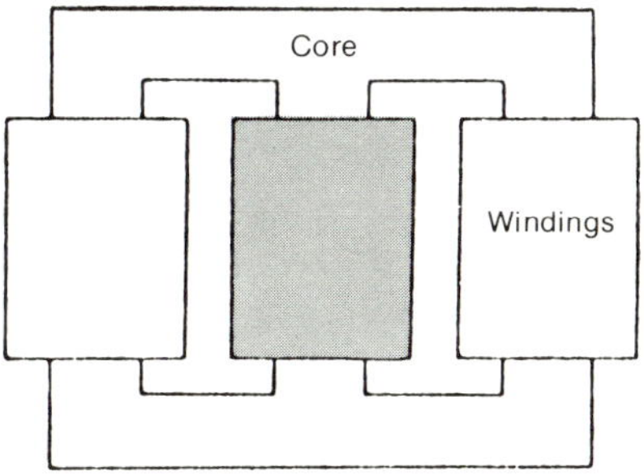

(A) Transformer Winding

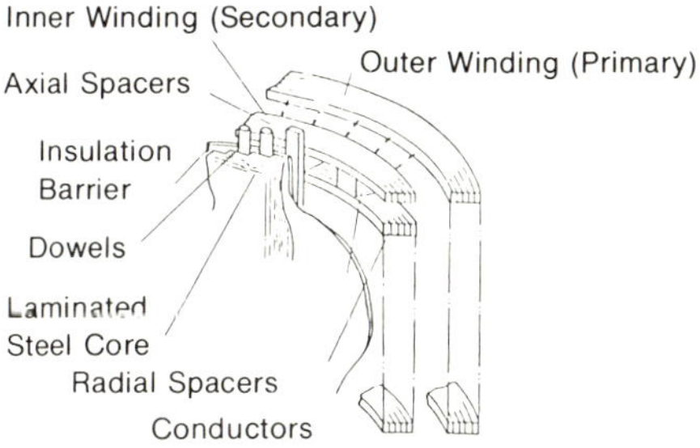

(B) Winding Configuration

Figure 1.36 - A typical transformer winding (A) and one type of winding configuration (B). The complete winding cylinder is illustrated in Figure 1.55.

Types of Winding Construction

A variety of coil types and arrangements have been used over the years. Detailed consideration is beyond the scope of this book.* The specific type depends on current and voltage requirements.

Most of the early transformer windings were rectangular in form. Nonetheless, examination of subsequent failures showed they should be circular in form. Since then, most windings have been of a circular design to help control mechanical forces.† Rectangular coils (Figure 1.37) are still in use but limited in practice to less than 10 MVA (and 69 KV).

Essentially, there are two principal types of coil arrangement for concentric core-form transformers—cylindrical (layer-wound, helical, spiral, or crossover) windings (Figure 1.38) and disk (single and double-wound) windings (Figures 1.39 and 1.40).

The layer wound (spiral type) winding is normally used for high currents (greater than 100 amperes).

The continuous disk winding is commonly used where the cylindrical or helical coil ceases to be economical for high voltage low current applications. A modified helical form (multi-layer) is also commonly used for voltages above 132 KV. Increased insulation may also be placed on both layers and usually at layer ends to increase creepage strength.†

The shell-form type construction generally uses interleaved rectangular pancake coils (Figure 1.41). This same design is employed for rectifier and furnace transformers where the mechanical stresses are quite severe.

Coil Clamping Construction

A transformer undergoes electrical stresses (conductor vibration, lightning, etc.) and mechanical forces (short circuits). Short circuits constitute a high percentage of all transformer failures.‡ Coil clamping

See Stigant, **J & P Transformer Book (1973) and Feinberg, **Modern Power Transformer Practice** (1979).*

†Chapter 2, Part 4.

‡Chapter 2, Part 3, Table 1.19.

is, thus an important factor in the design of *larger* power transformers.* As with core construction, clamping of the windings minimizes these forces and in larger units provides a means for taking up insulation shrinkage. In most types of windings adjustable coil support is necessary, and is generally provided for high voltage windings only.

High tensile strength copper alloy and oil impregnable top and bottom coil support insulating materials are also used in conjunction with the coil clamping mechanism (Figure 1.42).†

Figure 1.37 - A three-phase rectangular coil transformer assembly. (Courtesy of Westinghouse Electric Corp.).

*Stigant, ***J & P Transformer Book.***

†For larger windings it is common practice to use a resin-pattern-coated layer paper to increase bonding of the coil structure, thus increasing short circuit withstand strength.

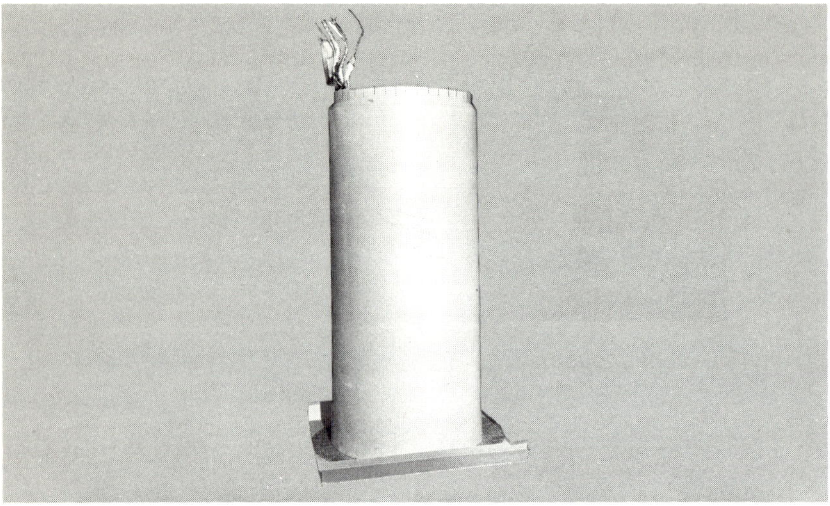

Figure 1.38 - A typical cylindrical type winding for core type transformers.

Figure 1.39 - A core type transformer power element assembly, complete prior to tanking with a disk-winding.

Figure 1.40 - A continuous disk coil being wound. (Courtesy of H.K. Porter Co., Inc.).

Figure 1.41 - Individual coils of an interleaved pancake winding for a shell-type transformer. (Courtesy of Westinghouse Electric Corp.).

Figure 1.42 - High strength, one-piece top coil support permits uniform clamping pressure on all coils and effectively transmits short-circuit forces to the clamping system. (Courtesy of General Electric Company).

During manufacture all windings of larger equipment should be progressively tightened during vapor phase (or other method of) drying, as well as when the completed assembly is finally placed in its tank. A locking device may be fitted to the adjusting screw to prevent any slackening.

Prior to final assembly, RTE-ASEA relies on pre-compressing four times the coils and spacers in a hydraulic press to pressures exceeding maximum forces the unit will experience.

The DYNA-COMP™* adjustable clamping system illustrates one of the most recent types of spring loaded dashpots (Figure 1.43). This design assures a tight coil throughout the assembly, shipping and service life of the transformer.

The coil springs on the dashpots, by providing a constant follow-up pressure, prevent any loosening of the windings.

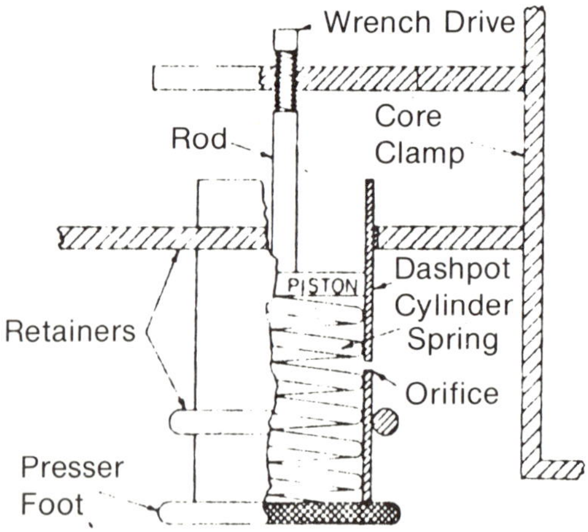

Figure 1.43 - Hydraulic dashpot, part of the new DYNA-COMP™ adjustable coil clamping system, assures a tight winding structure and prevents winding movement under short circuits.

**DYNA-COMP is a registered trademark of the General Electric Co.*

The Insulation System

Historical Survey

Modern transformers include various insulating materials which together form an insulation *system*.

- Pressboard (approximately 1/8"-1/2" thick)
- Kraft paper (approximately 5-20 mils thick)
- Manila and hemp paper
- High density particle-board
- Pressboard collars, and end insulation
- Laminated (plywood type) particle-board
- Enamels
- Inorganic and organic core lamination coatings
- Porcelain
- Epoxy powder coatings
- Maple wood structural forms
- Vulcanized fiber
- Cotton
- Plastics and cements, adhesive tapes, glass fiber bands, etc.
- Liquid dielectric fluid (except air-insulated or gas-insulated equipment)

This system isolates the transformer windings from each other and from the ground—to "insulate" the current-carrying parts of the transformer from the magnetic iron and structural steel parts. The insulation is thus more than "merely a mechanical means for keeping the wires apart" which was the only acknowledged purpose in the development of early equipment.

> Insulation materials and insulations were looked upon (in the 1890's) as a main division of the engineering work in the design of electrical apparatus.

Workmen were instructed in those early days in terms of:

- Cross section of iron (core)
- Physical dimensions of iron (core)
- Size of copper (windings)
- Number of turns on primary and secondary (windings)

That was it! What about insulation?

Well, that was at the workman's discretion! Figure 1.44 illustrates one common insulation "acceptance test" method. Given "A" and "B" as insulated windings, if the light bulb glowed, then the unit was *under-*insulated. Needless to say, this was a crude way of checking what we now know is the *most important* part of the transformer to maintain. The integrity of this insulation system spells the difference between equipment life or failure.

Physicist Joseph Henry is credited with developing the first transformer insulation (1832). Using strips of cloth from an old petticoat, Henry varnished them and wrapped them around a wire, thus forming a primitive insulation system.

In the late 1870's, uninsulated copper wire was laid as part of an incandescent lighting system. It was believed that a pressure of 100 volts would not require insulation, but the first attempt to transmit current in this fashion failed. Thomas Edison, in 1879, experimented with underground cables using hollowed out logs, cloth, tar, and resinous junk in the development of insulation systems. At length Edison brought forth a successful compound of tar mixed with oxidized linseed oil, paraffin, and beeswax.

Most of the early insulating materials used in the construction of electrical apparatus were thus manufactured for other purposes and merely adapted to electrical uses as they happened to apply. As another example, in the early 1890's, tapes originally intended for the millinery and dress goods trade were used for insulating tapes. On one occasion, samples of practically all the various types of fabrics obtainable from a large department store were tested for adaptability as insulation.

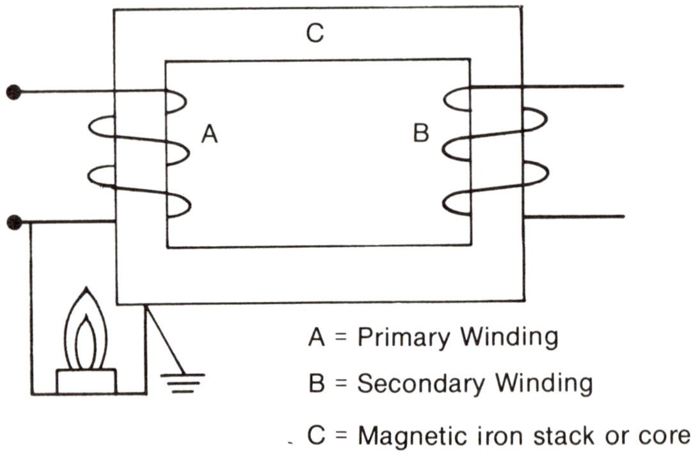

Figure 1.44 - Insulation acceptance testing in 1890.

The first transformers were sufficiently insulated using asbestos and low-grade pressboard (jute, cotton or a mixture of other vegetable fibers) in air. The introduction of shellac-impregnated papers in about 1900 was a tremendous improvement. Soon, however, shellacked paper no longer satisfied the higher temperature requirement of oil-filled transformers. In subsequent years, various forms of synthetic impregnating or thermally upgrading materials have been used.

While such detailed items have changed, however, the most widely used transformer insulation systems continue to use two *basic* items—a liquid insulation (transformer mineral oil, askarel or silicone, of which more than 90 percent are oil-filled) and solid insulation (Kraft paper, pressboard, wood; that is, cellulose products).

Functions of Mineral Oil

Surrounding the coils, core and solid insulation materials of most power transformers is the most common insulating liquid, mineral oil. This fluid serves three primary purposes:

- Provides dielectric strength
- Provides efficient cooling
- Protects the insulation system

Functions of Solid Material

A solid material "insulates" because it possesses two distinct properties. First, it has the ability to withstand both electrical and mechanical stresses due to the voltages used; secondly, it is such a *poor* conductor that a negligible but small current can flow through it and leak away. In other words, a good insulator will neither allow the current "to break through it" nor "to steal through it".

Consequently, a practical insulation system must contain material that performs these four major functions:

- The ability to withstand the relatively high voltages encountered in normal service (dielectric strength). This includes both impulse and transient surges.
- The ability to withstand the mechanical and thermal (heat) stresses which accompany a short circuit.
- The ability to prevent excessive heat accumulations (heat transfer).
- The ability to maintain desired characteristics for an acceptable service life period given proper maintenance.

Any weakness of insulation may result in the failure of a transformer. Insulation is deteriorated when it has lost a significant portion of its original dielectric, mechanical or impulse strength. The continuation of the deterioration process leads to inevitable mechanical and/or electrical failure.

Classifications of Solid Insulating Materials

The effect of temperature upon insulation materials is so significant that the thermal characteristic has become the basis for the classification of electrical solid insulation (Table 1.5).

Benchmarks descriptive of the various classes of insulating materials have been developed since the turn of the century for use in their identification.

The first thermal classification scheme consisted of only three classes: natural fibrous material (O), such as paper, cotton, linen; varnish gums and natural fibrous materials combined with oil (A); and inorganic fibrous materials (B), such as asbestos.

A seven-class international system replaced an earlier four-class system in 1958 as newer materials withstand temperatures above 130° C. Even this system is outdated, but it has proven very difficult to agree on a new one.

These temperatures should not be confused with the actual temperatures at which the insulating materials may be used in particular environments (air, oil, or gas) or with the temperatures on which specified *temperature rise* in equipment standards are based. These temperature classifications refer only to the thermal evaluation of insulating materials themselves or in simple combinations. As an example, some materials suitable for operation at one temperature in air may be suitable for a higher temperature when used in a system operated in an inert gas atmosphere or in oil. Likewise, some materials when operated in dielectric liquids may have lower or higher thermal endurance than when operated in air.

Even though the thermal characteristics are recognized as the most important, other factors such as mechanical strength and moisture resistance are required in varying degrees for the successful use of insulating materials.

TABLE 1.5

THERMAL CLASSIFICATION
OF ELECTRICAL INSULATION[1] (IEC)
INTERNATIONAL ELECTROTECHNICAL COMMISSION

Class Designation[1]	Maximum Temperature Permissible	Typical Materials
Class 90 (Y) or (O)	90°C	Unimpregnated cellulose, cotton silk
Class 105 (A)	105°C	Impregnated cellulose, cotton or silk; phenolic resin
Class 120 (B)	120°C	Cellulose triacetate
Class 130	130°C	Mica, glass fiber, asbestos with organic binder
Class 155 (F)	155°C	As in Class 120 with suitable binder
Class 185 (H)	185°C	As in Class 120 with silicone binder
Class 220	220°C	As in Class 185
Class over 220 (C)	Above 220°C	Mica, porcelain, glass quartz and similar inorganic materials

[1]The system based upon numerical values of maximum temperature rating is preferred to the letter symbols by IEEE. The International Electrotechnical Commission (IEC) is responsible for the preparation of world-wide electrical and electronic standards; yet is little known to the average American engineer. IEC is composed of 43 National Committees which represent 80 percent of the world's population producing and consuming 95 percent of all electrical energy. IEC standardization programs are developed by 190 technical committees and sub-committees that now span virtually every sphere of human endeavor in electrotechnology. (Chapter 2, Part 4, Figure 2.30).

Oil-Impregnated Cellulose

Cellulosic insulation and mineral oil still form the strongest known electrical-mechanical insulating system for power and distribution transformers, and will be our primary focus.

Power transformers are mostly or entirely insulated with various forms of paper—tapes, sheets or other pressed shapes. While there has been no basic change in the principal materials (oil and cellulose paper) used in transformers for at least 50 years, there have been improvements, both in the materials themselves and the method of manufacturing. But how did oil-impregnated cellulose become accepted as "the insulation cornerstone of the electrical industry"?

Historical Acceptance

Through the process of experimentation, paper fibrous materials became in a short time the accepted insulation standard. The outstanding features and benefits of cellulose include:

- Availability and low cost
- The relative ease with which it can be applied
- Flexibility and simplicity of handling and shaping when the working temperature is not high
- Moderate strength combined with lightness
- The facility with which these materials absorb impregnating material

Cellulosic insulations are rarely used without initial drying and impregnation with a second insulation material. Impregnation is basic to the effectiveness of the insulation system. Such materials are used in order to:

- Wet the fibrous components so as to maintain higher dielectric and chemical stability (mechanical consolidation).
- Seal the cellulose materials from moisture absorption.
- Fill any voids in order to eliminate air pockets in high voltage insulation that ultimately lead to dielectric breakdown.
- Prevent contact with oxygen.

In addition, the absorptive or "ink-blotter effect" of cellulose causes the absorption and breakup of air bubbles, whereas plastics and other synthetic materials have not thus far done this. A large air bubble caused during impregnation between windings could cause a flashover between them, resulting in a turn-to-turn failure. Cellulose alone aids in the prevention of this problem.*

Before considering the current practices in an insulation system, it is important to understand both the electrical and chemical structure of cellulosic materials.

Basic Electrical Structure

It is a fact that cellulose is the fundamental solid constituent of plant life; also the basic division of matter consists of atoms and molecules.

Atoms are constructed according to the plan of our solar system. The atom has a positively-charged central core or nucleus (corresponding to the sun) around which negatively-charged electrons of equal total charge are distributed in one or more shells or orbits, much in the same manner that Earth, Mars, Venus and the other planets revolve about the sun.

The ability of atoms to combine and form molecules is determined by these planetary electrons. The tendency in combining is to form arrangements in which the electron shells are completely filled.

When an atom is subject to an electric field in a current conductor, negatively-charged electrons and the positive nucleus are attracted toward opposite electrodes. In an insulating solid, however, the electrons are not really detachable from the atoms. If one surface of a glass plate, for example, is maintained negatively and the other positively, the electrons in the dielectric do not drift throughout the plate but undergo only a slight shift in *displacement*.

While it is difficult to free an electron from such a molecule, electrons do become free to form a conduction current when the insulation breaks down. The prime example of change in conduction current with deterioration occurs in cellulose which becomes contaminated with moisture.†

**Certain well known synthetic films impede oil circulation and may cause a disparity in stress distribution where there is a difference in permittivity between the oil and the solid film (Chapter 2, Part 2, Table 2.2).*

†Chapter 2, Part 5.

Basic Chemical Structure

Cellulose is one of a number of vegetable substances formed from repeated glucose units (recognized in everyday language as sugar). Most authorities recognize this basic molecular "building block" in the cellulose molecule as having the chain structure of Figure 1.45.

The molecular formula for cellulose is accepted as $(C_6H_{10}O_5)_n$. The degree of polymerization (that is, the number of repeated units making up the molecules, indicated by the letter n) varies widely depending on the source of the material and the method used for its formulation. The molecules of most insulations are composed of many atoms bonded in a complex arrangement. Typically, cellulose is made up of a chain of about 1200 glucose rings. Molecular weights as high as 1,500,000 have been reported where the molecules usually consist of a few thousand of these basic "building blocks."

Cellulosic Composition and Properties

Cellulosic materials used for electrical paper are ordinarily manufactured from the coniferous woods pulped by the Kraft process. This is usually northern spruce but some percentage of pine has been used only for pressboard. The Kraft process consists of boiling chips of wood with an alkaline solution of sodium sulfide and sodium hydroxide, using a sodium sulfate "make-up" solution from which the process gets its name.

While the number of atoms in a molecule is great, the types of atoms may be few. Cellulose consists of carbon, hydrogen and oxygen, while transformer oil is predominantly made up of carbon and hydrogen only. The length of the cellulose molecular chain appears to be limited only by its ability to withstand thermal and mechanical rupture.

Further observation of the cellulose structure (Figure 1.45A) indicates there are three hydroxyl groups (OH) in each glucose unit.

Figure 1.45B is an enlarged diagram of a single glucoside unit. Notice that one group is a primary hydroxyl and two are secondary hydroxyls. It is the presence of these secondary hydroxyls as potential water formers that is responsible for the objectionable characteristics of cellulosic paper. In addition, these hydroxyls can be considered as centers around which various polar molecules (for example, acids, alcohols and water) are attached by the hydrogen bonds. These two factors constitute natural "seeds of destruction."

The cellulosic molecule forms a variable mixture of crystalline regions in which there is some degree of order and amorphous regions in which the chains form a random condition. It is this amorphous type material with voids (air space between fibers) which accounts for the water absorption capacity of the cellulose fiber (Figure 1.46).

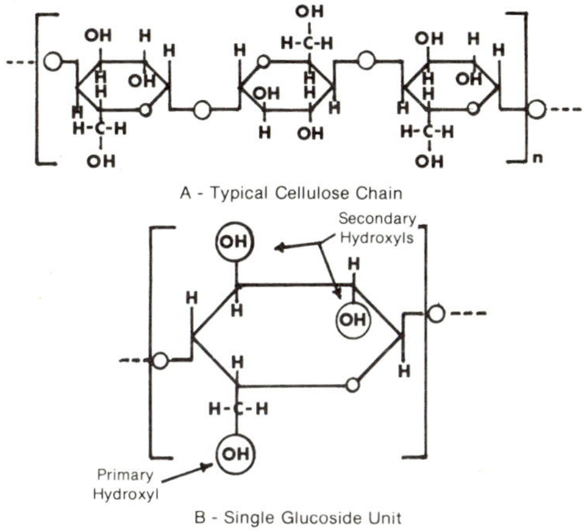

Figure 1.45 - Chemical structure of cellulose.

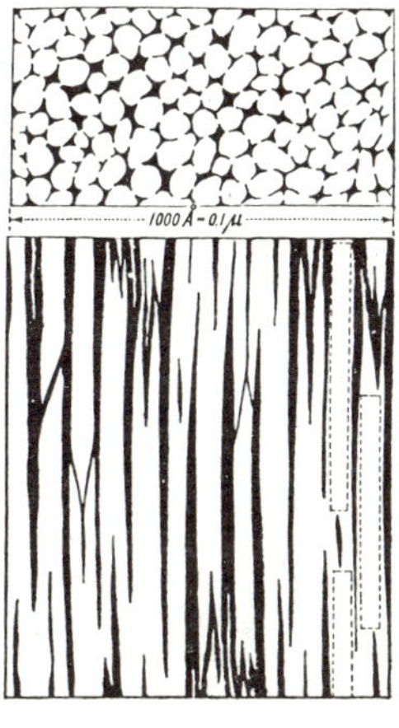

Figure 1.46 - Crystalline cellulosic structure. ***Frey-Wyssling****. The upper part is a transverse section, the lower a longitudinal view. (The longitudinal view is greatly reduced in size, the crystalline regions being actually much more than ten times as long as wide). The dotted lines signify the crystalline regions through which the long-chain molecules pass and the black lines signify the voids.*

Fibers of cotton flax and manila (or hemp) are also used, generally obtained from rugs and old cordage which are also broken down by boiling in alkali. The suspension of fibers is thoroughly washed and then "beaten" by passing between fixed and rotating blades or bars and finally matted on travelling gauze and dried on travelling felt.*

As existing in nature, cellulose is not pure but is found in a mixture with a variety of organic substances. The non-cellulosic but needed portion is called lignin (Figure 1.47). Even after such treatment, the paper contains up to as much as 7 percent impurities. Among the impurities is a certain amount of moisture that cannot be removed because of the colloidal nature of cellulose. Such a condition is found in peat; for example, this material may appear dry to the touch yet contain up to 75 percent moisture. This extreme example is given only to illustrate that there is no guarantee that cellulose is completely free of moisture, even during the manufacturing stage. However, there are characteristic differences between fibers from different sources—cotton fibers contain the highest proportion of pure cellulose, while manila fibers are among the strongest. Kraft-processed spruce from Norway, Sweden, Finland and Canada has proven to be the best overall material.

Finally, in contrast to plastics which are subject to the "crisis prone world" of petrochemistry, cellulose may be commercially in supply on a long-term basis since it is a renewable resource.†

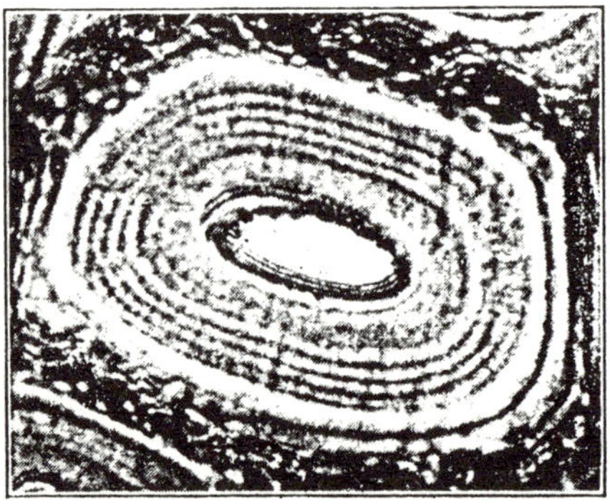

Figure 1.47 - Transverse section through wood fiber showing concentric layering of cellulose (light) and lignin (dark). (Courtesy of I.W. Bailey).

**With increased usage of nylon, manila is no longer available in sufficient quantity for use. As a result, virgin hemp fibers are now in use.*

†A pulp shortage took place in the early 1970's. Canadian sources indicate that a shortage may occur in the near future because of lack of mill capacity for unbleached Kraft pulp.

Criteria in Selecting Paper Insulations

The most significant factors in choice of paper insulations consist of:

1. Mechanical Strength
   Manufacturers have considered mechanical strength as the best measure of deterioration; however, they have disagreed as to the best particular criteria to evaluate the endurance (thermal stability) of an insulation paper.

   Four particular physical tests have been emphasized—tensile, bursting, tearing strength, and toughness.

   a. Tensile Strength (and elongation)
      The maximum stress a material subjected to a stretching load can withstand without tearing, expressed in pounds per square inch (psi) or kilopascals (KPa) per square meter.

   b. Bursting Strength
      A measure of the ability of a material to withstand pressure without rupture (breaking), generally expressed in pounds per square inch.

   c. Tear Strength
      The force needed to initiate or to continue tearing a sheet or fabric, generally expressed in grams.

   d. Toughness (folding paper or blending strength for pressboard)
      A measure of the amount of energy required to rupture (break) a material, expressed in inch-pounds.

   While one manufacturer most often has used the bursting strength criteria, the consensus by several authorities has been the use of tensile strength as the most meaningful and easily measured quantity. There are also those who have questioned the real meaning of any of these simulated laboratory test results as compared with actual results in an in-service transformer.*

   In any event, tensile strength has proven to be "the best handle" anyone has developed thus far to approximate material endurance.†

**Chapter 2, Part 4.*

†Tensile strength has also formed the basis for the loss of life and overload guides (Chapter 2, Figure 2.28).

2. Density (Compactness)
   The weight per unit of material volume generally expressed in pounds per cubic foot or grams per cubic centimeter, measured under defined conditions of temperature and physical condition.

3. Impermeability and Porosity
   The ratio of the volume of the cellular structure to the volume of its mass, generally expressed as a percent.

4. Uniformity
   Mechanical strength depends on the type of fiber and is usually greater the longer the fiber. High density and porosity, provided they are obtained in the right way, tend to increase the endurance or impulse strength. This can be achieved by more intensive beating and shorter fiber length.

   The cellulosic absorption of water weakens the interacting forces between the secondary hydroxyls (Figure 1.45 B). Thus, as the cellulosic structure becomes unstable, a rapid decrease in mechanical properties takes place.

Major Steps in Insulation Paper Improvement

Three major steps of cellulosic research and development to produce improved insulation papers have taken place: modification of the cellulosic fiber (Permalex®, TherMEcel®); use of stabilizing chemical additive (Insuldur®); and improvement in the paper-making process.* The first major improvements were made about the same time in the late 1950's.

In the Permalex System the more stable cyanoethyl groups replaced the less stable secondary hydroxyl groups. Though it proved to have two to three times the life in tensile strength (and therefore gave a 20° C temperature benefit), its ultimate phase out in usage was because of its higher cost.

The Insuldur® approach was the addition of an Amine-type chemical stabilizer which offered the advantage of somewhat lower cost.†

The second major "chemical modification" called TherMEcel® was a combination of additives chemically bonded to the cellulose (not mere

**Registered trademarks of General Electric, McGraw-Edison, and Westinghouse respectively.*

†Three chemical agents of which the primary active one was dicyandiamide.

mechanical holding); therefore, this advance material protected the Kraft fibers from the normal destructive reactions within a transformer.

Figures 1.48, 1.49, and 1.50 illustrate typical comparisons between these thermally upgraded papers and untreated Kraft paper. Each conceivably has proven its own advantage. However, data has also shown that consideration of only *one* property can be misleading; therefore, evaluation of any improved insulation system should be based on several factors rather than a decision based solely on one criterion.

Most recently, for cost measures and at the instigation of the paper manufacturers, the standard additive is now dicyandiamide which is added at the paper mill. Virtually all manufacturers now use this type of thermally upgraded paper. Recent laboratory tests, such as tube aging and accelerated heat-aging tests (Figures1.51 and 1.52 respectively) illustrate the contrast between standard Kraft and dicyandiamide-treated (Amine) Kraft papers. In-service use of these improved papers has proven superior to standard Kraft papers, giving 25 to 30 percent longer life in addition to better heat resistance.

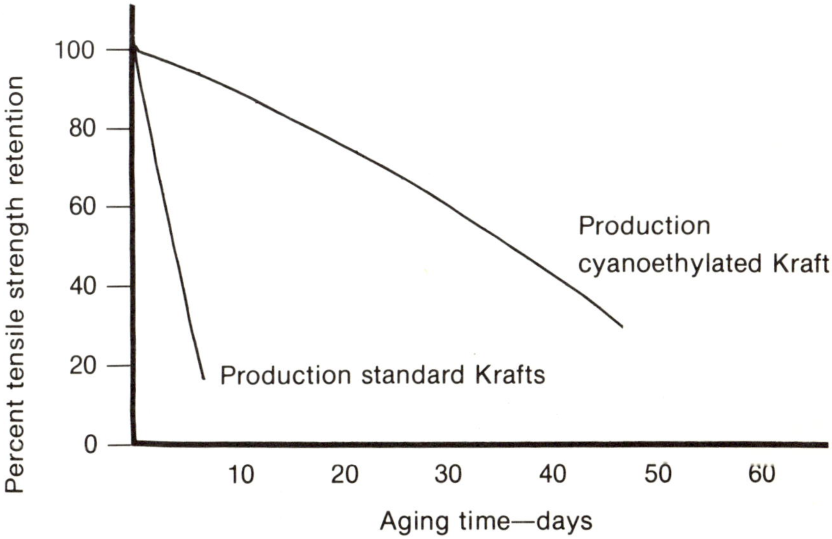

Figure 1.48 - Comparison of tensile strength aging characteristics of two production lots of cyanoethylated Kraft paper with standard Kraft paper. Paper aged in transformer oil at 175° C. (Courtesy of General Electric Company).

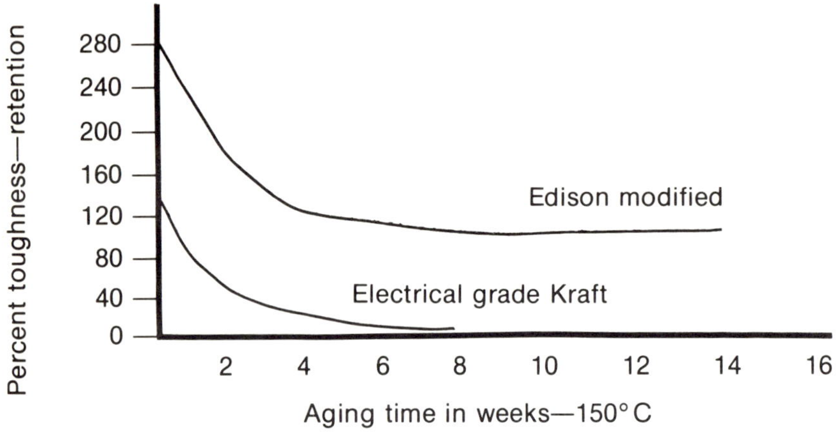

Figure 1.49 - Comparison of absolute values for toughness. (Courtesy of McGraw-Edison Company).

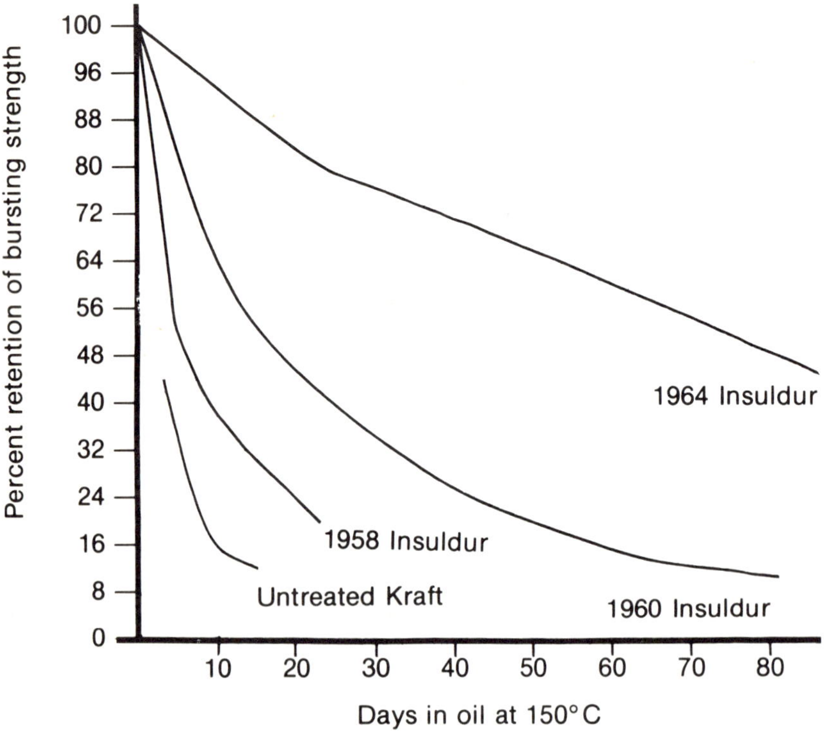

Figure 1.50 - Percentage of retention of bursting strength. (Courtesy of Westinghouse Electric Company).

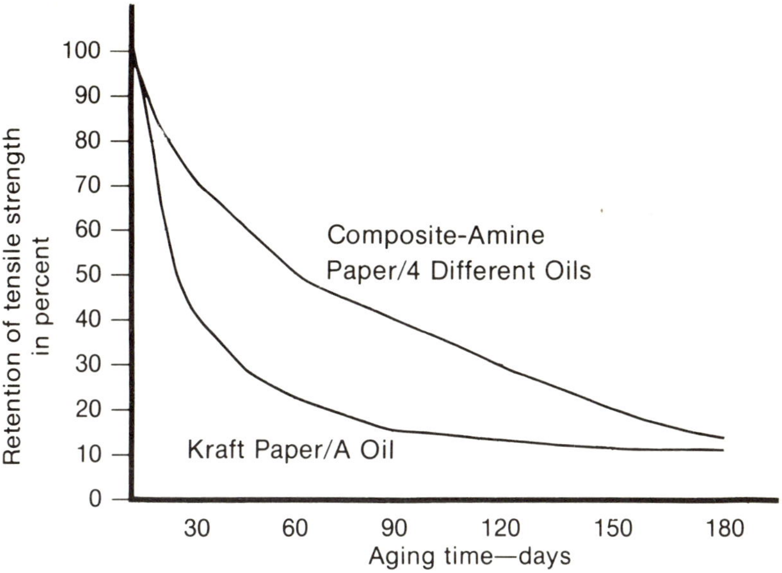

*Figure 1.51 - Heat-aging at 160° C in nitrogen gas.**

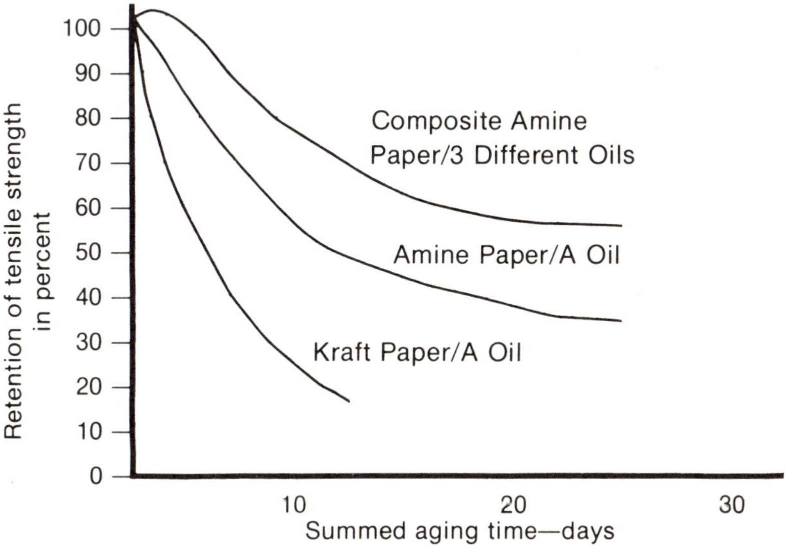

*Figure 1.52 - Accelerated heat-aging of transformer at 180° C hot spot temperature. Oil temperature 150° C.**

**Yasufuku, et al., EIC (1979).*

Current Practice for Transformer Insulation Systems

The usefulness of electrical equipment is dependent upon the *integrity* of its insulation system. Various components make up this system.

In an ideal system all components should give the same insulation life and performance at expected operating temperatures. However, in practice one insulation material will frequently out perform some of the others. Moreover, all the components should remain stable and compatible under all operating conditions. But again in practice, change in one may lead to damage of another material. Two such examples of system deterioration include:

1. A Dry Type System
   Asbestos paper and wire film insulation are compatible at normal temperatures, but at elevated temperatures and subsequent cooling a binder material in the asbestos paper might react with the wire insulation and cause its separation from the wire, with ultimate severe transformer damage.
2. An Oil-Paper System
   System-aging liberates free moisture from initially dry paper, which then reacts with film insulation and reduces its effectiveness severely.

In both instances, the insulation system components will individually perform satisfactorily under the expected conditions but in the system—in the combination of materials—chemical or physical reactions between the components occur which, if ignored, may lead to transformer failure.

Criteria in Determining the Insulation System

Selection of an insulation system includes at least five factors:

- Preferred core location
- Required end-use purposes
- Required heating and cooling
- Location of apparatus
- Desired basic impulse level (insulation class)

1. Preferred Core Location
   Recall that some manufacturers designate transformers by the location of the iron stack or the core in relation to the

windings—either core or shell type construction. When the core is the central portion surrounded by the copper windings, the transformer is a core type. When the core surrounds the windings like a shell, the unit is of the shell type. No sharp dividing line exists between the two common types; either may be used above or below the ratings mentioned. By contrast, other manufacturers only use core-form construction, regardless of size required.

2. Required End-Use Purposes

   a. Standard Size

      1) Distribution Transformers
      These limits cover ratings of 500 KVA or less. Every transformer must be protected against short circuits and other abnormalities; therefore, a minimum amount of insulation materials is required.

      2) Power Transformers
      These limits usually have ratings in excess of 500 KVA. However, as transformers are called upon to handle larger blocks of power, more insulating materials are required. New materials are being developed for greater mechanical strength to cope with lightning, more frequent and more severe short circuits, and transient surges.

   b. Special Function
   Transformers used for such purposes as furnace loads or rectifiers are generally classified as special type. They are normally regarded as distribution types, even though they may overlap into both standard categories.

      1) Arc-Furnace Transformers
      Electric arc melting processes require high amperage currents at low voltages. Electrode short circuits during the melting process make high demands on the short circuit strength. These short circuits are a normal operating occurrence rather than a rare fault condition as with a standard power unit. Although the equipment is similar to the ordinary insulated power unit due to the extreme variation in the load current, these transformers are heavily constructed with extra supports for the windings. These units also require

special bracing to withstand the mechanical forces caused by the high, widely-fluctuating currents and require extra insulation between turns to guard against high voltages. The insulating material must be carefully dried so that no loosening of the windings occurs in-service. These units thus have need for large connecting bus bars and such units are very difficult to seal tightly and frequently leak oil.

2) Rectifier Transformers
The older mercury vapor type rectifiers are being phased out by the use of silicon controlled rectifiers (SCRs). In both cases, they are designed to develop high dc voltage for short duration in the electrometallurgical and electrochemical fields. These units require greater mechanical strength to resist deformation from short circuits.

3) Other Special Purpose Units
Such devices as autotransformers, network, precipitation and converter station transformers (dc transmission stations) and voltage regulators are only a few of the numerous other types of transformers in use today.*

3. Required Heating and Cooling

a. Choosing a Coolant
A second classification by type includes the coolant employed in a transformer to limit the temperature rise of a unit. For transformers, the most common substances are air, oil and askarel. For environmental reasons, askarel or PCBs (polychlorinated biphenyls) are now being gradually replaced by substitute synthetics or by non-liquid cooling.

In the dry type unit, air or gas (such as SF_6) is the coolant whereas in the others, the windings and core are completely immersed in the liquid. Its surge voltage strength is only about half that of an oil-filled unit.†

*See Feinberg, ***Modern Power Transformer Practice*** (1979); and Pansini, ***Basic Electric Power Transformers*** (1976).*

†*Chapter 2, Part 1.*

The transformer nameplate will indicate a "class" designation (a group of letters) referring to the type of insulation medium and the method by which the transformer is cooled.*

b. Choosing a Method of Oil Preservation
Three basic methods of oil preservation are employed:

- Free breather (older units)—Open to the atmosphere.
- Conservator tank—A small separate tank connected by a small diameter pipe to the main transformer tank and mounted above it. Its purpose is to exclude moisture and reduce area of oil in contact with the atmosphere.
- Sealed type unit—The interior of the tank is sealed from the atmosphere but the top gas space is filled with air or an inert gas, usually nitrogen.†

4. Apparatus Location
Separate insulation classifications are required for units located "indoors" or "outdoors." Normally, dry type or fire/explosion-resistant liquid-filled equipment is used indoors. In contrast, indoor oil-filled equipment, as fire hazards, must be enclosed in a vault in accordance with the latest *National Electrical Code,* and state and municipal laws. Outdoors, the commonly used oil-filled unit would be subjected to greater overvoltages because of possible lightning strikes.

5. Basic Impulse Level
Traditionally, each transformer winding has had an insulation class. This class depends essentially on the unit's voltage rating. Since the operating voltages are relatively high, special attention is given to the insulation associated with various parts of the system. The insulation has to withstand not only the normal operating voltage, but the possible surge overvoltages caused by lightning or switching. A certain insulation class is then decided upon.

Over the years, however, the use of the term "insulation class" has decreased and the "handle" for a series of required dielectric tests has become the Basic Impulse Level or simply "BIL."

**Chapter 1, Part 2, Table 1.10.*

†For additional details, see Chapter 1, Part 2, Figure 1.59 ff.

The BIL is the crest value of the impulse voltage which a given unit is required to withstand without failure. These designated minimum levels may be from 2 to $3\frac{1}{2}$ times the normal operating voltage depending on the degree of reliability desired. Figure 1.53 illustrates an analogous situation between electrical and water flood control.

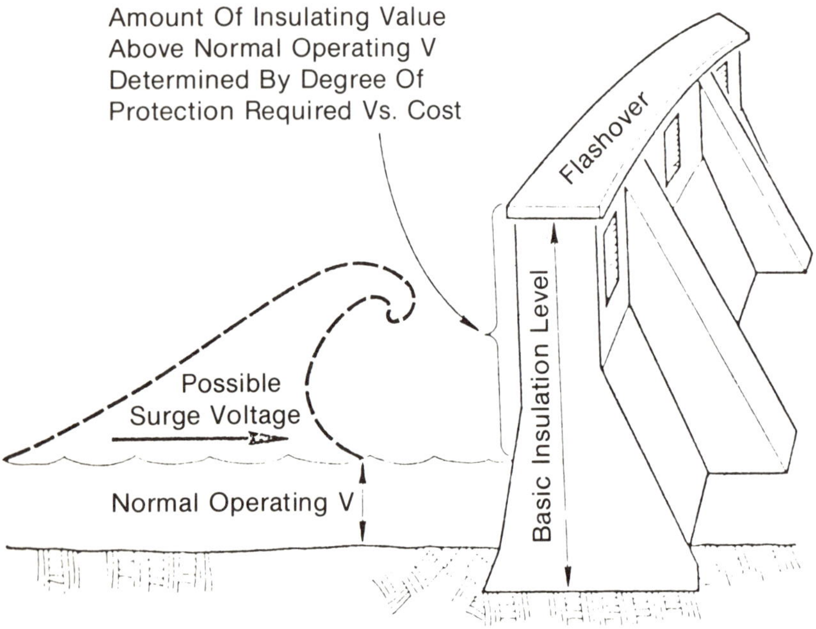

Figure 1.53 - The basic insulation level (BIL). Pansini, ***Basic Electrical Power Transmission*** *(1975), p. 102.*

Construction of Transformer Insulation Systems

Solid insulation is required in a transformer, wherever a difference in potential exists between two points. The selection of insulation is generally made in proportion to the anticipated overvoltages and with a safety margin to compensate for decreases due to normal service aging. Various components are designed to work best together and achieve which is called the *coordination* of *insulation* within the insulation system.

1. Insulation Components
Transformer insulation falls into one of the following three categories:

   a. Major Insulation
   The heart of the transformer's insulation, as defined by IEEE standards, consists of the insulation between the high voltage and low voltage windings in the *same* phase and from the windings to ground. This is commonly known to the electrical designer as "major insulation". (Examples: thermally upgraded Kraft paper cylinders and layer sheets; high density organic sticks bonded with synthetic resin; and pressboard sheets for phase insulation and collars).

   b. Minor Insulation
   The insulation between adjacent turns in a winding and between different sections of the *same* winding. (Examples: pressboard spacers, tape, and synthetic enamel wire).

   c. Phase-to-Phase Insulation
   The insulation between the windings of *different* phases. (Example: heavy Kraft paper or pressboard).

   These insulation systems are illustrated in principle in Figure 1.54.

   Constructional details of a Class 105 oil-filled cellulose system are illustrated in Figure 1.55.*

   A minimum of two wraps, three mils thick, are used on a conductor. *Lap* wrapped is when the paper overlaps the previous turn; *butt* wrapped butts or touches the previous wrap.

2. Kraft Cellulosic Paper
Electrical grade Kraft wood pulp papers or modified Kraft papers remain the most widely used of all insulating materials. It is applied as paper sheets and tapes. Kraft paper has been found to have greater thermal stability than other materials such as manila paper or hemp. Though Kraft insulation has a lower initial tensile strength than manila, Figure 1.56 illustrates how the effect of moisture on aging stability is less than that

**A simplified version of Figure 1.55 is also located in the Color Section, following Chapter 10.*

observed in the manila paper and thus explains why Kraft paper has been the standard insulating paper for more than 50 years. *

Kraft paper and electric grade dicyandiamide-treated Kraft paper are both standard materials used in modern transformers.

3. Composite Laminated Insulating Systems
The major insulation takes the form of enameled wire, coil winding cylinders, tubes, rods or insulated board.

   a. Resin-Enameled Wires and Conductors
   Resin-insulated wires consist of three types: enameled, fiber-covered, and multiple combinations. Of these, the enameled wire is the most widely used.

   Enameled wire not only acts as part of the dielectric material but this type of wire possesses an excellent space factor and may be used wherever the space requirements are critical, such as for small distribution units. Numerous types of enameled wires are available for different environments, temperatures and under voltages at all frequencies.

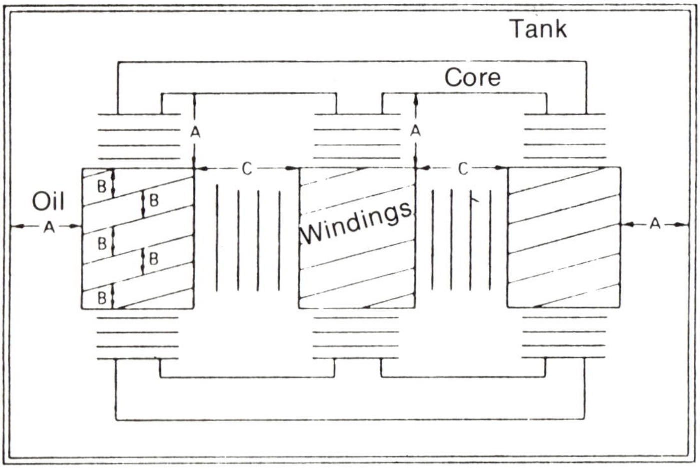

Figure 1.54 - Pictorial diagram of three classes of transformer insulation. (A) major insulation, (B) minor insulation, and (C) phase-to-phase insulation. IEEE Tutorial Course, ***Applications for Distribution and Power Transformers,*** *#76 CH 1159-3-PWR (1976), p. 28*

** Manila or hemp is still used as one layer on transposed cable (larger power transformers).*

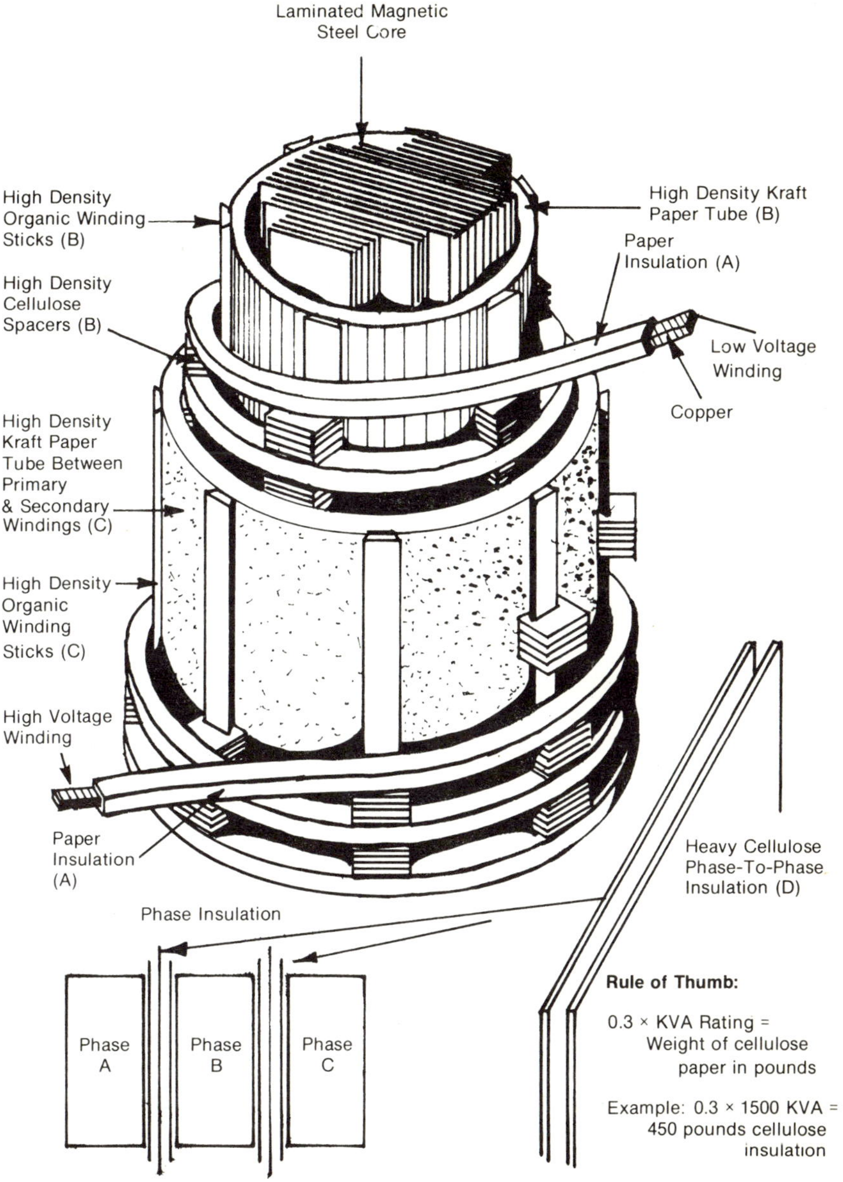

Figure 1.55 - Basic insulation system of a core type power transformer where (A) is insulation on wire (minor); (B) is insulation to ground (major); (C) is insulation between windings (major); and (D) is insulation between phases (phase-to-phase).

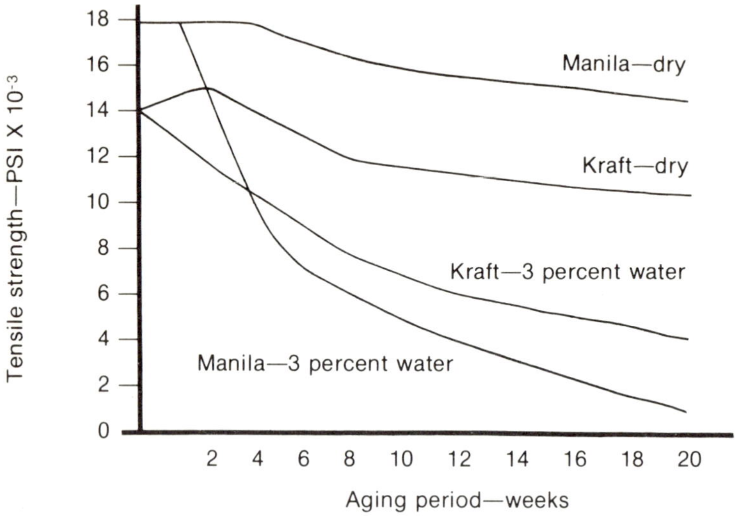

Figure 1.56 - Comparing the aging of 0.003 inch manila and Kraft cable papers at 100° C in sealed glass tubes from which the air has been displaced by nitrogen gas. Clark, ***Insulating Materials for Design and Engineering Practice*** *(1962).*

The first coils were in a dry transformer and were insulated with cloth tape and paper. For oil-filled transformers, manila and cotton wrap were long applied on the copper conducting members. Various combinations of paper wrapped enamel-coated wire have been developed in recent years (Table 1.6). Figure 1.57 illustrates the greater thermal stability between the old system, manila and cotton, and the new Formvar enameled wire as part of one thermally upgraded paper system. All seek to give higher stability and longer "Kraft" paper life. As a typical example for in-service transformers 200 KVA and up, especially where higher current capacity is required, bundles of copper conductors are used. The coils are double wrapped. Each strand is thinly-coated with Formvar enamel coating and then wrapped a second time with hemp and spruce pulp cable paper.

Through the years, a number of other materials have been tried and subsequently dropped either because they were found wanting in some respect or proved too expensive for general use:

- Thermoplastic materials, such as polyethylene and polypropylene (dissolved with heat)
- Mylar®* (limited temperature range)
- Teflon®* (too costly)
- Epoxy and acrylic enamels (inferior dielectric characteristic)

More recently, Formvar enamels are being modified to minimize solvent content that goes into the atmosphere; therefore, the solid content of these enamels was increased from 14 percent in 1958 to 26 percent in 1978. Unfortunately, in trying to meet environmental guidelines, loss of adhesion of the enamel to the wire has become evident; therefore, it appears that these long-used enamels may be on their way out.

An assembled wire structure has also been commonly impregnated with a resin such as polyester to furnish rigidity and to protect the unit against solvent and other environmental contamination. These insulated wires are most frequently wound into coil structures using high speed winding machines. Thus, enameled coating must be able to withstand this mechanical stress. The film thickness is generally less than 0.002 inch.

For years copper and aluminum conductor wires have been coated with solvent-based coating enamel but its future use is now limited. Epoxy powder coatings in which there is no loss to the atmosphere may solve the current dilemma. The use of electrostatic powder coatings has proven in pilot tests to be a reliable alternative for medium power transformers.

The epoxy powder insulation system provides many of the properties of wire enamels without solvent emissions. Epoxy coatings suffer less mechanical damage and are only up to $^1/_3$ as thick as paper. In addition, the epoxy insulation produces no water in contrast to the release of water through thermal decomposition of traditional papers. Also, while the paper shrinks during the coil drying process, the epoxy does not.

**Mylar and Teflon are registered trademarks of E.I. Dupont De Nemours & Co.*

TABLE 1.6

INSULATING COATINGS ON COPPER WIRE (OIL FILLED TRANSFORMERS)
Type of Enamel (Solvent-based)
Polyester-Amide, one of the first used
Acetal, used in many older units
Amine, 1958, thermally and hydrolytically stable
Polyester, thermally stable but hydrolytically unstable
Formvar, phenolic modified polyvinyl, early 1960
Electrostatic power spray; non-solvent type substitute

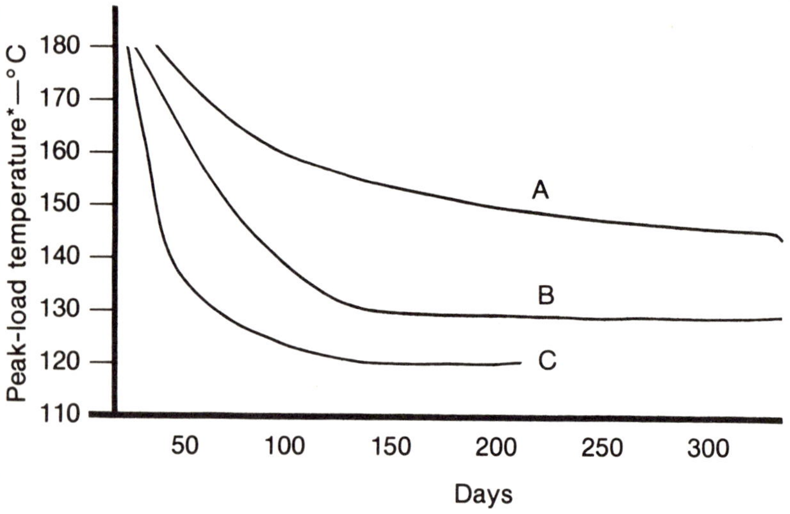

A—Permalex system Formex wire in both windings. Cyanoethylated Kraft paper insulation.
B—Prior system Formex wire in both windings. Untreated Kraft paper layer insulation.
C—Old system Manila paper and cotton turn insulation. Untreated Kraft paper layer insulation.

**In the range of about 130°-170° C these temperatures are approximately equal to percent load, based on ANSI loading guides.*

Figure 1.57 - Turn insulation dielectric breakdown. (Courtesy of the General Electric Co.).

The electrostatic method is well-suited for the continuous coating of wire. In this electrostatic cloud coating process, charged powder particles are coated on a grounded, moving wire by electrostatic attraction.

Recognize that insulation in the immediate vicinity of coil conductors is subject to the greatest thermal stress and is the hottest spot in the whole insulation system. The compatibility of multiple components becomes most significant in four areas:

- Temperature—The compatibility of multiple components becomes increasingly important as the temperature of such systems is increased. Various resins have maximum temperatures of general use ability.
- Moisture—The thermal stability of enameled wires can rarely be separated from the accompanying effects of moisture.
- Liquid insulant—Formvar has been incompatible with askarel.
- Selection of materials—If, during the design stage, the co-ordination of materials of the insulating system is not properly selected for compatibility, its effect may be the primary factor in the degradation of the enameled wire; therefore, when the purchase of a new transformer becomes necessary, it behooves the owner to buy units that have already demonstrated service reliability. Otherwise, the best maintenance procedures will be of no avail.

b. Resin-Laminated Structures

These resin laminated structures consist of two or more layers of reinforcing material (paper, wood, glass, asbestos paper or synthetic film) bonded together with an adhesive, usually by the application of heat and pressure. The exact composition of a structure varies as widely as the types and uses of transformers However, the phenolformaldehyde resin binder in combination with Kraft paper has been the most widely used of all the laminates. Unless the laminate contains a high resin content, it is susceptible to electrical degradation in moist atmospheres. Figure 1.55 pictures two Kraft composite tubes to ground and between the primary and secondary windings.

4. Comparison of Materials Used in Modern Transformers

The mineral oil-cellulosic combination is the most common, least expensive and most reliable system used world-wide, yet this insulating system remains the weakest link in the transformer. Therefore, our premise is that this part of the transformer requires early and prompt attention.

Normally, we think of the weight of a transformer primarily in terms of pounds of steel in the core and the coils, and so it is (Table 1.7). In addition, the amount of liquid in the unit comprises a somewhat large amount of known weight. Nonetheless, the surprise lies in how much paper is in a transformer, particularly compared to the weight of the insulating liquid. Again, Table 1.7 shows this comparison—and reveals that the weight of the paper in a transformer amounts to approximately ten times the weight of the mineral oil!* Indeed, we may presume that the life of the transformer could be best measured in terms of the life of the insulation system, particularly the cellulosic paper!

TABLE 1.7

COMPARING QUANTITIES OF MAJOR MATERIALS USED IN MODERN TRANSFORMERS

KVA	KV	Lbs. Steel (Core)	Lbs. Copper (Coils)	Insulating System	
				Lbs. Paper (Insulation)	Lbs. Oil (Coolant)[1]
3,000	13.2	9,000	3,500	1,000	100
10,000	115.0	34,000	10,500	3,540	330
16,000	115.0	40,000	11,000	4,050	480
20,000	132.0	58,000	13,000	5,760	600
30,000	154.0	79,000	21,000	8,020	730
40,000	230.0	116,000	16,000	10,600	980

[1]Assuming 8.25 pounds/gallon of mineral oil.

**Chapter 1, Part 3, Table 1.14. This ratio is increasing!*

Transformer Auxiliary Equipment

A transformer includes a number of auxiliary devices. Subsequent paragraphs will consider each of the various components (Figure 1.58).*

Tank Design (Core/Coil Assembly Enclosure)

The transformer case or tank, generally made of steel, provides mechanical protection for the complete core and coil assembly and a container for the liquid coolant/insulant, if one is used. Several different types of tank construction for oil-insulated equipment are found in-service. They protect the oil against air, moisture, and other contamination, and in some cases, provide for the expansion of the oil. This thermal expansion may range up to a 6 percent change in oil volume.

1. Free-Breathing Type or Open Type
   This is the older unsealed type with the air space above the oil at atmospheric pressure, which allows air to enter or escape from this space with pressure and temperature changes. This type may be equipped with an open, unidirectional, or dehydrating breather. Many arc-furnace transformers are of this type, with water-cooling coils inside the tank. The breathing device limits the intake of moisture by extracting water vapor from the air.

2. Conservator or Expansion-Tank Type
   Many medium or large transformers, especially in Europe, are equipped with an auxiliary expansion tank above the transformer which has a volume of from 3-10 percent of that of the transformer tank itself. The oil or other insulating fluid completely fills the transformer tank and the lower half of the expansion tank, as shown in Figure 1.59. These transformers breathe through the expansion tank filled with cold oil, usually through a dehydrating breather; but the oil in the main tank is sealed from the atmosphere. This reduces oxygen and moisture absorption and retards sludge formation.†

**See Chapter 6, Part 2 concerning transformer oil pumps.*

†In the U.S. there has been a marked increase in the use of the synthetic rubber bag type inside the conservator.

1. Primary bushings
2. Secondary bushings
3. Primary lightning (surge) arresters
4. Secondary lightning (surge) arresters
5. Liquid level gauge
6. Dial type thermometer
7. De-energized tap changer control handle
8. Instruction nameplate
9. Cooling system. External tubes for self-cooled (OA) or fans for forced air (FA).
10. Lower drain valve filter press (reclaiming) connection and sampling valve
11. Tank grounding provision
12. Upper valve for filter press (reclaiming) connection or sampling valve (for moisture content)
13. Pressure vacuum gauge
14. Pressure relief device

Figure 1.58 - Major auxiliary components of a 3750 KVA medium power transformer (6.7 KV primary/12.5-7.2 KV secondary).

3. Sealed-Tank Type
   In this type the space above the oil in the transformer tank is filled with inert gas under pressure (Figure 1.60). Some transformers of this type use an auxiliary gas tank above the transformer. Welded covers are generally used to eliminate gasket leaks, and a pressure-vacuum relief device is required to prevent the development of a high differential pressure between the inside and outside of the tank. These devices are set to maintain the pressure or vacuum within a range of -8 to +8 psi.

4. Gas-Oil Sealed Type
   In this type the interior of the tank is sealed from the atmosphere by means of an auxiliary tank to form a gas-oil seal operating on the manometer principle, as shown in Figure 1.61. It will be noted that the auxiliary tank is divided, with one section connected to the tank gas space, while the other is connected to the atmosphere through a breather. The two sections are isolated by an oil seal.

5. Automatic Inert-Gas Pressure Type
   In this type the space above the oil is filled with an inert gas, usually nitrogen under positive pressure supplied by a gas cylinder which is at the side of the transformer and is controlled by an automatic pressure regulator (Figure 1.62).

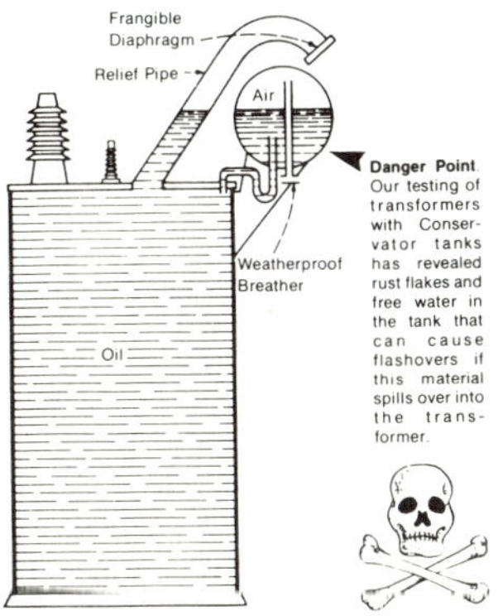

Figure 1.59 - Conservator method of oil preservation. The oil or other insulating fluid completely fills the transformer tank and the lower half of the expansion tank. (Courtesy of Factory Mutual).

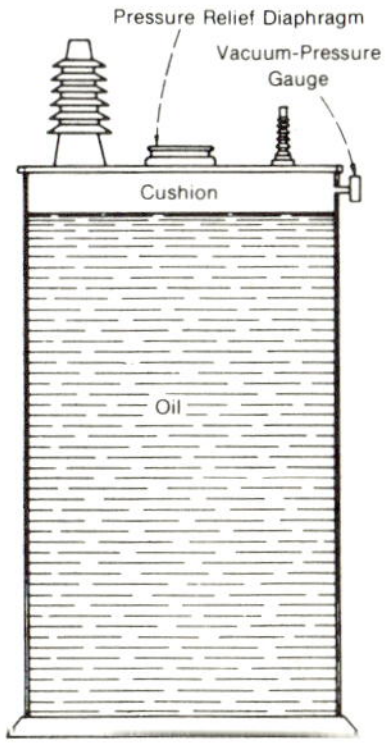

Figure 1.60 - Sealed-tank type transformer. The space above the oil in the transformer tank is filled with inert gas under pressure. (Courtesy of Factory Mutual).

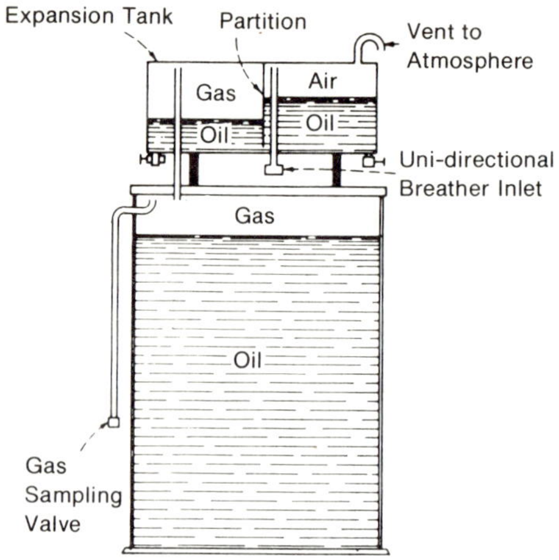

Figure 1.61 - Gas-oil sealed construction. The interior of the oil tank is sealed from the atmosphere by the auxiliary tank which forms a gas-oil seal operating on the manometer principle. (Courtesy of Factory Mutual).

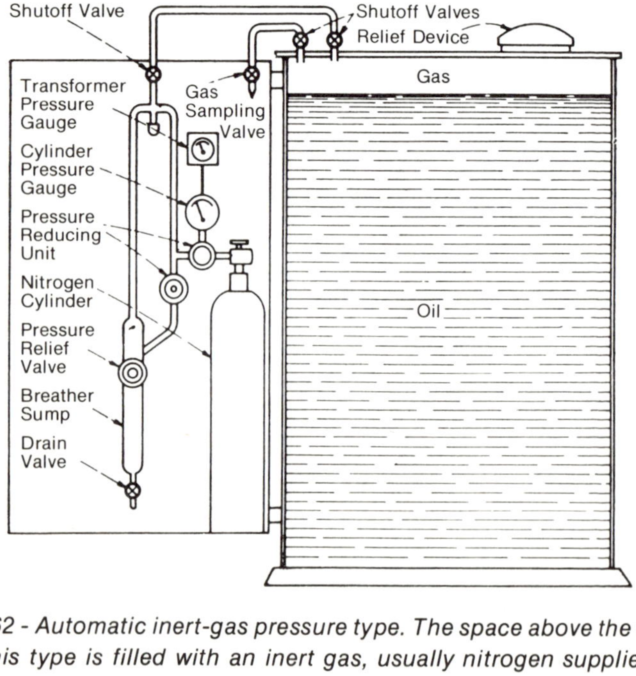

Figure 1.62 - Automatic inert-gas pressure type. The space above the oil in the tank in this type is filled with an inert gas, usually nitrogen supplied under positive pressure from a gas cylinder. (Courtesy of Factory Mutual).

Bushing Insulators

Conductors from the transformer windings must be safely brought through the tank structure in a protective terminal arrangement. The bushing insulator is the assembly that generally performs this function. The bushing is designed to insulate a conductor from a barrier, such as a transformer lid, and to safely conduct current from one side of the barrier to the other (Figures 1.63 and 1.64A). In addition to being a good insulator, bushings must be water tight, gas tight and oil tight, because they must keep moisture out.

Bushing insulators should be distinguished from standoff or suspension insulators, associated with transmission lines, bus bars, and so forth (Figures 1.64B and 1.63, as noted).

Several types of transformer bushings are found in-service and are distinguished by construction:

- Solid (high alumina) ceramic—(Up to 25 KV)
- Porcelain—Oil-filled (25-69 KV)
- Porcelain—Compound (epoxy) filled
- Porcelain—Synthetic resin bonded paper-filled (34.5 to 115 KV)
- Porcelain—Oil-impregnated paper-filled (above 69 KV, but especially above 275 KV rating)

As noted, the bushing class employed is based upon the voltage level it is designed to handle.

Transformer and (oil circuit breaker) bushings are generally cylindrical in shape and thus fall into the rod type of insulator (standoffs are usually cap-and-pin or barrel type, Figure 1.63, as noted).

Specialized bushings have been designed to collect only small pollution deposits and others have been made to give low electrical stresses on the surface. The bushing must provide sufficient dielectric strength to withstand the high voltages, particularly at the point where the bushing passes through the cover (Figure 1.64B). Externally, and for bushings used outdoors, the distance over the outer ceramic surface may be increased by adding "petticoats" or "watersheds" which increase the "creepage" distance from the line terminal and the tank. In turn, this modification can reduce potential flashover across the surface, due to moisture and particulate matter contamination. Nevertheless, proper design (and maintenance) is a learned art (not any single formula or graph), which depends especially on knowledge of the contamination flashover process, the environment, and available maintenance methods.*

Transformer bushings have traditionally been externally clad in porcelain because of its excellent electrical and mechanical qualities. Porcelain insulators are generally oil-filled beyond 35 KV in order to utilize the high permittivity and dielectric strength of oil, as compared with air.

For power apparatus (> 69 KV) capacitor-type oil-filled bushings are standard equipment. The condenser bushing has oil-impregnated special paper with aluminum foil layers between them inside the porcelain cylinder. These foil layers form a series of condensers in order to equalize voltage distribution between the outer and inner layers of insulation.

In addition to porcelain, the outer portion of a bushing exposed to the atmosphere may be clad with glass, epoxy, Teflon, a unique special synthetic rubber Hi-Lite™† or a unique silica compound (Polysil™). The insulating conductor is enclosed by clean dry material such as porcelain, toughened glass, fiberglass, epoxy, or some other cast organic material, as well as resin bonded paper or resin impregnated paper. The last two also provide an insulating envelope, in which case the intervening space may be filled with oil or another insulating medium.

The most recent outer material Polysil may have the potential of replacing porcelain in various applications, but conclusions cannot be made until the completion of a two-year test program.

The non-ceramic Hi-Lite suspension insulator introduced in 1977 does not contaminate as quickly and achieves a large reduction in weight over porcelain, but once it is contaminated it is more difficult to clean due to softness of material and design. In addition, it is not suitable for oil-filled bushing applications.

Porcelain insulators are likely to remain the primary material for some time to come, especially in power station installations where *energized* cleaning is mandatory.‡ In light of technological challenge, an extra-

**Chapter 8, Part 3.*

†Hi-Lite is a registered trademark of the Ohio Brass Co., Mansfield, Ohio.

‡Chapter 8, Part 3.

high-strength porcelain material has been developed. Slimmer design and less heavy units made from this improved porcelain are now available.

Figure 1.63 - A typical medium power transformer, highlighting transformer bushing and rod standoff insulator.

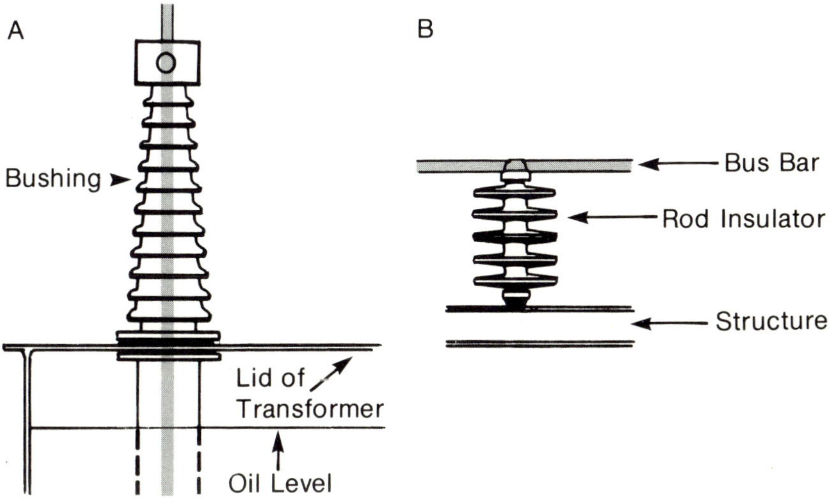

Figure 1.64 - Distinguishing between transformer bushing (A) and rod standoff insulator (B).

Tap Changing Equipment

Most transformers are equipped with a tap changer to permit small changes in voltage ratio.* The high voltage winding is the one generally constructed with taps. By changing the winding tapping, this device provides the way to vary the unit's turns ratio and thus the level of voltage (either the supply or load voltage with the other remaining constant). Typical applications include arc furnaces, electrolytic plants, chemical manufacturing processes, and so forth. Changes in connections may be done only when the transformer is de-energized (no load tap changing) or while the unit is connected to a load (on-load tap changing). Tap changers have been designed for either operation within an enclosure inside of the tank or externally mounted in a separate oil-filled housing bolted to the outside of the main transformer tank.

1. No-Load Tap Changing
   A transformer with off-circuit tap changing must not be energized until the tap changer is latched in proper position and locked. The traditional method uses a terminal board, making it necessary to lower the oil level, open the manhole, and use tools to make the change. Design practice recognizes that captive parts should be used in order to prevent loose nuts or washers from falling into the tank while tap positions are adjusted.

   Tap changing may also include manual turning of a handle protruding above the oil level or using a selector handwheel operated from outside the tank which changes connections under the oil.

2. On-Load Tap Changing
   The on-load tap changer contains the only moving parts connected into the transformer windings. The importance of the continuing reliability of this equipment cannot be emphasized too much.

   Load-tap changing may be performed manually or automatically operated. A wide variety of circuits is thus used in this process. Nonetheless, the basic principle of operation remains essentially the same:

For detailed description, see Stigant, **J & P Transformer Book (1973), and Feinberg, **Modern Power Transformer Practice** (1979).*

- Voltage adjustment under varying load conditions is required (supply voltage is constant).
- The taps to be changed are connected to switches.
- A transition impedance is used to limit the circulating current at the time when the adjacent taps are being connected.

The impedance can take either one of two forms, and thus the tap changer is classified as a resistor or reactor type. A high-speed resistor mechanism (40-70 milliseconds) is now generally used, although a reactor (inductor) type of gear is still used in the U.S., in spite of its size and cost.

Two types of resistor tap changers are in common use. The selector (pennant) switch combines tap selection and current transfer. For larger transformers, separate (flag) selecting and diverter (current transfer) switches are used.

Lightning (Surge) Arresters

Most transformer installations are subject to surge voltage originating from lightning disturbances, switching operations, or circuit faults. Some of these transient conditions may create abnormally high voltages from turn to turn, winding to winding and from winding to ground.

Lightning arresters are the principal means of providing this protection.* Lightning *surge* arresters are special lightning rods for additional protection against overvoltages, reducing these surge voltages before they reach the electrical system. The arrester should always be located as close as possible to the terminals of the transformer to be protected, preferably mounted directly on the transformer (Figure 1.58).

The basic arrester consists of an air gap in series with a resistor element. The most widely used type of arrester is the valve type. It consists of one or more gaps in series with a dielectric element, or so-called "valve material," which serves as the high resistance.

The valve arrester is similar to the relief valve on a steam boiler; that is, when the pressure becomes too high, the relief valve opens and discharges the excess pressure; then it closes when conditions are safe

For detailed description, see Pansini,* ***Basic Electrical Power Distribution**, Volume 1 (1971).*

again. In its normal state the arrester acts as an insulator. When a high surge voltage occurs, the arrester must cease to be an insulator and must turn into a short-to-ground in millionths of a second.

Arresters are available in three classifications for primary distribution system overvoltage protections: station (up to 684 KV); intermediate (up to 120 KV), and distribution types (up to 30 KV). The station class arrester is the most expensive and affords the best protection, while the distribution, the most widely used, provides less protection.

Two other types of arresters include the expulsion arrester which is on its way out and the metal oxide arrester, which gives even faster response time is in the development stage. The most effective surge protection is obtained when:

- The maximum voltage rating of the arrester is equal to or greater than the normal line-to-ground voltage of the transformer.
- The arrester connections are made as close to the transformer and as direct as possible.
- The arrester, tank grounds and secondary neutral, if any, are tied together to a common station ground connection.
- The resistance of the arrester ground connection is a maximum of 5 ohms and preferably 1 ohm where possible.
- The manufacturer's recommendations are followed.

Switchgear (Protective Devices)

When an electrical fault or short circuit occurs, protective devices isolate the particular area of the fault to minimize its effect on the larger power system. In particular, the device nearest a fault (a relay) will operate first to isolate the problem and minimize the disturbance by de-energizing that portion of the system (actuating a circuit breaker).

Switchgear is the name given to all the equipment designed to control and/or protect the electrical power system. It includes the following components:

- Bus bars
- Disconnecting switches
- Motor controllers
- Circuit breakers/fuses
- Oil-fuse cutouts
- Protective relays
- Measuring instruments
- Metal-clad cnclosures

As with the transformer's insulation system, the coordination of protective devices is a necessity for complete substation design. In fact, an electrical fault generally means that the insulation of some component in the power system has failed.* The only thing limiting a short circuit is the impedance that exists between the source of the electricity and the fault.

Our focus concerns *direct* and indirect protective devices (circuit breakers/fuses and relays). Keep in mind that what switchgear is protecting is not the device that has failed, but the power system.

Electrical protective devices are essential to isolate a fault promptly, to minimize electrical damage and interruption to production, and to help prevent fires that sometimes accompany electrical breakdowns in transformers. Both circuit breakers and relays are manufactured so that they can be set to respond to different fault currents and voltages.

1. Power Circuit Breakers/Fuses
   A circuit breaker (CB), the most important direct-acting high powered device, is used to interrupt circuits while overcurrent is flowing through it or during a transformer overload condition.

   Circuit breakers have two coils—one to close the breaker and one to trip (open) the breaker. In larger and modern breakers, a motor compresses a spring that provides most of the energy for closing and opening the breakers. Oil, air, vacuum, or gas can be the switching medium.

   In addition to protecting people and equipment, CBs ensure that short circuits do not cause damage in distribution lines, generators, or transformers. They ensure that a lightning strike on a power line results, for example, in only a barely noticeable flicker of consumers' light and not in a regional power failure.

   The making and breaking of contacts has been traditionally done for outdoor applications and older metal enclosed switchgear assemblies under oil. The oil serves to quench the arc when a circuit is opened.

   Air circuit breakers are now widely used. The arc is put out by a blast of compressed air. Air/magnetic devices have replaced oil units because of cost and the hazard of oil not being contained by the breaker in fault-clearing operations.

**Chapter 1, Part 3 and Chapter 2, Part 4.*

Many power breakers for large industrial plants and utilities now use sulfur hexafluoride, since SF_6 permits the handling of higher voltages than air, oil, or vacuum.

Modern electrical power systems generally use circuit breakers instead of power fuses. Whereas a circuit breaker can trip all three phases in a three-phase supply line, it is possible for only one fuse to "blow". Moreover, breakers can interrupt fault currents more than once. They need only be reset and periodically checked to make sure they are not damaged.* Naturally, CBs are more expensive than high voltage power fuses.

Drop-out style solid-material fuses have been applied to outdoor industrial substations in the transformer's primary circuit to protect the unit from secondary faults, as well as to protect the distribution system (up to 138 KV).

Because of the non-adjustability of the more basic devices, the low voltage power circuit breaker (up to 600V ac, 250V dc) is frequently used in critical locations of electrical power systems (as it allows for selective tripping in fault current or overload conditions). Oil may be used in the dashpots to provide desired time-delay characteristics. The low voltage molded-case circuit breaker is a current limiting device that is commonly used in residential circuits.

2. Protective Relays

   The relay, a low powered device, activates the associated circuit breaker (gives a trip command) during an overcurrent, undercurrent, or overvoltage condition. Nevertheless, relays do not operate on an overload condition! The circuit breaker should open during this condition. In effect, these relays are measuring instruments but are equipped with auxiliary contacts which operate when quantities flowing through them exceed or go below some preset value.

   The classic protective relay operates by electromagnetic induction. Many of these devices also have an instantaneous-acting attachment that operates on the principle of attraction.

   One element of the relay that is often overlooked is the target or indicator flag. This device can be calibrated in the field and can

Chapter 8, Part 1.

be serviced/monitored while energized. It functions when the relay attempts to trip the breaker. When the "red" target is showing, it means bad news, but the target is an invaluable aid in diagnosing a system fault!

Overcurrent relays on the *primary* side of a transformer will provide protection for the primary cable and the transformer windings and provide fast separation from the primary circuit in order to minimize the disturbance to the system and reduce damage to equipment.

Overcurrent relays on the *secondary* side of the transformer will prevent overloading and short circuiting of the windings in the event that the protective devices in the distribution circuit fail to function as intended.

Various other relays are associated with transformers:

a. Ground Fault Relays
   Ground fault relays can be arranged to isolate a transformer from the system if an internal fault develops into a fault involving ground. The type of ground fault protection depends upon the type of system grounding used. Ground fault relays connected in the neutral lead of a transformer can provide a backup for other system ground relays and transformer differential relays where provided. They provide the only ground fault protection for a transformer which does not have differential relaying. This type of relay should be provided in the transformer neutral of all grounded neutral power systems.

   Ground fault relays connected residually with phase overcurrent relays in a transformer secondary circuit can be set for very sensitive operation since current flows in the neutral connection only during a ground fault. They are suggested for transformer feeders on both solidly grounded and resistance grounded systems.

   When residually connected ground fault relays are used together with those in a transformer neutral, they can be arranged to differentiate between ground faults in the distribution system and those in the secondary circuit.

b. Differential Relays
   Differential relays are used to detect short circuits or faults to ground. The zone of protection formed by transformer

differential relays can be expanded to include circuit breakers and bus bars as well as the transformer windings and bushings. They provide the best form of transformer protection because of their simplicity, sensitivity, selectivity, and speed.

There are several types of percentage differential relays, some of which incorporate a harmonic restraint feature. Since each type is suited for a particular application, the manufacturer should be consulted as to the choice of relay that will provide the best protection for the particular application (transformer, generator, bus protection, etc.).

Differential relays should be provided for all transformers of 6000 KVA and above which have primary circuit breakers. All transformers of 10,000 KVA and above should be provided with differential relay protection. The primary should be provided with a circuit breaker for each winding into which current can flow if not already so equipped.

3. Factory Mutual Circuit Protection Guidelines*

> Circuit breakers with overcurrent relays are the preferred protection for the secondary side of transformers. The overcurrent relays should have a range of overcurrent settings above the permissible overloads and also have instantaneous trip elements that can be set within the transformer short circuit withstand rating...
>
> It should be recognized that the overcurrent protection ratings called for by the *National Electrical Code*, will not protect a transformer from overheating resulting from a sustained overload up to 250 percent of the nameplate rating. Even for short periods of time, operation of a transformer with hot spot temperatures in excess of its rating will *reduce* the normal life expectancy of the insulation. It should be noted that the time factor is not mentioned in the *National Electrical Code* requirements.

"Loss Prevention Data—TRANSFORMERS," Factory Mutual Engineering Corp., Bulletin No. 5-4/14-8, April, 1978, pp. 11-18.

Table 1.8 summarizes one of the ANSI standard capability tables for self-cooled and water-cooled transformers with moderate sacrifice of life expectancy for various equivalent loadings, before application of the overload.

The short circuit capability, as detailed in ANSI Standard C57.12.00 is another characteristic to be considered in developing the primary protection. Transformers shall be capable of withstanding without injury the thermal stresses caused by external short circuits provided the duration of the short circuit is limited to the values shown in Table 1.9.

TABLE 1.8

CAPABILITY TABLE FOR SELF-COOLED (OA) AND WATER-COOLED (OW) TRANSFORMERS WITH 0.25 PERCENT SACRIFICE OF LIFE EXPECTANCY

(Equivalent load before peak load—100 percent of rating)

Peak Load (hours)	Hottest Spot Temperature Reached (°C)	Allowable Peak Load Times Nameplate Rating in 30°C Ambient Temperature
½[1]	142	1.92
1[1]	134	1.69
2[1]	126	1.48
4	119	1.32
8	112	1.20
24	104	1.09

[1]This overload protection could be on the primary or secondary side of the transformer.

TABLE 1.9

SHORT CIRCUIT LIMITATION	
Times RMS Symmetrical Base Current in Any Winding	Time Periods (seconds)
25	2
20	3
16.6	4
14.3 or less	5

Alarms and Indicating Devices

1. Temperature Indicators
Oil- and askarel-insulated transformers (162.5 KVA) are usually equipped with a bulb and capillary-type (or dial-type) thermometer to measure the top liquid temperature. In addition, this device may be furnished with one or several contacts arranged to sound an alarm, activate a thermal overlight (red), control the fans for forced air cooling or trip a circuit breaker. A second thermometer should indicate the winding temperature.

2. Dehydrating Breathers
Many types of open- and conservator-type transformers are provided with either open or dehydrating breathers. The open type of breather permits air to enter and to escape from the space above the oil to take care of temperature changes and oil expansion. This breathing, in many cases, is through calcium chloride housed in a receptacle. The effectiveness of the calcium chloride in preventing the entrance of moisture depends upon keeping the chemical dry. The uni-directional breather or ventilator is of this type and consists of a pipe extending up through the oil to the space above the liquid, with a weatherproof open breather on the top of the transformer. This permits the air to be heated as it passes through the pipe in the hot oil and across the top of the oil and has been fairly successful in carrying off accumulated moisture.

Silica gel breathers, a later development, and known as dehydrating breathers, are now most commonly used and are replacing the open-type breather. They are used extensively on conservator-type transformers. The silica gel does not restrict breathing as does calcium chloride when saturated with moisture. Nonetheless, both types do involve maintenance for satisfactory performance.

3. Pressure-Relief Diaphragms and Mechanical-Relief Devices
   Electrical breakdowns in transformers cause decomposition of oil and the subsequent generation of gas. Thus, most transformers are equipped with some type of relief device.

   Relief diaphragms are usually of glass, bakelite, or thin metal and are mounted on the end of a large pipe connected to the transformer case above the liquid (Figure 1.65). They are usually designed to rupture at from 10 to 15 psi (69 to 104 KPa). *Mechanical-relief devices,* such as a diaphragm-loaded valve (Figure 1.66) are usually mounted on the top of the tank. The pressure-relief device on large power transformers may include a warning flag or alarm contacts to indicate operation of the device. The operation of pressure-relief devices, especially if some liquid is discharged, is indicative of internal arcing.

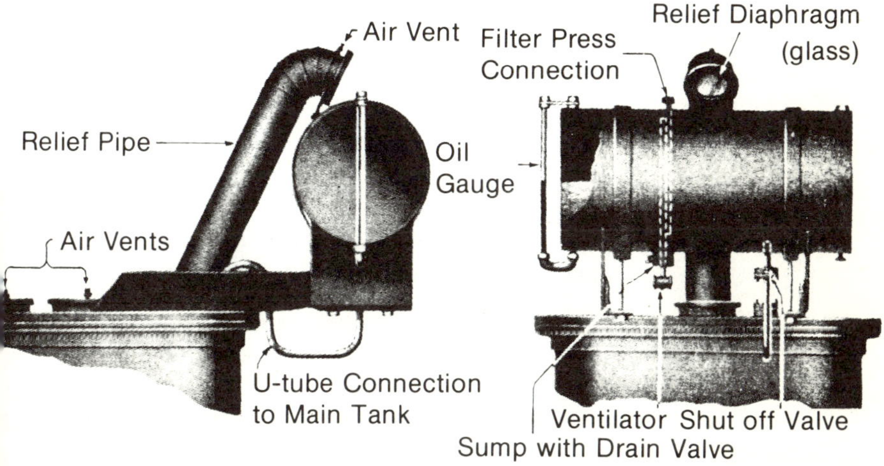

Figure 1.65 - Main features of oil-conservator construction, including relief diaphragm. (Courtesy of Factory Mutual).

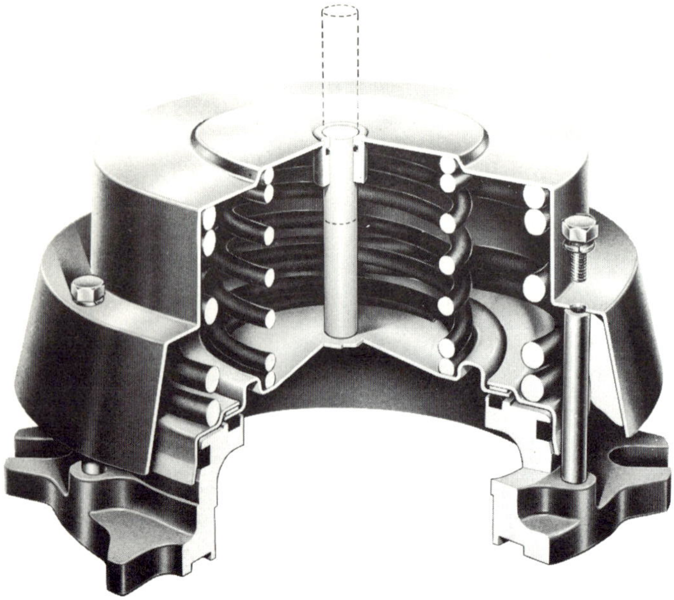

Figure 1.66 - One type of mechanical relief device to relieve excessive pressure in transformers. (Courtesy of Westinghouse Electric).

4. Other Transformer Alarms and Indicators
   Alarms associated with transformer malfunctions also include circuit breaker trip, accidental ground on an ungrounded system, and cooling water flow or overtemperature.

   Additional devices include circuit breaker "open-close" indicators, ground indicator lamps for ungrounded electric systems, liquid flow indicator, and most recently, an ultrasonic temperature sensor for transformer hot spot detection.

5. Fault Gas Detector Relays
   Fault gas actuated relays consist of two basic types—gas detector relays and sudden pressure relays.

   Conservator-type transformers may be equipped with a *gas-detector* (or Buchholz) relay. It warns of a slowly developing (incipient) fault* or localized overheating of insulating materials. When combustible gas exceeds a predetermined quantity of gas (usually 200 cc), a fault alarm sounds. Repairs

**Chapter 4, Table 4.3.*

can then be made before a major breakdown occurs. This relay is located in the pipe between the transformer and conservator tank and is designed to trap any evolved gas which may rise through the oil.

A new type of permanently mounted combustible gas-in-oil detector has recently been installed on medium and large power transformers. It measures the concentrations of various gases dissolved in transformer oil.

This device automatically extracts oil periodically (as desired) and determines the individual gas concentrations and the total dissolved gas concentration chromatographically. The data obtained may be stored on a paper chart recorder. These monitors are particularly valuable for large remotely located transformers or those critical to production but for which available downtime is minimal.

The second basic type, a *sudden pressure relay*, is actuated by the gas pressure that develops inside a transformer because of the decomposition of the insulating liquid when an internal fault develops. This type of relay operates on the rate of rise of pressure in the gas space above the transformer oil which, in turn, depends upon the power in the arc caused by the fault. The higher the rate of rise, the faster the relay operates. Sudden pressure relays act very quickly (within a half cycle of the fault) to limit damage resulting from a major fault.

Factory Mutual recommends that oil-insulated transformers of 10,000 KVA and larger should be equipped with gas detector relays arranged to sound an alarm in a constantly attended location.

Oil-insulated transformers of 50,000 KVA and larger should also be equipped with sudden pressure relays arranged to sound an alarm in a constantly attended location and isolate the unit from bus bars and other equipment in the event of an internal fault.

Understanding Transformer Nameplates

We have considered each of the major components of most transformers. The fact remains that every transformer is nonetheless unique in physical construction and electrical characteristics. Each unit is slightly different in some respect from other units—even from those of the same KVA rating.

Nameplate data, often overlooked or misunderstood, has played a major role in the specific design of all transformers (ANSI/IEEE Standard C57.12.00-1980). More importantly, this information gives necessary data required for the proper operation and maintenance of a given transformer (Figure 1.67).*

1. Phase (1)†
   Electrical power is most economically generated and transmitted by a three-phase internal connection, although much of it is used as single-phase (residential lighting). Phase can be related to the "push" and "pull" cycles of one cycle of a one-cylinder engine. Alternating current is generated in a sine wave (Figure 1.16), with half of the cycle above zero (push) and half below zero (pull). At the end of the stroke (between push and pull) the engine fly wheel goes through a position where there is no push or pull. To overcome this minimum energy situation two other pistons can be added to the engine 120° apart. Now the situation is such that at least one cylinder is either pushing or pulling at all times. Thus, the total energy of the system is never zero. With a transformer this can be accomplished by connecting three single-phase units into a "bank" of one three-phase unit.

2. Serial Number (2)
   The transformer manufacturer's identification number—year of construction—is also the owner/operator's key to much useful information available from the manufacturer, such as details of construction and factory test results.

3. Class (3)
   This nomenclature refers to the type of insulating medium and the method by which the transformer is cooled. Table 1.10 lists the most common classes.

R.L. Smith, "The meaning behind transformer nameplates,"* ***Power* *Magazine (1975).*

†The number following each nameplate item corresponds to the numbers in Figure 1.67.

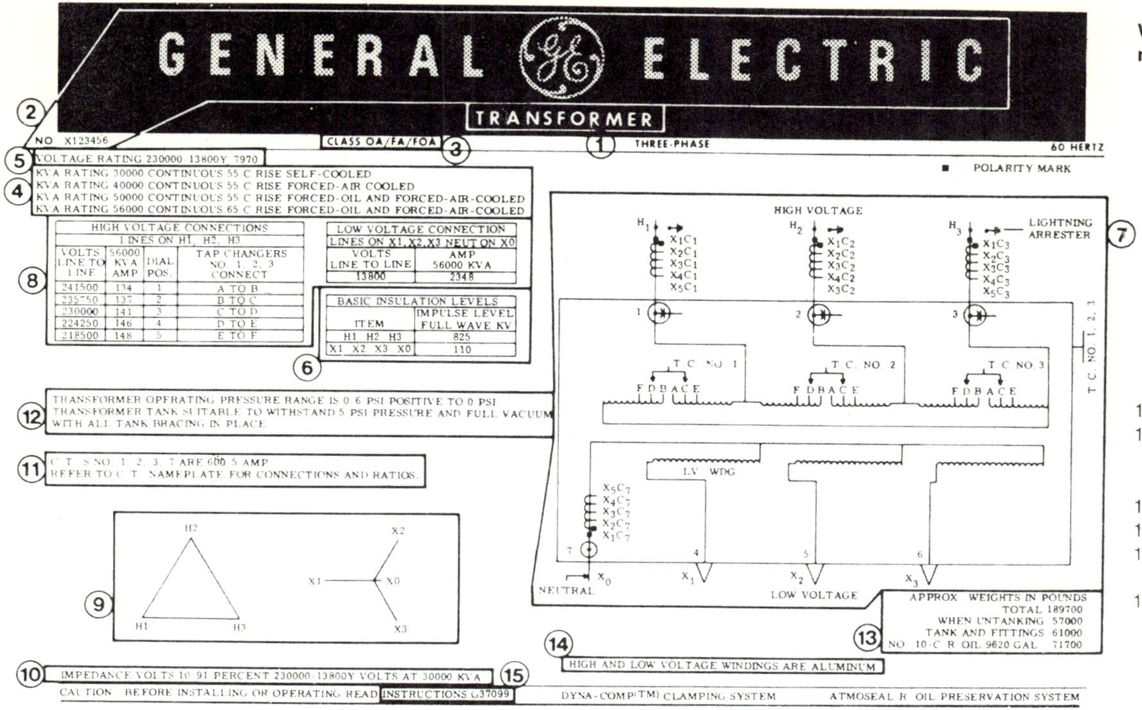

Figure 1.67 - Information on a typical nameplate for a three-phase power transformer. *(Courtesy of **Power** Magazine).*

What the numbers mean

1 Transformer
2 Serial number
3 Class
4 Temperature rise
5 Voltage
6 Basic insulation level
7 Connection or schematic diagram
8 Taps
9 Vector diagram or polarity
10 Impedance
11 Brushing current transformers
12 Pressure range
13 Weights
14 Windings other than copper
15 Instructions

TABLE 1.10

CLASSIFICATION OF TRANSFORMERS BY COOLING METHOD		
Class	Description/Cooling Method	Example (Figure 1.67)
OA	Oil immersed, self-cooled (natural circulation of insulating liquid)	30,000 KVA continuous 55° C rise
FA	Oil immersed, forced-air cooled (via fans)	40,000 KVA continuous 55° C rise
FOA	Oil immersed, forced-oil (via pump), plus forced-air cooled	50,000 KVA continuous 55° C rise
FOA/ FOW	Oil immersed, self-cooled plus forced-oil cooling via pump to circulate oil through heat exchanger	56,000 KVA continuous 65° C
AA	Dry type. Self-cooled (natural circulation of air)	
AFA	Dry type. Forced-air cooled (circulation of air or gas)	
AA/FA	Dry type. Self-cooled forced-air cooled	
OW	Oil immersed, self-cooled, plus water-cooling by pump through pipe/coil or heat exchanger	

4. KVA Rating (4)
   The kilovolt-ampere (KVA) or megavolt (MVA) rating of a transformer indicates the maximum full load power on which the unit is designed to operate under normal conditions.

5. Temperature Rise (4)

   The difference between the temperature of the transformer part under consideration (commonly the "average winding rise" or the "hottest-spot winding rise") and the ambient (surrounding) temperature. The *winding hottest spot temperature* is the highest temperature inside the transformer winding. It is greater than the measured average temperature of the coil conductors. The *average temperature rise* is the temperature above ambient due to the unit operating at full-load rating. Table 1.11 lists the hottest-spot temperature rises corresponding to various winding rises.*

   Keep in mind that the transformer temperature indicator gives an operating temperature, *not* a temperature rise. To determine the temperature rise in an in-service transformer, subtract the ambient temperature from the indicated temperature. For example, if a transformer rated for 65° C is operating in a 30° C ambient, the indicated operating temperature should be no greater than 95° C.

TABLE 1.11

CAPACITY LIMITS OF TEMPERATURE RISE FOR CONTINUOUSLY RATED TRAMSFORMERS (per ANSI C57.12.00-1980)

Transformer type	Winding rise (by resistance)	Hottest spot rise
55° C rise oil-immersed	55° C	65° C
56° C rise oil-immersed	65° C	80° C
55° C rise dry-type	55° C	65° C
80° C rise dry-type	80° C	110° C
150° C rise dry-type	150° C	180° C

Chapter 1, Part 3, Table 1.15.

6. Operating Voltage, Maximum Rated (5)
The highest root mean square (RMS) voltage at which a transformer is designed to operate, where:

$$RMS = \frac{V\ max}{2} = 0.707\ V\ max$$

In the upper left hand corner of Figure 1.67 (number 5) note the designation:

Voltage Rating 230,000 - 13,800 Y/7970

This data tells us that the primary phase-to-phase voltage is 230,000 volts. The transformer steps down the voltage to 13,800 volts phase-to-phase in a wye (Y) connection. Since the nameplate has the designation "three-phase" we can calculate the phase-to-ground voltage by dividing the phase-to-phase voltage by the square root of 3 or:

$$\frac{13{,}800}{\sqrt{3}} = 7970\ \text{volts}$$

7. Basic Insulation Level (6)
An indication of the ability of the transformer's insulation system to withstand lightning strikes and other system over-voltages.*

8. Winding Connection Diagram (7)
The schematic of internal interconnections between windings in the same phase, and/or windings between different phases in numerous ways to provide required voltages and operating conditions.

The most common transformer connections are delta Δ (windings connected in series to form a closed circuit), and wye or Y (the one end of each winding connected to a common neutral point and the other end to its appropriate line terminal). The four most common connections of a three-phase transformer

**Chapter 1, Part 2, Figure 1.53 and Part 3, Figure 1.72.*

bank are delta-wye (Δ-Y), wye-delta (Y-Δ), delta-delta (Δ-Δ), and wye-wye (Y-Y).* The "three-phase" designation indicates that all three phases are put in the same container (or tank). Many "tanks" consist of three single-phase units that are wired in a three-phase connection.

9. Tap Changing (8)
   An adjustment in turns ratio is accomplished by tapping the windings. All taps are tabulated on the nameplate by tap-changer letter, dial-position number, and tap connection. Some transformers with provision for tap-changing have automatic equipment that changes the turns ratio to compensate the voltage drops as system load increases. Other transformers have no-load tap changing to compensate for the variations in the incoming primary voltage.

10. Vector Diagram or Polarity (9)
   A vector diagram gives the phase relationship between voltages of various windings for three-phase transformers. For single-phase units, the polarity of the transformer is shown. The polarity is simply the way the coils are wound on the core and the terminals are brought out of the tank and labeled. Transformer polarity has *no relation* to the same term applied to a battery or to a dc generator meaning one terminal is plus (+), the other negative (-).

11. Impedance (10)
   Expressed as a percentage, the *tested* impedance is the transformer's opposition to the passage of short circuit current. Short circuits are frequently caused by insulation failure, equipment malfunction, or accidental contact between live parts of a circuit or device at different potentials, or between parts and ground.

   The current that flows on short circuit with full voltage maintained on a transformer is equal to 100 divided by the percent impedance of the transformer. Therefore, a unit with 4 percent impedance will have a short circuit current $\frac{100}{4} = 25$ times nor-

*For specific details of single and polyphase connections, see Stigant, ***J & P Transformer Book***, and Pansini, ***Basic Electrical Power Transformers***.

mal full load current. The higher the impedance, the smaller the short circuit current will be; in our example (Figure 1.67), the short circuit current will be $\frac{100}{10.9}$ or 9.2 times full load current.*

12. Pressure Range (12)
Liquid- and gas-filled transformers operate with a particular positive pressure in sealed tanks. When a tank is indicated as suitable for "full vacuum," vacuum filling is required. This factor is also significant later in the life of the unit when dehydration is required.†

13. Instructions (15)
Manufacturer's instructions are in two parts—installation and maintenance. Installation should be in accordance with the latest *National Electrical Code*. If this information is lost, obtain a replacement from the manufacturer.

Transformer Factory Testing

Transformers are designed to meet *required characteristics* and *test levels* to assure long life under *normal operating* conditions.‡Special tests may be specified by the purchaser.

Significant items of transformer quality control include:

- Ratio tests
- Polarity/phase relation tests
- Resistance measurements
- Losses—core, coil, and total
- Impedance
- Temperature rise
- Dielectric Tests
    - Impulse strength
    - Switching surge strength
    - Low frequency induced and applied voltage
    - Radio influence voltage (R.I.V.) or
    - Partial discharge (corona)

**Chapter 1, Part 3, Figure 1.74.*

†Chapter 7, Part 2.

‡Reference ANSI Standards, C57.12.00-1980, C57.12.90-1980, C57.1290 A-1978, and proposed IEEE Standard 262B-1977; proposed ANSI/IEEE C57.12.14/D7 (1980).

- Audible sound level
- Other specified tests
  - Short circuit withstand
  - Unit power factor at 20° C
  - GC analysis
  - TV interference
  - Thermal Evaluation
  - (Heat run—several days)

Our immediate purpose will be to work our way through the various industry standards to determine what is most useful for the transformer owner/operator. A certain amount of background is essential. All tests (factory and installation) are at best only insurance against the more important or more probable things that may have gone wrong in design/construction or against damage in shipping.* The only test which will prove that a transformer will last 20 years—or as we believe 50 plus years—is to operate *the unit* for 20, 50, or 75 years. Actual unit life will depend to a great degree on adequate maintenance and knowledgeable operation.

Transformer factory tests are broken down into *preliminary* tests (those made on the core and coil prior to final assembly and tanking) and *final tests* on a completely assembled unit. Further, the final tests are normally divided into routine acceptance and optional tests.†

Preliminary Factory Tests on New Power Transformers

1. Turns Ratio and Polarity
   These tests prove the correctness of the turns ratio of the windings and that the windings have been properly connected; and that any winding tappings have been made correctly. The turns ratio must be correct within ½ of one percent of nameplate markings.

2. Resistance Measurements
   Using the standard volt meter/ammeter method, comparison is made of the measured dc resistance value with the calculated value of each winding (copper losses and winding

Installation tests as* ***field tests* *are considered in Chapter 5.*

†*Design or type tests are another group; however, these tests are performed only on one or more prototypes of each new design.*

temperatures at the end of a heat run). Verify that there are no poorly brazed, crimped, bolted or open connections.

3. Impedance and Load Loss
   The impedance of a transformer winding can be measured by simply short circuiting one winding and circulating *rated current* through the other. The voltage required on one winding that will cause full rated current to flow in the other winding when short circuited is called the "impedance voltage". The ratio of the "impedance voltage" to the rated voltage is the impedance of the winding. When it is necessary to dry out a wet insulation of an "in-service transformer" this "impedance voltage" can be calculated and may be a useful tool in accelerating the dry out period.* Guaranteed value must be met (C57.12.93.500). The limiting value of load loss is established by the difference between the guaranteed total loss and the guaranteed core loss.

4. Core Insulation Voltage Test
   An alternating voltage test at 2000 volts is applied to the insulation on all leg and yoke bolts, clamping plates and core laminations in order to prove the insulation of the magnetic circuit.

5. Excitation (Core) Loss and Exciting Current
   These tests prove that the transformer at rated voltage, frequency and wave shape meets the guaranteed value. We believe that excitation current should be required in factory specifications.

6. Drying Out and Insulating Fluid Fill
   During building and storage, the materials used in transformer construction normally absorb 6-8 percent by weight of moisture. Moisture is insulation's primary enemy, even at this point.† Most of this moisture must be removed by some means of dehydration.‡

   Drying out is continued until a desired moisture level is achieved (unit monitored by insulation resistance and power factor).

**Chapter 7, Part 2.*

†Chapter 2, Part 5.

‡Chapter 7, Part 2.

Next, a vacuum is applied while still maintaining approximately 90° C heat. The unit, still under vacuum, is then filled with clean dry degasified transformer oil or other fluid and allowed to stand for several days under specified pressure so the oil can soak into the solid insulating materials. The passage of time ensures that any air bubbles will be absorbed by the degasified oil.

Routine Final Factory Tests on New Power Transformers*

Turns ratio, polarity, value of dc winding resistance, core and conductor losses, and impedance (all tap positions and all connections) are officially measured to verify the nameplate or contractual guarantees.

1. Temperature Rise Test
   The purpose of the test is to confirm that under the maximum KVA rating given on the nameplate, the temperature rise of the windings and oil (if any) will not exceed the guaranteed values (first in self-cooled OA conditions, then in every stage of cooling). Generally, duplicate units need not be tested.

2. Low Frequency Dielectric Test (ANSI/IEEE C57.12.90-1973)
   These traditional applied and induced voltage tests are performed to prove the adequateness of the insulation system design.

   The applied potential test stresses the major components of insulation and major insulation between the windings and ground. The value of the test voltage is approximately twice the system's highest voltage, as indicated in Table 1.12.†

   The induced potential test stresses interwinding insulation structures (interturn and ends), as well as portions of the major insulation. The induced voltage is a value proportional to the *low frequency* stress that the transformer will be subjected to during normal operation. This value is related to the unit's BIL. This test applies greater than rated volts per turn to the transformer, and therefore requires use of a higher frequency supply than normal (to avoid core saturation).

For details of all routine and optional factory tests see Feinberg,* ***Modern Power Transformer Practice* *(1979), and Stigant,* ***J & P Transformer Book*** *(1973).*

†A lightning surge arrester controls only transient impulse voltages, ***not*** *low frequency voltages.*

The test is normally run for 60 seconds or 7200 cycles; however, for frequencies above 120 cycles the time of duration is *reduced.*

$$\frac{60 \times \text{twice rated frequency}}{\text{test frequency}} = \begin{array}{c}\text{duration of test}\\ \text{in seconds}\end{array}$$

$$\frac{7200}{180} = 40 \text{ seconds}$$

3. Insulation Resistance
This test, using a megohm meter insulation tester, verifies that the state of dryness of the insulation on various windings and the core are of acceptable values (typical values at room temperature are not less than 1000 megohms).

   These test results are used as bench marks to determine the amount of insulation deterioration on subsequent tests.*

Optional Factory Tests on New Transformers

1. Insulation Power Factor
This test verifies the state of dryness and determines a figure of merit for the insulation used in a transformer. Recent trends are to guarantee that the power factor of the windings will not exceed 0.5 percent at 20° C. We believe that this is the only test sensitive enough to monitor the moisture content of a new insulation system and strongly urge its classification as a *mandatory* test rather than *optional.*

2. Additional Dielectric Tests
Overvoltages which a transformer must withstand consist of lightning and switching surges. A series of tests prove the adequateness of the design when confronted with these high frequency disturbances. The individual line terminals of a transformer are subjected to voltage impulses of specified amplitude (vertical height), duration (time), and characteristic shape.

3. Lightning Impulse Tests
Voltage impulses applied simulate these respective electrical disturbances (Figure 1.68 and Table 1.12):

**Chapter 5.*

- Reduced (50-70 percent) full wave (lightning strike hitting lightning surge arresters that replaced rod gap)
- Two chopped waves (flashover of external insulator)
- Full wave (lightning strike)

Oscillograms of a full wave, a reduced full wave and *two* chopped waves are compared with one another in the sequence given. Any inexplicable discrepancy between the two indicates a failure in the windings, major or minor insulation, bushings, or elsewhere.

On repaired transformers when the redesign is such that the voltage distribution is changed significantly or if the cause of failure was the result of impulse breakdown, it is advisable to have the repaired unit impulse tested.

4. Switching Surge Impulse Test
Simulated system surge voltages (such as those resulting from system switching and faults) produce the wave shape of Figure 1.69. These various impulse wave shapes illustrate vividly how transformer windings are changing. Dynamic systems are not static (with fixed "inherent" strength) as was once thought.

5. Audible Transformer Sound Level
Most of the noise measured in decibels (db) during unit operation at 120 Hz originates in the core steel. This value also is related to construction technique. Typical medium power transformers have a guaranteed sound level of approximately 58 db* (55 db is the maximum level recommended for any type of office). The noise level test is a significant factor for indoor transformers used in apartments or hospitals, and units located in residential areas.

   On repaired transformers, when a particular sound level is significant and much rework of the core is necessary, a repeat of the original sound tests is advisable.

6. Short Circuit Test
A transformer should be capable of withstanding high currents flowing between the generation source and the fault (from a ground phase or interphase fault). Because of the power requirements necessary today, this type test is usually performed

**Levels may range from 48 to 91 db.*

at a test laboratory, and as such, is performed on a unit of a given KVA and voltage. Similar designs are also presumed capable of withstanding short circuit forces.*

Today, it is recognized that all sizes of transformers need not have the same short circuit characteristics (four different categories). Magnitude of current (25-40 x base current) varies with unit KVA size for a duration of approximately two seconds.

7. Extended Thermal Evaluation (Heat Run)
A long term "heat run" of several days is occasionally specified for large units.

8. Dissolved Gas-In-Oil Analysis by Gas Chromatography (GC)
This is an increasing trend that may become mandatory on larger power units. A GC test on the oil may follow factory tests and "heat run," particularly if any gas bubbles are noted during tests at the oil surface. Air bubbles or "combustible gas bubbles" may be due to a variety of reasons; for example, bands on the core, magnetic steel used where no mag steel should have been used, or hot spots.

9. Insulating Fluid Tests (if liquid is used)†

10. Transformer Accessories
All transformer accessories should be tested separately prior to assembly to the main unit.‡

Future Test Standards

The induced low frequency voltage in its present form (2 x normal winding voltage) may be replaced by (1.5 x N) for one hour instead of 60 seconds maximum. This is because of the impracticability of increasing test voltages to evaluate higher voltage ratings.

The Corona Radio Influence Voltage (R.I.V.) test, introduced in the mid-1960's and used to detect partial breakdown of insulation during a low frequency test, may be required practice for EHV (> 345 KV) units. Maximum R.I.V. voltage level is 150 microvolts.§

**Chapter 2. Part 4.*

†Chapter 2, Part 2.

‡See Feinberg, ***Modern Power Transformer Practice,*** *(1979).*

§Proposed "Trial" IEEE Standard 262B-1977.

A proposed switching surge (phase-to-ground) wave shape (10 x 1000 microseconds vs 100 x 1000 μs) is also proposed to improve the reliability against field failures attributed to fast-front switching surges (Figure 1.69).*

The use of low voltage impulse testing and GC testing after field installation, as well as following factory testing, are gaining favor.

In the past, manufacturers included an appropriate number of hours of inspection service during installation as part of the contractual agreement. Today, that may be a separate contract.

It is clear from such new conditions that a greater responsibility will be placed on the transformer user in selecting appropriate levels and making sure that actual test results match nameplate data and required specifications.

Long transformer life is also not necessarily assured by proven performance during the foregoing factory tests when followed up with only *minimal* preventive maintenance. A power transformer should last 35 to 50 years when the unit is kept dry and free from oxygen, and also protected as much as possible from overheating, overcurrent, overvoltage, and transient and short circuit surges. Then and only then—after 50 plus years of use—will embrittlement of the insulation finally result in failure.

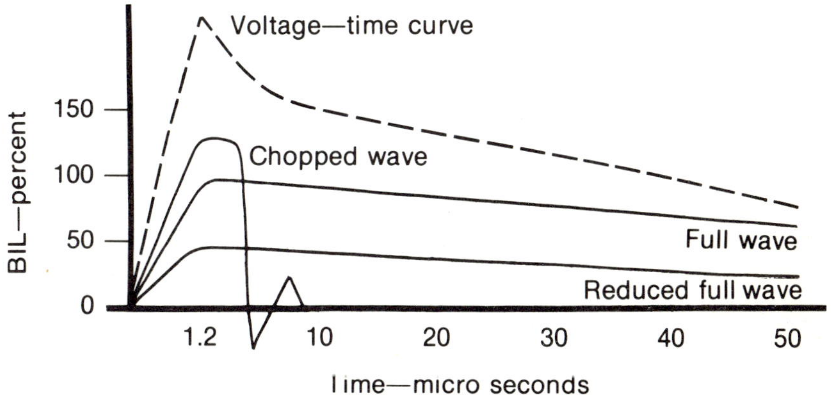

Figure 1.68 - Standard lightning impulse wave shapes at high frequency, or if subjected to fast transient voltages. Windings behave as capacitance networks (charging and discharging).

NEMA Standard No. 107-1964-R71.

TABLE 1.12

DIELECTRIC TEST LEVELS FOR OIL-IMMERSED TRANSFORMERS[1-2]				
Insulation Class (KV)	BIL and Full Wave (KV Crest)	Low-Frequency Test (KV rms)	Chopped Wave[3] (KV Crest)	Switching Surge Test (KV Crest)
15	110	34	130	75[4]
25	150	50	175	100[4]
34.5	200	70	230	140[4]
46	250	95	290	190[4]
60	300	120	345	235[4]
69	350	140	400	280[4]
92	450	185	520	375
115	550	230	630	460
138	650	275	750	540
161	750	325	865	620
180	825	360	950	685
196	900	395	1035	745
215	975	430	1120	810
230	1050	460	1210	870
260	1175	520	1350	975
287	1300	575	1500	1080
315	1425	630	1640	1180
345	1550	690	1780	1290
375	1675	750	1925	1390
400	1800	800	2070	1500
430	1925	860	2220	1600
460	2050	920	2360	1700

[1]IEEE Tutorial Course Text, **Application of Distribution and Power Transformers**, #76 CH 1159-3-PWR (1976), p. 20.

[2]Since the introduction of modern surge arresters, the front-of-wave test is seldom required anymore.

[3]Chopped wave minimum time to flashover is generally three microseconds.

[4]These voltages are not intended as test voltage values on high voltage windings, but are to be used for limiting the induced switching surge voltage in low voltage windings when a high voltage winding is tested.

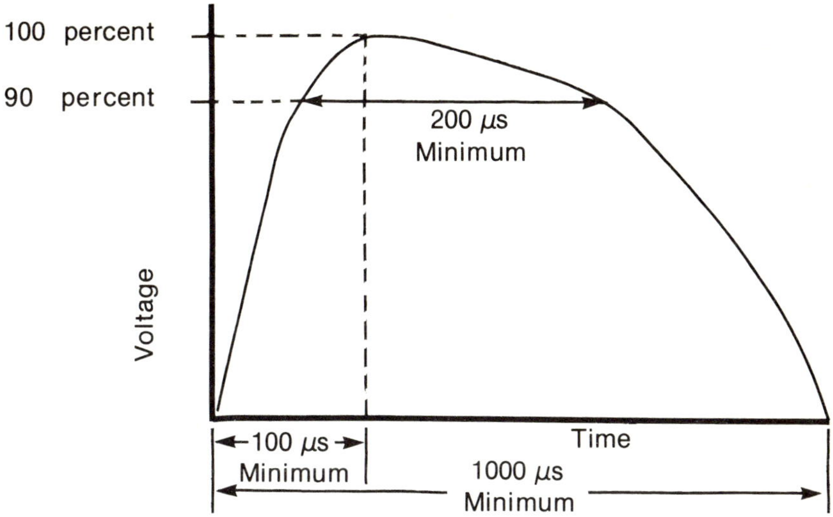

Figure 1.69 - Standard switching impulse wave.

Part 3—Defining Transformer Life Expectancy

Introduction

Numerous electrical authorities have suggested that *the life* of the transformer is *the life* of the oil and paper insulation system. Let's go one step further, and state even more precisely that the life of the transformer is the life of the *solid* insulation—the cellulosic paper, since the fluid may be reclaimed or changed during the service life of the transformer (except pole type distribution units).

Much data confirms that the majority of transformer breakdowns are attributable to the failure of the insulating system. This range has extended up to 85 percent of all transformer failures, depending on the particular source and date.*

The Paramount Importance of the Insulation System

Deterioration of the insulation's mechanical and dielectric properties is the primary factor that governs the length of useful transformer service life. Both the National Electrical Machinery Association (NEMA) and the American National Standards Institute (ANSI) have developed Transformer Loading Guides that recognize this fact. These Guides are viewed by many as being on the conservative side.†

Of the four major categories of materials utilized in the construction of transformers, the solid or cellulosic insulation has remained through the years the most difficult material to properly evaluate in terms of characteristics, reliability, and life.

The other three classes—namely, conducting (copper or aluminum), magnetic (iron), and structural (steel) materials—are all more readily evaluated with regard to those qualities desirable in the construction of electrical apparatus. The conductivity of copper, for example, can be measured to a fraction of a percent. Permeability, hysteresis loss, and eddy current loss in magnetic material can also be measured.

The study of electrical insulation has been the subject of exhaustive research by chemists, physicists and engineers for years. From the

**Chapter 1, Part 3, Table 1.19.*

†Chapter 2, Part 4.

earliest days of the electrical industry, it has been recognized that the *weakest* link in the harnessing of electric power has been the transformer's insulating system. Yet, the fact remains that little of this information has reached down to where it's needed *most*—to the man in charge of electrical maintenance.

With the advent of higher voltages, higher loadings and system interties, the question of a satisfactory insulation system to fit the various design applications is of paramount importance. This truth has been recognized since 1920 when it was originally stated in *The Electric Journal.*

Theoretical vs Practical Transformer Life

The transformer is a living, moving electrical device. Add to this the broad normal temperature range to which a transformer may be subjected:*

Hot spot temperature of -40° C to 160° C
Hot spot temperature of -40° F to 320° F

In practice, a constant winding temperature is seldom, if ever, achieved. Keep also in mind that the flash point of transformer mineral oil is 145°-150° C (294°-302° F)! This fact is just another reason why the cellulosic materials—Kraft paper, cotton, and pressboard—are the weakest link in the insulating system.

Under a proper preventive maintenance program these so-called "weak links" can have indefinite life. Controlled laboratory investigation has shown that the theoretical life of a small oil-filled distribution transformer (built before 1955) at a 95° C continuous hot spot temperature, could be up to approximately 400 years. Granted, the life of a power transformer would probably be considerably less than 400 years. Both design and field maintenance experience have shown, nonetheless, that a power transformer *properly maintained* should have a practical life of 50-75 years. A typical utility may work with a 20- or 40-year unit life expectancy, per plant accounting requirements. Insurance carriers generally consider transformer reliability to be questionable after 20 years. Therefore, the condition in which the insulation system is maintained largely determines the difference between 20 and 50 years of transformer life.

**250° C is the maximum temperature during a short circuit in the secondary winding for a few seconds.*

Much can be done to ensure full life. Preventive and predictive maintenance remain the keys.* An understanding of transformer life is a first step.

The Chemistry of Insulating Systems

Although we use the term "transformer life" as if it were something quite definite, transformers don't die of old age. Aging or gradual degradation is the result of several chemical reactions within the transformer. The solid insulation is one part of the transformer that has an aging characteristic that is irreversible. By contrast, mineral oil can be cleaned, making it as good as or better than new oils. When the paper is neglected, any loss in mechanical and electrical strength can *never* be regained. Therefore, the life of the transformer is the life of the cellulosic paper! In short, transformers do not die of old age, but are killed by neglect! Consider the two factors that contribute to eventual failure.

Failure to Recognize Basic Facts Concerning Cellulosic Degradation

The process of deterioration of oil-impregnated cellulose through the loss of mechanical and electrical strength begins at the paper mill, occurs during the manufacturing process and continues throughout its useful life.

Consider these specific effects of cellulosic deterioration on transformer life. Each can lead to premature transformer failure.

1. Cracking of the Cellulose
   Though a transformer may maintain sufficient dielectric strength, embrittlement nevertheless occurs as moisture is driven from the paper into the insulating fluid. Embrittled cellulose without excessive moisture is actually "quite good electrically." Failure usually occurs when the cellulose material falls apart or is mechanically distorted by vibration, short circuit, or switching surges.
2. Loss of Cellulosic Mechanical Strength
   When cellulosic insulation is heated for any length of time, a proportional decrease of mechanical strength occurs. Yet

**Chapter 6.*

strangely enough, when a transformer is overheated (moisture again driven out from the paper), the insulation resistance reading may actually increase! The result, however, is that the unit cannot handle high-current short circuits of, possibly, six times rated current, or even shock loading.

3. Shrinkage of the Cellulose
   Heat can cause embrittlement of cellulosic material, resulting in its shrinkage. In a unit such as a furnace transformer which is subject to great mechanical stress, shrinkage may permit movement by lowering the clamping pressure.

   The cellulose deterioration amplifies the normal tendency for the windings to grow progressively looser over the service life. The transformer no longer is a tightly compacted unit. The jack screws compressing the windings and superstructure become loosened (Figure 1.70). They were initially torqued to a predetermined setting that held the windings under pressure to keep the unit tight. However, as the cellulose shrinks, it causes or permits additional coil movement—either 120 cycle vibrations or surge voltages.

   Later, we'll consider in more detail several practical aspects of transformer life.* Mineral oil, likewise, starts to deteriorate from the moment it is introduced into a transformer. In other words, oxidation of the insulating liquid begins. Soluble contaminants from the solid materials are dissolved in the oil and may affect its insulating properties and life.† Nothing else but preventive maintenance can slow down this inevitable process.

4. Incomplete Maintenance Procedures
   Improper inspection because of negligence may fail to find, for example, a sealed transformer that permits oxygen to enter the transformer tank (due to an automatic pressure gauge that is malfunctioning), an empty nitrogen bottle that is not replaced, or developing gasket leaks.

Thus, these phases pinpoint the major issue—no recognized or agreed upon standard by which a problem is recognized. These facts have been evident for years but they have been largely ignored.

**Chapter 2, Part 5.*

†Chapter 2, Part 3.

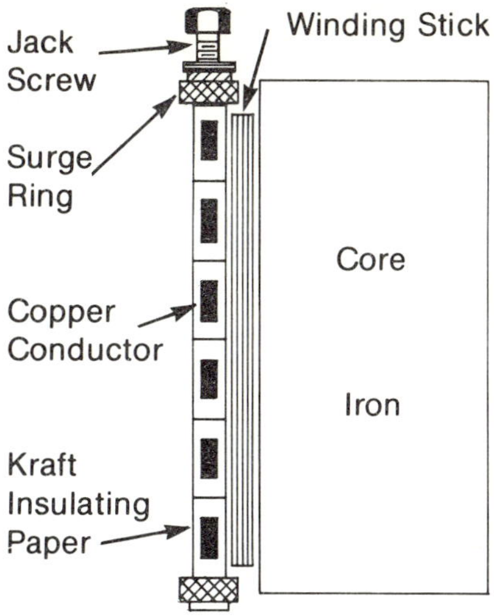

Figure 1.70 - A vertical cross section of major transformer components, focusing on jack screw clamping of insulated copper conductors. Initially the jack screw is torqued to a predetermined setting. However, thermal aging (heat) causes the insulation to shrink. Therefore, the conductors become loose; a sudden mechanical jolt will then bring about premature transformer failure. Recent transformers are thus built with greater bracing through additional jack screws and/or spring-loaded clamping devices.

Operational Accidents

A transformer with its insulating system properly maintained is designed to handle many severe problems. However, a transformer damaged through insufficient preventive maintenance cannot perform to its original design. As a result, these operating accidents can stress the insulation beyond its normal strength. The unit will then go off the line or even fail completely.

- Start up (re-energize spare unit)
- Switching surges
- Controlled short circuits
- Through-faults
- Lightning strikes
- Spike impulse voltages
- Physically moving transformer

When does a transformer reach the end of its life? It's impossible to predict! But here's one definition:

Consider the end of a transformer's life as "the length of time which will elapse until the insulation is so badly deteriorated that the probability of failure with *one* year is unreasonably high."

The very fact that this and any definition of transformer life is vague attests to the significance of the problem. The deterioration process through the loss of mechanical and electrical strength of the impregnated cellulose has its origin at the paper mill and continues throughout the cellulose's life.

Other Causes of Premature Failure

Causes of transformer failure, particularly those beyond the user's control, include:

1. Faulty Workmanship or Inferior Materials
   For example, mineral oil was discovered below specification after 14 months; the supplier did not notify the buyer and the buyer did not have adequate quality control.

2. Poor Shipping Technique
   Shock damage during negligent rail transit has resulted in extensive damage. One case history of 21 core type power transformers included these examples of internal damage:

   > Damaged core laminations, core displacement—both longitudinal and lateral movement; coil displacement—both axial and radial movement, destruction of some coil, insulation paper, loose core clamping straps and thrust jacks; coil jack screws; and much more—yet only minor external damage was observed.

3. Negligent Installation
   When everything else fails, read the manufacturer's instruction booklet! As a classic case history at a midwestern firm, three transformers that each required over 200 gallons capacity were received filled with only 10 gallons of insulating oil. This was not determined until reclamation of dirty oil was undertaken. It took less than two minutes of pumping to know something was wrong!

4. New Design Parameters
   In the mid 1960's, industry reports began to show an increasing number of transformer failures. The causes were pinpointed as

the result of rapid growth of EHV (extra high voltage) power and reduced through-fault capability (reduced BILs).

5. User Applications
   The transformer operator exceeds design specifications; for example, loading, number of line surges per year, and so on.

In all these instances, proper maintenance can at best minimize these five problem areas of which modern design practice may be the most significant. Its scope demands further attention.

Evaluating Current and Former Transformer Design

An IEEE paper has acknowledged that transformer design has been in a state of change for at least 20 years. Manufacturers have reduced short circuit strength during the 1960's and have in good faith made proposals to customers for actual repair/upgrading of these power transformers.

In the past, electrical industry designers focused on *minimum* volts per turn, *full* insulation, and *maximum* oil per KVA. Thus, prior to 1959, engineers estimated the theoretical life of a properly assembled and adequately serviced small oil-filled distribution transformer as 400 years.

Today, however, computer design optimizes the *minimum* use of materials and space with the maximum stressing of conductor, iron, and insulation.

> The manufacturers interviewed were unanimous in pointing out that they try to design and build a product that will deliver its intended performance for a satisfactory period of time. They recognize that the necessity to remain competitive in their field, against other domestic manufacturers and against increasing pressure from foreign competitors, requires computer design that optimizes the minimum use of materials and space and the maximum stressing of conductors, iron and insulation. As a result of this, they are certain that their product will perform according to its nameplate specifications, but it is acknowledged there simply are no built-in excesses that will tolerate operational or maintenance abuses.

On the other hand, manufacturers can point to the use of thermally upgraded paper insulation, as well as better sealing of the transformer tank and better gasketing, which minimize the entrance of undesired materials. While the computer does ensure a maximum speed and accuracy of calculations, it also depends upon the designer to imput correct information per customer specification. If the initial design is incorrect in any way, then the final program will also include these faults; therefore, no short cut to good design exists—careful evaluation of input data is still necessary. In computerese—"Garbage in equals garbage out."

Actual data show that the most critical years of transformer life are the first three and, secondly, the first ten years of equipment use. Yet keep in mind that field experience has demonstrated that a power transformer, properly maintained, may have a practical life of 50 years!

Increased Volts Per Turn

Relative electrical stress in a transformer can be determined by comparing values of volts per turn. This value has increased greatly over the past 50 years (Table 1.13). Reasons for the increase include better core steel, better insulation, less space requirements, and larger transformer size. Consider in Figure 1.71 the effect of reduced ampere-turns upon the electrical stress of the insulation system between 1902 and 1970 60 Hz equipment. Certain EHV transformers may have as much as 200 volts per turn. This increase in stress will cause "gassing" by breaking down the cellulosic insulation and create other problems with the oil in these high stress areas.

TABLE 1.13

TYPICAL TRANSFORMER CORE STRUCTURE ACCORDING TO TRANSFORMER DESIGN PERIOD

Year	Volt/Turn
1915	2-4 Volts/Turn
1932	8-10 Volts/Turn
1975	19-20 Volts/Turn
Present	200 Volts/Turn (EHV)

Reduced BILs

A trend that has now become the usual practice concerns the building of higher voltage transformers with greater reduction in BILs (Basic Impulse Levels); or simply stated, reduced insulating materials.

> It is evident that greater cost incentives are present for using reduced BILs at higher system voltages. Thus full insulation is used at essentially 69 KV and below, while several steps reduced BIL may be used in the EHV ranges.

Experience has already resulted in recognition of some limits for reduced BILs. The effect of this current practice is illustrated in Figure 1.72, in which several commonly used levels of transformer BIL are plotted against rated voltage.

> The BIL indicates the dielectric strength, and rated voltage is the steady state stress. The significantly greater proportional operating stress in the higher ratings is apparent.

Though lower BIL does drop the cost of transformer construction, it also improves the efficiency of the unit, *provided* lightning arresters and other fault protection equipment are properly specified and installed. Zinc oxide arresters are very efficient, and are more accurate to rating, giving a more desirable coefficient of ground.

Transformers located in areas where thunderstorms are not common may safely use much lower BILs (Figure 1.73). Nevertheless, all manufacturers basically agree that possible cost savings may not be worth the potential added risk to system security.

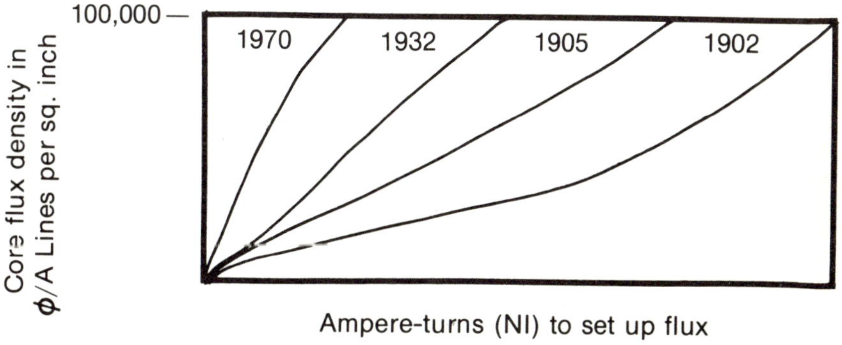

Figure 1.71 - Historical comparison of electrical stress upon the insulation stress in terms of reduced ampere turns and resulting increased volts per turn. In the NI relationship only the "N" can be varied.

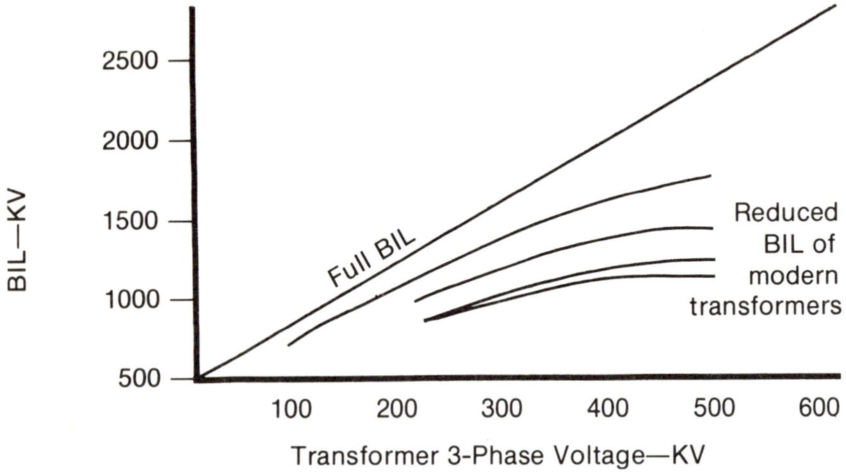

Figure 1.72 - Several commonly used levels of transformer BILs are plotted against rated voltage. IEEE Tutorial Course ***Application of Distribution and Power Transformers*** *(1976), p. 66.*

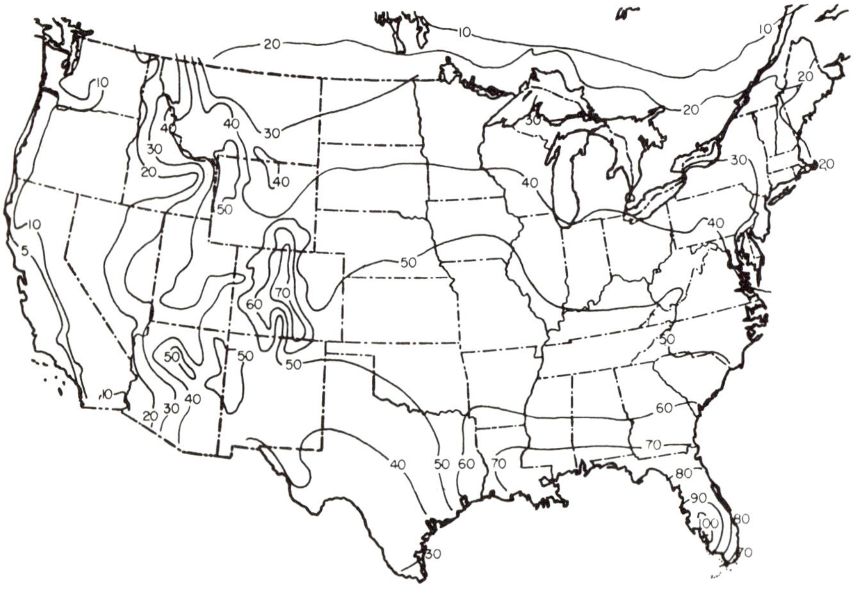

Figure 1.73 - Incidents of thunderstorms in the U.S. Lightning generally is more common in tropic and temperate zones than in colder climates. The numbers on the isograms indicate the average ***annual*** *days with thunderstorms for that area. Alaska averages 1 to 6 days and Hawaii 5 to 9. (Courtesy of* ***Plant Engineering****).*

Reduced Impedance

A lower impedance rating has simultaneously occurred with reduced BIL (Figure 1.74). This diminished ability to withstand higher short circuit current surges has resulted in the increased frequency of through-faults. On the other hand, this reduction in impedance has reduced the inductive reactance (lower losses), therefore, requiring less paper and permitting the application of UHV (ultra high voltage) transformers.

Improper bracing for the increased electromagnetic forces has been corrected through use of spring-loaded clamping. Therefore, the epidemic of through-faults during the last ten years should taper off in the years to come.

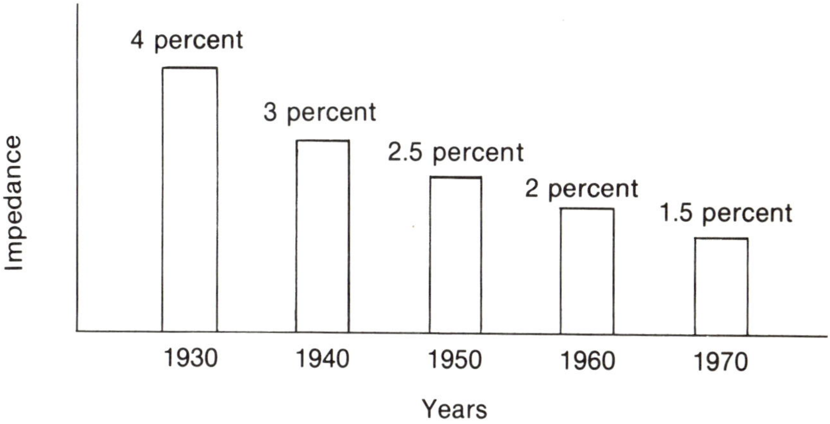

Figure 1.74 - Historical comparison of distribution transformer impedance. (Courtesy of the RTE Corp.-1973).

Reduced Physical Size and Insulation System

A fourth design factor has been the reduction in physical size, alleviating transformer shipping problems. A very large transformer in 1950 was about 100 MVA at 230 KV volts. Today, a "huge" unit would be 1300 MVA, at 765 KV. This great increase in capacity has been accomplished by improved core steel and cooling, thermally upgraded paper, lower BIL, form-fitting tanks and improved transit.

Table 1.14 illustrates how the size of transformers relate KVA rating to unit oil capacity. The insulating oil now provides 80 percent of the dielectric strength of a transformer. In prior years cellulose provided more of the electrical strength, but mini-construction has reduced this value. Less oil per pound of insulation also means less capacity of the oil to remove moisture from the cellulose paper when hot and less capacity to dilute the concentration of any soluble contaminants.

TABLE 1.14

TYPICAL TRANSFORMER OIL CONTENT ACCORDING TO TRANSFORMER DESIGN PERIOD	
Year	Gallons Oil/KVA
1915	2 Gallons/KVA
1930	1 Gallon/KVA
1945	1/2 Gallon/KVA
1960	1/3 Gallon/KVA
1975	1/6 Gallon/KVA
1977	1/7 Gallon/KVA
1979	1/9 Gallon/KVA
Some of the newer EHV transformers contain as little as 1/50 gallon of oil per KVA.	

Contrasting Transformer Design Criteria

The operating temperature is not uniform throughout a transformer. The unit part operating at the highest temperature is said to be the winding hottest spot temperature and naturally undergoes the greatest thermal stress.

Determination of this upper temperature limit makes use of three factors:

1. Average yearly ambient temperature (air temperature external to the unit).
2. Average winding rise above ambient due to full load.

3. Allowable temperature permitted for hot spot beyond top oil temperature (items 1 plus 2).

The sum of components 1, 2, and 3 equals maximum hottest spot temperature.

The former limit on temperature rise for oil-filled transformers was 55° C average winding rise (65° C hot spot—Table 1.15). However, since 1963 there has been a gradual and then a rapid transition from the 55° C standard to the 65° C average rise (80° C hot spot). Today, nearly all new units are 65° C average rise transformers.

TABLE 1.15

CONTRASTING CURRENT DESIGN CRITERIA WITH THOSE PRIOR TO 1963

Temperature Criterion	Past Design Criteria (before 1963) Temperature Degrees Celsius	Current Design Criteria (since 1963) Temperature Degrees Celsius
Ambient	40° C	30° C
Average Winding Rise Due to Load	55° C	65° C
Estimated Allowance For Hot Spot[1]	10° C	15° C
Hottest Spot Temperature	105° C	110° C

[1]The location of the hottest spot has proved most elusive.

> Many modern transformers are more sophisticated in design and are not "overbuilt" as were many of the older transformers in-service. Additionally, with the advent of 65°C rise transformers...such units are smaller and contain less oil per KVA rating. This means a higher ratio of solid insulation per gallon of oil (less dilution factor for contamination and deterioration), higher internal temperatures, and more insulating materials under maximum permissible electrical stress.*

In comparing past and current design criteria, it is a moot question whether one is superior to the other or merely compensating trade-offs; in both designs, reality points to the overriding importance of ongoing preventive maintenance.

The Importance of Dehydrated Insulation

The foregoing discussion of new transformer design clearly illustrates one fact—how important the elimination of moisture has become. *If* (and when) less insulation is used, the insulation must be maintained at a *higher* level of dryness.

Consider in this light the early history of the transformer insulation just from the factory to a plant site. A 30 MVA transformer at 154 KV has over four tons of paper (Table 1.16).

During the manufacturing process the paper will absorb moisture from the atmosphere up to 10 percent by dry weight of the insulation; that is, in our example, 802 pounds or 100 gallons of water. This water must be removed from the insulation before the transformer is shipped to a plant. Dehydration results in a 9 percent mechanical shrinkage in the insulation which must be offset by tightening the whole transformer under presses, use of jack screws, measuring devices, and other devices before the unit is oil-filled and tested.

How does the user know sufficient drying has been performed? Insist on an insulation power factor test no higher than 0.5 percent. This will be your only assurance that the windings have been sufficiently dried before the unit was oil-filled. Even after proper drying practice, residual moisture in the solid insulation will still remain in the range of 0.3 to 1.0 percent of the dry weight of the insulation.†

**E.L. Raab (retired from the General Electric Co., Pittsfield, Mass.).*

†*Chapter 6, Part 2 and Chapter 7, Part 2.*

TABLE 1.16

EXAMPLES OF PAPER IN TRANSFORMER					
KVA Rating	KV	Weight in Pounds	Pounds KVA	10% Initial Water[1]	
				Pounds	Gallons
3,000	13.2	1,000	0.33	100	12.5
10,000	115.0	3,540	0.35	354	44.3
16,000	115.0	4,050	0.25	405	50.6
20,000	132.0	5,760	0.28	576	72.0
30,000	154.0	8,020	0.26	802	100.0
40,000	230.0	10,600	0.26	1,060	132.5

[1]When the transformer is dried at the factory, there is a 9 percent shrinkage in the insulation.

Evaluating Transformer Faults and Failures

Both manufacturers and transformer maintenance authorities hold that a typical transformer does not die of old age—but is killed.

While "the chief contributing factor to failure of electrical equipment is inadequate maintenance," major sources of transformer failure also include:

- Manufacturer: design/workmanship
- Shipper: handling/packaging
- User: installation, operator error, overloads, excessive number of line surges, and improper or insufficient maintenance
- External Forces: earthquake, lightning, animals and so forth

At the same time, recognize the proper place of transformer *failure* in transmission and distribution outages, especially those where fire ensued (Table 1.17).* Fortunately, the "heart of the substation" is not

**Failure - A disturbance that causes complete coil/core and permanent insulation damage. It requires replacement with a new unit.*

among the worst offenders; nevertheless, when frequency of *faults* is considered, transformers do rate higher among the various links of a power system (Table 1.18).*

Since overhead transmission lines may extend over hundreds of miles, they are often exposed to numerous chances for breakdown—storms, falling external objects, damage to standoff insulators, and so forth. Fortunately, many of these faults are transitory in nature, and thus vanish within a few cycles—as would be the case when a twig falls across a line and a cross-arm and burns itself out or just falls down.

Though frequent transformer failures or faults do occur, "percentage-wise, they are relatively few," considering the hundreds of thousands of units in-service worldwide.

Analysis of Transformer Failures

Engineering consultants, utility associations and insurers annually survey their clients to determine the causes of transformer failure. Tables 1.19, 1.20, and 1.21 list the primary causes of transformer failure.

In recent years, a large percentage of power transformer failures have been due to through-faults and tap changers.† In spite of recent attention given through-faults, data indicates that the number of these failures still remain high (Figure 1.75).

Our laboratory data confirmed by internal inspection has revealed that a large number of combustible transformer faults caused by dissolved gas in transformer oil have their origin in the auxiliary tap changer.

Fault* - *A transient (short) disturbance or minor or major failure that impairs normal operation: for example, insulation failure; the transformer is thrown off-the-line. (*Minor Failure*** - A disturbance that requires removal from service and field repair.* ***Major Failure*** - *A disturbance that requires removal from service and factory detanking and repair).*

†*Disregarding "no comments" and "unknowns" in Doble's 1980 questionnaire, 17 percent were related to tap changers and 28 percent were through-fault failures.*

TABLE 1.17

FIRES OF ELECTRICAL ORIGIN (1965 to 1974 inclusive)				
Class of Electrical Equipment	Number	Amount[1] of Loss	Percent of Totals	
			Number	Amount
Wiring	1,184	$37,502,000	30.5	40.9
Motors	505	4,977,000	13.0	5.4
Resistance Elements	371	7,571,000	9.6	8.3
Switchboards	239	6,951,000	6.2	7.6
Lamps	208	10,476,000	5.4	11.4
Controllers	205	8,094,000	5.3	8.8
Automobile	173	2,623,000	4.5	2.9
Transformers	131	2,111,000	3.4	2.3
Electrostatic Equipment	130	3,832,000	3.4	4.2
Electronics	114	755,000	2.9	0.8
Circuit Breakers	90	1,595,000	2.3	1.7
Generators	28	372,000	0.7	0.4
Fuses	18	35,000	0.5	—
Miscellaneous	486	4,781,000	12.5	5.2
Total	3,882	$91,675,000	100.0	100.0

[1]Includes property damage and business interruption.

TABLE 1.18

FREQUENCY OF FAULT OCCURRENCE IN DIFFERENT LINKS OF A POWER SYSTEM[1]	
Equipment	Percent of Total
Overhead Lines	50
Cables	10
Switchgear	15
Transformers	12
Current and Potential Transformers	2
Control Equipment	3
Miscellaneous	8

[1]Ravindranath, **Power System Protection and Switchgear** (1978).

TABLE 1.19

MAJOR CAUSES OF POWER TRANSFORMER FAILURES (1961-1979)[1-4]	
Cause of Failure	Reported Failure Percentage
Insulation[2]	620
Through-Faults (short circuit)	302 (1969-1979)
Tap Changers	221
Lightning or Switching Surges	150
Manufacturing Defects	66
Core (Ground and Insulation)	57
Mechanical	80
All Other[3]	834
Total Failures Reported	2330

Table 1.19 continued on next page.

Table 1.19 continued.

[1]Annual survey conducted by Doble Engineering Company, Watertown, Massachusetts.

[2]Causes of insulation failures consist of winding, moisture, solid insulation.

[3]Other responses include operator error, overload protective system failure, vandalism, foreign article, ruptured tank, other equipment, unknown causes, no comments, and so forth.

[4]Furnace and rectifier transformers have failure rates approximately six times greater than the average rates for power and distribution transformers (Factory Mutual Insurers).

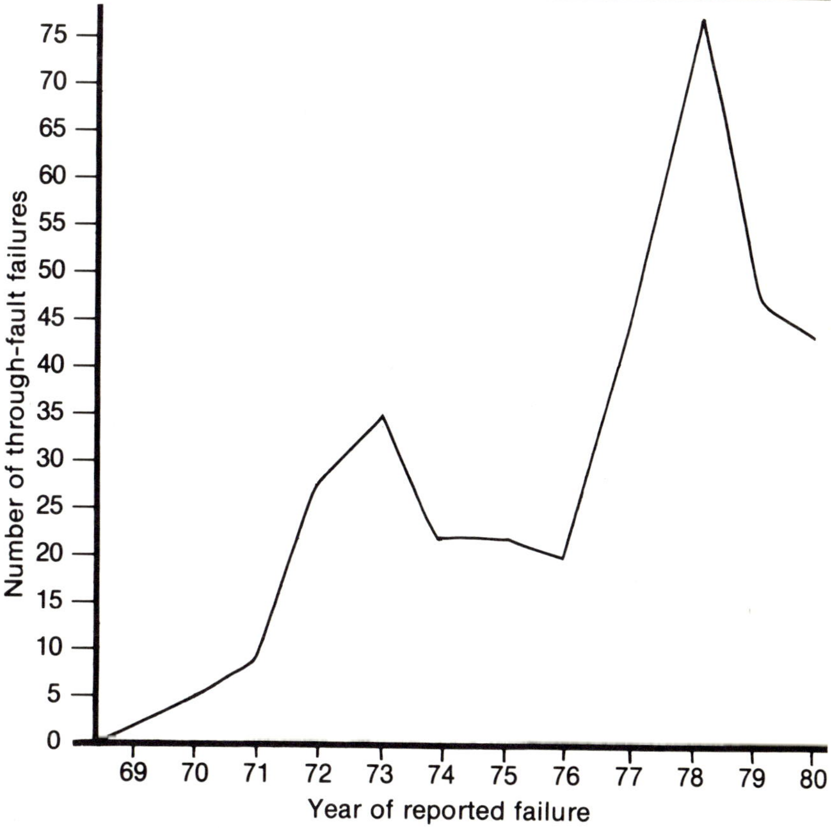

Figure 1.75 - Reported power transformer failures caused by through-faults. None were reported between 1960 and 1968. Reasons for the increase include the use of less iron, copper, and oil, as well as reduced BIL (Chapter 2, Part 4). (Courtesy of Doble Engineering Co., Watertown, Massachusetts).

TABLE 1.20

PRIMARY CAUSES OF DISTRIBUTION AND POWER TRANSFORMER FAILURES[1]

Cause of Failure	Percentage 1963	Percentage 1975[5]
Lightning surges	26.4	32.3
External short circuiting[2]	8.5	13.6
Poor workmanship (Manufacturer)[3]	7.3	10.6
Deterioration of insulation	10.2	10.4[6]
Overloading	4.5	7.7
Moisture	11.4	7.2
Inadequate maintenance and improper operation	6.4	6.6
Sabotage, malicious mischief	4.9	2.6
Loosening of connections	3.7	2.1
All others[4]	16.7	6.9
	100.0%	100.0%

[1]Hartford Steam Boiler Inspection and Insurance Company (1976).

[2]External short circuits include those caused by animals, broken lines or bushings, or flashover (not contamination).

[3]Failures occurred within the first year of transformer operation.

[4]All others include instances of misapplication, operator's error, owner's or repairer's workmanship.

[5]The average age at the time of failure from all causes was 9.4 years (the range of continuing service before varied from one month to 60 years).

[6]The average age of units failing due to insulation deterioration was 23 years. In all cases, the *actual* age of the insulation was used.

TABLE 1.21

INITIAL TRANSFORMER PART TO FAIL			
Transformer Part	1963[1]	1975[1]	1976[2]
High voltage windings	57.3%	58.0%	48.0%[3]
Low voltage windings	14.7	19.8	23.0
Bushings/Insulators	5.3	8.8	2.0
Leads	11.4	4.4	6.0
Tap Changers	5.3	3.2	——
Gaskets	——	——	2.0
All others	6.0	5.8	19.0
	100.0%	100.0%	100.0%

[1]Hartford Steam Boiler Inspection and Insurance Company.

[2]The Kemper Insurance Company, Long Grove, Illinois.

[3]A number of failures result from either 60 Hz or impulse creepage over the coil end insulation or between phases. Deposits of materials in these areas by magnetic attraction tended to promote this failure.

Time of Transformer Failures

Both surveys and transformer authorities acknowledge that the most critical years of transformer life tend to be the early years (that is, faults caused by manufacturing, shipping, and installation). These usually show up in a few days to one year after installation. Aging factors begin to take effect in seven to ten years. Table 1.20 (Notes 5 and 6) also gives some significant aging results.* All of this data, thus, confirms the familiar "Bathtub Curve" of Figure 1.76.†

**See Chapter 4, Table 4.26 for results arising from development of combustible gases in transformer insulating oil.*

†It is this data that verifies our conviction that occasioned the writing of this book—50 plus transformer years of life!

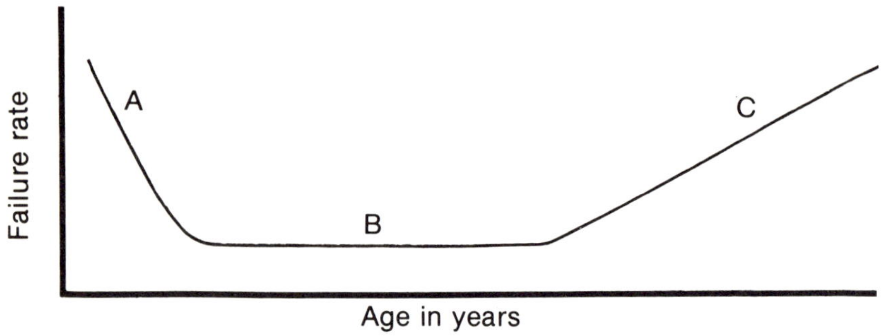

Figure 1.76 - Transformer failure rate "Bathtub Curve": (A) is failures due to infant mortality; (B) is constant failure rate (random); and (C) is failures due to aging. (R. Sahu, "Using Transformer Failure Data to Set Spare Equipment Inventories"-1980).

Importance of Transformer Maintenance

One clear guideline should be obvious. Every facility should have a good working knowledge of the failure rates of their major substations and should gladly participate in surveys that would give even more complete input into how to best plan an effective preventive maintenance program.

In today's inflated world economy, "zero-loss" of property or interruption of service must be the goal. No one can afford *not* to make this goal a reality! The only factors within the control of the owner/operator to prolong transformer life are proper maintenance and adequate substation protection. Recognized steps of maintaining this equipment in a safe and reliable operating condition are described in Chapters 3 through 10. Prior to this consideration, it is necessary first to understand the underlying causes of transformer failure.

The "field enemies" recognized by designer and consultant alike (and considered in Chapter 2) include:

- Moisture
- Oxygen
- Solid contamination
- Gas bubbles (from supersaturation or from oil decomposition in consequence of hot spots or electrical discharge
- Overcurrent
- Overvoltage (transient or dynamic)
- Overtemperature
- Short circuit (mechanical forces)

Chapter 2

The Insulation System—The Lifeline of a Transformer

Part 1—A Panoramic Survey of Insulating Media

Introduction

Earlier discussion focused on components of the transformer insulation system, giving special attention to solid insulating materials, particularly cellulose.* Our purpose now is to review the historical development of both liquid and gaseous dielectrics and their role as transformer insulants/coolants. These materials—transformer mineral oils, askarels, silicones, gas and air dielectrics—have served the electrical power industry well, and in most instances are likely to continue doing so. Transformer oils in particular are the most common insulating materials that have been utilized in the industry since the turn of the century.

Development and Use of Transformer Oils

The earliest transformers built between 1884 and 1886 did not contain any liquid insulation. However, the first decade of transformer use saw the need for a more effective insulating medium, such as mineral oil. Linseed oil, and various mineral and vegetable oils were tried unsuccessfully.

Even after Professor Elihu Thomson (Figure 2.1) had patented mineral oil for use in transformers in 1887, it was not until at least 1892 that the first insulating liquid—lubricating oil (Pennsylvania paraffinic type)—was employed for transformer use by the General Electric Company.

> It was not easy...to induce customers to employ oil-immersed transformers. They objected that the oil added to the cost, and constituted a fire risk should it leak through the case.

**Chapter 1, Part 2, Table 1.5.*

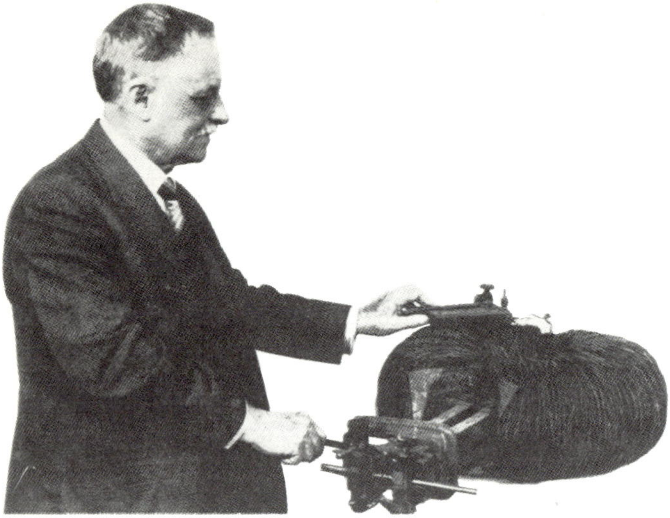

Figure 2.1 - Professor Elihu Thomson and his welding transformer. (Courtesy of the General Electric Co., ***Men and Volts*** *(1941).*

The unique value of transformer oil has since been recognized as serving four basic functions:

1. Providing sufficient dielectric strength
2. Providing efficient cooling/heat transfer
3. Preserving the core and coil assembly (by filling the insulating material voids)
4. Minimizing content of oxygen with cellulose and other materials susceptible to oxidation

When any one of these functions is impaired, both transformer efficiency and life is reduced. It is, therefore, understandable why the first effort was made to use a common paraffin-based lubricating oil.*

Lubricating oils will handle function number three very well, but leave something to be desired with regards to functions number one and number two. Lubricating oils contain undesirable additives which can lower the dielectric strength. Fortunately, in the early days, the ad-

Parraffin-based** crude oils are those which yield a paraffin wax (canning wax) as the final product of a distillation process (Chapter 2, Part 2).* ***Naphthene-based *oil is made from a crude which has as its final distillation product an asphalt compound.*

ditives weren't as many as in a modern lube oil, which may contain some 20 additives. In addition, the units were engineered with plenty of space between the windings so that in lower voltages, air alone was a sufficient dielectric. In other words, transformers were overdesigned with a safety factor to cover up the unknowns.

So long as the early units were located indoors, lubricating oils were good enough. However, when application was extended outdoors, weaknesses of lube oils became evident.

Paraffin-Based Oils

The year 1899 saw the first transformer oil manufactured to specifications. During the early 1900's, five more oils—all consisting of a paraffin base—came into use. Very little has been recorded of the properties of these early oils. However, some of them are known to have contained as much as ½ of one percent of water. Transformer owner/operators made no complaints unless the water content exceeded the ½ of one percent. The presence of water was detected by the "hot iron test". This consisted of plunging a hot iron into the oil. If a "crackling" noise was heard, the oil was too wet. Another name for this test is the "Crackle" test. Other oil developments followed, resulting in 12 different kinds of oil with varying properties.

As years went by, pressure began to mount to find a universal transformer oil with better properties that could be used in all types of electrical equipment. One of the properties that had to change was the pour point (-10° C). This was because larger transformers were installed outdoors where they were subjected to seasonal temperature extremes. Previously, most units had been housed indoors where a fairly high pour point was sufficient.

Naphthene-Based Oils

About 1925, a series of conservator freeze-ups at the pipe connecting the conservator with the main tank led to the application of a low pour point (-40° C) naphthene-based transformer oil to replace the paraffinic oils. These new crudes originated in the Gulf Coast area.

By now, many old transformers that were in service began to show the effects of little or no maintenance. This lack of transformer oil maintenance can be attributed to a combination of:

- Many kinds of oil in use

- Lack of standard specifications for manufacture of oil
- Lack of standards for determining oil quality
- Lack of understanding of what transpires inside a transformer as it ages

Thus, the 1930's witnessed many maintenance programs aimed at establishing criteria for in-service oil, accumulating test data, and finally, replacing transformer oil in units that did not meet testing standards.

While the newer naphthenic oils brought solutions to existing oil problems, they also created a new problem. For example, the pour point problem was solved, but was replaced by a sludging problem. It is interesting to note that sludge forms much faster in naphthenic oils than in paraffin-based oils, but sludge in the latter oils is generally insoluble and almost impossible to remove. Because oils were causing *sludge* in transformer units, more attention was paid to methods of solving such problems. Oils were being tested more often to see whether they were beginning to sludge. In some cases, they were being replaced to ward off the damage caused by sludge deposits.

Challenges to Oil Reclamation

Reclamation Suffers Loss

Recognition of the oil sludging problem led to the next improvement in the history of transformer oils—oil reclamation. Data had been collected on in-service oils between the mid-1920's and the early 1950's, during which period oils had been reclaimed. This approach, however, lost favor when it was discovered that the reclaimed oil deteriorated at a much faster rate than new oil. Probably the most significant breakthrough of the early efforts was the discovery that new oils contained "natural" inhibitors which either were consumed in-service or stripped off during the reclamation process.

Replacement of Natural Inhibitors

Twenty years of research supplied the answer to this problem by finding a suitable antioxidant which would replace the characteristics that were lost through reclamation.

In December, 1950, transformer oil—supplied with new transformers—began to contain up to 0.3 percent by weight of an antioxidant commonly known as 2,6-Diteriary-Butyl-Para-Cresole or

DBPC®.* Because of an extreme shortage of DBPC in early 1951, the industry returned to the use of uninhibited oil. When supplies of DBPC became plentiful again in 1953, the industry agreed to supply inhibited oil in distribution transformers and uninhibited oil in all others.

This additive, DBPC, has proven its value to industry, but transformer manufacturers and some large utilities even at this point in time have differing views on the subject. Therefore, there is no universal agreement on the use of DBPC. We are of the firm conviction that 0.3 percent by weight is a must if long transformer life is important.

Figure 2.2 compares the effects of inhibited or uninhibited reclaimed oils. While up to 0.3 percent DBPC by weight clearly improves the oil quality, further experiments have shown that beyond this content, the oil's impulse strength may be affected.

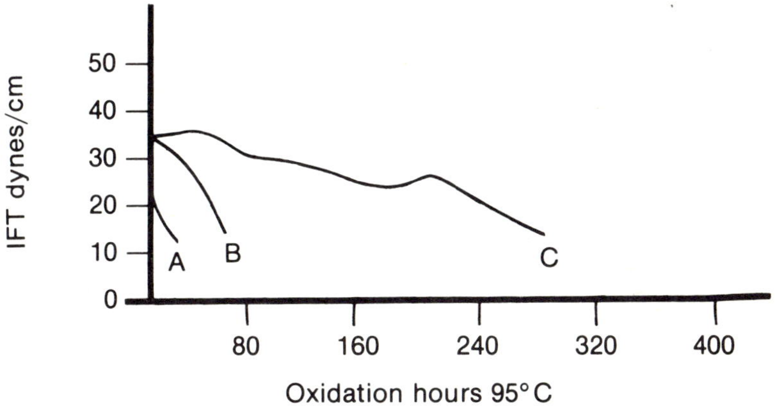

A—Reclaimed oil: uninhibited
B—Reclaimed oil: inhibited (0.1 percent DBPC by weight)
C—Reclaimed oil: inhibited (0.3 percent DBPC by weight)

Figure 2.2 - Comparing the effect on reclaimed transformer oil with and without oxidation inhibition. F.C. Doble, "The Reclamation of Insulating Oils," ASTM STP # 152, (1952).

**DBPC® is a registered trademark of the Koppers Co., Pittsburgh. This additive is also used as a preservative in food oils (crackers, bread, and cake mixes).*

Reclamation Regains Favor

As time went on, oil reclamation began to regain popularity as evidenced by the following excerpt from the American Society for Testing and Materials (ASTM) publication of June 23, 1952:

> The reclamation of insulating oils, including reconditioning, rerefining, and inhibiting, is of growing importance to power industry operators. Recent developments in a long studied power system maintenance program have furnished the basis for a new concept of the operating characteristics of insulating oils. The new concept is that insulating oils, when properly serviced, can be given practically unlimited extension of life free from the formation of sludge or excessive acidity due to oxidation even in free-breathing transformers.

Worldwide Oil Shortage

The latest phase in the history of oils is summed up in the word "shortage" which began to affect both oil and the antioxidant DBPC in the early 1970's. In 1973, the Western world suffered from an extreme shortage of oil partly due to the Arab oil embargo, which pushed oil prices from $0.30 per gallon before 1973, to at least $2.15 per gallon in 1980.

A recent problem facing the electrical industry is the shortage of the best crudes used in the manufacture of transformer oil (Figure 2.3). Of all the crudes available, only three percent are of the type used for transformer oil. What this means is that now different crudes are being refined. Moreover, environmental constraints have brought about different techniques of refining.* The industry has now gone back to the use of oil blends with paraffinics at concentrations up to 50 percent.

These factors may bring changes in the equipment in which oil is used. Other factors such as increasing costs, unduly restrictive transformer oil specifications, and government allocation requirements are forcing the industry to take a harder look at newer oil refining methods in order to meet transformer oil demands. Therefore, oil conservation and preservation methods, as well as other forms of insulating liquids, are being given serious consideration (Figure 2.4).†

**Chapter 2, Part 2, Table 2.3.*

†Chapter 7, Part 3, and Chapter 9, Part 3.

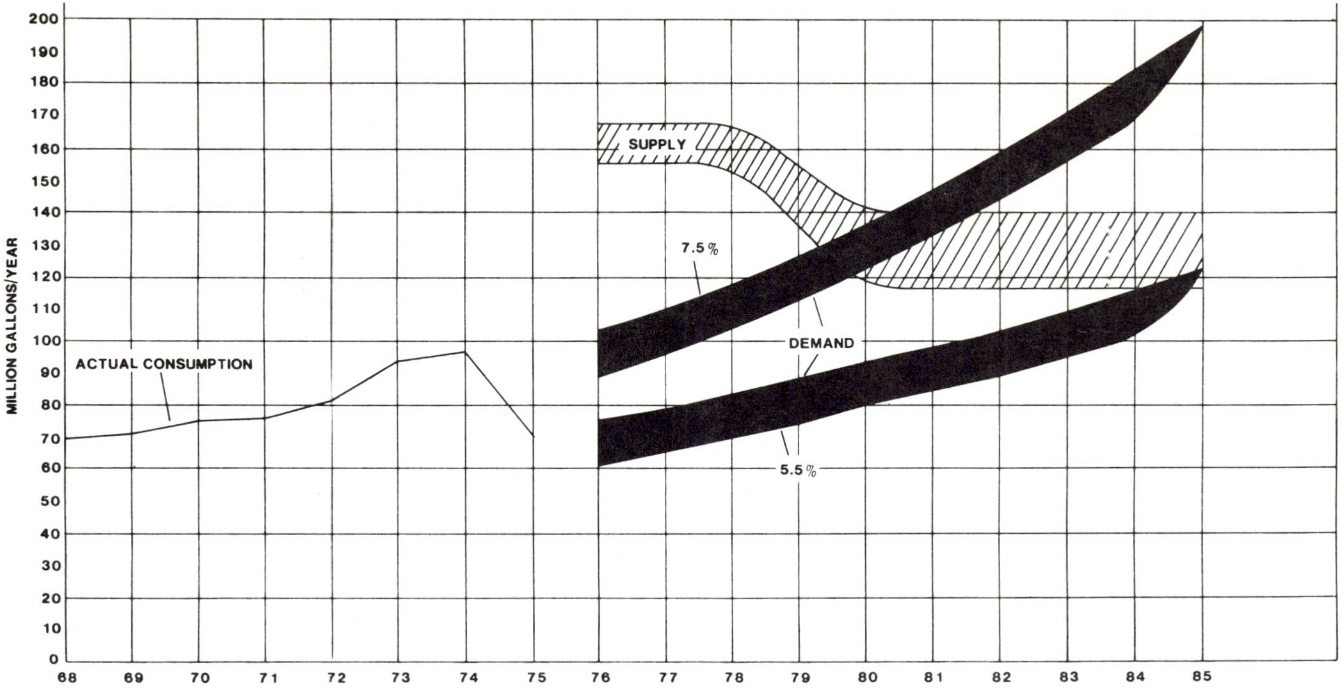

Figure 2.3 - Comparing U.S. transformer oil supply and demand 1968-1985. E.L. Raab, "U.S. Transformer Oil Supply and Demand 1975-1985 EPRI Report" (1976).

Figure 2.4 - The transformer oil shortage.

Alternatives to Oil-Filled Transformers

Research and development and a desire to overcome the limiting property of mineral oil constitute the story of ongoing transformer technology. The PCB story is the classic example.*

Immersed Askarel Transformer Insulation Systems

Throughout this volume, our primary focus is and will continue to be oil-immersed transformer insulation systems. Nonetheless, our treatment of these materials would be remiss if consideration were not accorded to askarel-based systems.

Askarel, developed in the early 1930's, is a generic term for a group of synthetic chlorinated aromatic hydrocarbons utilized in capacitors (1931) and transformers (1933). Use of askarel was developed primarily to eliminate the fire and explosion hazard commonly caused by mineral oils. Askarels, also known as polychlorinated biphenyls (PCBs), have come to be known by various trade names such as Aroclor®, Pyranol®, Inerteen®, Chlorextol®, Saf-T-Kuhl®, and so forth. PCB was invented and patented by Swann Chemical Company

**Chapter 9, Part 1.*

(1929). The company chlorinated the biphenyls to various degrees (for example, 54 percent, 60 percent, etc.). Samples were then sent to several companies for evaluation in an effort to find out if commercial applications were possible. The late F.M. Clark (General Electric Company) evaluated these materials and immediately filed for *application* patents for electrical equipment such as capacitors, transformers, cables, etc. Swann, however, continued to own the *material* patents. Shortly thereafter, Monsanto Chemical Company of St. Louis, Mo., bought out Swann, including the patents.

The first askarel-filled transformers were sold to Consolidated Edison Company of New York City in 1933 or 1934. The liquid composition of these units was 60 percent by weight of PCB and 40 percent by weight of electrical grade trichlorobenzene (TCB).

Over the years, Kraft cellulose-based paper has been widely used as the predominant solid insulating material with askarel. Moreover, selection of coordinating materials for an askarel-based system was more critical than for its oil counterpart. Such paper could not include any binder or varnish materials, soluble in askarel. Aromatic polyamide (NOMEX®)* papers have also proved most effective impregnated with PCBs.

Askarel-impregnated papers tend to have better low frequency ac characteristics and lower surge voltage capabilities than oil-based systems. Yet, even as askarel and oil differ in numerous characteristics, both lay claim to moisture as insulation's primary enemy.†

Though askarels (or PCBs) have been banned from sale and production as a threat to the natural environment, the existing 140,000 askarel units in the U.S. alone will be a maintenance concern for years to come. At this time, no transformer alternative (with *sufficient* in-service use) matches askarel's unique fire resistant characteristic. Moreover, it will take time to phase out these units through attrition, replacement (such as transformers designed utilizing vaporization cooling), or retrofill by available substitutes, such as silicone or high temperature hydrocarbons.‡

**NOMEX® is a registered trademark of E.I. Dupont DeNemours and Company.*

†Chapter 2, Part 5.

‡Chapter 9, Part 3.

Development of Silicone Fluids

In recent years, research attempts have been made to provide a fireproof liquid without environmental restrictions such as those placed on askarels. The result has been the introduction of silicone liquid, currently being used in at least 7000 transformers.

Silicone is a generic term for a family of linear polymer liquids between silicates and organic polymers. Basically, the fluid is "less flammable" than conventional mineral oil as classified by Underwriter's Laboratories, Inc. (UL).* UL defines one flammability scale as tested by ASTM (Table 2.1).

However, this is not the final answer. Insurance companies will have to help define this one. Although silicones do not oxidize and do not sludge, their BIL rating and voltage stress application limits have not been clearly established.

TABLE 2.1

FLAMMABILITY RATING (Underwriter's Laboratory)	
Water	0
R-113	0
Transformer Askarel	1 - 2
Silicone	4 - 5
Transformer Oil	20 - 30
Kerosene	30 - 40
Ethyl Alcohol	60 - 70
Gasoline	90 - 100
Ether	100

**Chapter 9, Part 1; Part 3.*

Why are silicones being seriously considered as an alternative fluid? Mainly because there is no environmental problem with them. Silicones can take more heat which will lead users to load transformers more, and which in turn, will lead to more research to combat the bad effects that higher heat will generate. Silicone has been used as a dielectric since 1956, but has not been seriously considered until lately. Probably the two biggest obstacles to the complete replacement of askarels with silicones are that silicones generate hydrogen gas as an arc-over product, and the high cost of the fluid ($15 per gallon circa 1980). Besides utilizing silicones in transformers, air or gaseous dielectrics have also been used as an alternative to other forms of transformer insulating systems.*

Air- or Gas-Insulated Transformer Systems

One common way of transformer classification is in terms of insulation medium liquid or nonliquid cooled.† The ultimate demise of askarel transformers has also brought about rising expectations for dry type apparatus. In considering equipment as suitable PCB replacements for certain applications, we are going full circle; returning to the unit we started with in 1884—the air-insulated transformer!

For dry type equipment, as for liquid-filled units, the life of the transformer is dependent on the temperature at which it is made to operate; therefore, insulation materials are classified worldwide according to their maximum safe working temperature.‡

The common "dry type" transformers are cooled with:

- Air
- Nitrogen
- Fluorogases

Nitrogen comprises 79 percent of air by volume, but what we are talking about here is a sealed, pressurized, nitrogen-filled unit. Air by itself is a poor dielectric and its presence in gases, including sulfur hexafluoride (SF_6), will lower the dielectric. Nitrogen is a better dielectric than air, and its dielectric strength increases with pressure. SF_6 is about 2½ times better as an insulant than N_2 under the same pressure.

**Chapter 9, Part 3.*

†Chapter 1, Table 1.1.

‡Chapter 1, Part 2, Table 1.5.

The obvious advantages of nitrogen over sulfur hexafluoride are the economics of the gases themselves. The disadvantage is size limitations. Only the smaller KVA rating and lower voltage units can be serviced by N_2. Also, nitrogen will not effectively quench an arc so the spacings between the coils must be greater. This aspect has resulted in the limited ratings available. Noise is also a problem with nitrogen-filled units.

1. Air-Insulated Transformers

   Both liquid and dry type transformers are constructed in much the same manner. The high-voltage and low-voltage windings are wound with either copper or aluminum conductors (foil wound or round wire) and placed over a magnetic core. The cores are constructed in a wide variety of arrangements (barrel, disk, sectional, and random sections).

   Most air-insulated transformers use a 220° C insulation system (formerly called an "H" system in the U.S.). The 220° C represents the ultimate temperature of the windings and is the summation of maximum permissible rise by coil resistance (150° C), the hot spot differential allowance (30° C), and the ambient allowance (40° C).

   The materials commonly associated with class 220° C include epoxy resin, mica, porcelain glass, quartz powder, treated fiber textiles and built-up mica using silicone varnish as a bonding and impregnating medium. Naturally then, the cost of insulation is a sizable portion of the apparatus' cost.

   While the coils are wound with class 220° C insulation, the same insulation is designed to operate at 80° C temperature rise, 115° C (U.S., or 125° C IEC).

   This is accomplished by increasing the conductor sizes, resulting in lower losses through lower temperature rise; however, if dry type apparatus is operated continuously at full load, it should be supplied with insulation rated for *only* an 80° C temperature rise (temperature limits on the casting materials). Recent studies have shown that air-insulated transformers are designed to be the most efficient when approaching full load.* Oil-filled transformers with the same losses are normally more efficient at 75 to 100 percent load un-

**Chapter 9, Part 3, Figure 9.18.*

less specifically designed for a certain load. This is a design requirement in the very large transformer for heavy industry and utilities.

The dry type transformer should also be used in areas not prone to lightning strikes.

Air-insulated units have certainly won their place in the electrical industry. Remember the four functions of an insulating liquid. You will notice that air will probably serve function number two the best. To help with function number one, since air is a poor dielectric, *two* compromises over a liquid are made—first, the gaps are kept large, and second, varnish is added to help the dielectric situation. This varnish, while enhancing function number one, will diminish function number three. Since the unit is open to the air, humidity becomes a problem. As long as the unit is energized, heat will keep the moisture out. If the unit cools, the varnish will absorb moisture and both deteriorate the insulation and lower dielectric strength.

Thus, ventilation and exposure to the atmosphere dictate the use of air-insulated dry types. While their size is limited because of the large spacings necessary, noise can also be a problem. Other units such as those insulated with epoxy plastic (compound-filled) or fluorogas dielectrics are also available.

2. Fluorogas-Insulated Transformers
Several gases are superior to air in terms of dielectric strength and have been used as transformer insulating materials. Their history somewhat parallels the history of mineral oils, but the fluorogases (those containing atoms of fluorine) only date back to 1956.

Reasons for choosing these gases over liquid dielectrics include the following:

- Electrical characteristics
- Nontoxicity
- Inert to other transformer materials
- Stability over applicable temperature range
- Reasonable price

The gases that have been considered for usage are SF_6, and fluorocarbons C_2F_6 (hexafluoroethylene or R-116), C_3F_8, and C_4F_8. Each has different properties and the key word to their use is "compromise".

At this time (1980), sulfur hexafluoride (SF_6) is found to be the most suitable gas in light of the foregoing criteria. The viability of the SF_6-insulated substation (GIS) at 765 KV has in fact been proven. The biggest advantage of an SF_6-insulated system is that it is nonflammable, nontoxic, and nonexplosive. Nevertheless, the gas has still two long-term unknowns—testing and maintenance. Vacuum heaters, leak detectors, and dry gas handling equipment required to service SF_6 can be problems. The most serious known contaminant in GIS systems is conducting particles, such as metal or carbon. However, its arc-formed products may contain toxic gases. Because SF_6 is a very heavy gas, it settles. This must be recognized if maintenance personnel enter equipment that has contained SF_6.

Currently, SF_6 seems to provide for the same protection to the insulation as does a liquid (functions number one and number two). The overall advantages of the system must outweigh the number three function (protection of the core and coil assembly)—once again, a compromise. This SF_6 gas-insulated system offers a solution when oil-filled is not acceptable because of fire risk, and when the air-insulated unit is not feasible because of either its size or its voltage rating. The air-insulated transformer is limited to system voltage of 15 KV, although it has been "rumored" that one utility has some large air-cooled transformers operating at higher voltages (over 115 KV).

SF_6-insulated substations are sometimes the only answer, especially where land or clearance space is limited (Figure 2.5).

Future Transformer Prospects

Among current research and development projects, two compromise insulation systems stand out.

A transformer combining the best of air-insulated materials with silicone dielectric fluid is now commercially available. A longer range system may use the combination of SF_6 gas and mineral oil.* Another approach would replace insulating oil altogether by use of inorganic insulating materials along with advanced coil-winding technology.

**Chapter 9, Part 3, Table 9.29.*

Figure 2.5 - The SF_6-insulated substation, resembling an oil refinery, is the only solution for certain applications, where availability of land is limited. (Courtesy of ***Transmission and Distribution****).*

Of several other liquids in use or under investigation, synthetic esters have some advantages over mineral oil.* While their fire point is higher and thermal and oxidation stability of specific esters containing inhibitors may be better, organic esters may have certain electrical limitations.

Therefore, even with all the research and development now going on, mineral oil crudes (naphthenic, paraffinic, or a mixture) should continue to be the most widely used insulant/coolant because of their ready availability and technical acceptability, despite rising costs, "for all but the most critical and specialist applications."

**A natural ester is castor oil.*

Part 2—Characteristics of Modern Transformer Oil

Introduction

It is generally recognized that the term "transformer oil" is misleading, in that over the years, this oil has been used in a variety of electrical apparatus other than transformers. Such apparatus includes bushings, circuit breakers, reclosers, disconnect switches, tap changers and oil-fuse cutouts.

Oil characteristics required for one type of apparatus can be markedly different from those needed for other equipment. For instance, oxidation resistance is of concern for pole-top distribution transformers, whereas, such a property is of less importance for circuit breakers. In the latter, the principal concern is the ability to rapidly quench an arc. Transformer oil happily accomodates the needs of transformer manufacturers as well as other apparatus manufacturers.

A clear understanding of the characteristics of transformer oil will help in the operation of the equipment that the oil serves. If we can understand what the building blocks of oil are, we can better realize the benefits derived from the oil.

Modern naphthenic transformer oil has been available since 1925.* Unfortunately, now that the industry has gone from 12 kinds of oil for electrical equipment to one universal type, and although this oil has served industry for over 50 years, the naphthenic crudes are running out. The crudes used to produce this oil consist of only three percent of the total crudes available. While the electrical industry has very little "clout" with the oil industry (requiring only approximately 0.03 percent of the annual U.S. refining capacity), it is of paramount importance that the industry has a functionally acceptable insulating oil to insure proper operation of electrical equipment.

If we cannot get new oil, and we don't preserve the 1.73 billion gallons of oil already in-service, then we will face an oil crisis. The prudent and economic thing, then, is to preserve and maintain the oil currently in

Chapter 2, Part 1 (historical account); and Chapter 9, Part 3, Table 9.27 (other dielectric media).

use in order to prolong the availability of naphthenic crudes. Hopefully, a new type of oil based upon paraffinic crudes will be suitable for use in new apparatus and will be compatible with existing oils in-service for use as replacement oil.

Chemical Properties

The chemical structure of transformer oil is very complex and, for the most part, the nature and number of the individual chemical components for even the most simple natural refined mineral oil are still unknown. What we currently know about the chemistry of transformer oils has been exhaustively dealt with in several studies. Transformer oils are composed of both hydrocarbon and non-hydrocarbon compounds. By definition hydrocarbons are chemical compounds containing only hydrogen and carbon.

Hydrocarbons

Hydrocarbon compounds form the major constituent of oil and can be divided into three major groups:

- Paraffins
- Naphthenes
- Aromatic Compounds

1. Paraffins are generally considered saturated hydrocarbons characterized by a straight chain structure as shown in Figure 2.6A. They are classified in organic chemistry as open-chain or aliphatic compounds.

2. Naphthenes are classified as closed-chain or ring compounds as well as alicyclic, that is, a non-aromatic ring structure of carbon atoms. Naphthenes can be monocyclic, (Figure 2.6B), bicylic (Figure 2.6C), and so on, depending on the number of rings.

3. Aromatic hydrocarbons (Figure 2.6D) possess one or more aromatic rings which may be joined with alicyclic rings. Figure 2.6E shows a *mixed* naphthenic-aromatic hydrocarbon structure. While some types of aromatics may function as natural oxidation inhibitors and provide stability, too much of an aromatic contribution reduces dielectric or impulse strength and increases the solvency action of the oil for many of the solid insulating materials immersed in it.

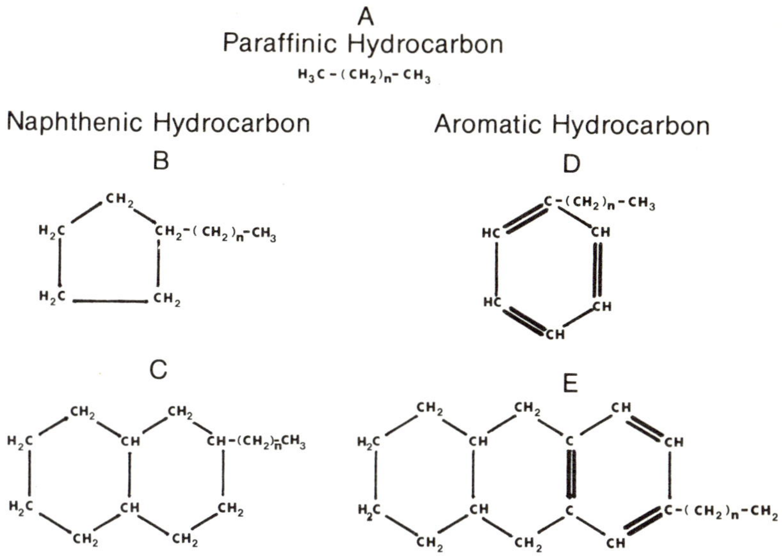

Figure 2.6 - Types of generic hydrocarbon structures, where n = 2 to 27.

Non-Hydrocarbons (Hetero Compounds)

Petroleum also contains non-hydrocarbons, which may contain a skeleton similar to hydrocarbons with the carbon atoms replaced by one, two, three, or more sulfur, oxygen or nitrogen atoms. Such a mixture is called a *hetero* compound. Non-hydrocarbons of transformer oil include naphthenic acids, esters, alcohols, tarry asphaltic compounds, as well as sulfur- and nitrogen-containing organic compounds.

All petroleum crudes contain sulfur and sulfur compounds (which vary in amounts from less than 1 to 20 percent by weight) as the crude comes from the ground. These sulfur compounds influence the properties of the petroleum crudes considerably, and govern the treatments used in the processing of such crudes. Naphthenics usually contain much less sulfur than paraffinics. The absence of sulfur and its unstable compounds is necessary to prevent the corrosion of copper in contact with the insulating oil (ASTM D-1275).

Petroleum crudes also contain relatively small amounts of nitrogen compounds. However, it should be noted that regardless of this small percentage (maximum 0.8 percent) of nitrogen compounds in petroleum products, they play a major role in the oxidation process.

Some compounds which contain amino acids and phenol groups together act as antioxidants in transformer oils and are thus desirable, as are some thermally stable sulfur-bearing compounds.

Naphthenic acids are also present in transformer distillates in considerable amounts. Most of them are, however, removed during purification of transformer distillates, so that their content in the finished transformer oil is as low as 0.02 percent by weight. Transformer oils may contain very small amounts of acids of the aliphatic and aromatic series, esters, alcohols, ketones, and peroxide compounds. New or fresh oils may contain very small amounts of iron and copper.

So you see—transformer oil as a finished product must be relatively free of all of the compounds mentioned above. It is a highly refined product for servicing the electrical heat transfer and arc quenching demands made upon it by the designers of transformers, breakers, bushings, and so forth.

Three other critical chemical factors are significant for new oils:

1. Neutralization (or Acid) Number
   A low total acid content of a mineral oil is necessary to minimize electrical conduction and metal corrosion and to maximize the life of the insulation system (ASTM D-3487).

2. Gas Formation Under Arcing
   Any significant increase in submerged gases formed under electrical stress (cm^3 per kilowatt-second at 25° C) could cause a problem in a transformer; for example, in mixing naphthenic, paraffinic or blended oils, any sign of *upward* change over that figure obtained with present oils could cause a problem. A value of 70 cm^3/KW-sec has generally been accepted (ASTM D-2300, Appendix).

3. Water Content
   A mineral oil of low water content is necessary to achieve adequate electrical strength, to maximize the insulation system life and to minimize metal corrosion (ASTM D-1533).*

Physical Properties

The physical properties of new oil include:

1. Viscosity

**Chapter 2, Part 5 (Moisture: Insulation's Primary Enemy).*

Viscosity is considered a measure of resistance to oil flow and is important for heat transfer and for oil-immersed moving parts. A thicker oil is said to possess a higher viscosity while a thinner oil is said to have a lower viscosity. Viscosity Index (V.I.) refers to the resistance of an oil to change in viscosity with temperature variation. An oil with a high V.I. is more resistant to temperature changes than one with a low V.I. Paraffinic oils generally possess higher V.I. values than naphthenic oils. However, at temperatures of 0° C and above (where most transformers operate), the viscosity values of transformer type paraffinic oils are generally slightly lower than those of naphthenic oils.

2. Specific Gravity
   This is the ratio of a given volume of a substance to the weight of an equal *volume* of water. Temperature correction is also required. The specific gravity of paraffinic oils is usually lower than that of naphthenic oils of corresponding viscosity. The test is useful as an aid in identifying types of new oil.

3. Pour Point
   The lowest temperature at which an oil will just flow is known as the pour point. A low pour point is particularly important in cold climates as it ensures that the oil will circulate and serve its function as an insulating and cooling medium and will not impede the movement of such mechanisms as circuit breakers. Such a property may help in identifying various types of base oils. For example, most paraffinic oils tend to solidify at higher temperatures (because of wax precipitation), which is undesirable, while naphthenic oils will tend to congeal at lower temperatures (-40° C or F), making them more desirable for use in transformer units in northern climates. In short, most paraffinic oils possess a solidification point, while naphthenic oils do not.*

4. Volatility
   While naphthenic and paraffinic oils may have comparable viscosity values and refining histories, the former oils are appreciably more volatile than the latter oils.

   Far more specific indications of the volatility of an oil are its *flash* and *fire points* determined under controlled conditions of

**The classic pour point test is not considered applicable for paraffinic oils in a practical sense.*

temperature, time, and flame size. The flash point is the lowest temperature at which sufficient vapor has been produced to ignite momentarily when a flame is applied to it. By contrast, the fire point is the lowest temperature at which sufficient vapors *continue* to be formed to sustain a fire for five seconds on application of the flame. The significance of these temperatures is an indication of the degree of safety in the environment in which the liquid is being used. Unfortunately, a blended oil (especially in Europe) may tend to have a *lower* flash point than a "straight cut" oil of the same viscosity.

5. Oxidation Stability
   Older paraffinic oils are generally more resistant to oxidation than newer naphthenic oils. It is possible for paraffinic oils to oxidize to higher levels of acidity without forming or precipitating insoluble oxidation products. Nonetheless, when *paraffinics* are badly oxidized, the result products are, in general, *more objectionable* than those of naphthenic oils. The paraffinic sludge formed is usually more adherent and of a more tarry nature.

6. Sludge Solvency
   Generally, naphthenic oils are better solvents than are paraffinic oils with respect to materials such as varnish, resins, asphalts, sludge, etc. This explains why a badly oxidized naphthenic oil is not as undesirable as a badly oxidized paraffinic oil!*

7. Interfacial Tension
   A high value for new mineral oil indicates the absence of undesirable polar contaminants, but its primary use is applied to service-aged oils as a sensitive indicator of the degree of oil deterioration and contamination of the oil by the solid materials immersed in it (ASTM D-971).

Electrical Properties

The primary electrical properties of new oil include:

1. Dielectric Strength (breakdown voltage)
   The standard dielectric strength of an oil (ASTM D-877, D-1816) is a measure of its ability to withstand electric stress

**Chapter 7, Part 3 (Aniline point temperature and its importance).*

without failure. It is the voltage at which breakdown occurs between two electrodes under prescribed test conditions. Normal acceptable values range from 38 to 55 KV depending upon the type and voltage rating of the equipment. Breakdown voltages for paraffinic oils are comparable with those of naphthenic oils.*

2. Impulse Strength
   This test of electric breakdown strength of an oil (ASTM D-3300) occurs under transient voltage or impulse stress conditions (switching surges or lightning strikes). As noted earlier, the aromatic hydrocarbon content has a notable effect.

3. Relative Permittivity/Dielectric Constant
   The relative permittivity is the ratio of the capacitance of an insulating material measured with a given electrode configuration to the capacitance of the same electrode configuration and spacing with air (or vacuum) as a dielectric (Table 2.2). Relative permittivity is dependent on temperature and voltage frequency. Typical values for new oil are 2.1 to 2.5 (90° C). Oxidation products tend to *increase* these values. Naphthenic and aromatic hydrocarbons have generally higher permittivities than paraffinics. This numeric relative permittivity is even more important in the utilization of *composite* insulations at their highest degree of efficiency (in the presence of a minimum amount of moisture).

Transformer Oil in the Petroleum Industry

Mineral oil liquids are obtained from crude petroleum, whose origin is said to have been formed from buried and decayed vegetable matter or by the action of water on metal carbides, such as calcium carbide. ASTM D-288 has defined petroleum as:

> A naturally occurring mixture, consisting predominantly of hydrocarbons which is capable of being removed from the earth in liquid state. Crude petroleum is commonly accompanied by varying quantities of extraneous substances such as water, inorganic matter and gas. The removal of such extraneous substances alone does not change the status of the mixture as crude petroleum. If such removal appreciably affects the composition of the oil mixture then the resulting product is no longer crude petroleum.

* *Chapter 3, Part 1, Table 3.1; Part 3.*

TABLE 2.2

PERMITTIVITY OF VARIOUS SUBSTANCES	
Substance	Permittivity
Air	1.0
Oil	2.25
Rubber	3.6
Paper	4.5 (average)
Porcelain	7.0
Water (pure)	81.0
Ice (pure)	86.4

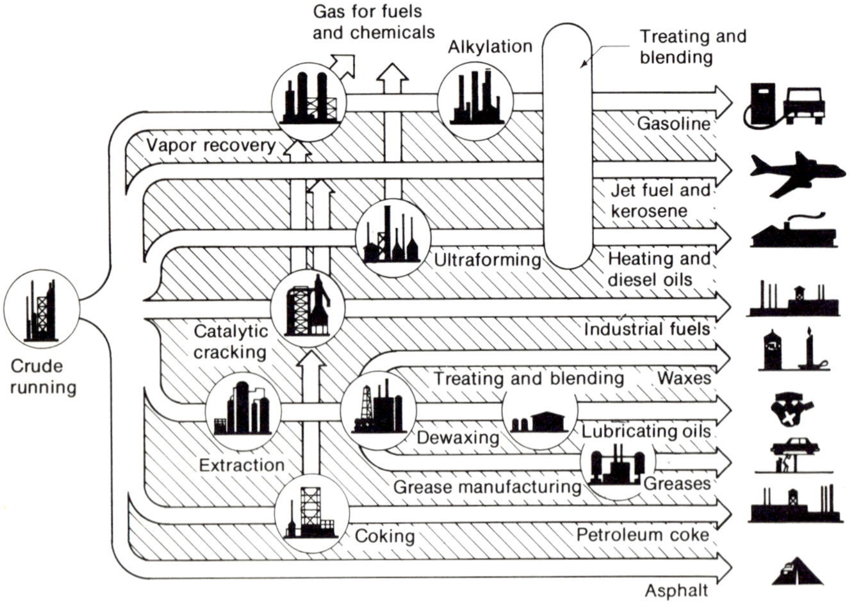

Figure 2.7 - Industrial fuels include transformer oil. (Courtesy of ***Plant Engineering***®*).*

The crude petroleum is then subjected to physical and chemical treatments to produce specially refined products. The end products of crude petroleum refinery include jet fuel and kerosene, heating and diesel oils, industrial fuels from which transformer oil is derived, lube oil blending stocks, petroleum coke and asphalts. Figure 2.7 illustrates the relative positions of various fractions which are given off in one type of crude oil refining process. During the distillation process, aviation gasoline is boiled off first, since it possesses the lowest boiling point. The residue from this vacuum distillation is further distilled under vacuum to give stock for the production of electrical and lubricating oils.

It is then followed by automobile gasoline, jet fuel, diesel fuel, and turbine and motor lubes. Transformer oil is cut at a point between the diesel and turbine oil fractions, that is, the middle fraction in the diesel-turbine oil area. Figure 2.8 gives a more detailed diagram of the conversion of crude oil end products that include industrial fuel from which transformer oil is obtained.

Modern transformer oil can be produced from four refining methods:

- Acid refining
- Solvent extraction
- Hydrogenation (hydrogen refined)
- Combination processes

Production (Refining) of Transformer Oils

The purpose of the refining process is twofold:

1. To select the liquid fraction with the most desirable characteristics, and
2. To remove or minimize objectionable constituents which can seriously affect the oil's stability to oxidations, its electrical insulating properties, and its low temperature fluidity.

These undesirable constituents include nitrogen and oxygen compounds, most sulfur compounds, tars, and unsaturated hydrocarbons as well as solid hydrocarbons, particularly amorphous and crystalline waxes.

Three principal methods of refining are used for the manufacture of modern oils. Each is diagrammed in Figure 2.9.

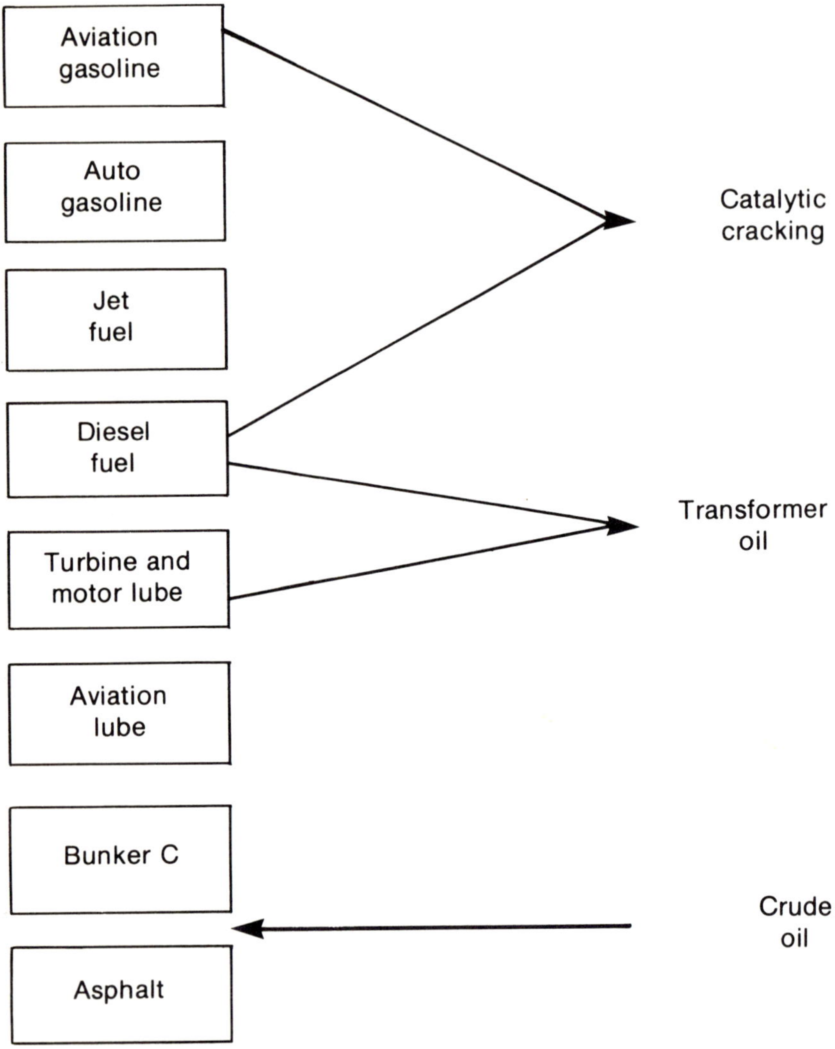

Figure 2.8 - Where transformer oil fits in the refining process. F.C. Doble, "A New Approach to the Insulating Oil Problem" (1954).

Acid Refining

This traditional method involves the treatment of the distillate with 93 to 98 percent strength sulfuric acid, followed by mixing, separation of acid sludge, alkali neutralization, washing with water, and clay treatment. The amount of acid used has varied between 2 and 20 percent

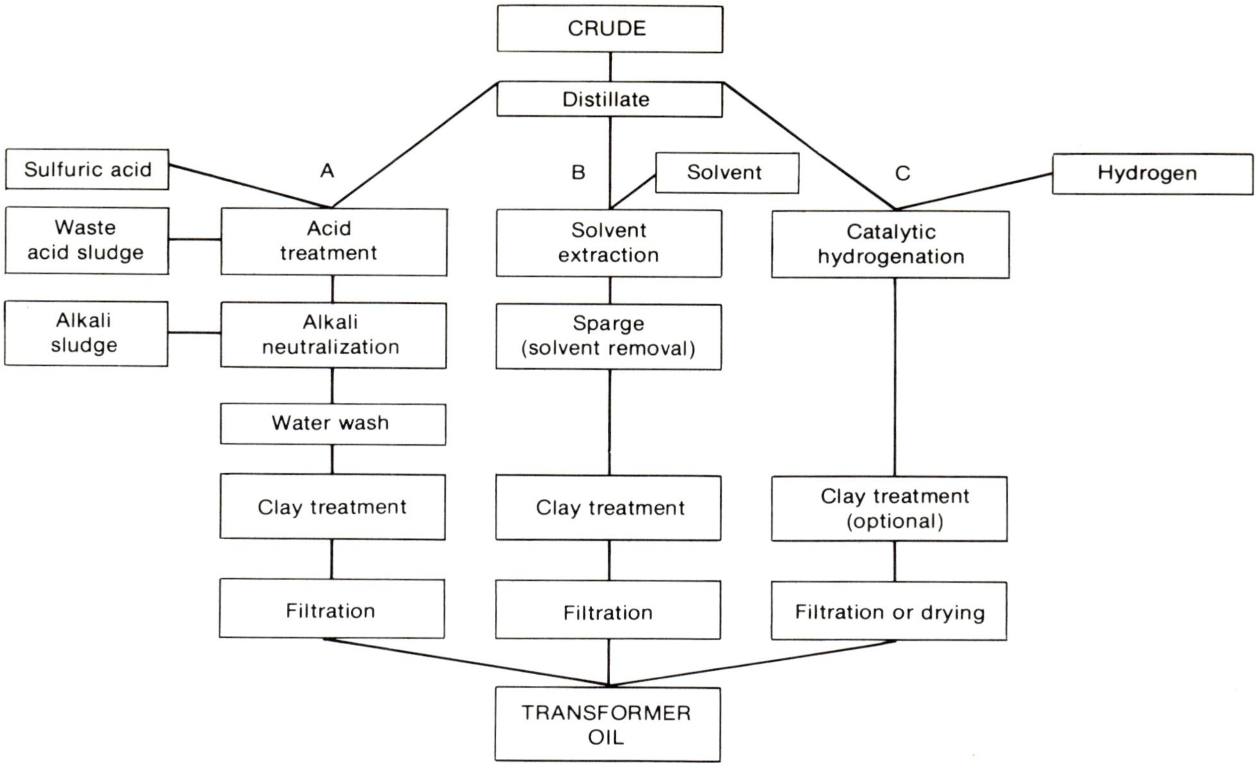

Figure 2.9 - Flow chart of basic refining processes where (A) is sulfuric acid, (B) is solvent extraction, and (C) is catalytic hydrogenation. D.R. Pugh, "Electrical Insulating Oil" (1974).

depending particularly on the degree of refining required. The best transformer oils may have resulted from this process; nevertheless, even before environmental constraints concerning disposal of acid wastes the method has had two major drawbacks:

1. Since sulfuric acid is insufficiently selective, valuable constituents are also removed along with objectionable constituents, resulting in a lower yield per gallon of distillate compared to some of the newer methods.

2. The sludge formed is worthless and expensive to dispose of in an environmentally safe manner. Thus, it has become necessary to develop other methods in order to produce more transformer oil for the industry.

Solvent Extraction

This process relies in principle upon the selective solubility of undesirable compounds in a chosen solvent. The most widely used solvent is furfuraldehyde (Furfural), although phenol also has been used.

Hydrogenation (Hydrogen-Refined)

This method is the most recent form of refining treatment. It involves the hydrogenation of the distillate under pressure and it raises temperatures over a catalyst. This is followed by steam stripping, and as with both acid and extraction, may include clay treatment.

The hydrogen-refined process is also the most versatile process. Light hydrogenation (hydrofinishing/hydropolishing) or severe hydrogenation (hydrofinishing) may be used to obtain the most desirable final product. Caution is the byword since, for example, if processing is too severe such treatment is reversible and promotes dehydrogenation with the formation of undesirable compounds.

The degree of hydrogenation required to produce a satisfactory product will vary depending upon the constitution of the crude distillate used. The refiner must carefully balance and control at least four variables, e.g., temperature, pressure, catalyst activity, and time. As the catalyst ages and loses some of its efficiency the other three values must be adjusted to compensate.

Combination Processes

A combination of any two or more of the foregoing methods is in-

TABLE 2.3

MODERN PROCESSES FOR REFINING TRANSFORMER OILS			
Process	Method Principle	Agent	Comments
Acid treatment	Chemical reaction	93 to 98 percent Sulfuric acid	Best quality oil; problem of disposal; waste
Solvent extraction	Dissolve	Furfural Phenol	Selective solvent of aromatics, resins, and sulfur compounds
Hydro-genation (Hydrogen refined)	Modify molecular structure	Hydrogen under pressure at elevated temperature over a catalyst	Most flexible; removal of sulfur and nitrogen by converting them to hydrocarbons. Least wasteful method; better yield
Combin-ation	Two or more stages	Many use clay treatment as one of the final stages	At least ten feasible methods; solvent extraction plus hydrogen refined is the most common

creasingly being used in purifying crude oils. The advantage of solvent extraction in contrast to acid treatment is that the extracted aromatics have valuable uses. Nevertheless, transformer oils are not the same as those obtained by the acid method. Logically, therefore, a combination process consisting of solvent extraction as the first state followed up by a form of hydrogenation or straight hydrogenation may be the way of the future.

Table 2.3 is a comparative summary of these four possible refining methods.

At least seven different combination procedures have been used. Nearly all complete the process by use of clay treatment. The most widely used "cleaning" material is fuller's earth.* This method uses the principle of adsorption. Unfortunately, while the clay treatment removes oil decay products, it can remove natural oxidation inhibitors that may be left in the oil. Thus, in some cases, a synthetic inhibitor has become mandatory. Depending upon the crude distillate type, some refiners do not use a clay treatment, but they usually add a small amount of inhibitor.

Distillates from crude oils are refined by various processes to yield mineral oil meeting standard requirements. Therefore, as long as the process utilized is properly developed initially and adequate quality control is maintained, satisfactory oil results comparable to acid refining can be obtained.

The Final Product

After oil has been distilled and then refined by one of the foregoing methods, nobody really knows what compounds remain in the oil. What we do know, however, is that transformer oil consists of an estimated 2900 hydrocarbons, and from 2 to 35 percent are aromatic.† The National Bureau of Standards in conjunction with manufacturers and users have analyzed transformer oil, yet still 90 percent of the compounds present remain unidentified because of the great cost in terms of dollars and time required to analyze each component. We do know the behavior and function of the total oil for its various applications and we know the function of the oxidation products of the oil in-service.‡

Even though there is this mystery, we also know enough that we can consistently produce batches of oil and make them to specifications. We know, for example, that there are "natural inhibitors" to oxidation in the oil.

Some of these are thermally stable aromatic sulfur compounds. We also know that when these compounds are exhausted oxidation takes place. This oxidation yields the following functional types of compounds:

Alcohols	R-OH
Acids	RC(=O)-OH

Chapter 7, Part 3.

†*Chapter 4.*

‡*Chapter 2, Part 3, and Part 5.*

Peroxides R-O-O-R

Ketones $R-\overset{O}{\overset{\|}{C}}-R$

Aldehydes RCHO

Soap R-O-Me+

Epoxies R-R (bridged by O)

Note that all these compounds have one thing in common—an oxygen atom. All these compounds act upon each other and upon the original oil, resulting in a product called sludge.*

Sludge is a resinous, polymeric type substance that is partially conductive, hygroscopic, and a heat insulator. If there is also water in the system (and water is a chemical by-product of oil oxidation), it will be attracted to the sludge.† If we could engineer our oil system so that all these reactions would not happen to it, we would greatly enhance the life of the insulating system. Nonetheless, oil oxidation is a fact of transformer life. The best we can hope for is to *control* the situation. Remember the third function of an oil mentioned earlier—to protect the coils. If we can protect the oil in use, we will certainly help the cellulose in existing transformers reduce the possibility of internal corrosion of metals and increase the heat transfer efficiency of the apparatus.

So, to start off right, where *new* transformer oil is needed, it is important that both the buyer and seller follow standard purchasing specifications. It is a fact that transformer oil is sold today that does *not* meet basic specifications.

Transformer Oil Specifications

Various purchasing specifications are published by different authorities, but all are very similar. Such specifications are used as a standard by both buyers and sellers to define their mutual obligations. Specifications include those established by the American Society for Testing and Materials (ASTM), National Electrical Manufacturers Association (NEMA), IEEE, and the Doble Engineering Company,

**Chapter 2, Part 3.*

†Chapter 2, Part 5.

commonly known as the Doble Transformer Oil Purchase Specifications (TOPS). At the international level the International Electrotechnical Commission (IEC) has also established its own specifications, most of which are similar to those established by American bodies. Each specification covers *new* mineral insulating oil of petroleum origin for use in insulating, cooling and arc quenching in *new* and existing power and distribution apparatus, such as transformers and circuit breakers.

It is of interest to note that on the average in the U.S., electrical equipment manufacturers purchase about 90 percent of all new transformer oil sold each year for use in new apparatus and in re-filling. Apparatus owners purchase about 10 percent of the new oil for makeup purposes. This can amount to as much as ten million gallons per year. Most of these purchases can be avoided by higher maintenance and suitable reclamation of existing oil in-service.

The following listing indicates the average new oil usage per year in the U.S.:

New distribution transformers	50 percent
New power transformers	31 percent
Makeup oil for in-service apparatus	10 percent
New circuit breakers	6 percent
Miscellaneous new apparatus (Bushings, generator coils, dashpots, etc.)	3 percent
Total	100 percent

In defining mineral oil, ASTM and NEMA classify such oil as Type I and Type II. Type I mineral oil is used for apparatus where "normal oxidation resistance" is required. Some oils may require that a suitable oxidation inhibitor be added in a small amount before such resistance is achieved. Type II mineral oil is designated for apparatus where "greater oxidation resistance" is needed. This is usually attained with the addition of a greater quantity of suitable oxidation inhibitor.

The most common specifications and test limits used in the purchase of transformer oils include:

- Dielectric breakdown strength
- Color
- Flash point
- Interfacial tension
- Pour point

- Specific gravity
- Viscosity
- Neutralization or acid number
- Water content
- Power factor

A comparison of the standards held by various authorities are given in Table 2.4. A comparison of transformer oil tests made by Doble (A), and a major transformer maintenance contractor (B), is shown in Figure 2.10.

A knowledge of these specifications and their interrelation is important to a prospective buyer who wants to keep his electrical apparatus in proper operating condition.

Doble Engineering is the only organization among the five that currently stipulates a "sludge-free life" (SFL) test among its properties.

Doble holds that its SFL laboratory test, as an additional bit of data, provides a basis for the comparison between test oils in an accelerated evaluation program and their comparison to oils of the past—oils with a known history and stability.

The SFL test uses a 300-milliliter (ml) oil sample in a large test tube into which air (oxygen) is atomized through a stainless steel jet in the presence of a copper catalyst (B of Figure 2.11). The tubes are maintained in a bath at 95° C. The test begins when "A" (test oil) reaches 100° C and is terminated with the first visible sludge.

The sludge-free life value of an oil is defined as the number of hours which have elapsed between the start of the test and the last sample which slowed a sludge precipitation. Historically, this value has continued on a downhill slide (Table 2.5). As a warning, however, the SFL test is a measure of the oil quality and is of *comparative* value only. It does not mean that an oil will sludge in a transformer 96 hours (58, 32, 16, or 8 hours) after it is energized. At the same time, knowing that oils which in prior years had a SFL of 96 hours are now down to 8 hours can be important, and might be a basis for separating large batches of oil both for storage and for mixing in use.

Evaluating Paraffin- vs Naphthene-Based Oils

The gradual depletion of sources of naphthene-based crudes has brought about the publication of numerous research studies, comparing naphthenic oils with paraffin-based crudes (the most likely replacement).

TABLE 2.4

A COMPARISON OF TRANSFORMER OIL PURCHASE SPECIFICATIONS ACCORDING TO VARIOUS AUTHORITIES[1]					
Property	Test Authority				
	ASTM Test Method	NEMA ANSI/ASTM 1979	IEEE 1977	Doble 1979	IEC 1969
Physical					
Aniline Point °C	D-611	63-78		63-78	
Color, maximum	D-1500	0.5	1.0	0.5	0.5
Flash point, minimum[2]	D-92	145		140	145
Interfacial tension, at 25°C minimum dynes cm	D-971	40	35	40	40
Pour point maximum °C	D-97	-40		-40	-30
Specific gravity 15°C/15°C maximum	D-1298	0.91		0.865	0.895
Viscosity maximum cST (SUS) at 40°C	D-445 or D-88	12.0		11.0	11.0
Visual examination	D-1524	clear and bright	clear	clear and bright	free from suspended matter or sediment
Electrical					
Dielectric breakdown voltage at 60 Hz:					
Disk electrodes minimum KV	D-877	30	30	26	30
Disk electrodes minimum KV VDF electrodes minimum KV either at 0.040 inch (1.02 mm) gap	D-1816	28	20	20	
Dielectric breakdown voltage impulse conditions 25°C minimum KV. Needle negative to sphere grounded 1 inch (25.4 mm) gap	D-3300	145		145	
Power factor at 60 Hz: maximum					
25°C	D-924	0.05	0.10	0.05	
100°C	D-924	0.30	1.0	0.5	
Chemical[3]					
Oxidation inhibitor content maximum percent by mass	D-1473	0.3		0.3	
Corrosive sulfur	D-1275, D-2668	←	Noncorrosive		→
Water, maximum ppm	D-1315, D-1533	35	25	30	40
Neutralization number, total acid number, maximum mg KOH/g	D-979	0.03	0.04	0.025	0.03
Sludge-free life (SFL) measured at 8 hour period, hours	Doble Procedure			40 ± 8[4]	
Resistivity, 50°C minimum ohm-centimeters	D-1169				1.0×10^{14}
Inorganic chloride and sulfate	D-878				none

[1]Type II oil specifications, including 0.3 percent by weight of ASTM approved oxidation inhibitor (DBPC or DBC).

[2]Until 1961, all naphthenic oil used in the U.S. had a 130°C maximum flash point. In 1961 the industry standardized a 145°C flash point in order to maintain the same safety margin as a result of the advent of the 65°C rise transformers.

[3]New oils should be free from PCB (polychlorinated biphenyl) contamination (0 ppm). This is not yet required by ANSI/ASTM but should be in the future.

[4]SFL characteristic has changed during the 1970's from 64 ± 8 hours. This change may possibly be the result of the different crudes available or the change of refining methods.

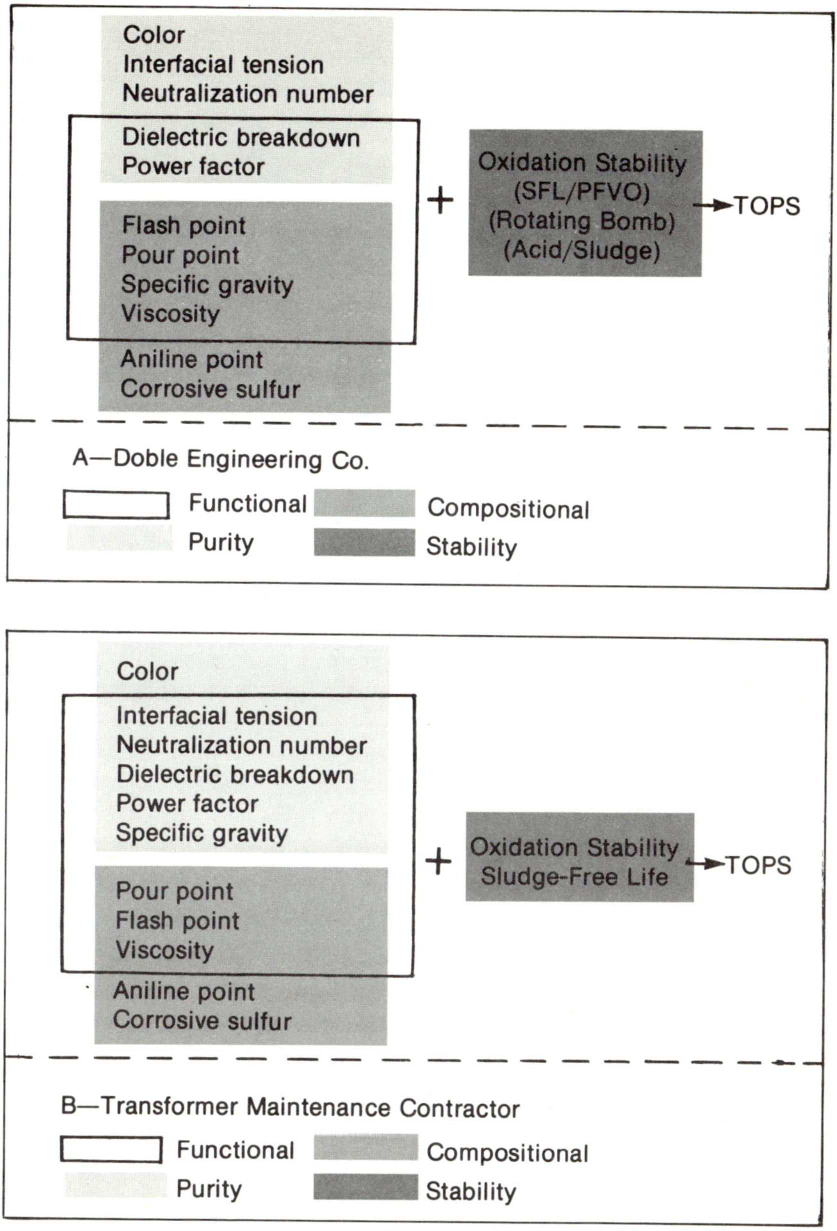

Figure 2.10 - Transformer oil tests classified according to major purpose by Doble Engineering Co., Watertown, Mass. (A) and a major transformer maintenance contractor (B). The oil is uninhibited.

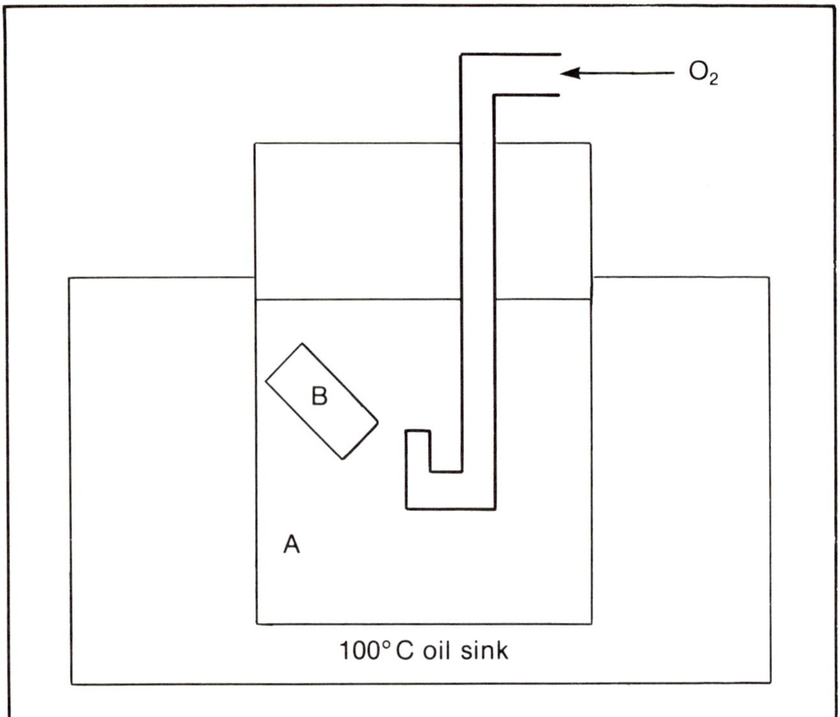

Doble Engineering procedure
A—Oil to be tested*
B—Copper strip
O_2—Oxygen metered through test oil
*Oil to be tested must be free of oxidation inhibitor (DBPC).

Figure 2.11 - The Doble Engineering sludge-free life laboratory test helps to further emphasize the differences in a group of inhibited test oils in an accelerated evaluation program.

An EPRI summary of results available at this time (circa 1980) concludes that "paraffinic oils appear to be a suitable alternative to naphthenic oils at usual ambient and operating temperatures." Nevertheless, some differences do exist:

1. Slow Carbon Contamination Settling Rate
   In paraffinic oils there is a delayed settling out of carbon generated by high power circuit breakers. This buildup of car-

TABLE 2.5

TRACK RECORD AND HISTORY OF TRANSFORMER OILS	
Year	Sludge Life
1930	96 hours
1945	96 hours
1965	58 hours
1975	40 hours
1977	16-32 hours
1979	8 hours
Warning: This test is a measure of the *comparative* quality only.	

bon could cause reduced dielectric strength in critical areas, possibly causing an internal phase-to-ground flashover. In naphthenics, carbon settles out in a relatively short time.

2. Low Temperature Behavior
   Unless very deeply dewaxed, a paraffinic oil can form wax at temperatures below 0° C (wax in oil solution is satisfactory but when precipitated out of solution it is a poorer insulator). When cooled to -40° C, naphthenic oils have served effectively over the entire required temperature range without special processing.

3. Acid Formation
   Evidence indicates that the type of acids formed by paraffin-type oils are "stronger" than those formed by naphthenic type oils.

4. Gas Evolution
   The undesirable tendency of paraffinic oils to *evolve* hydrogen gas from the oil in contrast to gas *absorption* by many of the naphthenic oils.

5. Reclamation and Aniline Point
   A higher aniline point temperature is natural for paraffinics (79°-94° C) compared to naphthenics (59°-82° C). Therefore, greater difficulty appears evident in oil reclamation.*

**Chapter 7, Part 3. This was the key original drawback with paraffinic oils.*

6. Cracks and Void Formation
These result from the inability of a paraffinic oil to flow easily as it contracts on cooling. This would reduce the dielectric strength of the total insulation system and result in partial discharge (corona). This formation can be minimized by the addition of a flow modifier, dewaxing, blending with other oils, or a combination of these.

7. Viscosity
Decreasing the temperature increases the viscosity. Again, a flow modifier can be added which decreases the viscosity, as it improves the oil flow.

Finally, two areas have given conflicting data. This demonstrates the complexity of the issue.

1. Deterioration of Cellulosic Insulation
Aging of paraffinic oils tends to produce more carbon dioxide (from the paper) than naphthenics. This would lead to the conclusion that the cellulosic paper would deteriorate faster. However, other data conflicts with this in that "aging of the solid insulation appears to be largely independent of the oil." Nonetheless, these *laboratory-scale* accelerated aging tests may not project long-term results of *actual* trial.

2. Effects on Dielectric Strength
Studies have shown that dielectric strength has been both affected and unaffected by the use of paraffinic oils. Paraffinic oils are currently being used in much of the world (Europe, Russia, Japan, etc.). In addition, from all the studies made to date, paraffinic oils in conjunction with judicious blending, dewaxing and flow modifiers can be obtained that are comparable with the traditional naphthenic oils.

Only a few oils have been tested, applications vary, and therefore, these particular results do not necessarily apply to all paraffinic oils. Additional studies of more representative samples of general paraffinic oil production and application should be made before any final conclusions are made. Most authorities recognize that the best way to test these oils is in-service. Thus, conclusions will not be complete for many years. These facts remind the transformer owner/operator that preventive maintenance of his present oils is the only sure way to go.*

**Chapter 6.*

Oil Disposal Practices

Once transformer oil has deteriorated to the point that it no longer can economically be reclaimed* or it cannot be used as a fuel or any other permitted use, it must be disposed of according to governmental regulations—federal, state, or local. In the U.S. waste oil can no longer be used as a sealant, coating, dust control agent, or dumped into a field or in any waterway—stream, pond, lake, or river, and so forth.†

**Chapter 7, Part 3.*

†Chapter 9, Part 2, (Table 9.11).

Part 3—What Makes a Good Insulating Oil Go Bad?

Functions of Mineral Oil

Many thousands of oil filled transformers, containing an estimated 1.73 billion gallons of insulating oil, are in service today. It is a fact that the mineral oil in these transformers is intended to:

- Provide dielectric strength of the transformer insulation system
- Provide efficient cooling
- Protect the transformer core and coil assembly from chemical attack
- Prevent the buildup of sludge in the transformer

When the properties of the oil have changed enough that the oil can no longer satisfactorily perform any one of these functions, the oil is said to be bad. Continuing to operate a transformer with bad oil significantly reduces the transformer's life expectancy.

Goals for Oil

Based on these functions of good oil, the goals to be reached include:

- Operating the transformer in the "sludge free range"
- Preventing premature transformer failure due to transformer oil oxidation products
- Conserving two very valuable resources—oil and transformers

The Chemistry of Oil Degradation

Origin of Oil Decay

"New" oil does not exist even in a newly installed transformer. Oil deterioration begins as soon as the oil is placed in the equipment at the factory prior to shipment. Traditional distinction has been made between deterioration and contamination.

1. Contamination—"In general, an oil will be considered to be contaminated when it contains moisture or other foreign substances that are not products of oxidation of the oil..."
2. Deterioration—"The term 'deterioration' will be used to denote only the effect of oxidation..."

Once again, in understanding what happens inside a transformer, we are in the midst of chemistry.*

General Chemical Reaction

The American Society for Testing and Materials (ASTM) has established that oxidation of an oil results from a process that begins when oxygen combines with *unstable* hydrocarbon impurities under the catalytic effect of the other materials in the transformer. Certain other factors accelerate heavily the sludge reaction.

> When a used oil, completely unfitted for further use in the transformer (worst case situation) is examined, it may be found that at least 80 percent of the hydrocarbons present in the used oil are unchanged and can be reused if the oxidized products are completely eliminated. It is correct to state that the oxidation of an oil is that of the impurities present in the oil and not that of the hydrocarbons. Pure hydrocarbons are not easily oxidized under normal conditions.

Basic Factors in Oil Deterioration

Oxygen

Oxygen is derived from the air inside the transformer. All oxygen cannot be removed, even by vacuum filling. At least 0.25 percent by volume oxygen still remains even in sealed units.

Oxygen is present in oil in solution as a component of dissolved air in a higher ratio, due to its higher solubility as compared with nitrogen (16 percent versus 7 percent). Here is another way to view oxygen content in the oil in contrast with the atmosphere:

Air Content 10-12 percent

Gas-in-Air (Atmosphere)			Gas-in-Oil		
N_2	=	79 percent	N_2	=	71 percent
O_2	=	20 percent	O_2	=	28 percent
Other Gases	=	1 percent	Other Gases	=	1 percent

*Chapter 1, Part 3.

Thus, it's clear that transformer oil has a special affinity for oxygen.

Degrading cellulose especially via heat also supplies a source of oxygen. Finally, the natural oxygen inhibitors in new insulating oil are gradually depleted with time, increasing the rate of oxidation throughout the in-service life of the oil.

As stated in the form of a chemical reaction, Figure 2.12 shows what makes a good oil go bad, where oxygen is the primary cause:

Unstable Hydrocarbons (in the oil) + Oxygen + Catalysts + Accelerators = Oxidation By-products

Transformer oil is adversely affected by oxygen and moisture (as catalysts) and heat (as an accelerator).

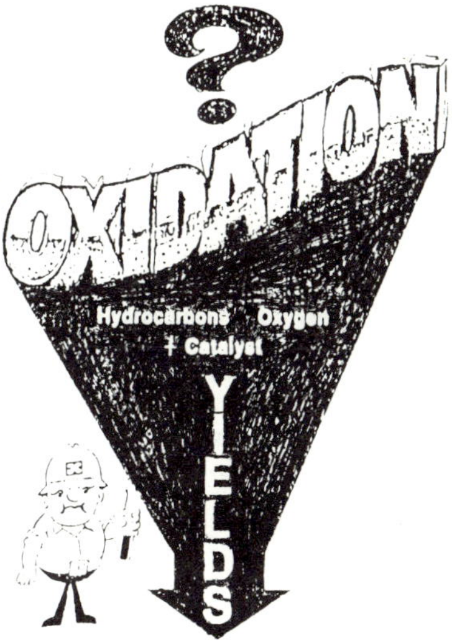

Decay Products

Figure 2.12 - What makes a good oil go bad...

Catalysts

The second grouping of factors affecting oil-life are catalysts—primary substances that increase the rate of oil degradation without being consumed in the process.

1. Moisture
   Moisture can enter the insulating oil externally via a leak, as con-

densation, or internally through the chemical process of oxidation. Water is a major catalyst in oil oxidation. All forms of water contain *additional* oxygen.

2. Copper and Oxygen
All of the preceding show the beginning stages of what is the inevitable process of deterioration. Fortunately, this is a long-term process, catalyzed also by copper (windings) and iron (core) and finally accelerated by several other factors.

Accelerators

Several *secondary* factors increase oil oxidation. These are called accelerators and consist of:

- Heat
- Vibration (120 cycle mechanical vibrations due to 60 Hertz power)
- Shock loading
- Surge voltages
- High electrical stress

1. Heat
Like most chemical reactions, heat is a *major accelerator*. As expected, time varies for oxygen to combine chemically with oil, depending upon the transformer operating temperature. At 75° C, for example, it takes approximately five days to combine with oxygen; in contrast, at 50° C, it takes several months.

2. Vibration and Shock
Electromechanical vibration and sudden electrical or mechanical shock also speed up a reaction.

3. Electrical Stress
The effect of cellulose-based insulation in the oxidation of the oil under electrical stress for higher operating voltages (especially *extra high voltage* equipment) is not as obvious.

   In newer transformer designs, because of weight and dimension restrictions, oil ducts are narrower and the electric stress (expressed in KV/cm) is increased.

   Laboratory studies tabulated in Table 2.6 show that oxidation proceeds at an accelerated rate even in the presence of low electric stress (49 KV/cm in contrast with *high* stress of 150-200 KV/cm).

Hydrogen and light hydrocarbons (methane and carbon oxides) are observed in relatively large amounts in the gaseous oxidation products.

Other observable phenomena of oil under *electrical stress*:

a. Visible particles in the deposits are much larger.

b. Accumulation of contamination deposits in zone of maximum intensity is characteristic.

c. Deposit is not uniform but forms separate elongated sections and is oriented in the direction of lines of force in the field.

Increased electrical stress thus:

a. Interferes with heat transfer.

TABLE 2.6

OXIDIZABILITY OF TRANSFORMER OILS UNDER ELECTRICAL STRESS[1]

Oil Test Method	Electrical Stress KV/CM			
	0	49	0	49
(Fatty) (heavy), acid number mg/KOH/gm	0.10	0.13	100%	130%
H_2O soluble acid mg/KOH/gm	0.032	0.049	100%	153%
H_2O weight in percent	0.003	0.017	100%	566%
Tan δ at 70°C[2]	5.5	10.7	100%	195%
O_2 absorption ml/100 g oil	28.5	48.5	100%	170%

[1]Lipshtein, **Transformer Oil** (1970), p. 68.

[2]Tan δ equals the dielectric dissipation factor of the oil.

b. Enhances aging of cellulose insulation through loss of mechanical strength.

c. Forms conducting "bridges" in transformer insulation leading to a decrease in electrical resistance.

Even the formation of additional water is accelerated under electrical stress.

4. Cellulosic Materials
Data has also shown that the cellulosic materials themselves exert an additive effect on the oil aging process. This is just one of the natural effects of cellulose taking place within the transformer.

Recall that each factor compounds the oxidation process until an actual "snowball" effect takes place.

Transformer Oil Decay Products

Much has been written about the chemistry of oil degradation, but not all that much is really known about the specific details. It is generally agreed that peroxides, acids, alcohols, ketones, and sludges are formed as the major stages in the deterioration process.

Early Deterioration Stages

Once a series of peroxides are formed, these unstable compounds start a chain reaction. The cellulose (paper, cotton, and so forth) reacts readily with peroxides, resulting in oxycellulose. This compound lacks in mechanical strength, with embrittled insulation materials as the result. Such embrittled insulation cannot withstand the shock produced by surge voltages.

As these decay products are considered, keep this fact in mind—because of the porous structure of insulating paper and board, they are considered *adsorbents*.

1. Peroxide gas
2. Water soluble acids
3. Low molecular weight acids
   a. Intensively adsorbed by cellulose
4. Fatty acids

5. Water
6. Alcohols
7. Metallic soaps
    a. Copper naphthenate
    b. Ferrous naphthenate
        1) More readily adsorbed than the naphthenic acids from which they were obtained
8. Aldehydes
9. Ketones
10. Lacquers
11. Sludges of asphaltene

Note that in the progressive sequence, fatty acids are unlike watery (mineral) acids. The fatty acid has a slimy feel, can be seen in oil, and though not as potent as mineral acids, nevertheless, eats up organic (synthetic) insulation.

In any case, the rate of acid buildup is accelerated as the acid itself acts as a catalyst for the formation of more acids. Field studies have shown that the critical neutralization or acid number is 0.25 mg KOH/gram of oil (Figure 2.13).*

Final Visible Stage (Sludge)

Unfortunately, the terminal stage of the deterioration process is sludge, the visible sign that the oxidation process has long been at work. Sludge is a resinous partially conductive substance which is moderately soluble in oil. In the laboratory, sludges have increased 18-20 percent even under very low electric stress (10 KV/cm).

Sludge comes about when the acids attacking the iron, copper, varnishes, paints, etc., and these materials come into solution and combine together. This sludge eventually precipitates out of solution and forms a heavy tarry substance which adheres to the insulation, the side walls of the tank, lodges in ventilating ducts, cooling fins, and so on.

One of the not so obvious but very serious effects of sludges is that this conglomeration of decay products has formed in the cellulose insula-

**Chapter 6, Part 1, Figure 6.2.*

tion itself, even before the sludge precipitates onto the interior surface. These sludges, once formed in the insulating system and deposited on the insulation, immediately go to work on the cellulose, resulting in shrinkage of the insulation. This shrinkage causes the transformer to lose its ability to absorb *shock loading*. As a result, coil movement under load brings about premature failure. In addition, a sludge deposit of 1/8 inch to 1/4 inch thick on the core and windings may increase the transformer operating temperature from 10° to 15° C. This can result in the necessity to derate the transformer in-service.

Even though the oil has deteriorated badly, relatively few of the estimated 2870 hydrocarbons present in the oil have reacted with oxygen. Further, and more important, the oil can be used again for its original purpose after the removal of the oxidation products.*

The Mechanics of Sludge Formation

An ASTM study has concluded that the process of oil oxidation consists of two main cycles of reaction:

1. The formation of oil *soluble* decay products such as acids, beginning as soon as the oil is put in operation.
2. The change of the oil soluble oxidation products into oil *insoluble* compounds.

Sludge, precipitated first on the cold parts and then onto the hot parts of the transformer, will continue to oxidize and eventually become insoluble. Sludging takes place periodically rather than continuously.

Figure 2.14 illustrates the cross section of a transformer coil with a resulting buildup of sludge by progressive layers. The layer in contact with the coil is relatively thin, and successive layers are progressively thicker. Now assuming that five successive layers of sludge have developed over a period of time, the first layer next to the coil, being thin and having good thermal contact with the heat source, will continue to oxidize and eventually become insoluble. As we progress from layer to layer, we will find varying degrees of hardness depending upon the extent of oxidation that has continued to take place since the sludge was laid down.

**Chapter 7, Part 3 (Reclamation).*

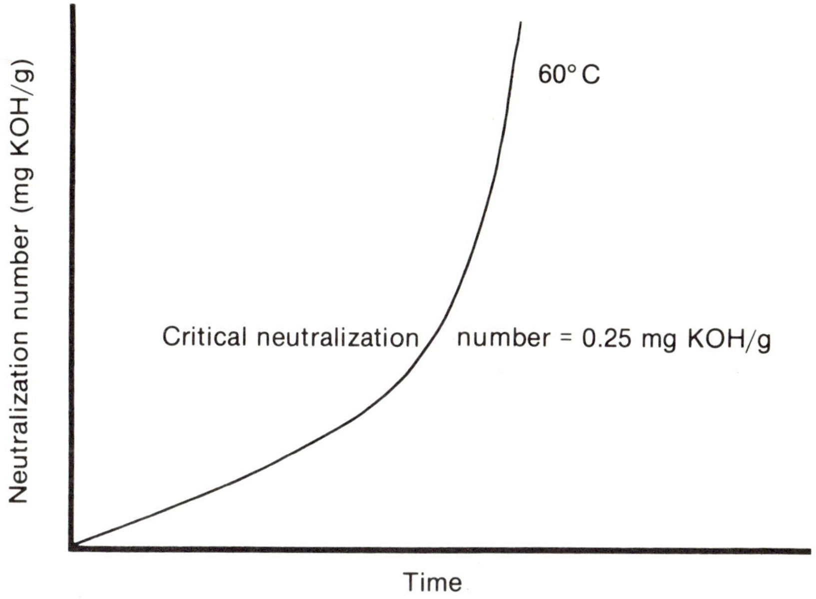

Figure 2.13 - The role of acid buildup, where the critical neutralization number is 0.25 mg KOH/gram.

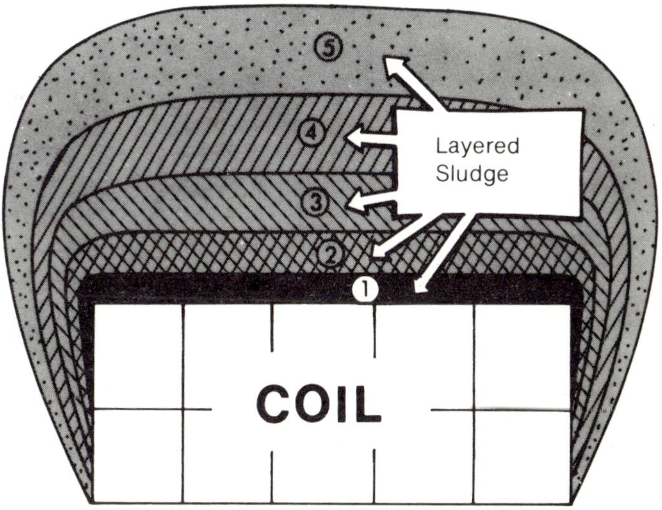

Figure 2.14 - Shows cross section of transformer coil with five successive layers of sludge buildup. Layer one has already solidified and is now a "permanent part" of the transformer.

Part 4—Cellulosic Insulation Life

Introduction

We concluded in Chapter 1 that transformers do not die of old age, but are killed! The primary factors that contribute to this eventual failure consist of poor maintenance, operational accidents, and overloading.

The practical life of a properly maintained transformer should be at least 50 years. But keep in mind that any suggested definition of transformer life is vague and therefore attests to the significance of the degradation problem.

Before we look at the problems of cellulosic insulation, we should understand the environment in which insulation in power equipment must work. The factors of influence are thermal, mechanical, and chemical, as well as electrical stresses; moisture and oxygen content; impact loading (short circuit duty); and degree of loading.*

The Insulation Environment

A transformer in-service is exposed to a variety of stresses—requiring both dielectric and short circuit or mechanical strength.

Dielectric Stresses

It is a fact that a transformer operates under 60 Hertz excitation. The resulting 120 cycle mechanical vibration of the entire structure under the influence of 60 Hertz power results in voltage stress. This stress is continuous and may be as high as 105 or 110 percent of rated voltage, depending on the tap range and the limits imposed by ANSI standards.

A second major dielectric stress involves lightning surges, arising from direct lightning "hits" to the electrical system or from indirect surges in the vicinity of lightning strikes. The result is a traveling wave on the transmission line which may "hit" the terminals of a transformer (Figure 2.15).

The third stress consideration involves switching surges. These impulses may be caused by a number of system operations or abnormalities but are often due to line switching.

**Chapter 2, Part 5, considers the primary enemies of cellulosic insulation—moisture, oxygen, and heat.*

The dielectric stresses are defined in terms of time in microseconds. Each stress presents greatly differing voltage-time relations (Figure 2.16). It would take a long graph to show these relationships in complete form. The essential parts, however, are shown by looking at the two ends of such a plot. For example, in the case of lightning surges, these are represented in laboratory studies as a wave (an undesirable disturbance which reaches its peak or crest value at 1.2 microseconds and decays to half value in 50 microseconds). The typical switching surge used reaches maximum value in 230 microseconds and decays to half value in 2000. In contrast, the 60 Hertz time requires 4167 microseconds just to reach its crest value.

Short Circuit Stresses

In time past lightning has been the primary stress. In recent years through-faults (that is, an excessive current flowing through a transformer) not only have been more severe, but they have also been more numerous.

The primary cause of transformer failure as the result of a short circuit is not because of direct thermal damage of the insulation but rather the mechanical forces produced in the windings. As the large KVA sizes of transformers have increased in recent years, so have the mechanical forces that are stressing the integrity of their insulation systems. The interaction between windings results in both horizontal and vertical electromagnetic forces.

The vertical or axial force causes the low voltage and the high voltage windings to shift with respect to each other, a condition called "telescoping" (Figure 2.17). These forces make the windings take positions that will increase the magnetic flux of the system. If two windings are in series, the electromagnetic force varies as the square of current. For example, a short circuit current 20 times normal will produce $(20)^2$ or 400 times the normal stress.

The vertical force between primary and secondary windings results because it is impossible to exactly balance the low and high voltage electrical center lines. The only way the interaction of the two windings could be eliminated would be to have the coils occupy the same space at the same time—an impossibility! Therefore, the resultant force between the two coils can be minimized only when the manufacturer is diligent in his design and workmanship. Because of the complexity of this problem in manufacturing, no two transformers are alike, each having its own characteristics. Although the horizontal force (radial or hoop) is the major force, (Figure 2.18), it is the vertical component that is *most difficult* for which to design and manufacture.

Unfortunately, all three stresses—dielectric, mechanical and thermal—often occur simultaneously.

> A lightning strike can create a fault near a transformer when it is loaded and hot, just as well as when it is relatively unloaded and cool. Indeed, in a severe lightning storm it may be overloaded since other equipment may be (already) out of service....Thus, an overheated insulation system may be subjected to a lightning strike, followed almost at the same time by severe mechanical shock with accompanying heat from the high fault current.

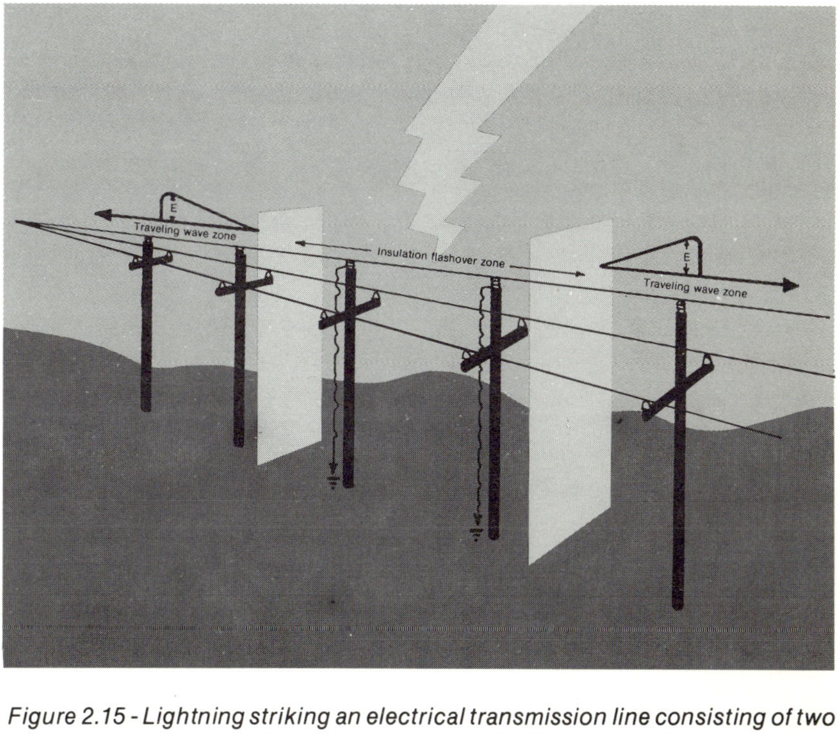

Figure 2.15 - Lightning striking an electrical transmission line consisting of two parts—structures experiencing flashover and adjacent traveling wave zones where insulation flashover does not occur. (Courtesy of The Ohio Brass Company).

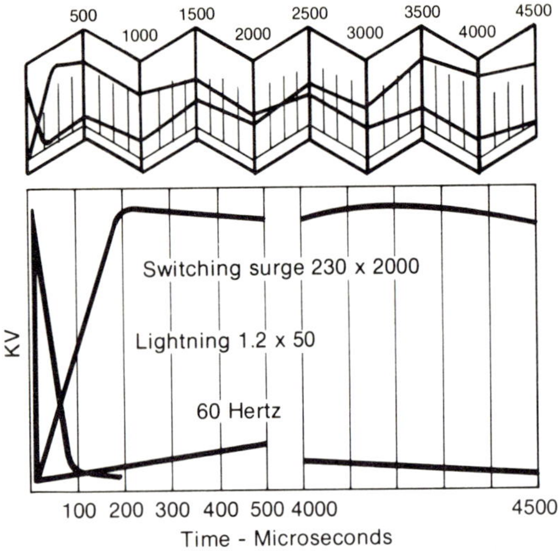

Figure 2.16 - An in-service transformer must be designed to withstand various dielectric stresses. Different voltage-time relationships are required for each of these three stresses. (Courtesy of The Ohio Brass Company).

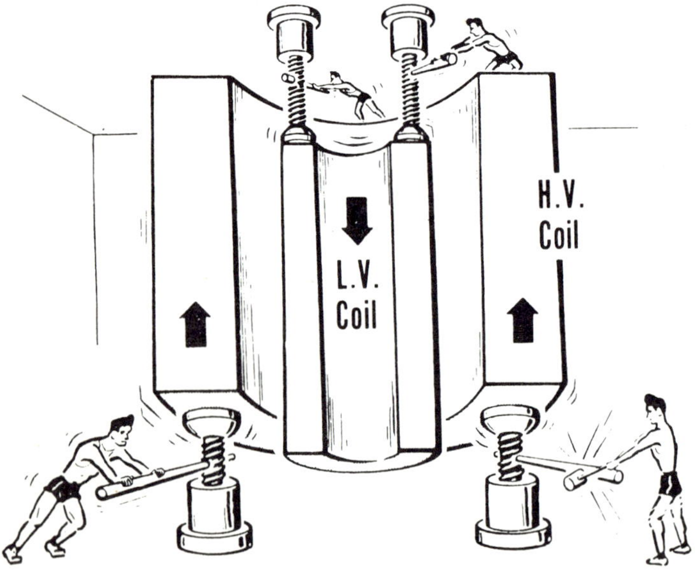

Figure 2.17 - Vertical (axial) forces between high- and low-voltage coils in core-form transformers in a through short circuit. (Courtesy of Doble Engineering Co., Watertown, Mass., and Westinghouse Electric Corp.).

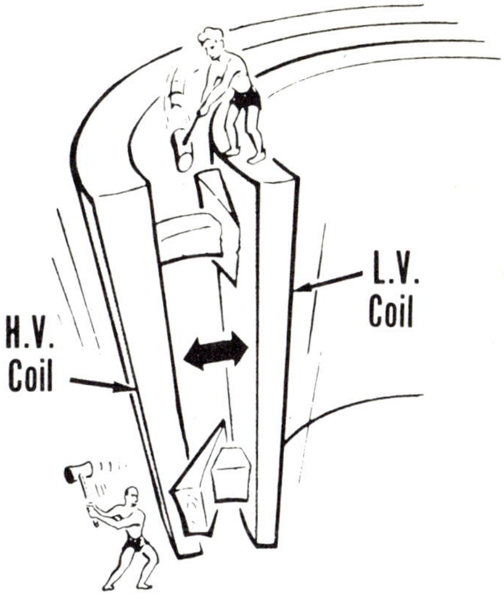

Figure 2.18 - Horizontal repulsion (radial) force between high- and low-voltage coils in core-form transformer in a through short circuit. (Courtesy of Doble Engineering Co., Watertown, Mass., and Westinghouse Electric Corp.).

This fault current can cause a hot spot temperature of 500° F due to the I^2R losses. If the transformer does not fail, but continues to operate, this temperature will reduce dramatically inasmuch as the rapid rise in hot spot temperature is not accompanied by a corresponding Btu input.

This is the situation, therefore, under which a transformer must operate through proper design, manufacture, quality control testing, and preventive maintenance. Rectifier and furnace transformers operate on "short circuit current" or are subject to many short circuit operations, and therefore, require greater mechanical strength to resist deformation from short circuits.

Indeed, the transformer does more than hum! A designer must recognize these various stress effects.

Effects of Stress on the Insulation System

60 Hertz Normal Excitation Voltage

The principal concern is both the major and phase-to-phase insulation. Failure in-service (or in a test) can result from:

- Inadequate insulation through error in design or manufacture.
- Improper installation.
- The presence of partial discharges; as the term implies, a partial breakdown of the insulation structure within a transformer. This is sometimes referred to as "corona", but the proper terminology is partial discharge.* Such discharges are undesirable as progressive tracking or formation of gas may result.
- Long time deterioration as the result of thermal and voltage stress.
- Improper or no maintenance.
- Excessive loading and/or overloading.

60 Hertz Transient Overvoltage

Normal operation for three-phase transformers results in a phase-to-ground stress that is 58 percent of the phase-to-phase stress.

However, this relation is altered substantially under single phase fault conditions. The stress on the major insulation may be higher than normal by 30 percent for a grounded system or 73 percent for an ungrounded system. This transient stress should normally be no problem, but in the case of design or manufacturing error, it is possible that insulation puncture, once initiated, would persist after returning to normal operating conditions.

Lightning Impulses

Failure during a lightning impulse (no more than 50 microseconds) involves an outright puncture or tracking of the insulation. This point of failure may occur anywhere in or on the impulsed winding or in another winding because of the oil stress produced by the voltage oscillations. The very rapid increase and collapse of voltage principally stresses the bushing, winding leads and the first few turns of the winding. If the puncture is through major insulation or phase-to-phase insulation, failure is instantaneous, whereas if the puncture is in the minor insulation, final failure may be caused by the resultant 60 Hertz follow current.

Refer again to Figure 2.16. Note that a non-linear initial voltage distribution followed by a series of oscillations occurs until a linear

**Chapter 2, Part 5.*

voltage is achieved. This places considerable stress on the *minor* insulation.

It is such considerations that illustrate *insulation coordination*—the correlation of the various discharge characteristics of protective devices and basic insulation levels.

Switching Surges

In contrast to the lightning strike, the front of this wave is slow enough so that the voltage distribution is approximately linear. The impulse is transferred to other winding terminals approximately in proportion to the turns ratio between the two windings. This means that any resultant failure will most likely occur in the major or phase-to-phase insulation (Figure 2.19).

Within any one phase, two other critical areas exist:

- Surge can creep around the major insulation and cause failure, (Figure 2.20).
- The high voltage can puncture the primary layer insulation, (Figure 2.21).

This is why furnace transformers require extra insulation between turns to guard against high voltage caused by arcing.

Short Circuit Through-Faults

A well-built transformer has been likened to a tough prize fighter who is hit hard and often—yet manages to come back for more. The "hits" a unit absorbs are in the form of short circuits. For distribution transformers approximately 450 "hits" occur during a normal lifetime, or as many as 54 through-faults of varying magnitudes in one year! Power units are subjected to even more numerous and more severe faults.

By their very nature transformers seem compelled to shake themselves apart or tear themselves to pieces. The more common vertical or axial forces (Figures 2.17 and 2.22), operating on the high and low voltage windings, tend to pull one another apart; that is, telescope. Figure 2.23 illustrates how the major and minor insulation are affected.

The horizontal repulsive radial forces tend to make the low voltage coil buckle inward and the high voltage coil buckle outward (Figures 2.18 and 2.24). The major insulation would be primarily affected, while the minor insulation only secondarily.

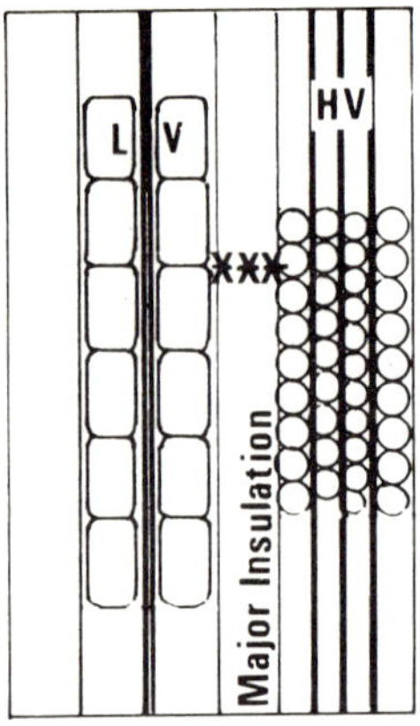

Figure 2.19 - Failure due to puncture of major insulation between high and low voltage windings.

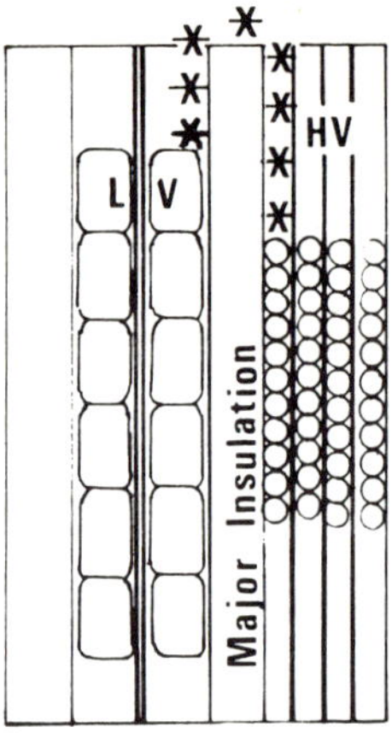

Figure 2.20 - Failure due to creepage around major insulation.

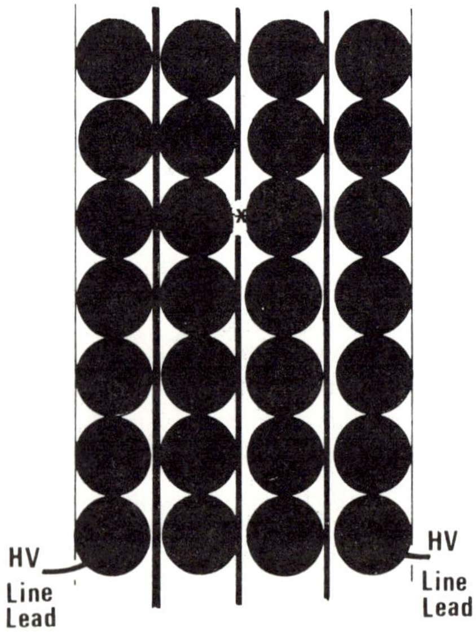

Figure 2.21 - Failure due to puncture of primary layer. (Courtesy of RTE Corp.).

Figure 2.22 - Coil failure due to shifting of low and high voltage coils. (Courtesy of RTE Corp.).

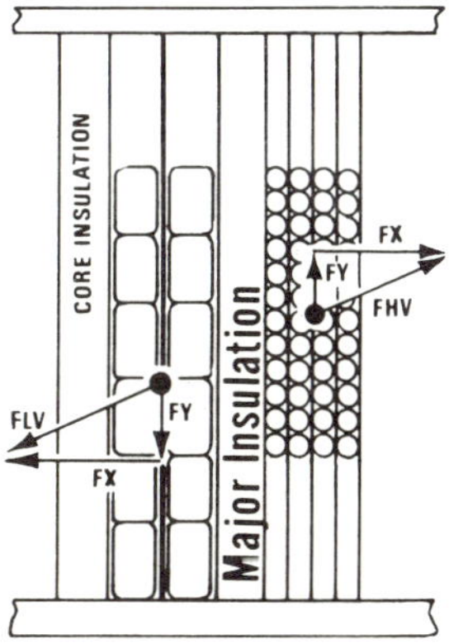

Figure 2.23 - Unbalanced forces between primary and secondary coils.

Figure 2.24 - Radial failure of coils. (Courtesy of RTE Corp.).

Insulation System Reliability

Insulation system performance is not predictable with any great degree of accuracy. Nevertheless, the principle of coordination of system insulation does involve simulating tests which approximate by a suitable margin the actual voltages to be expected in-service.

In-Service Voltage and Current Limitations

The actual in-service voltages to be encountered are limited as follows:

- The 60 Hertz line voltage to ground is inherently limited to slightly over 100 percent of normal operation stress on a continuous basis or to 130 or 173 percent on a short-time basis during single-phase fault conditions.
- The lightning impulse and switching impulse voltage levels are *limited* by the surge protective characteristics.
- Short circuit duty (made of operation) includes the magnitude of the current in all windings; the duration of the current during a fault, and the frequency of occurrence of faults. Existing standards have provided an incomplete definition, but subsequent publications have discussed the matter in more detail.

Simulated Test Procedures

Much of the data concerning insulation aging at different temperatures has been obtained in laboratory and model tests in which the decrease in mechanical and electrical strength has been measured.

The first tests were made on sample materials only in sealed tubes. The proven technique (on motors) of testing by using models of complete equipment was used then, being more representative of actual transformer conditions. A transformerette is a scaled down model transformer (0.75 KVA), containing a core, windings, and insulating oil, enclosed in a sealed steel container. The relation between these tests and actual in-service conditions is only a rough approximation.

More recently for large power transformers, a series of functional tests have been made under controlled conditions on production type transformers (25 KVA). Each part of the insulation system performs its normal function during the test.

The simulated system tests on sample materials tend to be inexpensive and lead to fairly quick results; so they are particularly useful as screening tests for new materials or for comparison to present systems. Actual equipment tests are more expensive and time-consuming, and therefore tend to be more useful as a final proof test of a previously screened test. Both types of tests are quite commonly performed by transformer manufacturers, and some simulated system tests are occasionally made by producers of insulation components. Figure 2.25 shows the testing of the proposed metallic glass core transformer.* Transformer Loading Guides based on such tests are developed and tempered by sound judgment through operating experience of industry.

The latest tube aging studies involve comparison of paraffinic and naphthenic oils as part of the insulation system (Figure 2.26). Whereas changes are clearly evident between use of Kraft versus thermally upgraded papers, paper aging was not noticeably different between the different oils.†

As new insulation systems are developed and accepted by the electrical industry, new or revised test standards and Loading Guides will be established in order to realize all their benefits.

*Figure 2.25 - A 15 KVA transformer being tested. This unit has a toroidal metallic glass core. (Courtesy of **Science** Magazine).*

**Chapter 1, Part 2, Figure 1.25.*

†Chapter 1, Part 2, Figures 1.51 and 1.52.

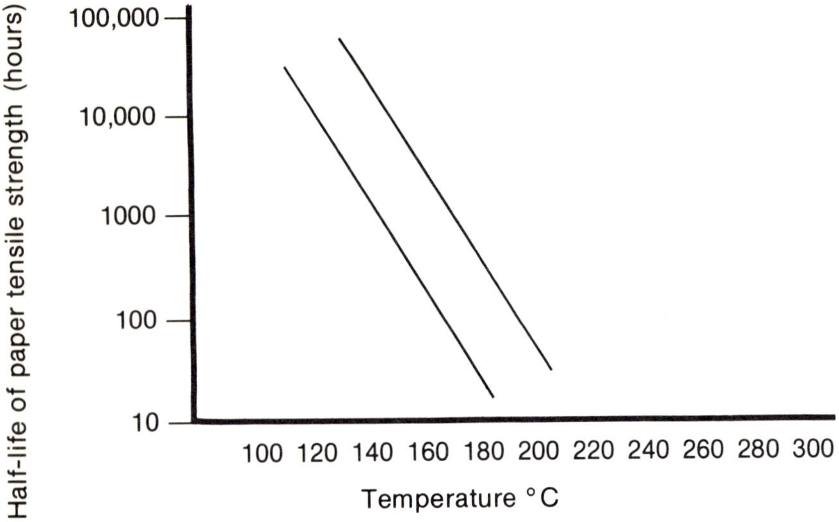

Figure 2.26 - Results of paper aging in a sealed tube. T.O. Rouse "Evaluation of Alternative Insulating Oils ...(EPRI-1980).

Measuring Transformer Insulation Life

The complex factors of mechanical and electrical stress, thermal aging, and most recently gas generation at elevated temperatures might make it seem impossible to establish any meaningful calculations of unit life expectancy. However, both insulation aging laws and test procedures provide guidance in the proper loading and maintenance of distribution and power transformers.

Insulation Aging Laws

Chemists, physicists, and electrical engineers have wrestled (and continue to wrestle with) the problems of insulation aging. Improved understanding of the insulation aging problem has required their joint efforts. These contributions have been made over a span of nearly 70 years.

Aging as a function of both time and temperature eventually renders the cellulosic insulation unfit to provide sufficient mechanical and electrical strength.

The earliest theory, nonetheless, held that aging does not begin until a definite temperature (90° C) had been exceeded.

By the early 1930's, V.M. Montsinger had clearly established that aging goes on at all temperatures, even room temperature, the rate increasing as the temperature increased. Indeed, a small amount of aging even occurs while a transformer is at no load. In addition, Montsinger believed that the end of life criterion was purely the mechanical failure of the insulation. In fact the dielectric strength generally increased until the material actually cracked open. "It is hopeless to judge...the rate of deterioration of insulation by its electrical strength. This leaves...only the mechanical strength to judge the (cellulose) insulations."

Numerous authorities, therefore, have considered the mechanical strength as the best criterion of insulation life. Consensus opinion holds that the most meaningful quantity is tensile strength measured in pounds per square inch (psi) or kilopascals per square meter (KPa/m^2).

Nine years of data concerning the relationship of tensile strength of paper, aged in oil and in air, provided the basis for what has become known as "the 8° C rule". The thermal life of class 105 insulation is halved for each increase of 8° C or conversely doubled for each decrease of 8° C.

The rationale for this guideline is found in the so-called "10° C rule" which has been widely applied to evaluate the rate of chemical reactions.

> By this it is meant that the rate of a specific chemical reaction can be expected to double for each 10° increment of temperature increase, all other factors remaining constant.

In the late 1940's T.W. Dakin and G. Malmlow pointed out that aging is the result of one or more chemical reactions and that the rate at which chemical reactions proceed varies with temperature. Dakin reasoned that the relation of insulation deterioration to changes in time and temperature is therefore assumed to follow an adaptation of the Arrhenius chemical reaction rate law. This theory states that the logarithm of insulation life is a function of the reciprocal of absolute temperature.

$$\text{Log}_{10}\text{Life (hours)} = A + \frac{B}{T}$$

Where:

- T = absolute temperature in degrees Kelvin (HST + 273); HST = the hottest spot temperature in degrees Celsius.

- A and B = Constants determined by the selected stresses imposed on the insulation (A) and by the material used in the insulation system (B).

Additional experiments have shown that this relationship of life and temperature has proven acceptable for many types of insulation structures up to 140° C. Dakin's use of the Arrhenius chemical rate phenomena is still generally accepted as the best available basis for thermal aging.

Various investigators today, nevertheless, still do not agree on the length of life at any given temperature. The International Electrotechnical Commission (IEC) believes that over the range of 80° C to 140° C the rate of useful life of the insulation on transformers is doubled for every temperature increase of approximately six degrees Celsius. While the 8° C rule is well known, research now has evidence that heat may even be more critical and tends toward the 6° C rule.*

Thermal Evaluation Testing

The second method of evaluating the relationship between transformer life and temperature consists of three types of test procedures. Each method continues today as an ongoing research and development activity.

1. Insulation Material Evaluation
   The first procedure involves the relative life of a specific material in a particular environment. As an illustration, one concept of measuring insulation life used the relationship between temperature and loss of tensile strength.

   Even though numerous studies have been made, the results are fairly consistent, considering the possible variations in the technique. Figure 2.27 summarizes these results.

   First, this plot projects how long it takes for paper to lose 20 percent of its tensile strength and also 80 percent of its tensile strength. But of what practical value is this? As an example, life at 100° C is approximately 18 months when 20 percent of strength is reached and 12 years for 80 percent loss of strength.

**Keep in mind that the oils in U.S. equipment have been refined from naphthenic crudes while paraffinic crudes are used primarily outside the U.S.*

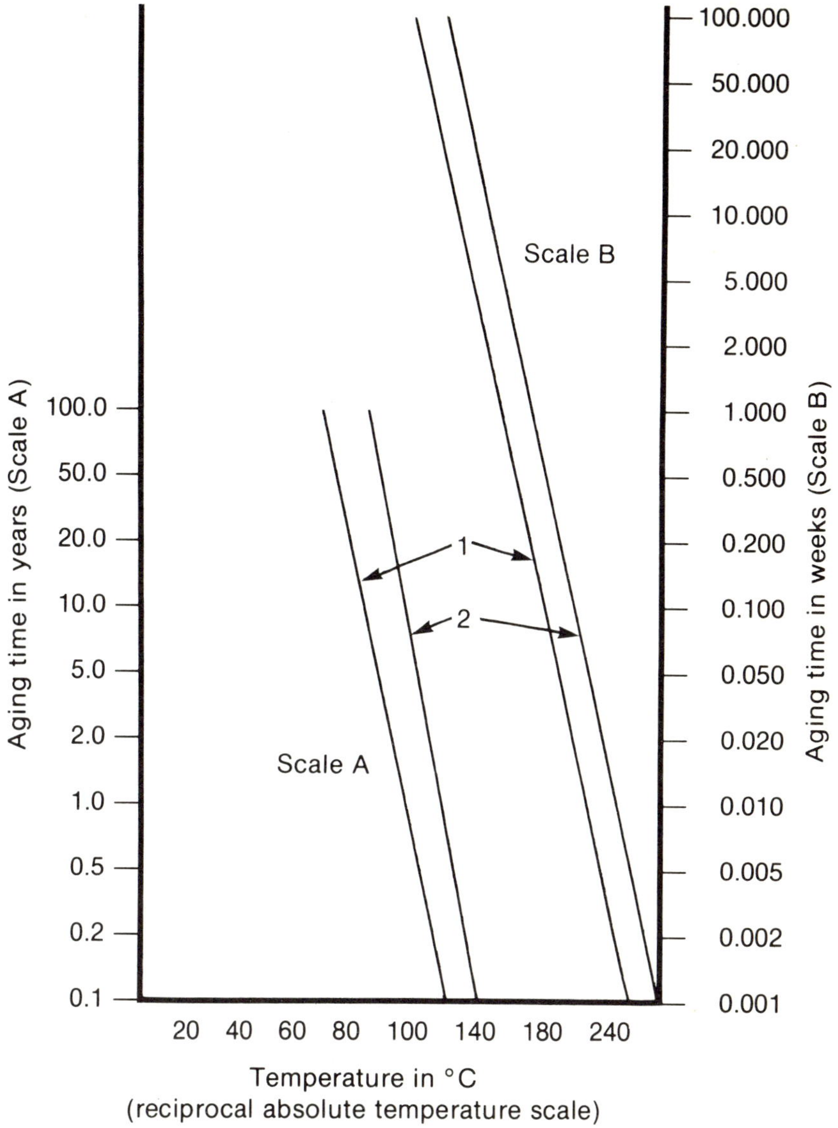

Figure 2.27 - Aging time of manila paper in oil, based on tensile strength. W.A. Sumner, "Life Expectancy of Oil-Immersed Insulation Structures" (1953). A.M. Lockie, "Loading Distribution Transformers Beyond Nameplate Rating" (1976).

Secondly, the composite plot gives an index as to the time limit at which transformers may be safely operated at various temperatures; for example, when the life at 95° C is 32,000 hours (80 percent tensile strength), then the life at 115°C is 2900 hours, and in one percent of this time, or 29 hours, one percent of the life would be lost.

2. Insulation System Evaluation
   In this procedure small models of complete apparatus are used to evaluate the compatibility of various insulation materials.

3. Functional Transformer Testing
   Both insulating material and System Evaluation Testing have proven useful, but, *only* simulated tests of "full-sized" equipment can provide thermal endurance comparisons of various insulation systems and form the basis of guides for loading in-service filled transformers.

   One accepted correlation of insulation life versus temperature for power transformers is shown in Figure 2.28. This curve is based on a a series of tests on *small* distribution transformers carried out in the 1950's and 1960's, modified per the latest end of life criterion. *A loss of one-half (50 percent) of the initial tensile strength of the cellulosic insulation is now the accepted end of life criterion.*

The Unique Effect of Elevated Temperature on Aging

In spite of the complex scope of thermal degradation, its unique effect apart from mechanical stress had not been fully realized until recent years.

The increasing practice of recurring overloading of transformers beyond the nameplate rating, as well as *higher average* loads is due to improved technology and business economics. The effect of thermal aging appears to have even greater significance for larger MVA units. As a consequence, the role of thermal aging has finally been recognized as a prominent factor in life expectancy.

An experimental program has demonstrated the interrelated effects of thermal aging and mechanical stressing on the integrity of transformer conductor insulation.

- Insulation which was aged as received from the vendor (after drying);
- Insulation which was dried and aged on conductors;

- Insulation which was aged on the conductor and subjected to repeated mechanical compression.

Figure 2.29 shows that the insulation which was aged in the conductor appears to have undergone greater degradation than that which was aged as received. Some of the degradation may be the result of minor mechanical damage which occurs during the wrapping process. Some may also be explained by the test procedure!

Though these tests were conducted on small turn segment models, necessitating follow-up research on larger transformer life structures, the relationships developed have the definite potential to be used for calculations of expired functional life.

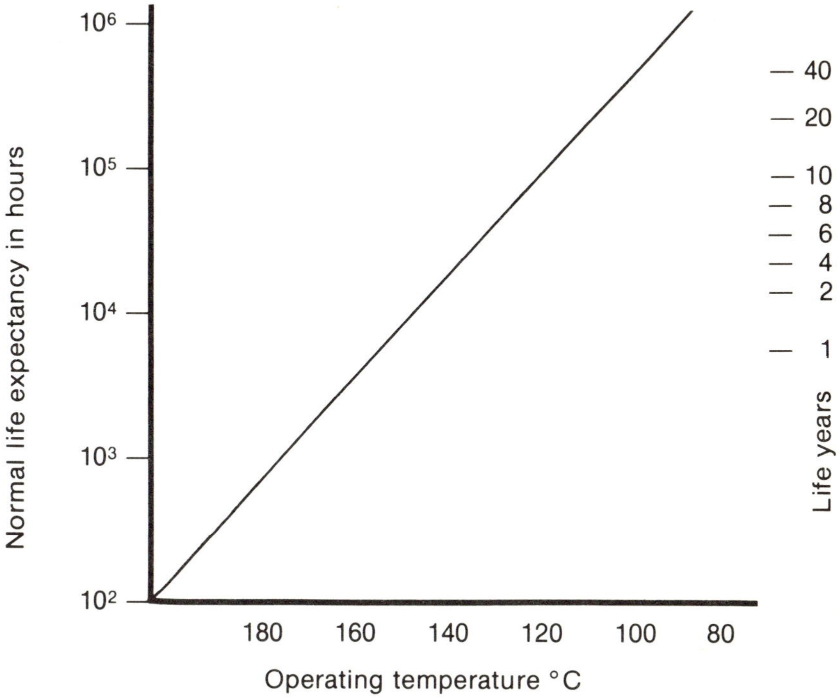

*Figure 2.28 - Life of winding insulation versus temperature, 65° C rise designs.**

**NEMA TR-98 1971 Transformer Loading Guide.*

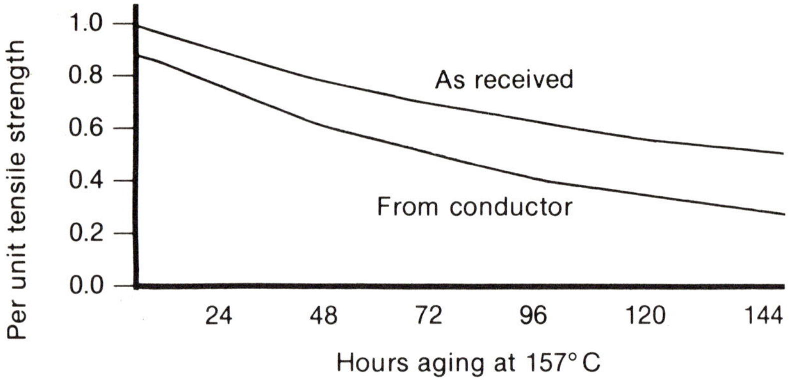

*Figure 2.29 - Tensile strength of conductor insulation which has been thermally aged as received, thermally aged on conductors, and mechanically stressed after aging on the conductors.**

Loading Guides for Oil Immersed Transformers

Based on the foregoing assumptions and test procedures, industry (manufacturers and users) has developed Transformer Loading Guides. The primary criterion for early Loading Guides (1937, 1948, 1962) had been that deterioration of mechanical properties *alone* was the governing mechanism for determining life expectancy, and the 8° C rule was set as the end of life criterion. The latest proposed revisions recognize that functional transformer life is influenced by *both* thermal and mechanical stress and is an application of the Arrhenius-Dakin formula for life of a chemical reaction.

The Loading Guide is a voluntary document, not a mandatory standard, and is necessarily conservative. The guides translate data of the Life Expectancy Curve (Figure 2.28) into a more practical handbook form for the power system-manager.

The NEMA† TR-98, 1971 Loading Guide provides the best U.S. approved power transformer life assessment. Reflecting even more recent technology, IEC Loading Guide Publication 354 was approved in

**W.J. McNutt, "The Combined Effects of Thermal Aging and Short Circuit Stresses on Transformer Life" (1976).*

†The National Electrical Manufacturer's Association.

1972. The latest ANSI/IEEE revision, Project 507, may be approved in the early 1980's. The American National Standards Institute, (ANSI), the top U.S. Standards organization, is a federation of organizations involved in the preparation of consensus standards, covering the complete standards spectrum. Figure 2.30 charts the relationship between various world-wide standardizing organizations. "Common membership of committees makes for even closer liaison."

Figure 2.31 shows the relationship between a power transformer's hot spot operating temperature and the total in-service usage of the unit in hours or years.* The standards for both distribution and power transformers (500 KVA ratings) are given for comparison. For example, the NEMA Standard and the proposed ANSI/IEEE Loading Guide suggest that a 65° C average winding rise power transformer operating continuously in a 30° C ambient at its 100 percent design rating (110° C hottest spot temperature) should only have a normal life expectancy of 65,000 hours (or 7.42 years). Fortunately, this life expectancy of 7.42 years does not seem near to reality. Several factors extend transformer life beyond this limited age.

- The average annual ambient temperature usually does not exceed 20° C by more than a few degrees (for example, 5° C in winter and 29° C in summer).
- Power transformers do not operate at a constant 100 percent load.
- Insulation tensile strength of 50 percent may not be the end for the transformer. Recall that insulation hardly ever dies, it is usually "killed".

On the other hand, the 1972 IEC Guide has established a specific temperature limitation: That in no case a hot spot temperature of 140° C in the windings is exceeded. Various authors have stated that above 140° C the Arrhenius Law is not completely applicable.

Recent studies have confirmed the presence of gaseous products of insulation aging reaction in transformers overloaded beyond 140° C, which has led to premature dielectric failure. These outages had previously been thought to have resulted from surges of fault failures. Therefore, cellulosic paper decomposition rather than an oil mechanism appears to be the primary source of such unit bubbling. This appears to be true, regardless of the type of cellulosic insulation. One confirmed result has been reduction in a transformer's impulse strength.

**IEEE Project Number 507, proposed revision of C57.92 (1962).*

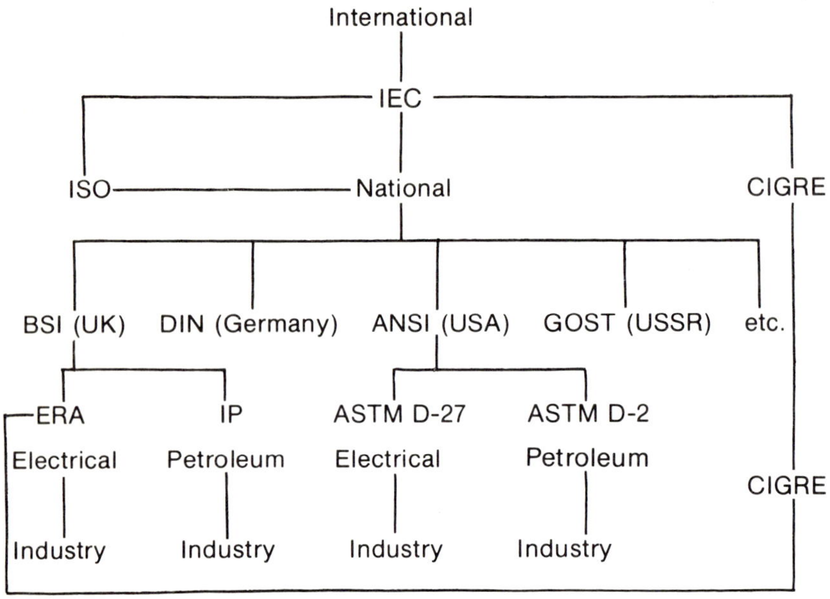

Figure 2.30 - Flow diagram shows how the various international standardizing organizations relate to one another. ISO (International Standards Organization) concerns itself with non-electrical standards. CIGRE is the international equivalent of U.S. IEEE Power Engineering Society.

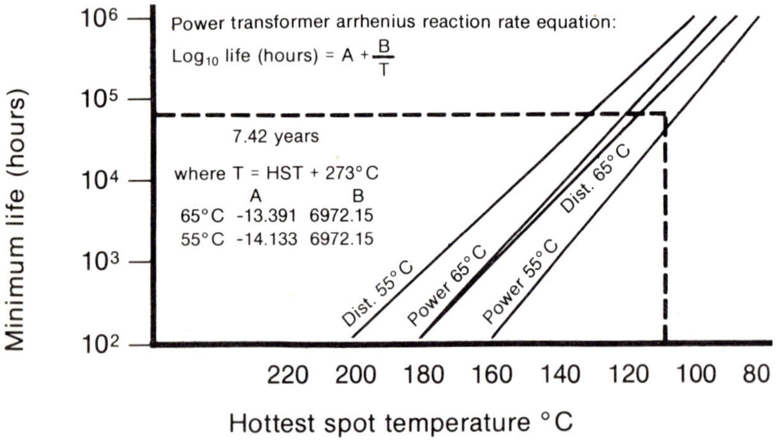

Figure 2.31 - Life expectancy curves—minimum life vs. hottest-spot temperature. IEEE project Number 507 to replace ANSI/IEEE Standard C57.92 (1962). B.D. Lahoti, "Evaluation of Transformer Loading Above Nameplate Rating," (1980).

In addition, no specific temperature limit has been assigned to metallic parts outside of the windings; rather, the general statement has been given that these parts "shall not attain excessive temperature rises". However, especially for units above 100 MVA rating it is now recognized that stray electromagnetic flux can produce dangerous hot spots in metallic parts, such as the core, core clamping members, tank, etc.

Among other variables that the user should consider:

- Variation in transformer design characteristics and therefore differences in stress effects for the same nameplate rating (check with the manufacturer concerning maximum overload capability). Model tests were conducted on a *normal cyclical* basis, whereas in recent years much continuous overloading and increased average loading has taken place.
- Influence of exterior finish (dark color is the worst; light color is best for overloading).
- Influence of corrosive atmosphere (rusting tanks, deterioration of gaskets).
- Influence of excessive vibration.
- Influence of difference in elevation (the higher the altitude, the more serious the overloading).
- Influence of solar radiation and wind velocity.

Keep in mind that while the manufacturer has improved the ability of the cellulosic paper to withstand heat by the addition of dicyandiamide to the pulp, the amount of oil per KVA (the fixed ratio) has decreased over the years. Therefore, reality points to the importance of controlling overloading of transformers, both in older and newer designed units. Even though ANSI/IEEE has published criteria for the cellulosic insulation, no equivalent guideline exists for the insulating oil. Since oil ages at *lower* temperatures than cellulose, we believe that mineral oil may be the limiting factor of a unit's insulation system. And yet, others say that the oil may not be the limiting factor because the oil can be sampled and tested, whereas the cellulose cannot.

Finally, though a Loading Guide for cellulose is available, it's at best only a "rough assessment" of the effect of transformer overload.

Value of Transformer Loading Guides

Much discussion in recent years has developed over the value of the Loading Guide.

While Figures 2.28 and 2.31 give the best insulation life data currently available, it is not directly applicable to power transformers because the stresses, both mechanical (short circuit) and dielectric are greater in power units than in distribution units. Since no model tests of power transformers have been made because of prohibitive costs, the power transformer curve was selected to be more conservative than those used for distribution units.

At this point experience indicates that normal life is "some tens of years."* A more precise estimate cannot be stated because it may vary even between identical units because of differing operating factors.

It is obvious that for economic and resource conservation purposes, justification exists for a joint manufacturer-user-governmental research and development project, involving an extensive series of thermal aging tests, beginning in the medium power unit range (1-100 MVA ratings). One suggested vehicle through which such an evaluation could be made is the American Electric Power Research Institute (EPRI).

In summary, since we know that any Loading Guide is only a "rough assessment" of the effect of the transformer load and overload, tensile strength can't be measured in the field, and no agreement on the effect of temperature on the oil exists, even the latest Loading Guide is at best only a good intellectual guess!

Therefore, prevention of severe oil and cellulosic deterioration is the only sure practical approach.†

**Circa 1981.*

†Chapter 3, Part 2; Chapter 4; and Chapter 6.

Part 5—Potent Enemies of Cellulose Insulation

Introduction

A transformer with its insulating system properly maintained is designed to handle many day-to-day problems, such as switching surges, limited short circuits, lightning strikes and through-faults. But a transformer that has its solid insulation system damaged through improper maintenance cannot. Keep also in mind that with many modern transformers, the insulating system is under greater stresses than older units.*

Much, however, can be done to assure long life. Preventive maintenance is the key, plus an understanding of the basic characteristics of transformer cellulosic insulation. Combining this knowledge will enable the owner/operator to recognize and quantify imminent problems.

The destructive effects upon the insulation system may be subdivided into inherent and external causes.

Inherent Problems of Cellulosic Insulation

When cellulose is dry, free from gas, and immersed in oil, it is the toughest physical insulating system available, even in this space and computer age. It is, however, the weakest link in the transformer insulation system. The "seeds or self-destruction" are inherent in the cellulosic insulation.

The Electric Journal of April, 1920, states that the arch enemies of solid insulation are moisture and temperature. This is still true today! The only difference would be one's perspective—the chemist would say heat is the number one problem, while the electrical engineer would say moisture is number one. F. M. Clark has written:

> Residual moisture weakens the resistance of the cellulose to thermal degradation resulting in undesired chemical, mechanical and dielectric change...(In fact) the basic chemical changes which are caused by the presence of moisture in the cellulose insulation when heated are more easily and more reliably observed than are mechanical and dielectric degradations which ultimately result from these chemical changes.

**Chapter 1, Part 3.*

Figure 2.32 illustrates how an aging electrical transformer acts similar to a chemical reactor. This development of compounded cellulosic deterioration begins with heated cellulose in combination with a trace of moisture. These reactions continue until proper steps are taken to interrupt the process or transformer failure terminates operation.

Moisture in combination with heat will destroy the solid insulation system long before the dielectric strength of the oil has given a clue that anything is going wrong.

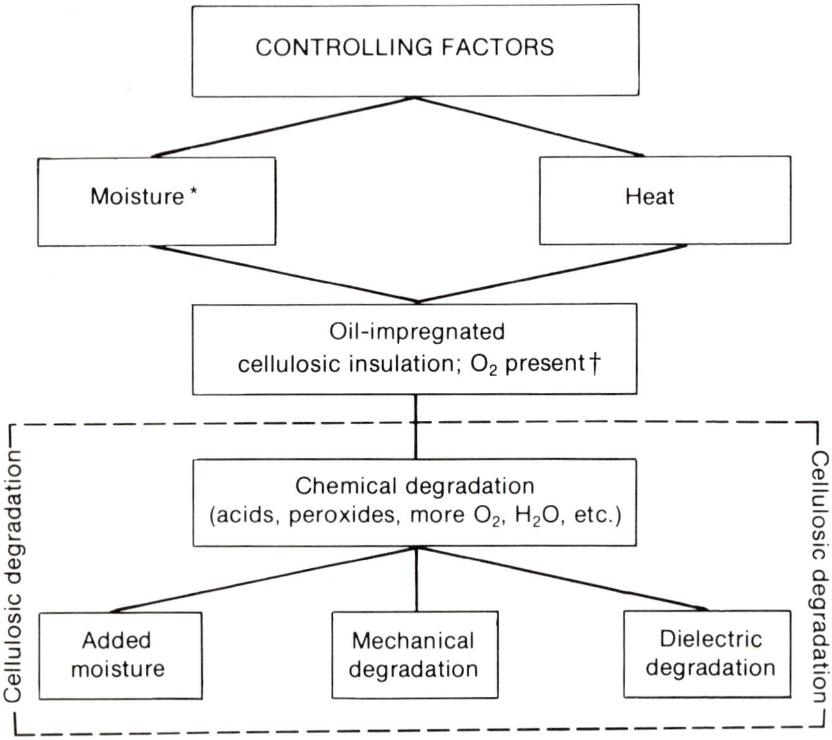

Figure 2.32 - The action of heat and moisture in the development of compounded cellulosic degradation.

* *Moisture is available to cellulose primarily from two sources: (1) residual absorbed moisture left in by the manufacturer (0.25% to over 1% by weight) and (2) moisture formed by thermal degradation during service life.*

† *The accelerated formation of oxygen is consistent with the effects observed even in the* ***total absence*** *of externally supplied oxygen.*

Three cellulosic characteristics that spell out its Achilles' heel are:

1. Affinity for water and oil decay products.
2. Adverse reaction to oxygen.
3. Vulnerability to excessive heat.

These three impose operating limitations on the transformer. Fortunately, the degradation can be effectively controlled by the proper recognition and respect for the limits of oil impregnated cellulose insulation. Nevertheless, this degradation can only be limited—never eliminated.

The Problem of Moisture

Cellulose has a strong affinity for water (hygroscopicity) and thus will not share the moisture equally with the insulating liquid.

> A very simple demonstration of the preferential absorption for water is to set a little beaker nearly full of oil, alongside a similar beaker containing water. If a strip of filter or blotter paper is bent into a horseshoe shape and one end immersed in the oil and the other into the water, it will be seen that the water will creep up the paper and will push back the oil that tends at first, to creep up from the oil filled beaker.

F.M. Clark found that moisture will divide between paper and oil in a definite ratio in a final state of equilibrium in which the moisture content of the cellulosic paper is *hundreds* of times as great as that of the oil. This is an important point to recognize and is a good clue as to why the dielectric strength test of oil is of little practical value. The paper insulation absorbs the water from the oil; in effect, it dehydrates the oil and stores the water in the worst possible place—in the paper and in the area of highest electrical stress!*

Recall the glucoside structure of the cellulosic molecule.† This demonstrates that it is possible to liberate hydroxyl groups (OH) and form moisture in a heat catalyzed process.

> When dried cellulose is exposed to atmospheric moisture, water is taken up both chemically and physically. The

**Chapter 3, Part 3, Table 3.6.*

†Chapter 1, Part 2, Figure 1.45.

moisture is ADSORBED *physically* on the surface and in the capillary structure of the cellulosic structure. The result, in some instances, is manifested in a swelling of the cellulose fiber. Chemically, moisture is taken up by cellulose to form a weakly bonded molecular structure through the hydroxyl groups of the cellulose. The moisture so ABSORBED presents the chief problem in the drying of cellulosic insulation for use in electrical apparatus.

To assist in understanding the difference described, an *adsorbent* reaches out and grabs something else, like a magnet attracting metal filings. By contrast, the *absorbent* involves only the soaking of one material in another, such as the soaking of water in a sponge.

This hygroscopic nature of cellulose insulation constitutes an ever-present difficulty both in the manufacture and maintenance of transformers which are so insulated.

The presence of moisture increases the aging rate. Figure 2.33 shows that insulating paper with a one percent moisture content ages ten times faster than one with only 0.1 percent. Therefore attempts to substantially reduce these objectionable changes due to the presence of moisture in the solid insulation have been the motivation of a substantial engineering effort for many decades.

Although methods of drying the insulation are beyond the immediate context, recognize for now that it is impossible to have a transformer completely void of moisture.*

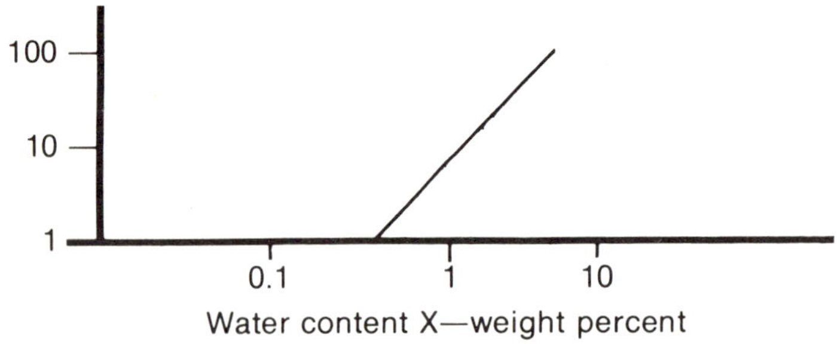

*Figure 2.33 - Influence of moisture in paper on aging time of paper: a multiple of the aging time of oil-impregnated paper with moisture content X compared to that of a well-dried, oil impregnated paper with 0.3 percent residual moisture. H.P. Moser, **Transformerboard**, (1979).*

**Chapter 2, Part 5, and Chapter 7, Part 2.*

The removal of moisture from cellulosic insulation using dry heat alone necessitates exposing the insulated equipment to high temperature—usually in the range from 100° C to 120° C. Nonetheless, temperature should be only as high as possible without breaking down the insulation chemically.

Moisture, therefore, can only be *safely* removed as the temperature reaches the limit of 100° C. Dr. Clark has concluded, "The *complete* removal of moisture from cellulosic insulation *without* causing chemical degradation is a practical impossibility."

The Problem of Oxidation

Oxidation is normally associated first with the deterioration of insulating oil. However, oil oxidation, heat, and moisture work in combination with one another, or, are synergistic. Compare:

Liquid Insulation (oil):	Oxidation and MOISTURE as arch enemies with *heat* as the primary accelerator.
Solid Insulation (paper):	Heat and MOISTURE as arch enemies with *oxidation* as the primary accelerator.

Oxidation can be controlled but not eliminated. Oxygen comes from the atmosphere or is liberated from the cellulose as a result of heat. Oxidation of the cellulose is accelerated by the presence of certain oil decay products called polar compounds, such as acids, peroxides and water.

The Kraft paper, being extremely porous (probably 85-95 percent air), will *absorb* (or soak up) oil up to about 10 percent of the transformer oil fill. Therefore, the process of cellulosic degradation begins immediately, starting at the initial oil fill of the unit. In addition, both paper and pressboard have pronounced selective *adsorption* for the volatile organic acids formed as the initial oil decay products.

The first decay products—peroxides and water soluble and highly volatile acids—are immediately adsorbed by the cellulose insulation up to its saturation level. In the presence of oxygen and water, these "seeds of destruction" have a potent destructive effect on the cellulosic structure. This deterioration may begin long before any oil test data indicates a problem. The acids of low molecular weight are most intensively adsorbed by the cellulosic insulation in the *initial* period; later, the rate of this process slows down. The oxidation reaction may attack the cellulose molecule in one or more of its molecular linkages. The end result of such chemical change is the development of more polar groups and the formation of still more water.

F.M. Clark also found that the effect of oxidation on the mechanical life of cellulose insulation becomes more pronounced in the presence of moisture and higher temperatures. This brings us to specific consideration of the second arch enemy of cellulose insulation—heat.

The Problem of Heat

Approximately 90 percent of cellulosic deterioration is thermal in origin. Elevated temperature accelerates aging by reduction in the mechanical and dielectric strength (Figure 2.34). Secondary effects include paper decomposition (DP or depolymerization), and production of water, acidic materials, and gases. If any water remains where it is generated, it *further* accelerates the aging process (Figure 2.32).

Recent evidence indicates that transformer overloading beyond 140° C hottest spot temperature causes the formation of gas bubbles. A gas bubble will lower the dielectric and will precipitate a flashover and a premature failure. Excessive heat accelerates cellulosic circuit withstand strength, or what is now termed "short-time failure".

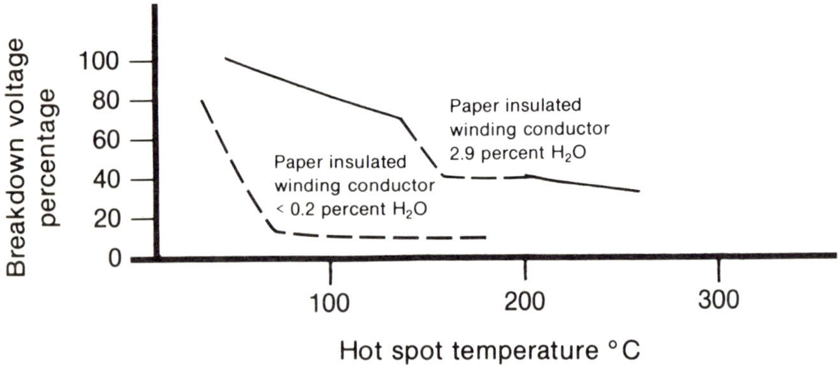

Figure 2.34 - Sixty Hertz breakdown voltage in percent of 25° C strength of dry paper (< 0.5 percent H_2O). The "wet" data is represented by a (dotted) line to indicate that the precise shape and location of the curve are uncertain because of limited data available. W.J. McNutt et al., "Short-Time Failure Mode Considerations Associated with Power Transformer Overloading," (1979).

Thermal degradation is a function of time, temperature, and initial system dryness. Unfortunately, the temperature is not uniform throughout a transformer.

As a rule of thumb, the hottest spot temperature is two-thirds up the coil stack and one-third the way into the high voltage winding. Approximately two-thirds of the heat is located in the coil and one-third in the core.

When the oil immersed cellulose is subjected to heat, the cellulose deteriorates, resulting in formation of water, acids, carbon dioxide, and carbon monoxide.*

The circled hydroxyls are "water formers" when loosened by heat. They are normally held together by "weak" or secondary bonds in contrast to the primary bonds that cannot normally be broken.

Whether cellulose is heated in the presence or absence of oxygen, the application of high temperature accelerates its deterioration (Figure 2.35). Heating severs the linkage bonds within the cellulose (glucose) molecule, breaking down with the formation of water. This resulting water causes continuous new molecular fission, and weakens the hydrogen bonds of the molecular chains of pulp fibers.

Data continues to indicate the stability of cellulosic insulation for continuous operation up to temperature of 105°-110°C for typical Class 105 (or Class A) oil impregnated cellulose.† These temperature levels are based on the three criteria of Table 1.15.‡

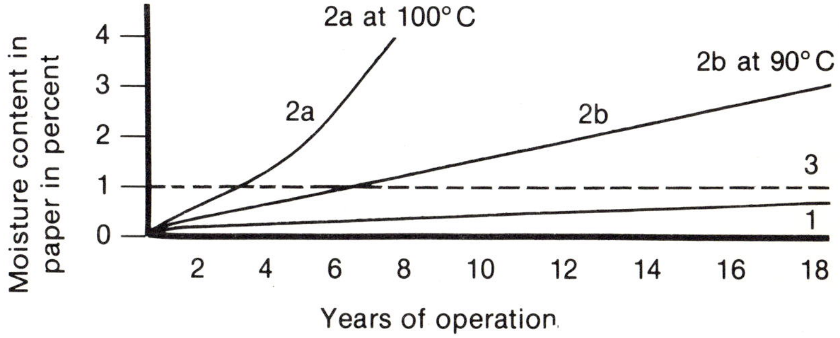

Figure 2.35 - Moisture content in the paper of a 200 MVA transformer. H.P. Moser, ***Transformerboard****, (1979).*

**Chapter 1, Part 2, Figure 1.45.*

†*Chapter 1,Part 2, Table 1.5.*

‡*Chapter 1, Part 3.*

Determination of the ultimate limit to which cellulose can be safely heated—for the purposes of dehydrating either a new or service-aged unit without affecting its mechanical and electrical properties—continues to be a major problem for transformer designers and manufacturers. The rate at which these changes take place, that is, the breakdown of the structural chain, depends primarily on the hottest spot temperature.

External Contaminants in the Insulating Oil

One outside contaminant may be in the form of "dandruff" present in the manufacturing process and not properly filtered out during the oil purging procedures. Lint particles may also have broken off the cellulose while in-service.

Another contaminant is polychlorinated biphenyl (PCB) which reduces the impulse strength of transformer oil.*

Immersed Askarel (PCB) Based Systems

Primary consideration of insulation life and characteristics has been in terms of transformer mineral oil.

Although askarel units are being phased out because of environmental regulations, mention should be made of moisture's significant effect on these PCB-based systems.* Dielectric properties of askarel are reduced by both free and dissolved water in the liquid. Such moisture usually originates in the components of the insulation system, such as aging paper insulation or from the breathing of external air through defective transformer tank sealing. Askarels are also contaminated by the various solid insulation materials which may be liberated by the PCBs solvent action.

What Happens When Cellulosic Insulation Deteriorates?

Aging of Insulation

Earlier, cellulose aging was defined as deterioration of insulating materials through the loss of mechanical and electrical strength.† Ag-

**Chapter 9, Part 2.*

†Chapter 1, Part 3.

ing is a phenomenon involving both chemical and physical changes. The more subtle chemical reactions foreshadow the *gross* physical manifestations (Figure 2.32).

1. Chemical Changes
   Manifold chemical changes occur in insulating systems with time under conditions of elevated temperature and applied voltage.

   a. Oxidation
      This chemical change occurs in apparatus operated in air or in oil in the presence of a catalyst such as copper. Oxidation may take several forms.

      A low oxygen content in oil is desirable in order to retard aging and the formation of aging products. Under the condition of low oxygen content occasional overloads will thus result in less loss of life (Figure 2.36).

      Under the same temperature conditions an oxygen-free insulation system vs. an oxygen-loaded system yields a ten times longer life period; or what might be a more practical consideration, an oxygen-free system can withstand a 20° C higher temperature without harmful effect.*

      The most common form of oxidation contamination introduces acid groups into the solid or liquid insulation. The acids brought on by oxidation splits up the polymer chains (small molecules bonded together) in the cellulosic insulation, resulting in a decrease of tensile strength. It also embrittles the insulation materials, such as wire polymer coatings.

      The presence or absence of oxygen in the space above the transformer oil will affect transformer aging. Tank constructions which minimize the admission of air and moisture to the unit, while they are used primarily to preserve the oil, also have a bearing on the solid insulation.

      The oil in oil-immersed equipment normally retards aging of the solid insulation by hindering the oxidation process and by carrying away heat that accelerates the rate of decomposition. If the oil itself becomes oxidized, however, peroxides, acids, and more water are produced, all of which

*Chapter 7, Part 2.

accelerate the cellulosic aging process. *Ergo*, a snowball effect! Peroxides in turn produce oxycellulose ("bad"); acids form hydrocellulose (also "bad").

The design philosophy that results in today's higher operating "hot spot" and average winding rise temperatures* neglect the effect of higher heat on the oil. As mentioned earlier, the tendency is for the chemical reaction rate of oil to double for each 8°-10° C increase in operating temperature (Arrhenius rule). Proper insulation system coordination requires consideration of the effect of the hottest spot temperature on the oil, as well as the cellulose.

For manila paper the effect of oxidation reaches its maximum at approximately 120° C, Figure 2.37. Kraft paper or thermally upgraded papers would not be affected as severely as manila paper.† Keep in mind also that the effect of oxidation on the mechanical life of cellulose becomes more significant in the presence of moisture. Again, we're back to the arch enemies—heat and moisture.

Another effect of heat-induced decomposition is hardening and embrittlement of the insulation. The introduction of oxygen increases the cohesion between some neighboring chains while, at the same time, decreasing and breaking the chains to make them shorter. This has the net result of increasing the hardness and decreasing the resistance to bending or stretching.

Ultimately, oxidation results in the production of volatile products that actually evaporate, and decrease the thickness or volume of the insulation.

Frequently all three chemical effects occur at once, but usually one of them plays a dominating role in each particular situation, depending upon the nature of the insulation and its environment.

b. Depolymerization

The result of high temperature chemical decomposition of cellulose is depolymerization, that is, the breakdown of the

**The higher operating temperatures were brought on by the advent of thermally upgraded papers and/or concern for continued utility service or industrial production at the expense of transformer life.*

†Chapter 1, Part 2, Figure 1.56.

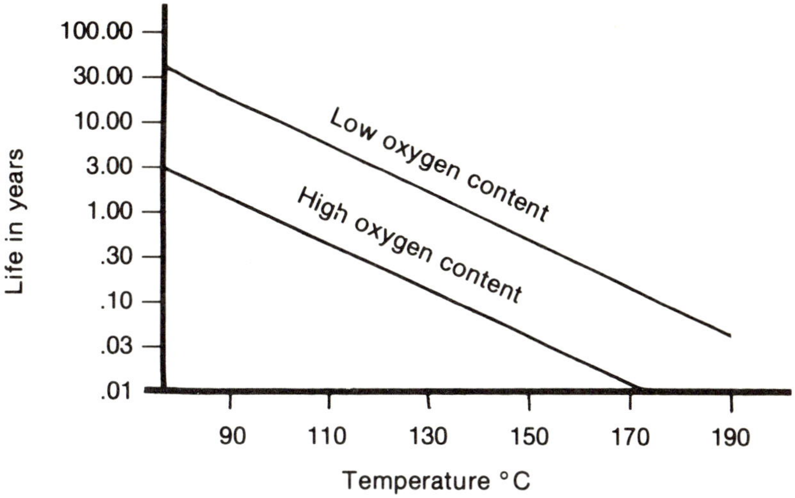

Figure 2.36 - Life of cellulose. Insulation vs. temperature. End of life defined as DP = 200 (degree of polymerization). W. Lampe et al., "Continuous Purification and Supervision of Transformer Insulation Systems in Service," (1978).

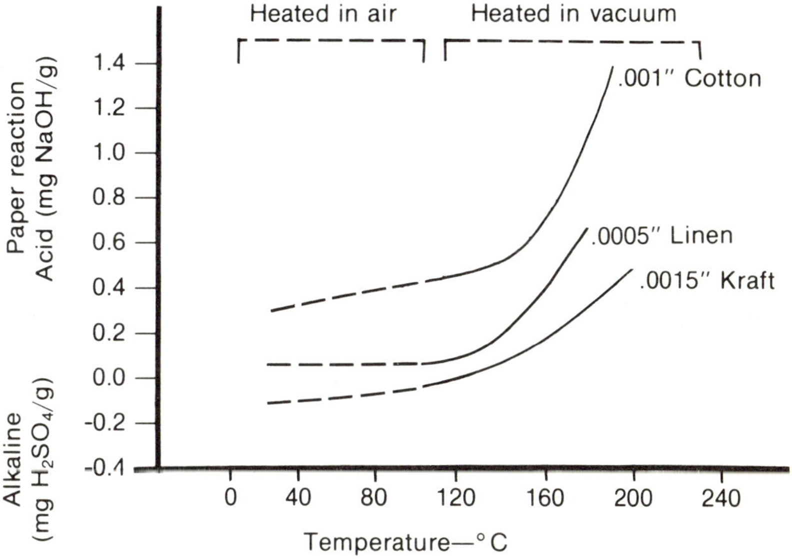

*Figure 2.37 - The effect of heat on the reaction of undried cellulose papers. The papers are heated as indicated in a tightly wound roll, 2 inches wide and 1½ inches in diameter. The heating period is 44 hours. F.M. Clark, **Insulating Materials for Design and Engineering Practice**, (1962).*

cellulosic structure in its chemical makeup.* The polymer chains actually break into shorter units at elevated temperatures even in the *absence* of oxygen. This reaction is generally a much slower one than oxidation, but at *high* temperatures it becomes an important factor. The net effect of this break in chemical bonds will be to cause the insulation to evaporate, flow, or otherwise change its shape, strength, or other desired property.

c. Hydrolysis
This form of chemical change is the chemical reaction of water with the cellulose that also results in a form of depolymerization. It is also true in the case of polyester resins, especially in sealed systems, and is further catalyzed by acids or bases.

d. Miscellaneous Effects
Many other chemical reactions can deteriorate insulating materials. Among these, for example, are electrolytic and spark or corona-induced reactions, and the decomposition of chlorinated materials to produce hydrogen chloride, etc.

This discussion is necessary to help maintenance personnel visualize these characteristics—evaporation, flow, shrinkage—applied to the solid insulation in order to realize the seriousness of the effects of heat and deteriorating service-aged oil on the life of the transformer.

2. Physical Changes
These physical changes are the result of the following factors:

a. Normal occurrence without need of chemical reaction. Melting or flow of solid insulation. Key Factor—elevated temperature and lower molecular weight. The solid insulation, usually resin insulation, is under tremendous pressure. This pressure plus heat results in melting or flowing of the insulation.

b. Occurrence with or without chemical reaction—softening. Key Factor—elevated temperature. Volatilization (a slower process). Key Factor—lower molecular weight.

c. Direct result of chemical reaction. Hardening and ultimately cracking. Key Factor—excessive mechanical stress; strain from thermal expansion and contraction; embrittlement.

**Chapter 1, Part 2, Figure 1.45.*

3. Cellulosic Evaluation by Chemical Criteria
   The cellulosic aging process can be evaluated by various chemical criteria, such as the degree of depolymerization of the cellulose, the viscosity, the copper and acid numbers, and also by the decrease in the mechanical strength of the cellulosic materials.

   When Kraft insulating papers are heated in the presence of moisture, change resulting in molecular depolymerization can be promptly demonstrated by measurement of the viscosity and copper number.

   The measure of the ability of the cellulose to reduce a standard copper solution is defined as the copper number. With a properly prepared insulating paper, the reducing action of the cellulose is of low order. With degraded paper, the copper number rises.

   Figures 2.38 and 2.39 show reduction in the viscosity and an increase in the copper number under heat treatment. A maximum value of approximately 0.6-0.75 has been found to be satisfactory.

   The amount of moisture present (Figure 2.39) ranges from zero to approximately five percent by weight of the dry paper. The obvious need is to remove the moisture. This and other tests have proved that reducing moisture content will substantially increase life expectancy.

   F.M. Clark found conclusively that moisture, *not* air, is the contaminant most to be feared in the use of cellulose insulation. It is the moisture that primarily reduces the stable life of cellulose.

   A closely related influence on aging is the presence in the insulation structure of materials which restrict the free movement of water through the insulation, such as metal stress-grading shields, aluminum sheet around the secondary winding with paper between layers, or in older units, heavy varnish films. These may prevent the removal of water during processing, which increases the rate of aging still further.

More Practical Effects on Transformer Life

Transformers are like people—they are all the same but they are different. Therefore, scientific evaluation of the behavior of deteriorating cellulose insulation is not that simple. Nonetheless, various studies have shown those four factors affect the cellulosic insulation and consequently the transformer life:

- Reduced tensile strength
- Reduced impulse strength
- Reduced short circuit withstand
- Reduced winding compactness (shrinkage)

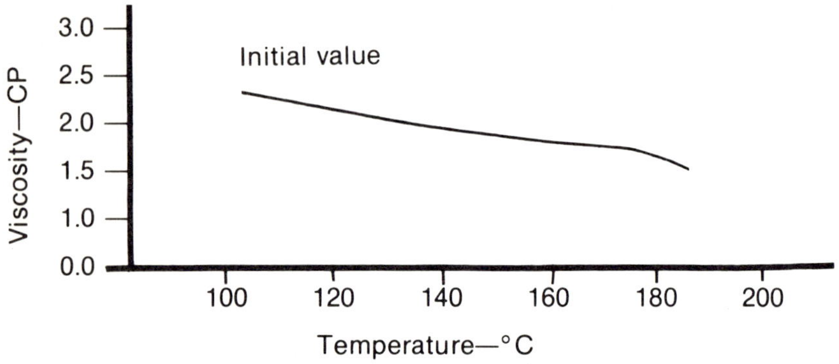

*Figure 2.38 - The change in the viscosity of cuprammonium solutions of cellulose as the result of four hours of heating at the temperatures shown.**

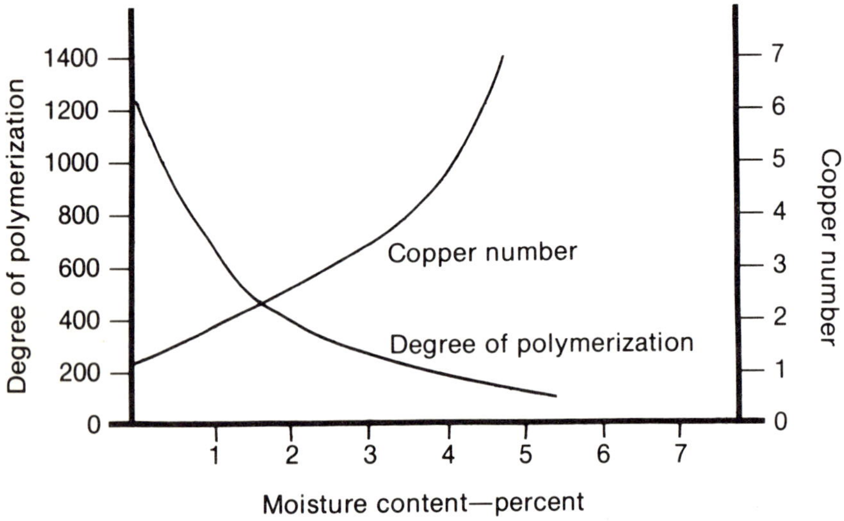

*Figure 2.39 - The effect of moisture on the properties of Kraft insulating papers when heated for 18 days at 140° C.**

F.M. Clark, **Insulating Materials...(1962).*

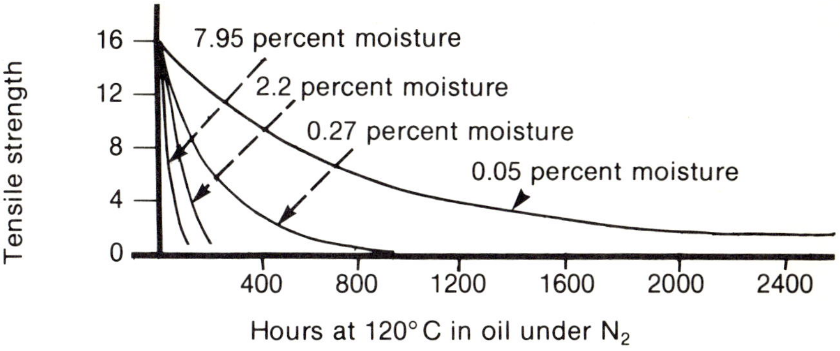

Figure 2.40 - Effect of moisture on loss of tensile strength of manila paper with aging. F.M. Clark, "Factors Affecting the Mechanical Deterioration of Cellulose Insulation," (1942).

Reduced Tensile Strength

When oil-impregnated cellulose is heated for long periods of time, loss of mechanical strength occurs. The tensile strength of the cellulose insulation has become the most accepted measure of transformer life.

Moisture, temperature, overheating, and oxidation are factors that, over a period of time (aging), have been found to affect the mechanical stability of the cellulose insulation. Let's consider each of these individually.

1. Moisture

   Small amounts of moisture—even microscopic amounts—accelerate deterioration of cellulose insulation. Studies made by F.M. Clark (Figure 2.40) and M.F. Beavers (Figure 2.41) show more rapid degradation in the strength of cellulose with increasing amount of moisture—even in the absence of oxidation.

   Such mechanical deterioration of cellulose insulation is not limited to aging temperatures above 100° C. Figure 2.42 shows that *manila paper* is sensitive to the effects of moisture even at temperatures as low as 75° C. While it is true that the results with Kraft paper would not be as pronounced as manila, the effect may, nevertheless, be substantial.

   Finally, moisture absorption by enameled wires can reduce dielectric strength. Most cases are temporary: for some enamels, a form of hydrolysis occurs; for others, embrittlement becomes evident.

Clearly, therefore, the presence of even very small amounts of moisture does actual harm to the solid insulation. Dehydration of the insulating system can reduce the rate of loss of mechanical strength, but the previous loss is never recovered.* Therefore, great wisdom lies in keeping transformers dry. This should be done before the moisture can be detected by the old tried (but untrue) methods of yesteryear!†

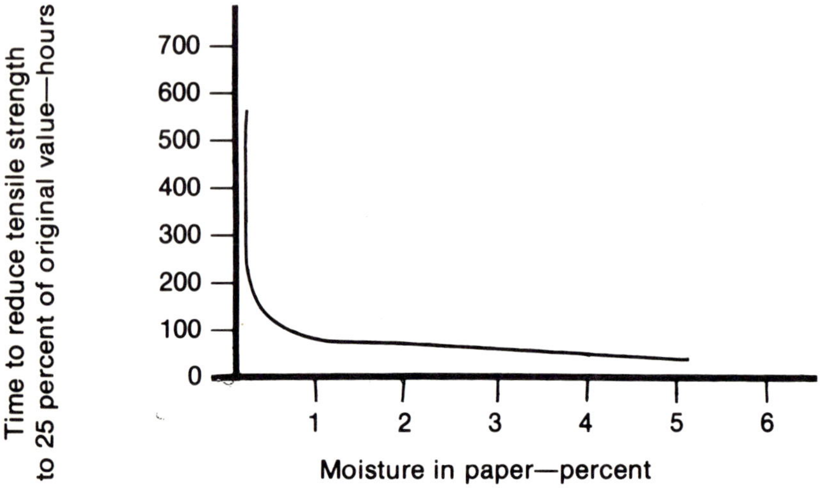

Figure 2.41 - This shows the effect of moisture on the aging period at 150° C to reduce the tensile strength of 0.003 inch manila paper to 25 percent of its original value! M.F. Beavers, "Distributing Transformers Surpass ASA Guides," (1959), and A. Lockie, ***IEEE Tutorial Course***, (1976).

2. Temperature
   Transformer users today are loading their equipment, especially distribution units, in excess of nameplate rating, often on a *daily* basis. Studies have shown the rapid decrease in tensile strength of cellulose paper in oil even while under nitrogen (absence of oxygen) at *elevated* temperature, (Figure 2.42). F.M. Clark found a similar relationship using the same conditions of aging.

   The unusual and unexpected conditions of aging are most important in affecting the deterioration rate. For example, Dr.

**Chapter 7, Part 2.*

†Chapter 3, Part 3.

Dakin has cited the example that with the addition of insulation pressboard and cotton tape sealed in the tube with the paper tape, the decrease in tensile strength was less rapid (Figure 2.43). That means, the more solid insulation used in a transformer, the greater the resistance to aging. Unfortunately, the trend today is toward *reduced* insulation and therefore increased rate of aging.* As an owner/operator, therefore, this becomes an important concept to factor into the preventive maintenance program for your plant. Newer design transformers need more work and much earlier than the older designs.

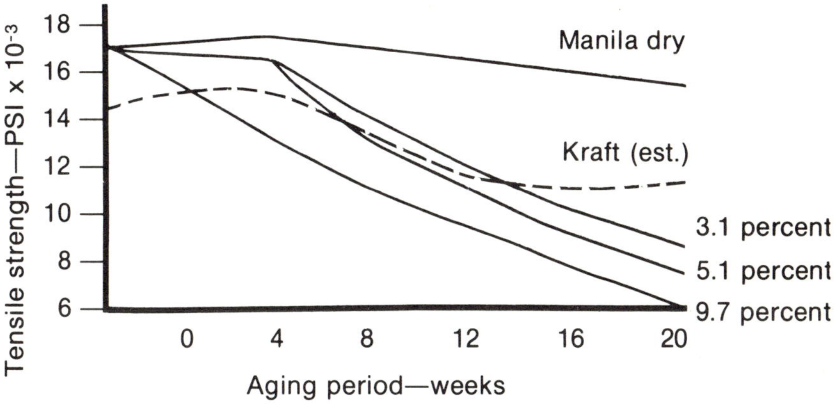

Figure 2.42 - The aging of manila cable paper (Kraft paper) as affected by the presence of moisture when sealed in glass tubes and heated at 75° C in the presence of nitrogen. The moisture contents of the papers tested are indicated on each curve in percent by weight. F.M. Clark, ***Insulating Materials...***, *(1962).*

Figure 2.44 illustrates the effects of the combined factors of high temperature and 7 percent moisture. The 12-week aging characteristic of *dried* insulation is substantially free from degradation at 75° C and 100° C. By contrast the *undried* insulation has lost 50 percent of its initial tensile strength after an equal period of aging at 100° C. Thermally upgraded Kraft paper loss would again show much less drastic reduction in tensile strength. Once again the data presented illustrates only a general relationship and not the kind of results that would

**Chapter 1, Part 3, Figure 1.72.*

provide exact numbers today. Much study is obviously needed and, fortunately, is even now being vigorously pursued by research and development groups worldwide.

This effect of high temperature (though certainly dated) still serves as a warning against the use of excessive temperatures during the drying and impregnating treatment of the insulation at the factory.

Again recall Figure 2.32. Values derived from this study and with data taken from the ANSI Loading Guides (C57.92-843) result in the "Loss of Life with Temperature Chart," Table 2.7. When the same data is plotted as Figure 2.45, it reveals even more vividly the accelerated reduction in transformer life as the temperature is increased above 85° C.

Finally, at today's higher operating temperatures and lower oil use per KVA rating, moisture becomes that much more detrimental to the insulation system. Figure 2.46 shows the effect of increased moisture content when manila paper has been aged at 75° C and 120° C.

3. Oxidation
   The third factor that reduces the mechanical strength of cellulosic insulation is acidity caused by oxidation.

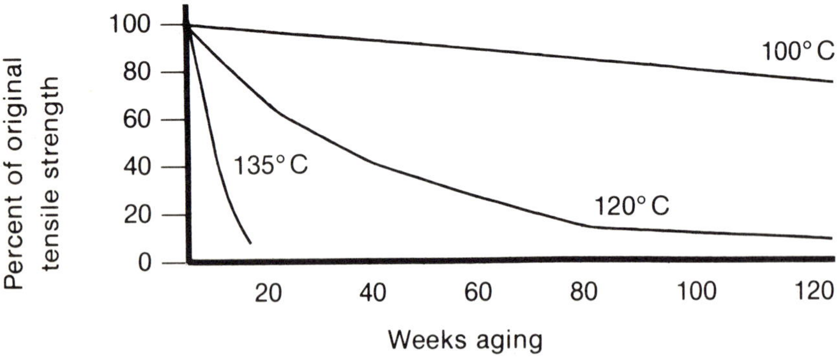

Figure 2.43 - Decrease in tensile strength of cellulose paper in oil under nitrogen with time of aging at elevated temperatures. T.W. Dakin, "Electrical Insulation Deterioration Treated as a Chemical Rate Phenomenon," (1948), and A. Lockie, ***IEEE Tutorial Course****, (1976).*

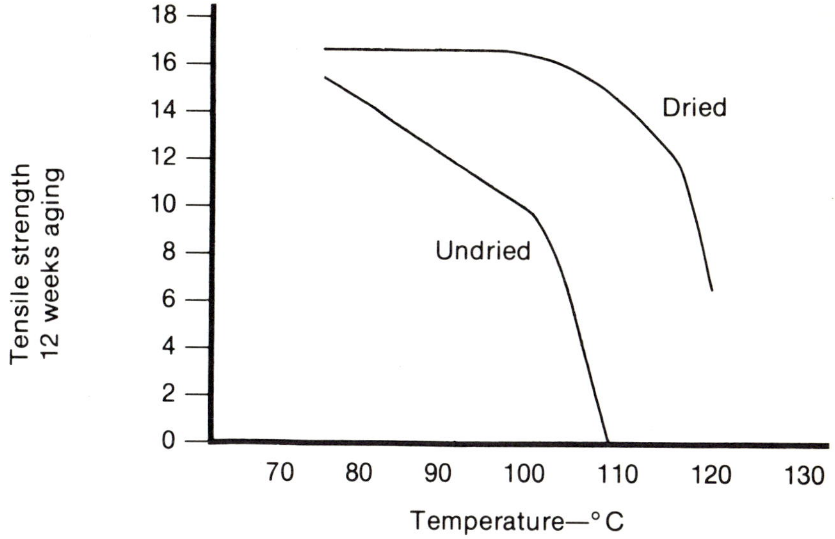

Figure 2.44 - The aging characteristic of oil-impregnated manila paper in sealed glass containers, as affected by the presence of 7 percent moisture. F.M. Clark, ***Insulating Materials...****, (1962).*

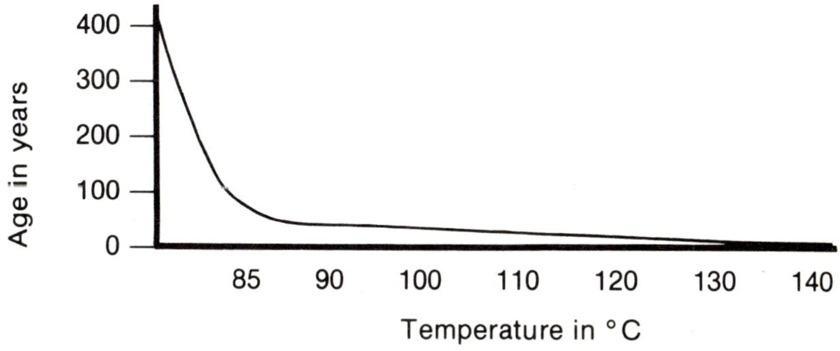

Figure 2.45 - This curve relationship has been plotted from data in Table 2.7 and extrapolated to end of life.

TABLE 2.7

INSULATION AGING OR LOSS OF LIFE WITH TEMPERATURE[1]		
	Time to use up 1 percent of insulation life if the end point of life is a loss of tensile strength of:	
Temperature °C	80 Percent	20 Percent
95	2620 hours	308 hours
99	1400 hours	168 hours
104	610 hours	86 hours
109	340 hours	47 hours
115	202 hours	27 hours
119	143 hours	19 hours
124	67 hours	12 hours
134	20 hours	3 hours
142	10 hours	1.4 hours
150	5 hours	.5 hours
175	4 hours	.06 hours

[1]R. Bean et al., **Transformers for the Electric Power Industry**, (1959).

The curve relationship in Figure 2.47 relating tensile strength of manila paper to oil acidity shows the devastating effect of high acid numbers. The effect would not be as drastic with Kraft or thermally upgraded papers as with manila, but nevertheless, it remains substantial.

> While we are prone to look at the worst condition (that is, 50 percent loss of life at an oil acidity of 0.30 mg NaOH/gram) the loss of life at an oil acidity of 0.10 mg NaOH/gram shows a reduction to 65 percent of the original value.

Further study has indicated that although the oil-formed acids have a degrading effect on the mechanical strength of cellulose insulations, the presence of oxygen at the same instance becomes a powerful one-two punch and certainly creates conditions that accelerate the aging effect.

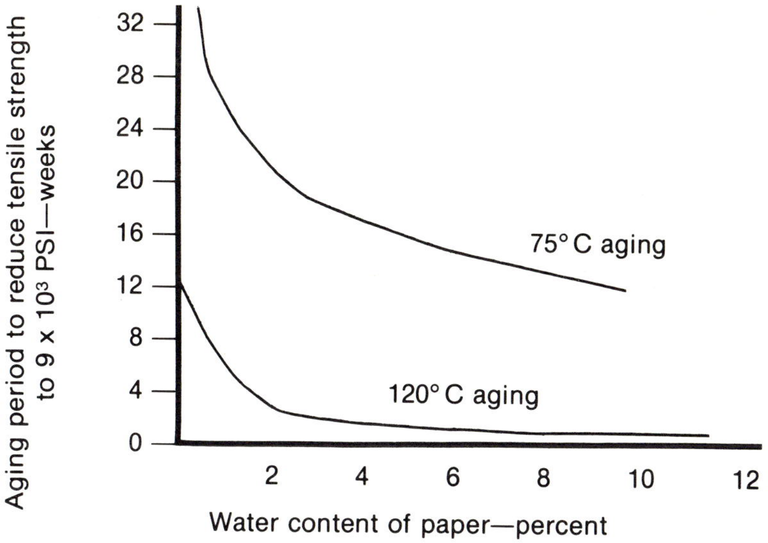

*Figure 2.46 - Showing the effect of moisture on the period of time necessary to reduce the tensile strength of manila cable paper to 9 × 10^3 psi when aged in nitrogen filled sealed glass containers at 75° C and at 120° C.**

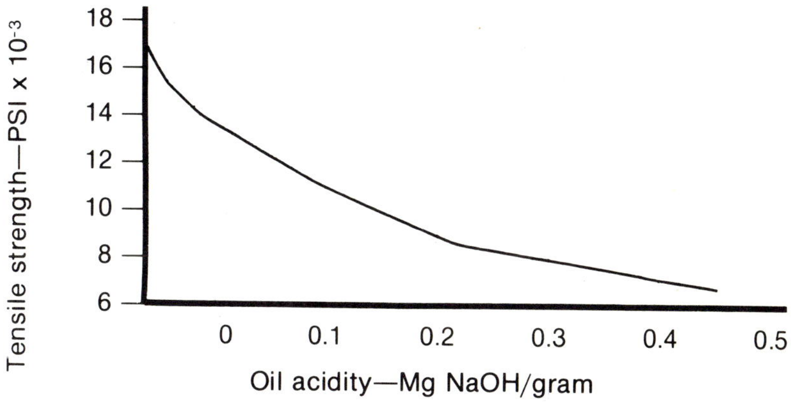

*Figure 2.47 - Illustrating the relationship between the tensile strength of vacuum dried and oil-impregnated 0.003 inch manila paper and the development of oil acidity as a result of oxidation in open containers with a free supply of air at 100° C.**

F.M. Clark,* ***Insulating Materials...**, (1962).*

Recent studies concerning new oils vary in effects of oxidation on cellulose degradation. Keep in mind that newer oils will likely contain greater amounts of paraffinic crudes.*

Transformers now also function under higher operating temperatures—a factor that is inherent in the newer designs of transformers and may not be something the user can control. Therefore, it behooves the user to minimize the factors of moisture and oxidation, which are within his control and control the temperature of the transformer using every means at his disposal.

Reduced Impulse Strength

"Impulse strength" relates to the ability of the insulation system to withstand a surge voltage and retain its original characteristics within described bounds. The impulse strength is 2-2½ times the normal electrical stress under operating voltages.

Increased temperature, thermal aging, and moisture drastically affect the impulse strength of the transformer, and shorten life expectancy.

1. Temperature
   Studies under service-related conditions have shown that the impulse strength decreases as the operating temperature increases. The "hot spot" temperature was selected as the *most* significant indicator of the thermal state of the unit, since the hot spot temperature is the weakest link in the chain. Therefore, limiting the hot spot temperature to a more realistic figure than manufacturer's design is one way of preserving the unit's impulse strength.†

2. Thermal Aging
   Deteriorating cellulose insulation due to aging (high heat plus long time duration) will still show good dielectric strength, but as it ages the insulation becomes brittle. A surge voltage comes along, cracks develop in the cellulose structure, and the game is over. The unit cannot handle the impulse voltage and it will fail.

**Chapter 2, Part 2.*

†*Chapter 7, Part 4.*

3. Moisture
   Moisture is also a factor in the reduction of basic impulse strength. The moisture content of oil-impregnated pressboard has a curve relationship to the insulation power factor (Figure 2.48). In Figure 2.49 the impulse strength in turn varies inversely with the moisture content. In addition, Figure 2.50 shows that a decrease in voltage impulse strength is clearly evident as soon as the moisture content of the paper is substantially greater than 0.13 percent.* So you see that water even in minute amounts is clearly trouble. Moreover, running the classic D-877 dielectric test is not the proper yard stick to measure the problem.†

Reduced Short Circuit Withstand

While modern arresters protect equipment against high *voltage* lightning strikes, less protection exists against high surges of *current*. Therefore, reduced short circuit withstand may be considered today a more critical factor than impulse strength.

1. Temperature
   During a switching surge or short circuit of one to two seconds, a mechanical jolt occurs in the transformer. The temperature may go as high as 500° C for a short period of time as the result of a short circuit. Since insulation deterioration is cumulative, and mechanical losses can't be regained, the conclusion is obvious.
2. Thermal Aging
   Again, *but even more important* than it is for impulse strength, a transformer cannot continue to handle short current and switch surges, unless proper maintenance steps are taken.

Reduced Winding Compactness

Keep in mind that heat creates two problems: embrittlement of cellulosic material, and shrinkage of cellulose.‡ This results in a loose transformer structure, free to move under impulse or through-fault—resulting in damage to the insulation and ultimate failure!

**A moisture content of 0.13 percent is especially low, even for new U.S. large power transformers.*

†Chapter 3, Part 3.

‡Chapter 1, Part 3.

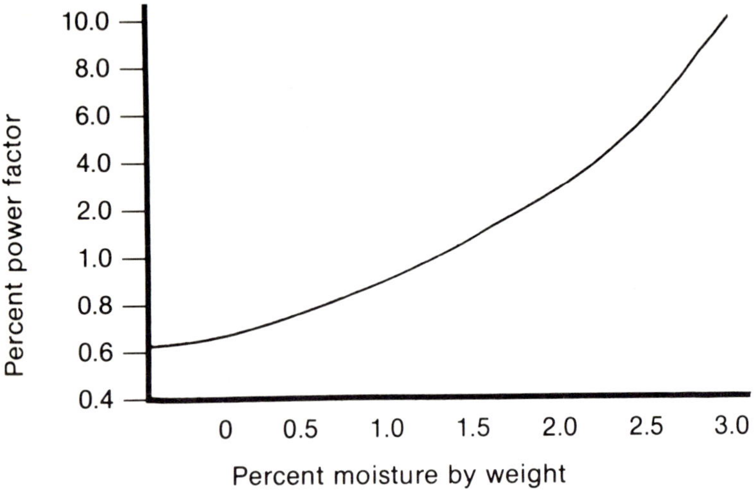

*Figure 2.48 - Variation of power factor with the moisture content of oil-impregnated pressboard.**

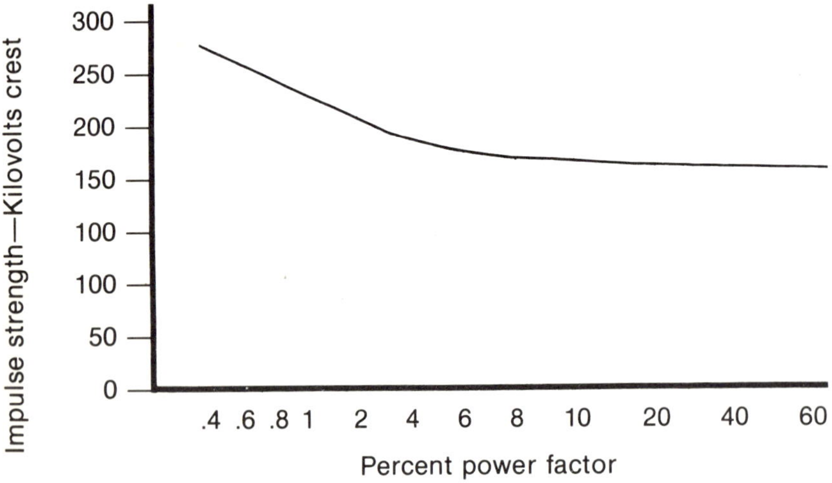

*Figure 2.49 - Impulse puncture strength of oil-impregnated pressboard.**

**W.L. Teague, "Dielectric Measurements on New Power Transformer Insulation," (1952).*

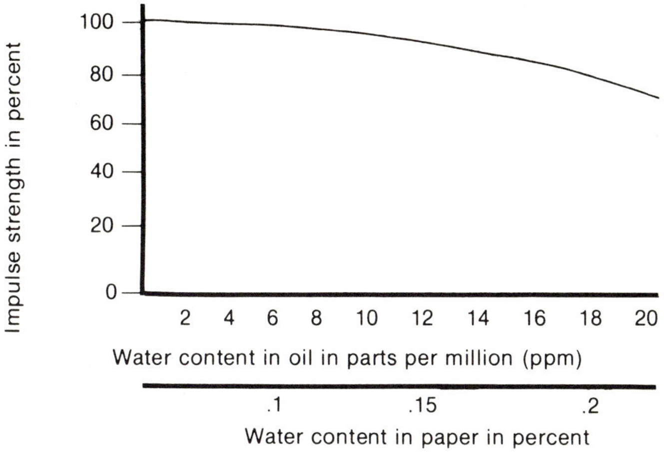

Figure 2.50 - Impulse voltage strength of paper. H. Moser, ***Transformerboard,*** *(1979).*

Where Does the Water Come From?

Introduction

Operating experience with transformer insulation over many years has shown that moisture in microscopic amounts—*not* gallons—is the cause of more electrical breakdowns than any other impurity.

Recognition of the importance of extremely *small* amounts of moisture has grown immeasurably with the increase in voltage stress and reduced BILs. This truth was first published in a 1902 AIEE paper.

Indeed, moisture "constitutes a hazard not only to the dielectric performance of the oil itself but also to insulations that are immersed in the oil." Thus, we should all agree that water is "enemy number one." What is not known or generally accepted is that a water problem may begin before ASTM standard dielectric test D-877 reveals it. The late F.M. Clark has also stated that:

> The electrical user...recognizes the dielectric hazard which is presented by the presence of water but rarely does he recognize the equally disastrous effect which *traces* of moisture have on the long-time usefulness of the insulated equipment, even when the amount of moisture is *not* sufficient to cause dielectric difficulties.

Factory and Installation

It may take up to two years to build a large transformer and during that time the unit may collect up to ten percent moisture by dry weight of the insulation. Therefore, initial drying is required prior to assembly and ultimately shipping.

Consensus opinion says that the paper insulation of a new transformer leaving the factory may contain between 0.3 and 1.0 percent water by dry weight. Thus, a certain amount of water *is always* present and can be readily accommodated by the transformer. Therefore, the purchaser of the new equipment should not tolerate an upper limit of 1.5 percent.

Indeed, it is recognized that a transformer will perform to all its design criteria and nameplate rating with a moisture content as high as 1.5 percent by weight. Nevertheless, this higher water content does affect the longevity of the transformer. An analogous situation would be like adding a 50 pound weight on the back of a world-class marathon runner. When he drops out of the race at five miles and cant't finish the 26 mile, 385 yard race, the reason is obvious—too much handicap! It's the same with 1.5 percent moisture content in a transformer. It won't finish the race.

Figure 2.51 illustrates how critical small quantities of moisture are to the dielectric strength of transformer mineral oil, as compared with silicone and askarel (polychlorinated biphenyl or PCB).

Table 2.8 shows that the water in solution in the oil even for 20 ppm is still less than one-tenth the residual water in the paper for very large (EHV) power transformers.

To minimize the entrance of moisture during transit from the manufacturer, a new transformer tank is filled with dry gas (such as nitrogen) under pressure. Moisture content should remain essentially the same as at the time the unit was shipped. If the gas pressure and oxygen content are not the prescribed values, high probability exists that the transformer has been contaminated with atmospheric air or moisture. In such a case, as part of the installation procedure, a dry-out becomes necessary.*

**Chapter 7, Part 2.*

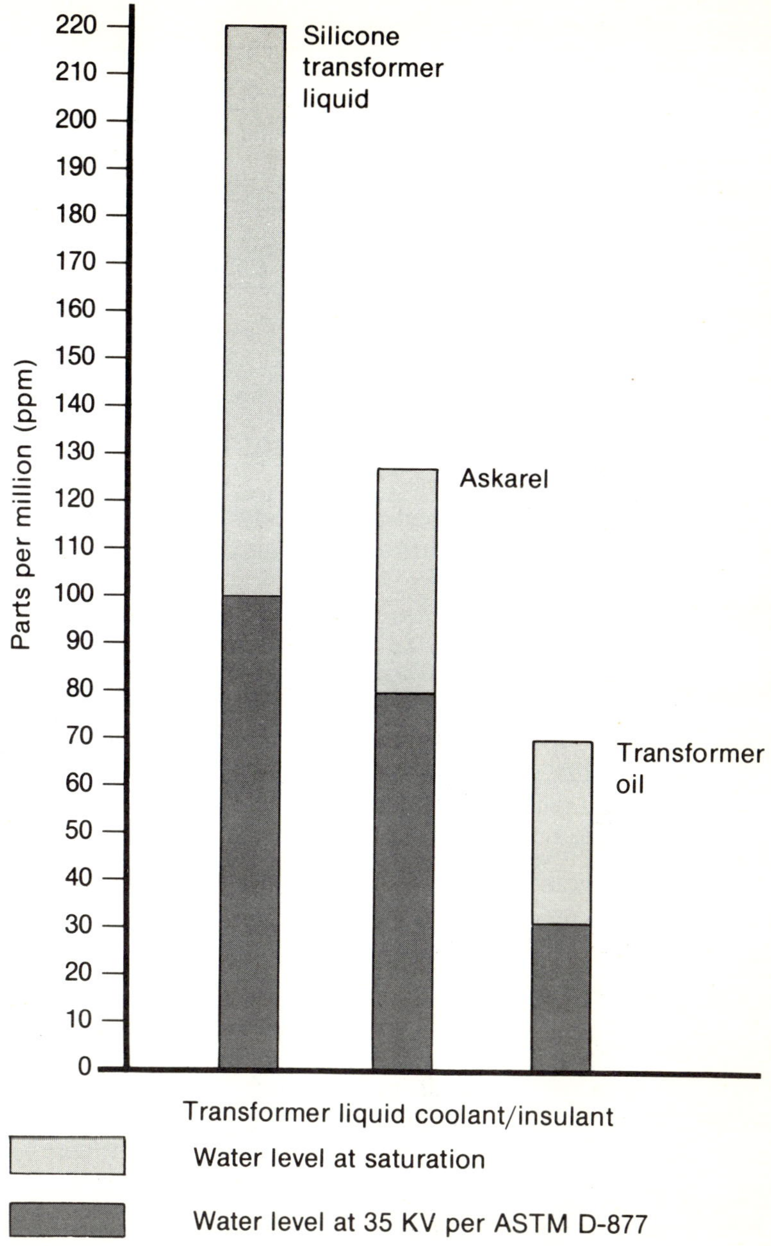

Figure 2.51 - Acceptable limits of moisture content in various dielectric fluids (medium power transformers). Dow-Corning Corp. ***Silicone 561 Instruction Manual*** *(1979).*

TABLE 2.8

COMPARISON OF WATER IN OIL AND IN CELLULOSE INSULATION[1] (Very Large Power Transformers)

EHV Power Transformer	Cellulose Insulation (pounds)	Oil (gallons)	Residual Water in Cellulose Insulation		Water in Oil Parts Per Million	
			0.1 percent (gallons)	0.2 percent (gallons)	5 ppm (gallons)	20 ppm (gallons)
A	15,000	10,000	1.8	3.6	0.05	0.18
B	25,000	15,000	3.0	6.0	0.07	0.27
C	30,000	20,000	3.6	7.3	0.09	0.36

[1] P.L. Bellaschi, Doble Engineering Co., (1967).

When a large power transformer leaves the factory, hopefully it has been dried sufficiently so that the residual moisture is typically 0.5 percent of one percent. Nonetheless, moisture can enter into the insulating system in these three ways:

1. External Sources
   - Opening the transformer tank during installation or maintenance which results in exposing the unit to the atmosphere.
   - Accidental leaks in the tank—around lids, bushings, or inspection cover gaskets.
2. Partly External Sources
   - The precipitation of water from the atmosphere or a workman's breath on inner tank surfaces (lid and side walls in the air space).
   - Moisture condensation on the tank, dripping back on a terminal board, tap changer, leads, or through the oil onto and into the cellulosic paper insulation.
   - Temperature changes in transit plus leaks or loss of nitrogen because cellulose is an excellent desiccant (moisture absorber).
3. Internal Reactions
   We have already established that as the insulation system thermally ages, water is produced. Nothing can prevent this gradual deterioration process:

- The decomposition of the cellulose—the aging of insulation resulting from heat—as the combination of hydrogen and hydroxyl (OH) groups.*
- As one product of oil degradation—the chemical reaction of the oxygen present with hydrocarbons.†

Several laboratory studies have shown that when cellulose degrades, moisture is produced. An applicable everyday analogy is what happens to old newspapers used as a shelf-liner—embrittlement occurs after several weeks, even at room temperature.

Degradation may be expressed in terms of the degree of polymerization. Recalling that the deterioration rate of the paper depends mainly on the temperature, Figures 2.52 and 2.53 then give a picture of what happens when *heat* liberates water from the cellulose. Clark, Fabre, and Pickon made this startling observation—water liberated from the paper accelerates the rate of cellulosic degradation, otherwise dependent solely on temperature. Analysis has thus shown that water content may fall in a range between 0.5 and 2.5 percent.

When oil oxidizes, it also produces water as a degradation product together with acids, sludge, and other polar compounds. The relative proportions of these products depend on the oil and the aging conditions. If it is supposed that a typical transformer contained ten tons of paper and pressboard and 8000 gallons of oil, then in oxidizing to 1 mg KOH/gram the oil would produce between 2 and 20 gallons of water. This quantity would increase the water content of the paper between 0.1 and 1.0 percent by dry weight. Therefore, the minimum quantity of moisture from the oil and paper together totals between 0.6 and 3.5 percent.

The significance of this compounded effect is that at 2 percent total water content the rate of aging is between 6 and 16 times and at 4 percent 12 and 45 times that at the initial 0.3 percent water content.

As a typical example, during a 25-year period the moisture accumulation may be 3.3 percent by dry weight of insulation (Table 2.9).

**Chapter 1, Part 2, Figure 1.45.*

†*Chapter 2, Part 3 and Part 5 (Figure 2.32).*

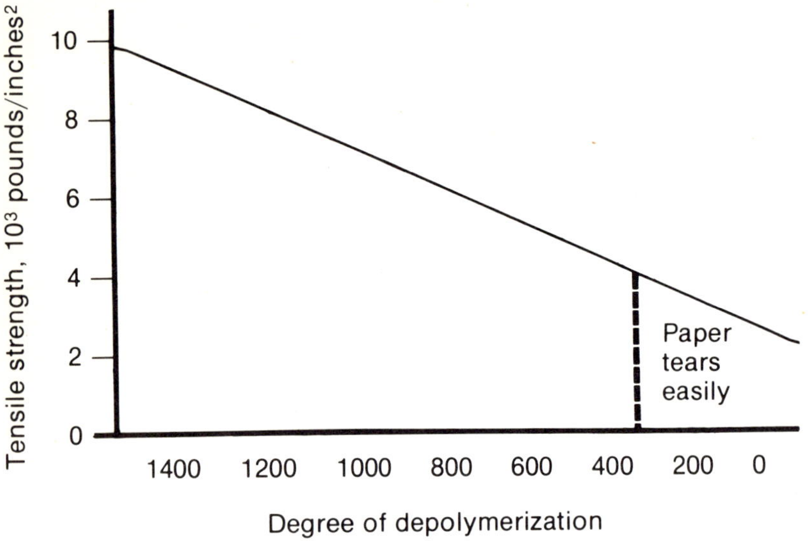

*Figure 2.52 - Relationship between tensile strength and degree of depolymerization.**

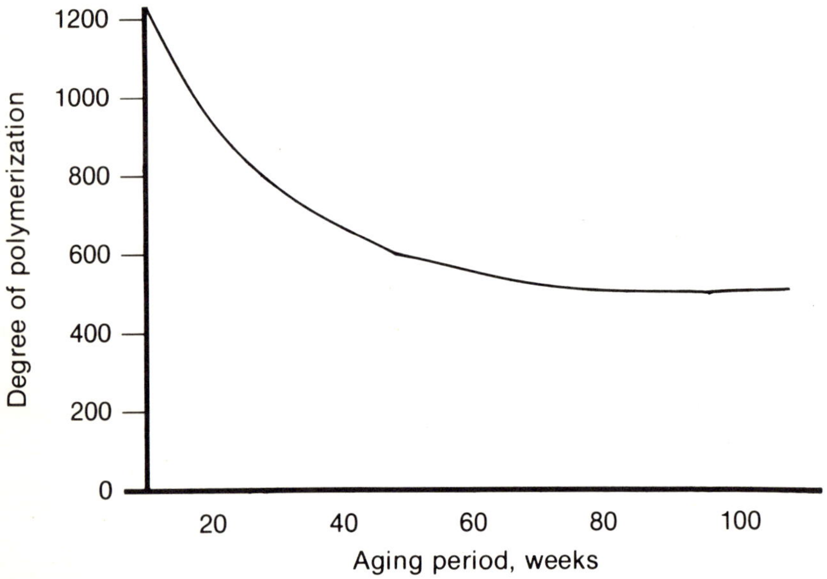

*Figure 2.53 - Aging of paper in model transformers at 85° C top oil temperature (paper at about 100° C).**

A.W. Stannett, **Electrical Times (1965).*

TABLE 2.9

TYPICAL MOISTURE ACCUMULATION IN A MEDIUM POWER TRANSFORMER	
Water Source	Percentage Dry Weight
Left in by Manufacturer From paper insulation[1] From transformer oil[2]	0.3 percent[3] 1.0 percent 2.0 percent
Total	3.3 percent water by dry weight

[1]Product of aging.

[2]A product of oil oxidation (with two oil changes).

[3]If the manufacturer had not dried the transformer sufficiently, so that residual moisture content was 0.3 percent, the total water content from all sources would be 4.5 percent—a complete disaster!

Characteristics of Water Affecting Insulation

What are some of the general characteristics of water in oil or paper that may affect the life of transformer insulation? Keep these in mind:

1. Water is always present in commercial insulations. In view of the sources this is to be expected.
2. Water (with air a standard of comparison) is a polar liquid having a high permittivity or dielectric constant, varying between 79.5 and 81 (pure water only). By contrast the dielectric constants of most commercial insulations range from 2.0 to 7.0.* Therefore, water is attracted to areas of strong electric field. It is not uniformly distributed throughout the insulating system of the transformer, but in fact may concentrate in the most dangerous parts of the system.

Chapter 2, Part 2, Table 2.2

3. Water is strongly electropositive and attracted to negatively charged electrodes and repelled from positively charged electrodes.
4. Water's electrical conductivity is largely due to the presence of polar contaminants, such as salts and acids.
5. Water is a universal solvent except for fats and sulfur. This solvent property of water can be used to advantage. The freshly formed oxidation products are water-soluble rather than oil soluble. To what extent? Better than 10 to 1, so that water removed will tend to wash out these fresh acidic volatile materials.
6. Water-soluble acids produced by oxidation of the oil will combine with water left in or with oil to assist or promote corrosion to exposed metal parts in transformers; as for example, on the underside of manhole covers. Rusting also occurs on the metallic parts of a transformer, even though covered with oil. Thus, it is of paramount importance that water be removcd regularly.
7. Water is a catalyst (a substance that accelerates a chemical reaction without itself being consumed) for almost all reactions. Nearly all reactions require at least a very small amount of water.
8. Water forms corrosive nitrous and nitric acids in combination with the gases released by the ionization of air during partial discharges.
9. Oxidized oils contain water-soluble polar groups which orient themselves on a water surface so the hydrophilic (meaning love of water) group is in contact with the water. The result of this inner action weakens the strength of the film at the interface of the water and oil. The reduction in the interfacial tension (IFT) of a transformer oil is an indication of the early stages of the oxidation of an oil.*
10. The cellulose has greater affinity for water than oil (hundreds of times). Water will replace the oil in oil-impregnated cellulose (moisture migration). Thus, the cellulose is sucking up the moisture because cellulose is hygroscopic while the oil is get-

**Chapter 3, Part 1, Table 3.1.*

ting rid of the water because it is hydrophobic; that is, the oil repels water.

Oil and Water *Do* Mix!

A familiar cliche says, "Oil and water do not mix." Undoubtedly, you have heard it many times but..."it ain't necessarily so"! It utterly fails to be true in the engineering sense. Oil and water *do* mix and in a number of ways. The full scope of moisture consists of free water, suspended water, dissolved water, and "chemically bound water."

1. Free Water
   Free water is generally present in many shipments of new, clean insulating oil whether in drums or tank cars. It is also frequently found at the bottom of transformer tanks and is relatively harmless at this point. The water can be seen with the naked eye when drawing a sample, provided the sample valve is located on the exact bottom of the unit.

   While the transformer is clean and has a high IFT (30-50 dynes/cm), water entering the unit will show up very rapidly in the bottom of the transformer because the water is heavier than the oil. If the free water did not have access to the insulating system of the transformer it would be relatively harmless in this location. (Murphy's Law says that some free water will end up in the cellulose. This signals the onset of an inevitable moisture problem.) While it is true that the oil in immediate contact with the free water on the bottom of the transformer will be water saturated at the lower temperature, it still is relatively harmless to the transformer. This remains true so long as the water goes down the sides of a unit rather than splashes over the top of a transformer's core and coils.

2. Water in Suspension (Emulsified)
   Moisture in oil free from solids in suspension is not a serious problem. When an oil is new and free from contaminants, water will quickly settle to the bottom of the tank.

   The problem of moisture becomes significant in the presence of cellulosic fibers and polar contaminants dissolved in the insulating liquid. When contaminants including oxidation products, fibers, or other debris are present in oil, water tends to cling to such contaminants or be absorbed by them. This water in suspension is closely associated with oil decay products. Under this circumstance water and water-soluble acids will combine with acids and sludges in an oxidized transformer oil.

Water in suspension cannot be seen with the naked eye and is not detected with a flat disc D-877 dielectric test.

If an oil is oxidized to any extent, any water coming into the transformer will partially be absorbed into the oil decay products. As the decay products build up in the oil, the surface tension of the water or the interfacial tension between the water and the oil is lowered dramatically. These oil decay products align themselves at the water-oil interface, reach into the water, and draw the water into existing decay products; they are thus hydrophilic (love of water) and will continue to suck the moisture. This heavier decay molecule will then recirculate throughout the entire transformer and will find its way into the cellulosic insulation, or into areas of high electrical intensity, thus causing a reduction in the insulation resistance. Here is another effect of Murphy's Law: the water saturated oil decay molecule has a preference for the coolest part of the transformer (bottom and fins) and areas of highest electrical stress. Aged oil holds much more water in solution than new oil and hot oil more than cold; therefore, hot aged oil contains more water than cold new oil.

The necessity to identify this problem very *early* in the history of the transformer is most apparent. Running of a power factor test on the oil is one method of detection.* A high power factor test on transformer oil will indicate a *problem*, and one of the causes could be water trapped in oil decay products; that is, water in suspension.

Consider now the most degrading form of moisture—water in solution or dissolved water.

3. Water in Solution (Dissolved Water)

   a. The established fact
      That oils could actually dissolve moisture has still not been completely recognized, even though this fact has been established for over 60 years.

**Chapter 3, Part 1, Table 3.1.*

F.M. Clark stated that the presence of *dissolved* water in oil "constitutes a hazard not only to the oil itself but also the cellulose immersed in the oil." The amount of water that may be dissolved is primarily a function of the temperature and the condition of the oil. This reminds us that the primary conditions which affect solid insulation are temperature and moisture.

It might prove interesting to study Figure 2.54 to follow the moisture in cellulosic insulation under heated conditions. It shows the variation of the insulation resistance with temperature. The increased resistance of a given path through the dielectric at lower temperature...is exaggerated by the presence of *free* moisture...

> If a piece of insulation is gradually heated and the moisture is free to escape, the resistance *decreases* at first, because of the moisture being vaporized and again *increases* because of the vapor being expelled at the higher temperatures. After the expulsion of the moisture the material begins to change in nature and ultimately begins to carbonize and behaves as a high-resistance conductor. The resistance then falls rapidly to a very low value...While this curve is typical, the exact temperature at which the minimum and maximum resistance values occur depend on the area of the electrodes, the amount of moisture content and the facility with which it may escape and on the chemical and physical changes that occur due to the temperature.

Relationships between the dissolved moisture content and each of the associated media—cellulose, oil and air—are normally stated in terms of temperature and parts per million (ppm). An analogy may assist in trying to picture in your mind the quantity in ppm of dissolved water in a given insulating medium: one part in a million is comparable to one minute in 1.9 years, or one hour in 114 years!

Figure 2.55 shows typical plots of the solubility of water in new, unaged insulating oil. Note line A stands for today's U.S. oils that contain 25-35 percent aromatics in contrast to line B, European oils.

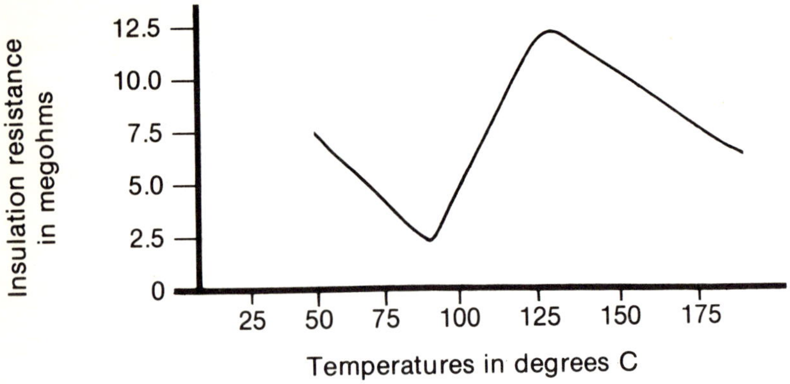

Figure 2.54 - Classic relationship—fibrous materials. E.C. Reed, ***The Electric Journal*** *(1918).*

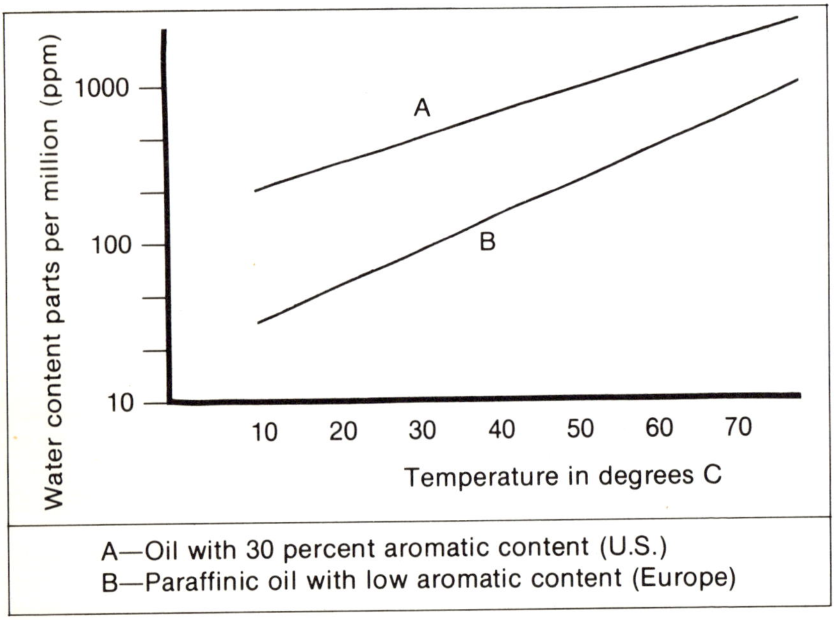

Figure 2.55 - Typical plots of solubility of water in new insulating oil against temperature. R.W. Sillars, ***Electrical Insulating Materials and Their Application*** *(1973), p. 214.*

b. Water saturation in transformer oil

A significant relationship between new oil and dissolved water has been found to exist. The plot of Figure 2.56 is called the "Cloud Point Curve." Three general statements can be made regarding this curve:

- All oils dissolve more moisture at higher temperatures than at lower temperatures.
- Cooling the unit gives oil the opportunity to unload its water.
- The greater the extent of refining operation, the less the solubility for water.*

The curve also indicates "soluble" water on the right of the curve and "free" water to the left, respectively. As an example, curve "A" shows that at 50° C oil has the ability to hold 100 ppm of water "in solution." If the temperature of the oil with 200 ppm of water is raised to 60° C, the oil would still be sparkling clear, and a standard dielectric (D-877) conducted at this temperature would give a high reading somewhere in the range of 40 or 50 KV. However, if the temperature of the 60° C oil would be lowered to 20° C, as the temperature drops to 50° C and lower, the oil would go from a "clear sparkling" oil to a "cloudy" oil. Therefore, the point at which the transformer oil goes from the soluble water area (clear oil) to the free water area (cloudy oil) is called the "cloud point" (ASTM Test Method D-1524); *ergo,* Cloud Point Curve!

1) Hot transformer oil holds more moisture

   Let's consider a transformer with free water in the bottom. The layer of oil next to the free water will be saturated with water. As the transformer heats up due to the load (especially above 40° C), so will the "water-saturated oil" heat up and at the new temperature, the oil will become unsaturated. This oil with water in solution will rise up through the transformer past the cellulosic insulation and eventually flow into the cooling fins.

2) Cooling the transformer

   As this oil and water combination comes down through the fins and cools, the water will precipitate out,

**Chapter 2, Part 2, Table 2.3.*

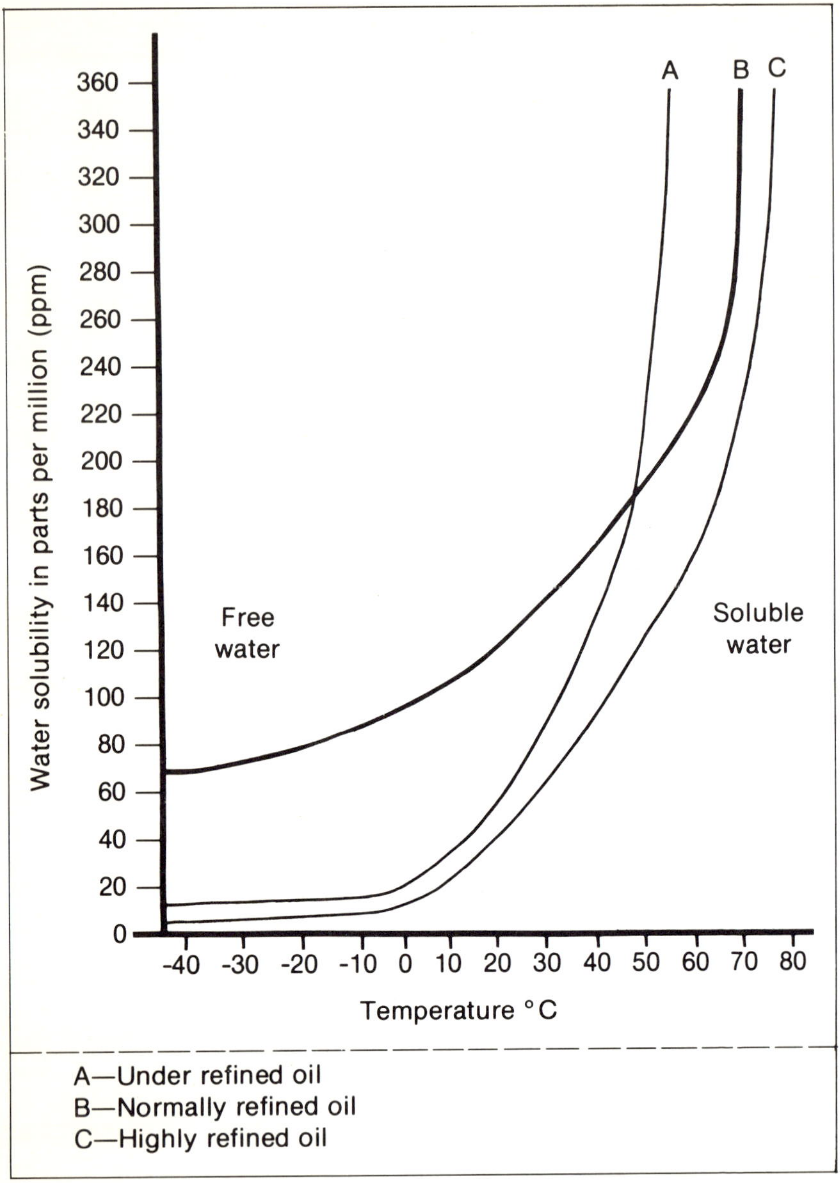

Figure 2.56 - Effect of temperature on solubility of water in new transformer oil ("Cloud Point Curve"). F.M. Clark, ***Insulating Materials for Design and Engineering Practice****, (1962), and Westinghouse Electric Co.,* ***Installation and Maintenance of Power Transformers****, (1975).*

recondense on the sides of the fins, and run down to the bottom of the transformer again.

> It is apparent if a hot transformer oil contained an appreciable amount of water in solution, this water (in solution) would have no (appreciable) effect on the overall power factor of the transformer (in solution) at the time of test. However, if the transformer was tested later, after the oil had cooled off, the overall power factor might be higher.

This is due to water phasing out of solution, being absorbed by the insulation. Free water brought on by cooling of the oil will end up in the insulation and also as water in suspension. The power factor of the oil and of the insulation will be seriously affected by the moisture. Inasmuch as decay products of oil have an affinity for the coolest part of the transformer and the decay products also attract water, the inevitable result is internal rusting that eventually results in leaks. Therefore, 95 percent of leaks in transformer cooling fins are from the inside of the fins to the outside. This statement would not necessarily hold true, however, when a transformer is located in an area of severe chemical fallout.

3) The effect of oil refinement
The higher the degree of refining, the less water can be held in solution for any given temperature. Generally, a more viscous oil dissolves more moisture than a less viscous oil, for example, lubricating oils dissolve more moisture than insulating oils (Figure 2.57).

Chemically Bound Water

It is a little known fact that a *very small* amount of water is a virtue in order to maintain the mechanical strength of cellulose. The cellulose contains a certain amount of "chemically bound" or absorbed water in the OH and H groups of the glucose molecule which may be released at high temperatures.* The present day trend toward transformer operation beyond nameplate ratings makes it necessary to keep this fact in

**Chapter 1, Part 2, Figure 1.45.*

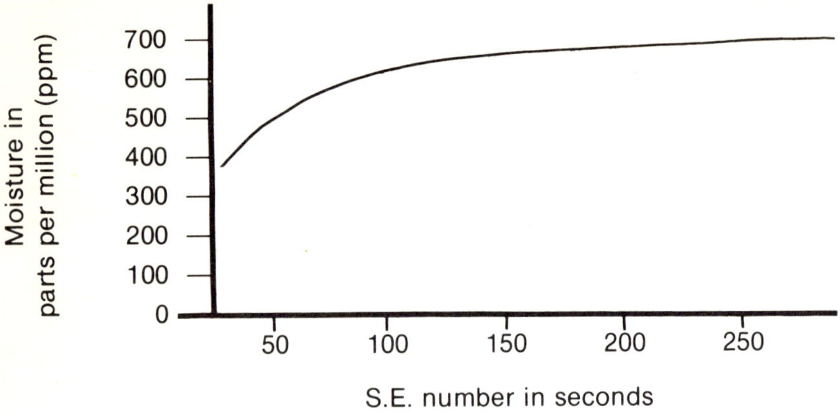

Figure 2.57 - Relation of steam emulsion number and water solubility in oil at 100° C. A.E. Flowers, "Some Effects of Moisture in Insulation Oils" (1939).

mind. As far back as the early 1940's the possibility was recognized that the beneficial "chemically bound" water would be released through excessive temperature and then lose its virtue altogether. This "chemically bound" water ends up as free water or water in solution.

Nonetheless, some moisture is required in the environment during the manufacturing of coils; if the humidity is too low, cracking and breaking of some materials may result.

Dissolved Water and Insulation Failure

1. Equilibrium and Imminent Danger
   Consider now the distribution of dissolved water between the insulating liquid and the cellulose, particularly at the point of equilibrium. This occurs in two substances. In a transformer these two materials are the transformer oil and cellulose insulation. The point of equilibrium is reached when the moisture content remains constant for at least 12 hours. No further change in quantity or distribution of the water takes place, regardless of the humidity or the number of days of exposure at a given temperature.

   Figure 2.58 shows what happens to the water as final equilibrium develops between various insulating liquids and air. F.M. Clark found that the three fluids exhibit a water solubility relationship which is proportional to the relative humidity of its environment.

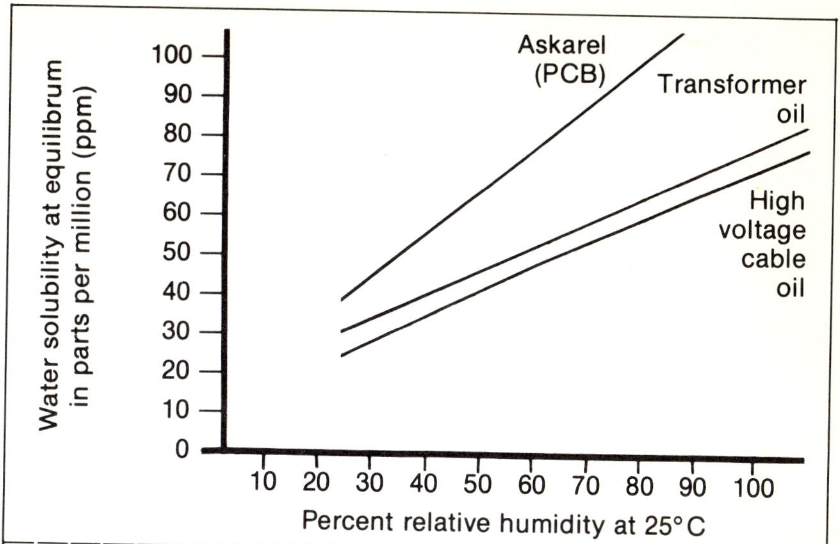

The maximum amount of water absorbed by transformer and cable insulating liquids exposed to humid air.

Figure 2.58 - Insulating liquids and humid air. F.M. Clark, "Water Solution in High-Voltage Dielectric Liquids" ***AIEE Transactions*** *(August 1940).*

The absorption of moisture by the cellulose insulation involves two stages:

First stage—The rate at which mineral oil absorbs moisture from the surrounding atmosphere depends on the intimacy of contact between the oil and the atmosphere. Figures 2.59 and 2.60 again show a proportional relationship between water content and exposure of transformer oil and askarel to humid air under these conditions:

Volume of liquid	7500 cc
Surface area exposed	375 sq cm
Depth of liquid	20 cm
Temperature of exposure	25° C

Second stage—When water is present in the oil or the cellulose insulation of a transformer at the relatively low temperature of 25° C, it tends to reach a common equilibrium, that is, moist oil gives up water to the dry cellulose (Figure 2.61) or wet cellulose gives up water to dry oil (Figure 2.62).

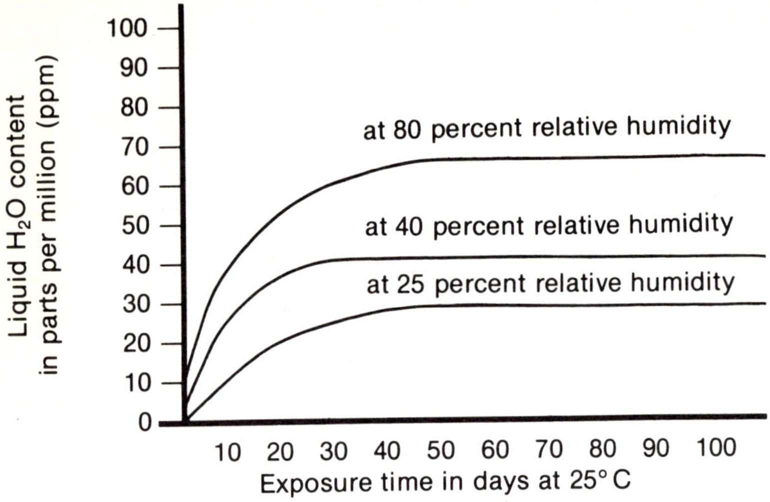

The absorption of H_2O from humid air by mineral transformer oil.

*Figure 2.59 - The absorption of water by mineral transformer oil.**

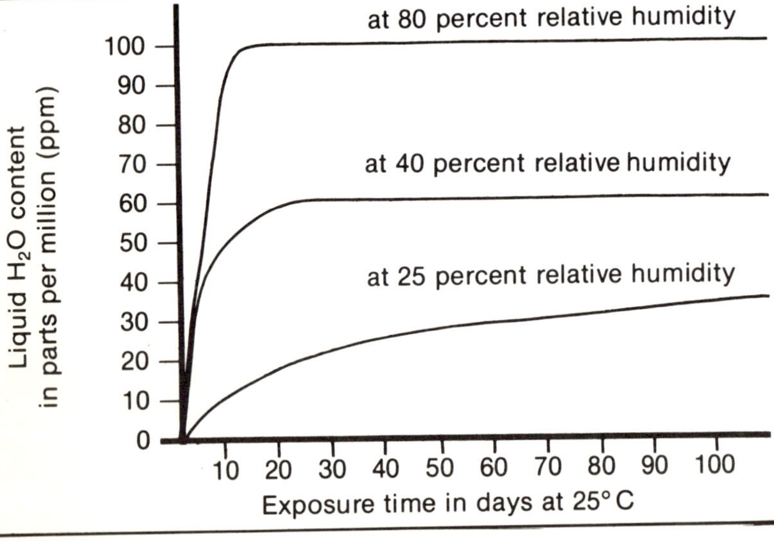

The absorption of H_2O from humid air by askarels (PCB).

Figure 2.60 - The absorption of water by askarels.†

F.M. Clark, "Insulating Materials for Design...**" (1962); †Clark (1940).*

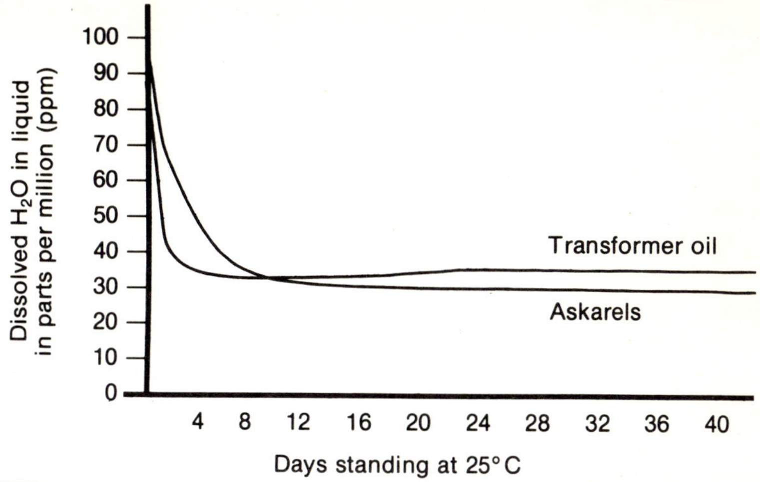

The dehydration of wet insulating liquids as a result of moisture absorption by dry cellulosic insulation immersed therein.

*Figure 2.61 - Drying liquids with dry insulation.**

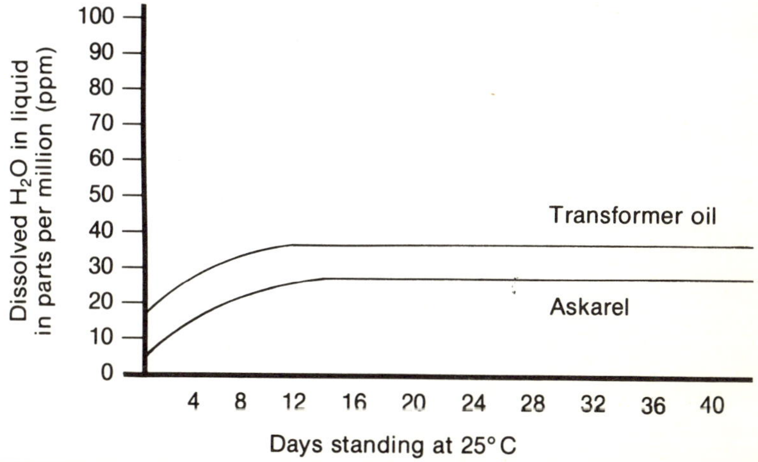

The absorption of water by dry insulating liquids in contact with cellulosic insulation previously exposed to humid air.

*Figure 2.62 - Water absorbed by oil.**

F.M. Clark,* ***AIEE Transactions* *(August, 1940).*

In a sealed container, once the equilibrium condition is reached between the water content of the cellulose and that of the insulating liquid, this equilibrium is not easily changed.

In this context, please note that in the manufacture of electrical equipment, this point of "equilibrium is rarely if ever achieved." These experiments of F.M. Clark and others were controlled life tests under laboratory conditions. We should note also that Clark reaffirmed the validity of these earlier tests (1940) in 1962.

> A dynamic situation is the order of the day. For example, in the case where temperature is increasing, the system moisture transfers from paper to oil, and when the temperature drops the system moisture transfers from oil to paper.

Again, we recognize the significance of the cloud point condition. Clark also found that equilibrium remains substantially unchanged when each set-up is heated from 25° C to 84° C liquid temperature.

All this has been said in order to focus attention on the *danger*, (Figure 2.63) which results when moisture is allowed to be dissolved by the insulating liquid. Thus, Dr. Clark speaking of the problems involved in drying out wet transformer insulation, concludes:

> When once the moisture is absorbed by the impregnated cellulose unless it is present in large amounts, it is released again only by drastic treatment, which involves the application of temperatures in the range of 95° C to 100° C and higher.

Recall the limiting temperature for Class 105 impregnated cellulose is 100°-105° C.* Clark's data was obtained using Kraft materials immersed in oil.

2. Equilibrium and Vapor Pressure
An equilbrium situation for the paper and oil system can also be expressed as a condition in which the water vapor pressure in the paper is equal to that in the oil at any given temperature.

The pressure of the saturated vapor in equilibrium with a particular medium is the water vapor pressure in that medium. This water vapor pressure expressed in millimeters of mercury (or

**Chapter 1, Part 2, Table 1.5.*

Torr), should be kept at a minimum. Normal atmospheric pressure is 14.7 pounds per square inch absolute, equal to 29.92 inches mercury absolute, and equal to 760 mm mercury absolute. The classic reference for moisture content of Kraft insulation as a function of temperature is the "Piper" chart, Figure 2.64.

The use of inert gas, such as nitrogen in enclosed transformers, has become the standard. This gas is used to minimize the exposure of the oil and paper from the moisture and oxygen of the air. An equilibrium condition develops between the moisture content of the fibrous materials and that in the gas space.

In this regard a modified Piper equilibrium chart has proved useful for predicting and interpreting the behavior of the moisture within transformers containing cellulosic material and a gas space. In using the chart, keep in mind that the vapor pressure is not the same as the pressure of the gas, but is only the partial pressure of the water vapor in the gas.

3. Ionization and Dielectric Breakdown
Ignoring dissolved moisture opens the door to additional chemical degradation. Such deterioration of the cellulose

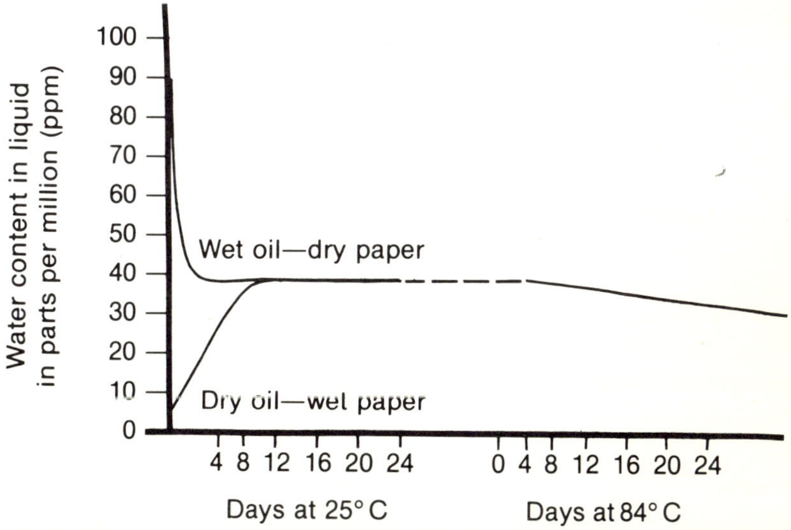

Figure 2.63 - The characteristics of the moisture equilibrium set up between mineral transformer oil and the impregnated cellulose insulation (1940).

results in the formation of additional moisture and acidic polar materials. The presence of these materials leads to ionization under electric stress.

Recall Figure 2.32 and note in Figure 2.65 several additional items in the block diagram. Not all the deterioration which leads to the failure of insulation can thus be ascribed to contamination. The second broad category involves ionization or partial discharge within the dielectric structure.

Thinking in terms of the cellulosic atomic structure, the freeing of one electron from a stable atom or molecule constitutes ionization. This usually refers to the progressive ionization involving a large number of electrons.

The term ionization does not necessarily require any form of electrical discharge whatsoever. Nevertheless, as the voltage stress is further increased, that is, as the number of free electrons continue to be released, an electron avalanche begins. Dielectric degradation already underway leads to more moisture evolution, dielectric shrinkage, and gas formation, such as carbon dioxide and carbon monoxide within the dielectric structure. This results in unequal dielectric stress distribution with greater concentration of stress within the void. The cellulosic air space becomes "ionized",* and at some point the ionization phenomena merges into a destructive type of a glow discharge, which is universally recognized by the term "partial discharge" or "corona". This process damages many insulating materials, evolves gas such as *ozone*, and ultimately leads to dielectric breakdown.

All this has been said to demonstrate the truth that for continued transformer operation cellulosic insulation *must* be maintained in a dry condition.

**Chapter 1, Part 2, Figure 1.46.*

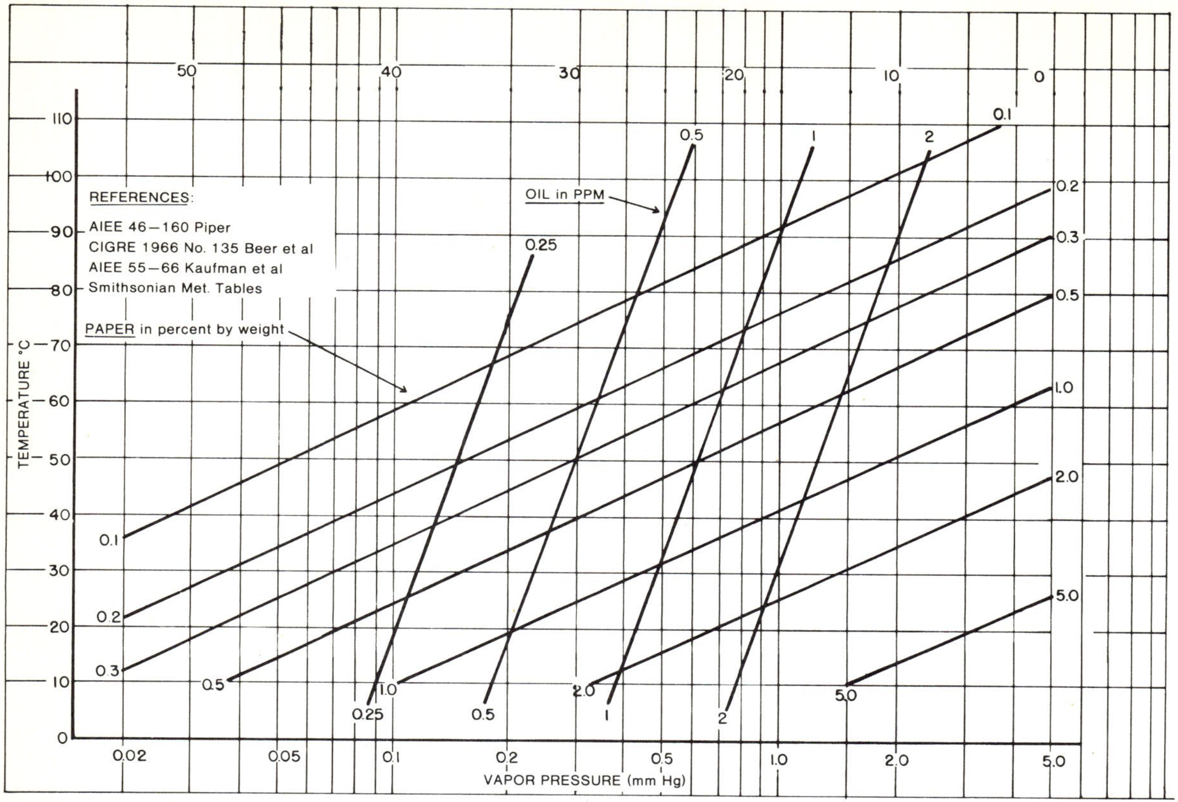

Figure 2.64 - "Piper Chart." Theoretical vapor pressure equilibria for cellulose, oil, and air.

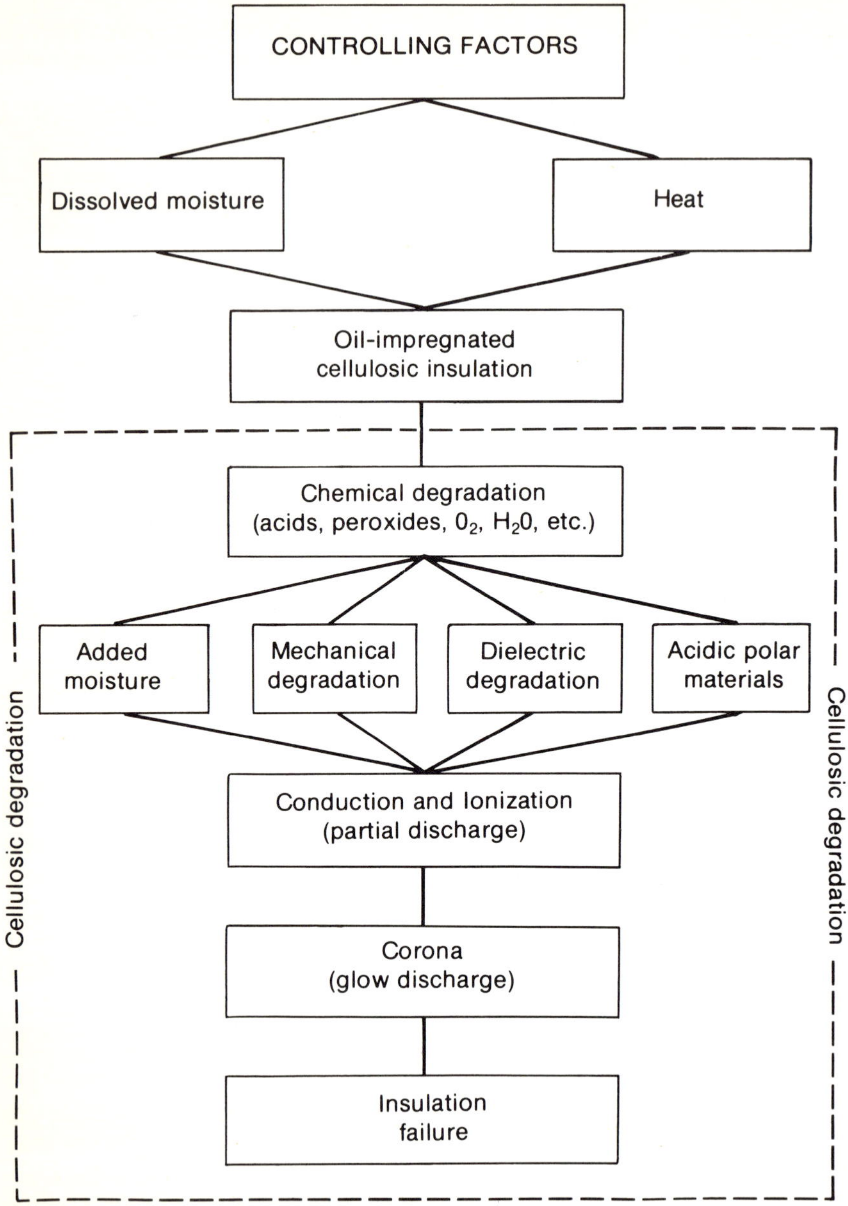

Figure 2.65 - The progression toward insulation failure.

Chapter 3

Quality Evaluation of Transformer Insulating Oil

Part 1—Nine Useful In-Service Oil Field Tests

Major Issues

It is a fact that the life of a transformer is the life of the insulation system. Knowing also what can happen to service-aged transformer oil, the "issue" at hand is a two part question: (1) What needs to be done? and (2) What tests yield the most meaningful information?

Since transformer oils are a part of the insulation system, they also require tests.* Thus, the first matter is relatively easy—yearly intervals are the most frequent examination period, but the second issue involves a greater difference of opinion.

In particular, the critical issue can be thus stated: how to recognize an unavoidable oil deterioration problem before it gets too severe in a system that is enclosed and not a subject for visible inspection.

The concern for testing only has received the wrong emphasis. Everyone recommends tests for new oil *before* it is put into service. Yet it is very unusual for new oil not to meet specifications (except for dielectric strength). Once this oil is put in service very few people are concerned enough to require on-going testing. And of those that do—only a few tests out of the 55 possible are run (ASTM Publication D-117).

Dielectric-breakdown voltage is by far the most common test made on insulating oil, according to the 1977 EPRI survey of 89 large electric utilities. The vast majority of the utilities reported that they tested all samples for breakdown voltage, but fewer indicate that they use neutralization number, appearance, power factor, interfacial tension and moisture in their test programs. Sixty-eight percent test both in the field.

**Insulating oil can be sampled without endangering the life of the transformer, whereas this is not always true with the solid insulation (Chapter 5).*

field and back-up these results with laboratory tests; 26 percent of the respondents use field tests alone, and five percent test only in the laboratory.*

These figures apply to utilities and indicate that many gallons of oil have not been looked at or, at best, have had only a cursory examination. If these large power units are being neglected, it is safe to conclude that industrial transformers are receiving even less testing.

Why the Reluctance to Test?

Many reasons account for why transformer owners/operators do not do what should be done. Insufficient knowledge concerning effects of deteriorated oil, lack of equipment, lack of time, or uncertainty as to what is a "good" or "bad" test result.

Lack of time in an industrial plant can be traced to two causes. Production department schedules won't allow equipment taken out of service except at turn-around time. Then, everyone is "too busy" to do anything (wrong priorities).† Secondly, since plant electricians maintain all types of electrical equipment, they are again, "too busy" to get around to transformers because this equipment does not give them much trouble.

While some of the oil testing equipment required is relatively expensive, it is not nearly as expensive as the damage that can be traced to the effects of oil deterioration. On the other hand, a testing program can be overcome by using a contract laboratory service for transformer oil testing. Other common responses include:

- "Do transformers have oil in them?"
- "Electricians won't go in an energized substation."
- "Unwillingness to rethink what's been done for years."
- "Why should I?"
- "What tests should I run?"
- "We filter press every year."
- "What do all those funny numbers mean, anyhow?"
- "I have too many other problems right now."
- "We tested 8 years ago."
- "We never had a transformer fail."

**P.E. Flanagan, "Utilization of Insulating Oils by North American Utilities - A Survey in 1977."*

†A company may have dozens of transformers, but ***thousands*** *of motors!*

- "I'm retiring next year, and I don't want to make any waves."
- "My transformers are in good shape - I tell production to lighten up when my units run hot."
- "My transformers are all new."

All of these responses are excuses, not genuine reasons for testing or not testing consistently.

What Oil Tests Are Essential?

The American Society for Testing and Materials (ASTM) lists 33 properties and 55 test methods in its Standard D-117, *Standard Method of Testing and Specifications for Electrical Insulating Oils of Petroleum Origin.* Practically, only the 10 tests in Table 3.1 serve as monitors for "in-service" transformers and no *one* test by itself is a reliable indicator.

We have already discussed the insulating oil tests in general. We have seen what the characteristics of *new* oils should be.*

For a true picture of a service-aged oil's condition and an insight into conditions inside the transformer, results of the 10 tests in Table 3.1—especially the first four—should be considered in combination. Moisture content in parts per million (D-1533) and gas-in-oil analysis by gas chromatography (D-3612) are also most significant.†

Of the 55 possible tests listed in D-117, several lose their significance when the oil leaves the refinery. Such tests as saponification number, corrosive sulfur, flash point, pour point and others are useful in determining that new oil meets ASTM specifications.‡ Once it is established that the oil contains the characteristics that these tests monitor, it is seldom necessary to recheck them. Several other tests may be useful if a complete oil analysis is justifiable. A consensus on typical screening tests performed in the field and in the laboratory is given in Table 3.2.

Time hasn't really changed this ranking although the latest authoritative consensus recommends dielectric, acidity and water content as principal in-service tests.§

**Chapter 2, Part 2, Table 2.4.*

†*Chapter 3, Part 3 (D-1533) and Chapter 4 (GC).*

‡*Chapter 2, Part 2.*

§*Feinberg,* ***Modern Power Transformer Practice*** *(1979).*

TABLE 3.1

THE TEN MOST IMPORTANT ASTM TESTS FOR IN-SERVICE TRANSFORMER OIL

ASTM Test Method[1]	Criteria for Evaluating Test Results - New Oil[2]	MTL or Lab[3]	Information Provided by Test
D-877/D-1816 Dielectric Breakdown Strength	New oil should not break down at 30 KV/25 KV or below respectively.	MTL	Conductive contaminants present in the oil such as metallic cuttings or free water.
D-974 Neutralization (or acid) Number (NN)	Milligrams of potassium hydroxide required to neutralize 1 gram of oil (0.03 or less for new oil)	MTL	Acids present in oil.
D-971 Interfacial Tension (IFT)	Dynes per centimeter (40 or higher for new oil).	MTL	Sludges present in oil or excessive polar contaminants from the cellulosic materials.
D-1524 Color	Compared against color index scale of 0.5 (new oil) to 8.0 (worst case).	MTL	*Marked* change from one year to next indicates a problem.
D-1533 Moisture Content Parts Per Million (ppm)	Below 25 ppm-new equipment.[4]	Lab[5]	Reveals total water content.
D-1298 Specific Gravity	Specific gravity of new oil is approximately 0.875.	MTL	Provides a quick check for presence of contaminants.
D-1524 Visual Examination	Good oil is clear and sparkling, not cloudy.	MTL	Cloudiness indicates presence of moisture or other contaminants.
D-1698 Sediment	(New oil) slight/moderate/heavy.	MTL	Indicates deterioration and/or contamination of oil.
D-924 Power Factor at 25° C	Power factor of new oil is 0.05% or less.	MTL	Reveals presence of moisture, resin varnishes, or other product of oxidation in oil or of foreign contaminants such as motor oil or fuel oil.
D-3612 Gas-In-Oil Analysis by Gas Chromatography (GC)	See Chapter 4, Table 4.26.	Lab	*Laboratory* test revealing ppm of combustible gases dissolved in the oil.

[1]For comparative purposes, specification of new insulating oil can be obtained from ASTM publication Part 40 D-3487, "Standard Specifications for Mineral Insulation Oil in Electrical Apparatus." Available from American Society for Testing and Materials, 1916 Race Street, Philadelphia, PA 19103.

[2]Criteria for evaluating test results—used oil—are listed in Tables 3.4, 3.5, and 3.6.

[3]Field mobile test laboratory (MTL).

[4]EHV units are even more critical, no more than 20 ppm and may be as low as 15 ppm (as proposed in IEEE Project 637, April, 1980).

[5]Field equipment has only recently become available.

TABLE 3.2

RANKING OF FIELD AND LABORATORY OIL SCREENING TESTS PERFORMED BY NORTH AMERICAN UTILITIES			
	ASTM Test Number	Importance in Field[1]	Importance in Laboratory[1-2]
Dielectric Strength	D-877, D-1816	No. 1	No. 1
Color	D-1500, D-1524	No. 2	No. 6
Power Factor	D-924	No. 3	No. 4
Moisture Content	D-1533	No. 4	No. 5
Interfacial Tension	D-971	No. 5	No. 3
Acidity	D-974	No. 6	No. 2

[1]Doble Engineering Client Survey, 1977.

[2]Electric Power Research Institute (EPRI) Survey, P.E. Flanagan, "Utilization of Insulating Oils by North American Utilities in 1977."

Quality of Oil Sample

Regardless of what test is specified, it is imperative that it be run on a representative liquid sample. Accurate sampling is the prerequisite to any standard test method. ASTM thought it was so important that they developed a standard for sampling (Part 40, D-923).

> As used oil may change appreciably in storage, samples should be tested as soon as possible after removal...and dates of sampling and testing should be noted.*

Therefore, oil tests performed on-site are the most representative and most informative (Figure 3.1).

**ASTM Book of Standards*, *Part 40, D-974, Note 7, p. 333 (1980 Edition).*

Figure 3.1 - Numerous transformer authorities recommend on-site transformer oil testing. This requires a mobile test laboratory (MTL®).

Cleanliness and quality control are essential to obtaining reliable test results; the oil sample must not be handled in dirty containers or subjected to contact with dirty test equipment. It is especially important that the sample be protected from contact with light, air, and moisture. The sample should be drawn quickly, kept in a sealed container, and *tested within hours*. It must not be exposed to the ultra-violet rays of sunlight; exposure to sunlight for a matter of days will accelerate aging and the sample will be wholly unrepresentative of the condition of the oil in the transformer.

ASTM Precision Statement

Whatever the test procedure, a statement as to the precision of the method should be known to the transformer owner/operator.*

- A *repeatability* statement (same technician) intends to answer the question, "How well can I expect to agree with myself?" (The *same* laboratory, *same* field test equipment)
- A *reproducibility* statement intends to answer the question, "How well can I expect to agree with *other* laboratories?"

Precision statements are only for laboratory tests; since there are also field tests, the best approach is thus to run (where possible)* ***lab* *tests in the* ***field****.*

Dielectric Breakdown Strength Tests

The first recognized problem in transformer oil maintenance was moisture. A typical early test for water consisted of heating the oil in a frying pan with a plumber's torch. If moisture were present in the oil, bubbles formed as the oil was heated and would rise to the surface and burst with a crackling sound.* Clear oil free from moisture would swirl smoothly in the pan with no visible bubbles or sound—a crude test, but quite effective for 1911. Up to two percent moisture (20,000 ppm) in the oil was considered satisfactory in those earlier days. Today, for even medium power transformers, the requirement is down to 20-30 parts per million of water in the oil.

Yet, because of tradition, availability of the equipment, and the simplicity of the method, the dielectric strength of an oil still ranks as the most common field and laboratory oil screen test. In many facilities the dielectric test is the only test applied to transformer oil. The significance of the test is that it will only detect free water, dirt, and conductive particles, not acids or sludges. Authorities now agree that careless technique has been the source of 99 percent of "bad" dielectric readings. All too many wet oils have been nothing more than the result of a test technician's sweaty hands!

Instrumentation available now for the dielectric test is quite compact and efficient (Figures 3.2 and 3.3). The test set applies an ac voltage at a controlled rate to two electrodes immersed in the dielectric fluid. The gap is a specified distance. When the current arcs across this gap the voltage recorded at that instant is the dielectric breakdown strength of the liquid. For new oils this is considered to be 35,000 volts or above. Used oils would not be acceptable below 25 KV.

Two standard test cups fit most of the commercial dielectric testers. One is the flat disk (ASTM D-877) and the other is the VDE (ASTM D-1816) spherical variety.† The flat disk cup has been the work horse of the industry since 1921, but is not as sensitive to moisture as the VDE cup.

**The "crackle test" is still used for liquids that are not as "water critical" as transformer oils.*

*†Details of D-877, D-1816 and other subsequent tests are outlined in **ASTM Book of Standards** Part 40. Make sure you consult the most recent annual standard.*

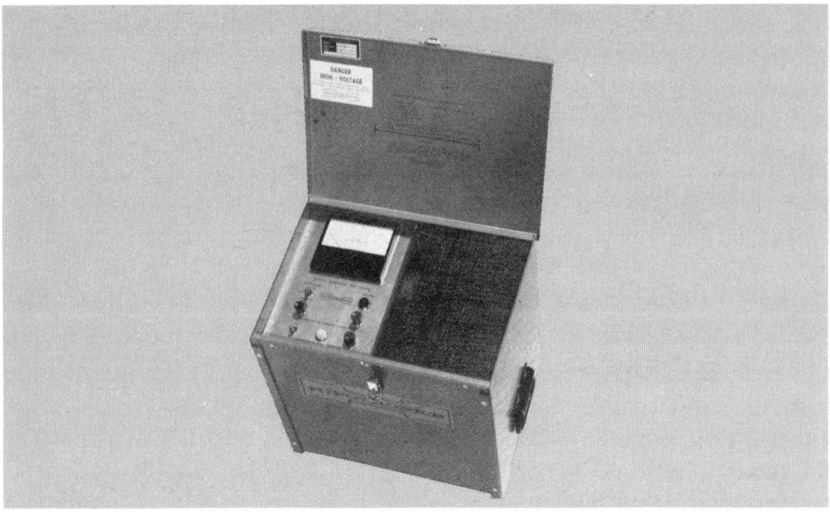

Figure 3.2 - Portable oil dielectric breakdown strength tester (Model OC-60). (Courtesy of Hipotronics, Inc.).

Figure 3.3 - TC/DE: ASTM test cell with disc electrodes (oil cup). Adjustable 1-inch diameter electrode and standard 0.100-inch gap gauge. (Courtesy of Hipotronics, Inc.).

The VDE (Verband Deutscher Elektrotechniker) electrodes were adopted in the U.S.A. in 1960 and are now used by an increasing number of laboratories. A small, motor driven, nylon impellor stirrer is immersed in the oil to stir the liquid (Figure 3.4). The flat disc cup used a 0.1 inch gap and a voltage rise of 3000 volts per second (RMS). The VDE cup has two gap settings 0.040 inches and 0.081 inches. The rate of voltage rise is 500 volts per second, RMS. The 0.040 inch gap can be used with the 35 and 50 KV test sets. The 0.081 inch gap needs an 80 or 90 KV test transformer. ASTM is currently reviewing this situation and is leaning toward endorsing the smaller gap.

Both methods have their limitations.* The flat sharp edges of the D-877 test cup electrodes do not approximate the conditions found in a modern transformer. Present day high voltage transformer construction emphasizes smoothly contoured metallic parts rather than sharp edges.

ASTM says D-1816 should not be used on an unprocessed oil but the practice is occurring and will probably become standard. Air and lint also affect dielectric strength and the transformer owner/operator needs to know that effect for practical information to arrive at valid conclusions.

By the same token D-877 is not sensitive enough to test the oil in units above 69 KV, where very dry oil is required. It is generally concluded that D-877 is insensitive to moisture in concentrations up to about 30 ppm.

A third type of cup is available for use with the commonly used dielectric test transformers. This cup is made for testing gases. One electrode is flat, the other electrode is spherical.

Probably the most confusing thing about dielectrics are the values that are considered "good" (Table 3.3).

Next, look at gas dielectrics:

• SF_6	19.4 KV
• $SF_6 + 0_2$	14.6 KV
• N_2	12.2 KV
• Air	8.1 KV

Thus, the transformer owner/operator must have more than just test data. Experience and practical judgment will determine when a value is

**Chapter 3, Part 3, "But My Dielectric Was Good...".*

TABLE 3.3

A WIDE VARIETY OF "GOOD" DIELECTRIC STRENGTH TEST RESULTS[1]		
ASTM D-877	ASTM D-1816 (VDE)	
0.1 inch gap	0.040 inch gap	0.081 inch gap
35 KV	Oil as Received—30 KV	40 KV
46 KV	Processed Oil (Vacuum treated and filtered)— 41 KV	Too high to measure
46 KV	Processed Oil Resaturated with Dry Lab Air—30 KV	40 KV
[1]No known relationship exists between readings or the two methods (Figure 3.14).		

just a number and when it has meaning. Fortunately, we do have a more objective method of measuring water content. This is Karl Fischer chemical titration employing ASTM Test Method D-1533. Traditionally, this has been a laboratory test only. Recently, however, equipment utilizing an automatic coulometric titrator has been developed for field application.*

Neutralization Number

The second traditional liquid field screening test is also called the acid number. ASTM Test Method D-974, the most common of four methods, measures acidity (from oil oxidation) in transformers. These (fatty) organic acids are detrimental to the insulation system and can induce rusting of iron when moisture is also present. An increase in the neutralization number is an index to the rate of deterioration of the oil (Figure 3.5).† Sludge will be the inevitable product of an acid situation that is neglected. The acid content is expressed as the number of milligrams of potassium hydroxide (KOH), a base, that it takes to neutralize the acid in a one gram sample of oil.

**Chapter 3, Part 3, (Figure 3.18).*

†Chapter 2, Part 3, Figure 2.13; Chapter 6, Part 1, Figure 6.2.

The D-974 laboratory acid test method is based on a color-indicator titration (Figure 3.5). This method is relatively simple to do and very precise. The test consists of mixing a known amount of sample with a suitable solvent, adding an indicator, and titrating (or neutralizing) with KOH to a colored end point.

The laboratory method is good for both new and used oils. The precision is such that you can monitor the deterioration and see trends being established. The repeatability of this test (duplicate determinations by the same operator) is 0.05 mg in the range of 0.1 to 0.5 acid number. The reproducibility (results from two different labs) in the same range of acid numbers is 0.04 to 0.08 mg KOH/g. By contrast, field test methods do not reveal the transformer is in trouble soon enough because they are not sufficiently sensitive to determine acid numbers up to 0.15 mg.

The only problem with test D-974 is that it is difficult to detect a color change in dark colored oils. A trained test operator can devise methods to do this, such as changing the mix of oil and chemicals, and observing the foam rather than the oil.

A spot test kit for field use (D-1902) was marketed by Allis Chalmers some 20 years ago. The acidity spot test was made by applying drops of an acidity buffer solution to a piece of filter paper and allowing it to soak in. Then two drops of oil were added to the same place. This was followed by a one drop addition of an indicator. This caused a color change which was compared to a standard color chart. The ratings were:

Green and yellow	=	0.3 mg KOH/gm
More green than yellow	=	Less than 0.3 mg
More yellow than green	=	More than 0.3 mg
Orange or brown	=	High acidity
Blue	=	Alkaline

The only problem with test D-974 is that it is difficult to detect a color change in dark colored oils. A trained test operator can devise methods to do this, such as changing the mix of oil and chemicals, and observing the foam rather than the oil.

A similar type test is the *approximate* acidity test D-1534. This field type test is run by placing a 20 milliliter sample of oil in a graduate cylinder and adding an indicator solution. KOH is added to the oil and the solution is shaken. After it is allowed to settle the color change is noted. If three milliliters of KOH are added and the color changes, the acid number is 0.3. Each one milliliter of potassium hydroxide represents a 0.1 acid number. A drop wise method would be more accurate at 1 milliliter = 10 drops (commerically available ampule kits 3, 6, etc.). ASTM does not publish a precision statement on D-1902 or D-1534, as they are field tests.

Figure 3.4 - TC/VDE: ASTM test cell with VDE electrodes and motor-driven circulating system. (Courtesy of Hipotronics, Inc.).

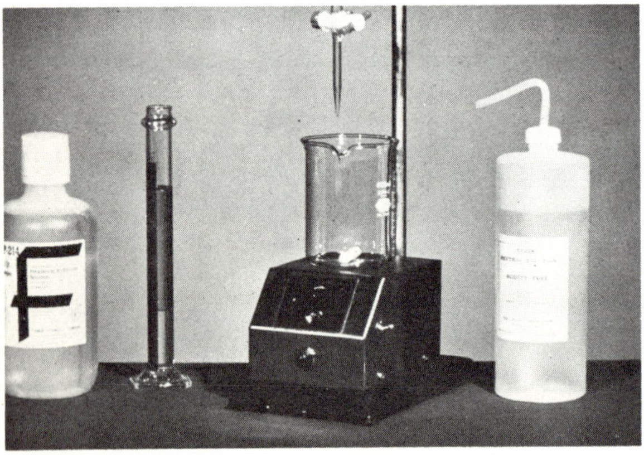

Figure 3.5 - Laboratory equipment (50 ml. graduate cylinder, burette and beaker) and titration solutions are required to perform ASTM D-974 neutralization (acid) number test. If a technician adds too much KOH the solution will look like tomato juice. Back titration using hydrochloric acid is then required to neutralize the basic mixture.

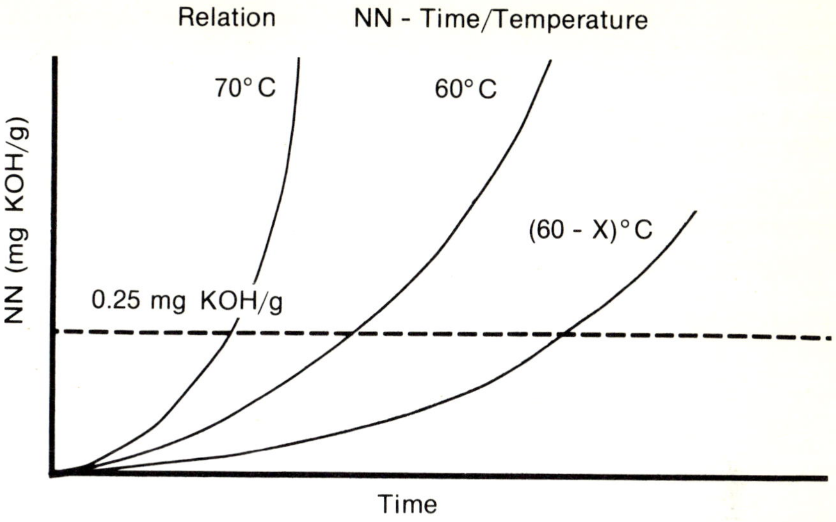

Figure 3.6 - Typical pattern of increase in NN as a function of time is exhibited by curve for transformer oil operated at 60° C. The exponential rise in NN at the critical point results from the catalytic action of acids and the depletion of the oxidation inhibitors. Definition—Critical Acid Number. The acid condition is of such magnitude that the acid acts as a catalyst to accelerate the formation of oil decay products, and the oil oxidation inhibitor is essentially deplete. This results in "run-away" oil oxidation as witnessed by the sharp rise in the curves shown.

A final method for acid testing is ASTM D-664 which uses a potentiometric titration. This calls for a rather elaborate piece of laboratory equipment, an involved lab procedure to prepare the solutions and the sample, and a good bit of maintenance to the indicating and the reference electrodes.

The results are determined from a plot of the increase of the potential of the electrode against the volume of KOH used (Figure 3.7). This is not a precise method. The precision drops off in direct proportion to the change in shape of the titration curve.*

**TMI does not recommend this procedure and it is questionable if it is used anywhere to monitor "in-service" transformer oils.*

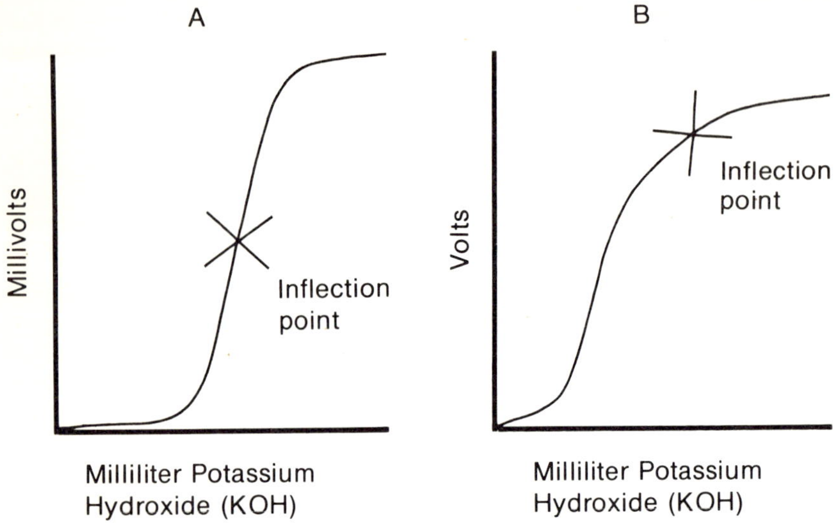

Figure 3.7 - Potentiometric titration. The end point is the mid point between the two inflections. In sample A, the end point is much easier to determine than that of sample B.

Interfacial Tension Test

The next test in the on-site series is the interfacial tension test (ASTM D-971), commonly referred to as the IFT Test (Figure 3.8).

The IFT test measures the tension at the interface between two immisible liquids, oil and water, and is expressed in dynes per centimeter (millinewtons/meter). This test is extremely sensitive to the presence of oil decay products and soluble polar contaminants from solid insulating materials.

Picture in your mind a needle floating in a cup of water. Adding detergent to the cup will break the "surface tension", and the needle will sink. Good clean oil will normally lay on top of the water (Figure 3.9) and will yield an IFT of 40 to 50 dynes/cm. Oil oxidation contaminants will lower the IFT even further. These contaminants are hydrophilic which means they have an affinity for water molecules, as well as for oil molecules. At the interface the hydrophilic materials extend across to the water so that a vertical linkage is established and thus the lateral linkage (which makes up the surface tension) is weakened. The interface is now less distinct, and the tension at the interface is reduced. The greater the concentration of contaminants, the less the tension. IFT test results are thus related to the degree of the oil's oxidation. The very *first* sign of sludge in solution may be in the range of 27-30 dynes/cm while a badly deteriorated oil has an IFT of 18 dynes/cm or less.

The ASTM D-971 test method is suitable for use on a mobile test laboratory. The test procedure calls first for a check of the water (surface tension).* Then the oil under test is added while the platinum ring is submerged in the water. After waiting for an aging period of 30 seconds the interface is broken by the platinum ring by lowering the table, using the left hand, while the ring is held in position by increasing tension on the torsion wire connecting it to the indicating dial (Figure 3.10). Once the break is achieved, the scale reading is subjected to a calculation to convert it to a true IFT reading. Read the vernier scale to the nearest tenth.

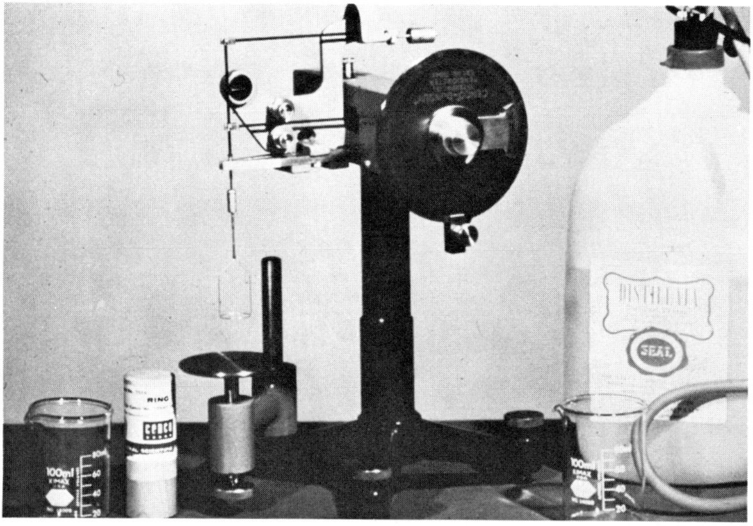

*Figure 3.8 - Interfacial tension is **the** test to determine the degree of sludging in a transformer's insulation system. Since oxidation products have an affinity for water molecules, this test is a logical one. ASTM D-971 laboratory test uses the Du Nouy tensiometer, a platinum ring, a glass container, and distilled water. Once the instrument is set up and calibrated, the test can be conducted in a few minutes. The term "tensiometer" refers to the instrument devised by Dr. Pierre Du Nouy. The ring method was known before the U.S. Civil War. Du Nouy's contribution was the development of a specialized form of torsion balance which greatly simplified the test and the calculations.*

The surface tension of pure distilled water is approximately 72 dynes/cm.

Figure 3.9 - At the interface each liquid exhibits its own surface tension, the molecules of one having no great attraction for those of the other. To break through the interface it is necessary to rupture both surface tensions. This is what happens as oil oxidation byproducts are added to the oil.

Figure 3.10 - A transformer maintenance technician carefully conducting an interfacial test on an oil drawn from the transformer only hours earlier.

The repeatability precision standard of the test is two percent of the mean value, and reproducibility is five percent of the mean.

The field test (D-2285) for IFT consists of a determination of the volume of water (a drop) that the oil will support. The larger the drop of water the higher the IFT. A high IFT number results in the drop of water clinging to the needle. This tendency is lowered by hydrophilic compounds lowering the surface tension of both liquids. The procedure calls for a hollow needle by which water is introduced into oil. A trial drop of water is expelled to determine the approximate size of the drop. Next, a drop 3/4 of the trial drop is expelled. This drop is then aged for 30 seconds in the oil. Then this drop is enlarged by metering more water into the oil. The amount of water in the final test drop is directly proportional to the IFT. The larger the drop, the higher the IFT number. Again, ASTM does not publish a precision statement for this method as it is a field test.

Oil Power Factor

The ASTM D-924 Power Factor test has been a traditional field test. However, this test is considered a "negative" screening test.* It measures the leakage current through an oil, which is a measure of the contamination or deterioration. Unfortunately, *the power factor is not specific in what it detects*. It does tell you the presence of polar materials, but other tests must be made to determine *what* polar compounds are present.

When water is held in solution no appreciable effect on the oil's power nor dielectric strength may take place. The ASTM D-924 power factor test may pick up the sludges. However, since oil oxidation further increases the oil's ability to dissolve moisture, some types of sludge have very little effect on the overall power factor. Years of experience have shown that sludge can thus best be detected by the D-971 interfacial tension test (See Table 3.14; Figure 3.11).†

When the power factor exceeds 0.5 percent an investigation is indicated. If a 100° C reading is more than 7 to 10 times the value at 25° C, then this usually indicates a soluble contaminant in the oil other than water. A high oil power factor is not necessarily harmful, however,

** Such a test is used to screen a large number of oils to be tested. The oils that do not pass the screening test are then subject to further lab tests.*

† See Color Section, following Chapter 10.

although it may be reflected in somewhat higher transformer insulation power factor. In addition to transformers, the power factor test is applicable to liquids in cables, oil circuit breakers, and other electrical apparatus.

Oil Color Comparison Test

The next in the standard series of oil tests is the color test. This test, meaningless in itself, is significant when there is a *marked* change. ASTM has two available standard methods—a lab test (D-1500) and a field test (D-1524). The lab test is conducted by comparing a sample of oil to some color standards. Unfortunately, the lab test is for all types of oils, not just transformer oil.

The field test is specifically for used mineral oils that have been used in electrical apparatus. The test is similar in that the oil is compared with a series of standard colored disk filters in a specially built comparator. Naturally, there is no precision listed for this field test. The lab test should be repeatable by 0.5 numbers.

Visual Examination of Oil Sample

This next test in the transformer oil series—visual condition—is also the subject of ASTM standard field method D-1524. As it is used in determining the color classification, the same is simultaneously checked for cloudiness, turbidity, (metallic contaminants), particles of insulation, carbon, or other suspended materials. Good oil should be clear and sparkling. Cloudiness indicates moisture, carbon, and/or sludge.* If the presence of carbon is indicated, the same arcing or partial discharge should be suspected. A GC test on the oil should be performed.†

Sediment in Transformer Oil

In the related visual test ASTM D-1698, "Sediment and soluble sludge in service aged insulating oils," the oil sample is filtered (centrifuged) to separate the sediment from the oil.‡ The upper portion of the sample

**Unusual odor (smell) of an oil is sometimes significant (Chapter 6, Part 2).*

†Chapter 4.

‡Chapter 7, Part 3

is then decanted off and saved. The sediment containing portion is dried, weighed, ignited, and then reweighed. The remainder is inorganic, and the loss in weight is organic. Inorganic sediment indicates contamination, such as rust; organic sediment indicates deterioration, such as sludge or contamination.

The portion of the sample that was poured off will contain any soluble sludge. Soluble sludge is determined by dilution with n-pentane, a hydrocarbon, to precipitate the sludge. The sludge is not soluble in n-pentane. Another test for organic sludge is to see if it dissolves in acetone.

Specific Gravity of Oil

ASTM D-1298 specific gravity is the final essential property that should be checked. This is simply a ratio of the mass of a given volume of liquid (oil) to the mass of an equal volume of water at specified temperatures. The test is normally conducted by floating a hydrometer in the liquid and taking the reading at the meniscus. If the reading would be over 1.0, then the oil probably contains some contaminant—quite possibly askarel (PCB) or silicone; if less than 0.84, then the oil may be a paraffinic oil. Nevertheless, specific gravity is useless for detection of accidental mixing with lubricating, hydraulic or fuel oil, or pot head compounds as the specific gravities fall into the same range. More meaningful is a test for flash point (D-92) and viscosity (D-88/445) if such contamination is suspected.

All of the foregoing tests apply to mineral oils. Askarels are tested for dielectric and acid number using the same standard methods as oils. Several of the other tests use different standards.* Interfacial tension is not tested on askarel as PCB does not oxidize to form hydrophilic or polar compounds.

The specific gravity of askarels is measured with ASTM D-1810. This method is basically the same as the method for oils except the range of the hydrometer is higher (1000 to 2000). If the askarel is thick, the test can be run at 194°F rather than the standard 60°F.

Supplementary Transformer Oil Tests

A knowledgeable physician would use up to 22 blood tests in his com-

**See Chapter 9, Table 9.15 for details concerning askarel testing. An acid value exceeding 0.02 mg KOH/g may well indicate that arcing has taken place.*

plete physical examination of a patient. Situations occur where a complete transformer oil evaluation is also imperative. As one example, these tests would serve as the basis for determining transformer reliability—whether to condemn, continue in-service or repair a given unit.

Such a matter as determining if a transformer retains its original oil is necessary to find out the basic source of these additional tests. We will consider these tests in terms of physical, electrical, and chemical properties.

Physical Tests

1. Aniline, Flash and Fire Points, Pour Point, and Viscosity Tests
   These were considered in Chapter 2, Part 2. Such properties do not markedly change during the aging process unless the oil has become grossly contaminated by some other material or liquid.
2. Coefficient of Thermal Expansion (D-1903)
   This is generally used only in the design of enough space for oil in a transformer tank.
3. Refractive Index (D-1218)
   This may be considered "the fingerprint" of an oil—a second method of determining where the oil has come from. Determinations made on the same liquid after periods of service may form a basis for estimating any change in composition of the degree of contamination.

Electrical Tests

Resistivity (D-1169) of transformer oil monitors the amount of conductive contaminants, such as metal salts, water, aldehydes, ketones and alcohols. Its importance is recognized in the IEC standard (Table 2.4). Aging tube studies have indicated that the resistivity decreases most in the first aging period and remains relatively constant thereafter.

Chemical Tests

1. Copper Content (D-2608)
   This is useful to assess the condition of service-aged oils from contact with copper metal during refining or service.*

This test is to be **distinguished from the Copper Number of Paper, D-919, Chapter 2, Part 5.*

2. Inorganic Chlorides and Sulfates (D-878)
   These may affect the dielectric properties of the oil and thus its ability to function properly under service conditions. This is still normally important only at the refinery.

3. Oxidation Inhibitor Content (D-1473)
   This is an important factor in maintaining long service life of inhibited insulating oils. Oxidation inhibitor is not efficient beyond 120°C (that is, high operating temperatures).

4. Peroxide Number (D-1563)
   This number measures the quantity of oxidizing constituents present.

5. Corrosive Sulfur or Silver (D-1275)
   This detects the presence of objectionable quantities of certain thermally unstable sulfur bearing compounds in new oil. When present, these compounds may result in corrosion of copper and especially silver parts on tap changers or leaks from some pot heads or bushings.

6. Atomic Absorption
   This measures metallic contamination such as iron, copper, or aluminum in oil in ppm levels.

7. Power Factor Valued Oxidation (PFVO)
   As developed by the Doble Engineering Company, this test periodically measures the power factor of an uninhibited oil while it is being aged at 95°C in the presence of copper and air. The test is run concurrently with the Doble Sludge Free Life Test which measures the time until the oil forms sludge.*

Future Prospect—Transformer Oil Testing

Even when various testing and reclaiming standards and specifications are complied with (ANSI/IEEE, ASTM, IEC), it cannot be assumed that oil within a transformer can be ignored for long periods of time. Thus, it is evident that decisions have to be made concerning *how* often periodic tests are made, and *how* to *evaluate* the test results (Part 2).

**Chapter 2, Part 2, Figure 2.10.*

In addition, while the foregoing tests are currently applicable, new and additional test methods must be evaluated and made available. One of these is *now* currently in use. It is identified as ANSI/IEEE C57.104 (ASTM D-3612) and concerns how to analyze combustible gases in mineral oil.*

A second document describes an improved method of measuring the acid number of transformer oil, using infrared spectroscopy. As oils degrade, two absorption bands are formed in the carbonyl region of the infrared spectrum. By determining these values and calibrating against oils of known oxidation, a good estimate of acid number can be obtained. Among other advantages, it requires a small sample which is especially beneficial for certain situations, such as monitoring accelerated oxidation tests, or evaluating field samples of limited quantity.†

**Chapter 4.*

†Gedemer, "An Improved Infrared Method for Determining Acidity in Transformer Oils," (1978).

Part 2—Interpreting Transformer Oil Test Data

Introduction

Insulation is the weakest link in any electric power distribution system—and the oil in an oil-filled transformer is an integral part of the insulation system. IEEE, transformer manufacturers, and insurance carriers agree that annual testing of insulating oil from energized transformers is a minimum requirement for ensuring a dielectrically acceptable unit.*

Multiple Benefits of Annual Testing

A program of annually performing specific oil tests will deliver numerous benefits, in addition to confirming the oil's dielectric strength. A comprehensive test program can:

1. Indicate the Interior Condition of the Transformer
   Presence of sludge—dissolved or deposited—can be detected, and the sludge can be purged before it can precipitate out onto the windings and other interior surfaces of the transformer. Operation in the sludge-free-range will help assure maximum *transformer* life, not just oil life. Oil test data can also frequently indicate the condition of the cellulosic insulation.

2. Prevent an Unscheduled Outage
   Internal oil problems can be detected early enough to permit corrective action to be scheduled when disruption will be minimal.

3. Provide Data for Plotting Deterioration Trends
   The naphthenic insulating oils used today have been used for more than 55 years. These oils deteriorate in predictable fashion; data is available to permit comparing normal and abnormal deterioration. Although a single set of test results will reveal the presence of sludge and other decay products, attempting to detect a trend toward abnormal deterioration on such a basis is like trying to plot a graph with only one reference point.

**Exceptions are arc and rectifier furnace transformers, other critical units, and all very large power transformers. See Chapter 6, Part 2 (Predictive Maintenance); Chapter 10, Table 10.1.*

4. Classify Oils
Reliable recommendations may be made concerning degree of servicing required—whether energized preventive or corrective maintenance or de-energized detanking is necessary.*

Too often we hear the terms "good" or "bad" oils. What we must define is how "bad" is "bad"—establishing a viable method for evaluating and classifying transformer oils. Secondly, such an oil classification should also indicate the condition of the inside of the transformer.

Establishing a Practical Oil Classification System

Conflicting Standards

Various authorities have established standards for service-aged oils. Table 3.4 illustrates the problem of *conflicting* standards. Who is right and who is wrong? In addition, the *multitude* of local manufacturer's service shops have a *multitude* of different standards!

IEEE has taken one formal step in bringing some objectivity into this confusing picture. They have defined these four groups of in-service oil.

Group I — Oils that are in satisfactory condition for *continued use.*

Group II — Oils that require only *reconditioning* for further service.

Group III — Oils in poor condition. Such oil should be *reclaimed* or disposed of depending upon economic considerations.

Group IV — Oils in such poor condition that it is technically advisable to *dispose* of them.

IEEE and EPRI have updated their acceptable test limits for each of the four classes (Table 3.5). In addition, IEEE has proposed suggested test limits for in-service insulating oil by voltage class (Table 3.6).

While Tables 3.5 and 3.6 meet the first criterion on defining how "bad" an oil is, they do not tell us anything concerning the interior condition of the transformer.

**Chapter 6 (Preventive Maintenance); Chapter 7 (Corrective Maintenance).*

TABLE 3.4

RECOMMENDED TEST VALUES FOR CONTINUED SERVICE OF TRANSFORMER INSULATING OIL

Authority	NN mg KOH/g Maximum	IFT Dynes/cm Minimum	Dielectric D-877 KV Minimum	Power Factor % at 25°C Maximum	Moisture Content PPM Maximum
Westinghouse (1977)	0.15	21	28	1.0	—
General Electric (1979)	0.20	24	26	0.65	—
Kemper (1979)	0.36	21	24	1.0	25
Factory Mutual (1978)	0.25	18	23.5	—	—
American Nuclear Insurers (1979)	0.20	22	23.5-31[1]	—	50
EPRI Utility Survey (1977)	0.32	23	27	0.80	34
IEEE - Project 637 (1980)	0.1-0.2[1]	24-30[1]	26	—	20-35[1]
National Fire Protection Association (1977)	0.40	—	22	—	—
International Electrotechnical Commission (1973)	0.50	15	30-50 KV[1]	—	20-30 mg/liter
U.S. Permissible Range	0.1-0.4	18-30	22-31	—	20-50

[1]Specific values depend on voltage classification from below 69 KV; up to 288 KV; and over 345 KV.

unit. These relationships between IFT, NN, and condition of oil and its effect on the transformer have been available for years, but never acted upon until recently.

Unique IFT-NN Relationship

Several studies in the 1940's have shown that a definite relationship takes place between acid number and IFT (Figure 3.11).* An increase in NN should normally be followed by a characteristic drop in IFT. Field tests have shown IFT to be a powerful tool for determining how an insulating oil has performed and how much life is left in the oil before some correction is needed to prevent sludge.

This inverse relationship between these two laboratory findings practically eliminates "mistakes" when evaluating oils. The NN test by itself becomes questionable in the high acid ranges because it rests on the skill of the technician. The IFT test, however, conducted on precision lab equipment provides an excellent back up test for the neutralization number. If the test results do not fall on the curve or in proximity to it (such as Points A and B), the results are suspect. Questions should be asked as to service conditions and further tests should be run. A low IFT not accompanied by correspondingly increased acidity and color indicates polar contaminants which have *not* come from normal oxidation of the oil.†

A second bit of data involves the findings of a ten year study on 500 in-service utility transformers.

Over the period from 1946 to 1957, ASTM sought to determine the correlation between IFT, NN, and transformer sludgings. Units selected for the study all had been in service for some time in different industrial environments. Each unit had its oil tested and was visually inspected for the presence of sludge periodically during the ten year study. Results of the study are summarized in Table 3.7.

**Figure 3.11 is located in the special Color Section, following Chapter 10. This data has been based on napthenic oils. When paraffinic type oils come into future use, the relationship between IFT and acidity and sludging may have to be re-established or may change as test data and experience are accumulated.*

†Although a low IFT with a low acid number is an unusual situation, it does occur because of contaminants such as solid insulation materials, compounds from leaky pot heads or bushings, or from a source outside the transformer.

TABLE 3.5

ACCEPTABLE TEST LIMITS FOR THE FOUR IEEE CLASSIFICATIONS OF TRANSFORMER OIL[1]

Property/Test	Oils For Continued Service	Oils Needing Reconditioning	Oils Needing Reclaiming	Oil After Reclaiming	Oils to be Scrapped
IEEE Classification	Group I	Group II	Group III	—	Group IV
Dielectric Strength D-877, KV, Minimum	27	24	23	30.4	19
Dielectric Strength D-1816, 0.1 cm (0.4" gap), KV, Minimum	24	—	—	—	—
Neutralization Number mg KOH/gm oil	0.30	0.32	0.45	0.10	0.85
Interfacial Tension dynes/cm., Minimum	24	23	20	33	14
Power Factor, 60 Hz 25° C, Maximum	.8	1.1	1.36	.26	2.2
Moisture Content, PPM, Maximum	34	40	66	30	91

[1]Flanagan, "Utilization of Insulating Oils by North American Utilities-Survey in 1977", Electric Power Research Institute (EPRI).

TABLE 3.6

SUGGESTED TEST LIMITS FOR SERVICE-AGED OIL BY VOLTAGE CLASS[1]				
Property/Test	Limits			
Voltage Class	69 KV and Below	Above 69 KV Through 288 KV	345 KV and Above	ASTM Test Method
Dielectric Breakdown Voltage 60 Hz KV minimum	26	26	26	D-877
Dielectric Breakdown Voltage KV minimum (.040 inch gap thin)	23	26	26	D-1816
Neutralization Number, mg KOH/gm oil, maximum	0.2	0.2	0.1	D-974
Interfacial Tension dynes/cm minimum	24	26	30	D-971
Power Factor, 60 Hz 25° C, maximum percent[2]	0.65	0.39	0.31	D-924
Moisture Content,[3] PPM-maximum	35	25	20	D-1533
Gas Content when specified, maximum percent	See note 4			D-831 D-1817 D-2945 D-3612

[1]IEEE Insulating Fluids Committee, Project 637, April, 1980.

[2]EPRI, Utility Survey - 1977.

[3]Values do not pertain to free breathing transformer or compartment.

[4]Some transformers are equipped with diaphragms to prevent the introduction of air. The dissolved gas content in these transformers should be maintained in accordance with the manufacturer's recommended list.

TABLE 3.7

HISTORICAL DATA BASE ESTABLISHING CORRELATION BETWEEN NEUTRALIZATION NUMBER—INTERFACIAL TENSION AND SLUDGE FORMATION IN OIL FILLED TRANSFORMERS		
Neutralization Number vs Sludge (A)		
NN mg/KOH/g	Percent of 500[1]	Units Sludged
0.00-0.10[2]	0	0
0.11-0.20	38	190
0.21-0.60	72	360
0.60 and up	100	500
Interfacial Tension vs Sludge (B)		
IFT Dynes/cm	Percent of 500[1]	Units Sludged
1) Below 14	100	500
2) 14-16	85	425
3) 16-18	69	345
4) 18-20	35	175
5) 20-22	33	165
6) 22-24	30	150
7) Above 24	0	0

[1]ASTM - 11 year test on 500 transformers (1946-57).

[2]Realistic value of 0.03-0.10.

In time past, neutralization numbers were allowed to exceed 0.3 mg KOH/g before any maintenance was considered necessary. As a result of such long range studies, it is clear that problems began much sooner than was originally thought. In fact, as the acid number rises beyond 0.2, the problem of maintenance extends into the transformer itself (Table 3.7A).

All 500 transformers exhibited visible sludging by the time NN had risen to 0.60 mg KOH/g or more, or IFT had dropped below 14 dynes/cm.

Combining the IFT-NN relationship Figure 3.11 with the ASTM field study (Table 3.7) establishes seven categories of oil condition (Table 3.8).

Oil Quality Index System

High IFT means that a transformer oil is relatively sludge-free and, therefore, purer than an oil with a low IFT. Conversely, a high neutralization number indicates high acid content and a badly deteriorated oil. Dividing the IFT by the NN provides a pure numerical value that is an excellent means of evaluating oil condition. This value is known as the Myers Index Number (M.I.N.) or Oil Quality Index Number. A new oil, for example, has an M.I.N. of 1500:

$$\text{M.I.N.} = \frac{\text{IFT}}{\text{NN}}$$

$$= \frac{45.0 \text{ (typical new oil)}}{0.03 \text{ (typical new oil)}} = 1500$$

This relationship provides a further basis for classifying oils into the seven categories according to their condition, (Table 3.8).

Examination of Table 3.8 reveals that there is some overlap of M.I.N. ranges in the first three categories (Figure 3.12). The reason for the overlap is that oil should meet the criteria for both minimum IFT and maximum NN for the category, in addition to falling within the range given for M.I.N. For example, an oil with an M.I.N. of 318 would be classified as marginal—rather than Proposition A or good—if the IFT were 27.0 or less or if NN were 0.11 or more.

TABLE 3.8

TRANSFORMER OIL CLASSIFICATIONS*				
1. Good Oils	NN 0.00 - 0.10	IFT 30.0 - 45.0	**(Pale yellow)**	M.I.N. 300-1500
2. Proposition A Oils	NN 0.05 - 0.10	IFT 27.1 - 29.9	**(Yellow)**	M.I.N. 271 - 600
3. Marginal Oils	NN 0.11 - 0.15	IFT 24.0 - 27.0	**(Bright yellow)**	M.I.N. 160-318
4. Bad Oils	NN 0.16 - 0.40	IFT 18.0 - 23.9	**(Amber)**	M.I.N. 45 - 159
5. Very Bad Oils	NN 0.41 - 0.65	IFT 14.0 - 17.9	**(Brown)**	M.I.N. 22 - 44
6. Extremely Bad Oils	NN 0.66 - 1.50	IFT 9.0 - 13.9	**(Dark brown)**	M.I.N. 6-21
7. Oils in Disastrous Condition	NN 1.51 or more		**(Black)**	

**See Table 6.10 in the Color Section, following Chapter 10.*

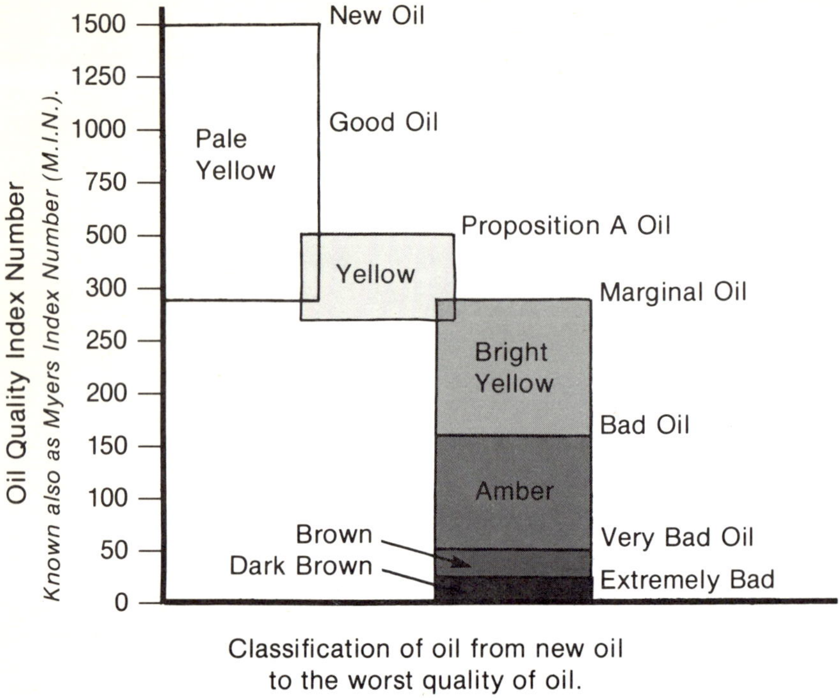

Figure 3.12 - Quality Index System showing full scale of oil condition from new oil down to the lowest category. To fall within a particular classification, an oil must meet all the criteria for IFT, NN, and M.I.N.

Using the Oil Classification Chart (Tables 3.8 and 6.10)*

Consider these seven categories and their ability to perform their four functions (cooling, insulation, protection against chemical attack, and prevention of sludge buildup).

The first category is Good in which these functions are efficiently provided. Classes 2 and 3 are in the "Preventive Maintenance" range. All functions are still maintained in Proposition A range; nevertheless, a drop in IFT to 27.0 may signal the beginning of sludge in solution. At this point, the oil is less than good; therefore, servicing is recommended.

**See Table 6.10 in the Color Section, following Chapter 10.*

A plant having eight transformers, for example, may have one in the Proposition A range, and all others in the third (Marginal) or a lower category. The most economical proposition would be to service all eight transformers at the same time, rather than to wait until the oil in the Proposition A category has deteriorated to marginal condition.

Oil in the Marginal category is not providing proper cooling and winding protection. Fatty acids are beginning to coat the winding insulation; sludge in insulation voids is highly probable. Significantly, case histories record transformer failures in these two categories. For example:

Consider a 250 gallon transformer with NN of 0.08 mg KOH/gram and IFT of 28.7 dynes/cm. Therefore, M.I.N. $= \frac{28.7}{.08} = 360$ (Prop. A). Inspection disclosed visible sludges on the superstructure. Consider also a 2500 gallon unit with a NN of 0.12 mg KOH/gram and IFT of 27.1 dynes/cm. Therefore, M.I.N. $= \frac{27.1}{0.12} = 226$ (Marginal).

The unit failed. Upon inspection (non-visible) sludges were layered under the insulation next to copper—in the form of an arc 28 inches long, 3/4 inches wide, and 1/8 inch thick. For this reason servicing should not be neglected beyond this point.

The next three classes stand in the "Corrective Maintenance" range and are "bad news" items requiring prompt servicing. In the Bad Category sludge has already been deposited in and on transformer parts in almost 100 percent of the units.

Insulation shrinkage and blocked cooling vents with higher operating temperatures characterize the Very Bad and Extremely Bad Categories respectively. The last category is "Disaster City". At this stage the concern should be for how much life remains in the transformer, not just the oil condition.

A *marked* change in color between oil testing periods has significance.* As a rule of thumb, once transformer oil changes from the yellows into the ambers and browns, the oil has degraded to the point where the insulating system has been seriously affected. Radical color changes may be caused by:

- Electrical problems
- Pot head or bushing compounds
- Uncured varnishes or polymers

**See Figure 3.11 and Table 6.10 in the special Color Section, following Chapter 10.*

- "Free breathers"
- New oil in dirty transformer
- Oil has been reclaimed

A situation where NN and IFT were bad, but the color was light may indicate contamination rather than oxidation would be the cause of the problem. On the other hand, bad color coupled with good acid and IFT readings would suggest an overheating or an arcing condition and indicate that a GC test is needed.* This classification chart has proven to be a valid criterion for Preventive Maintenance in evaluating and recommending servicing for used transformer oil. Experience has shown that with annual testing oils have remained consistently in the Good range year after year, especially when they were initially serviced in either the Proposition A or Marginal Oil categories.

Evaluating Results of Essential Transformer Oil Tests

We believe that NN and IFT are the most important tests, and tell the most about the conditon of the oil and its effect on the transformer. However, a proper oil-testing program recognizes the importance of all other tests in Table 3.1.

Our primary objective at this point is a realization that transformers *can* go bad. Once this fact of life is recognized, establishing an inspection and oil testing program is the first positive step. Then a realistic evaluation must follow.

The third step is to *prevent* or minimize the factors that inevitably cause transformer failures.† Finally, *corrective* steps must be scheduled to right any condition that has need of attention.‡

Additional tests are especially indicated if a sickeningly sweet odor is detected while the oil sample is being drawn. Carbon articles may also be noted in the oil. This odor can be caused by degradation products released by arcing. However, oil with a high acid content will usually have an acid odor that can mask the odor of the arc-over products. If there is reason to suspect that dissolved combustible gases are pre-

**Chapter 4.*

†Chapter 6.

‡Chapter 7, especially Part 3.

sent, the oil should be tested by gas chromatography.* Keep this fact in mind—though oil tests can detect dramatic changes in oils, they cannot detect incipient faults in the insulation system!†

If further tests of the oil indicate the possibility of damage to the transformer, other than sludging deterioration, a battery of de-energized tests—including tests for transformer turns ratio, polarization index, insulation resistance, insulation power factor, and excitation current—should be run on the transformer.‡

Criteria for Reuse of Reclaimed Oil

Inasmuch as use of reclaimed oil will likely become common practice in the years to come, Tables 3.9 and 3.10 list suggested requirements for reclaimed oil.§

**Chapter 4.*

†Oil test results can indicate certain dangers to the cellulosic or solid insulation system; for example, high water content with normal IFT and acidity; or noticeable solid insulation or carbon particles in the oil.

‡Chapter 5.

§Chapter 7, Part 3.

TABLE 3.9

SUGGESTED PROPERTY REQUIREMENTS OF RECLAIMED OIL FOR TRANSFORMERS[1]		
Property/Test	Limit	ASTM Test Method
Physical		
Flash point, minimum, °C	140[2]	D-92
Pour point, minimum, °C	-40[3]	D-97
Specific Gravity, 15°/15°C, maximum	0.91	D-1298
Viscosity, maximum, cST at 40°C (mm^2/sec)	12.0	D-88 or D-445
Color, maximum	1.5	D-1500
Visual examination	Clear	D-1524
Interfacial tension, minimum, dynes/cm (mN/m)	35	D-971
Electrical		
Dielectric breakdown voltage 60 Hz, minimum KV	30	D-877
Power factor at 60 Hz, 100°C, maximum percent	1	D-924
Chemical		
Neutralization number, maximum, mg KOH/g	0.05	D-974
Oxidation Inhibitor, maximum percent by wgt.	0.3	D-2668
Water, maximum, ppm	35	D-1533

[1]IEEE Insulating Fluids Subcommittee (proposed April, 1980) Project 637.

[2]145°C flash point may be desired for 65°C rise transformers.

[3]In certain sections of the United States and Canada, it is common practice to specify a lower or higher pour point, depending upon climatic conditions.

TABLE 3.10

SUGGESTED TEST LIMITS FOR RECLAIMED OIL IN TRANSFORMERS AND REACTORS BY VOLTAGE CLASSIFICATION[1] (After Filling But Before Energizing)				
Property/Test	Limit			ASTM Test Method
Voltage Class	69 KV and Below	Above 69 KV thru 288 KV	345 KV and Above	
Dielectric Breakdown Voltage, 60 Hz minimum, KV	30	35	35	D-877
Dielectric Breakdown Voltage, 60 Hz, .040 gap, minimum, KV	26	26	26	D-1816
Desirable, KV	30	30	30	D-1816
Neutralization number maximum, mg. KOH/g	0.05	0.05	0.05	D-974
Interfacial tension, minimum, dynes/cm (mN/m)	30	30	35	D-971
Water maximum, ppm	35	20	15 (10 desirable)	D-1533
Gas content when specified, maximum percent[2]	2	2	2	D-831, D-1817 or D-2945
Color, maximum	1.5	1.5	1.5	D-1500
Condition-visual	Clear	Clear	Clear	D-1524

[1]IEEE Insulating Fluids Subcommittee (proposed April, 1980), Project 637.

[2]Some transformers are equipped with diaphragms to prevent the introduction of air. The dissolved gas content in these transformers should be maintained in accordance with the manufacturer's recommended limits.

Part 3—The Measurement of Insulation Moisture

Introduction

Consider the basic relationship in an oil-cellulose-water-air insulation system.* It is fact that in a state of equilibrium with air, or any gas, cellulose will acquire a water content which is proportional to the relative humidity of its environment; the relationship (Figure 3.13) takes the form of a gentle "S" curve. Insulating oil behaves in a similar manner, but here the relationship is linear.

For oil-impregnated cellulose exposed to a humid atmosphere, the water content of the solid-liquid phase is again dependent on the relative humidity of the surrounding air. Therefore, false reasoning concludes (at physical and thermal equilibrium) to know the water content of the oil is to know the water content of the cellulose.

One major problem arises due to a false premise—that equilibrium between oil and cellulose is a practical normality. The fact is that equilibrium is an unnatural state in a transformer and is achieved only under closely controlled conditions. A *dynamic* situation (where the water is migrating from one medium to another) is the order of the day. Whatever method is used to determine moisture content, keep in mind that "the water content of the oil is not constant but rather continually changing with the transformer temperature and immediate past history." Therefore, in such a dynamic system, a top oil sample of the oil is required to best evaluate the moisture content of the cellulose.†

Even then, there is really no known way to determine the exact "weather map" inside an operating transformer. It is like when it is raining on the golf course, yet less than a half mile away the sun is brightly shining. No universal test exists that gives all the answers concerning the insulation environment.

A number of techniques must be used to make good engineering judgments concerning insulation of transformers. The two primary classes of insulation—liquid and solid—are subdivided as to respective quality and quantity of moisture present in the equipment.

**Chapter 2, Part 5, p. 250.*

†Chapter 6, Part 2.

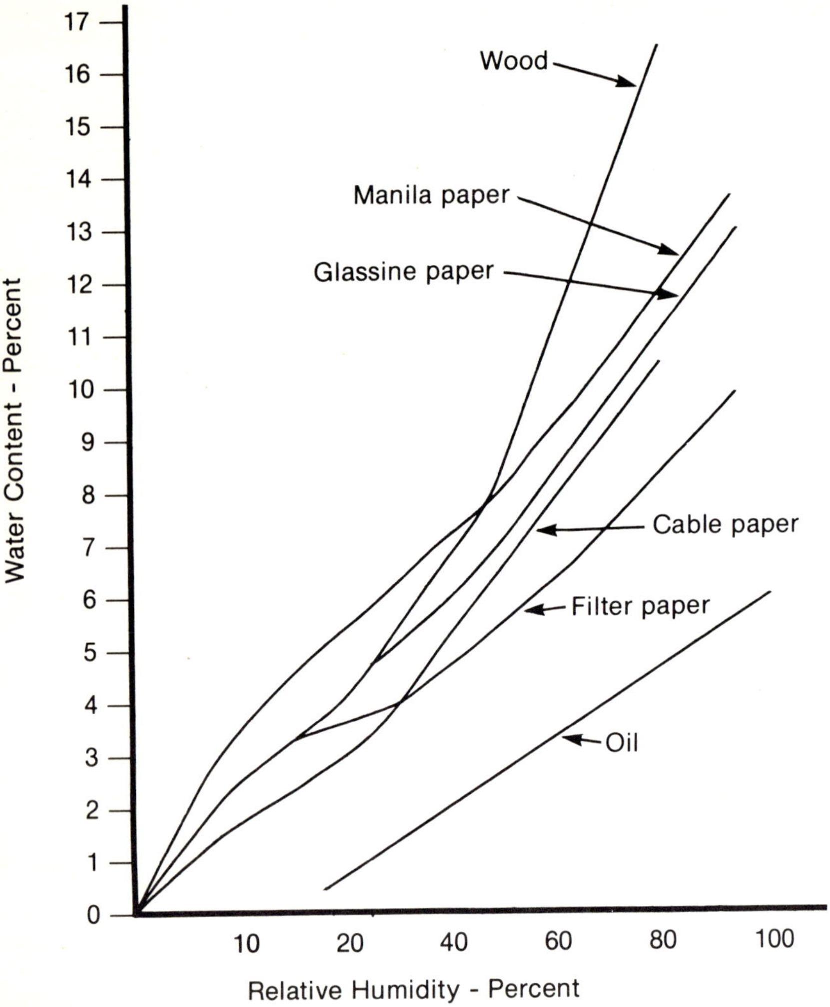

Figure 3.13 - The rate of moisture regain as a function of the relative humidity of the air with which the cellulose is in contact. F.M. Clark, ***Insulating Materials for Design and Engineering Practice****, (1962).*

Liquid Qualitative Measurement

Dielectric Strength Tests

1. Mineral Oil
   The "old work horse" of transformer oil quality testing is the

ASTM Method D-877. An oil sample is subjected to 60 Hz electrical stress between square-edged flat disc electrodes, one inch in diameter and spaced 0.1 inch apart.*

The Flat Disc cup was designed to be effective in detecting the dielectric breakdown voltage in transformer oil having a water content between the ranges of 30-80 parts per million (ppm) free water in the oil. This test is still the only universally accepted test for use on transformer oils that have been in operation and contain oxidation products.

A serious drawback to the Flat Disc test is lack of sensitivity to oil moisture contents below 60 percent of the saturation value or below 30 ppm. A more sensitive method (ASTM D-1816) is being used increasingly for in-service testing of oil.

A comparison of the sensitivities of these two methods for determining moisture content is shown in Figure 3.14. Even though there is no known mathematical relationship at this time between D-877 and D-1816, an effort is now being made to investigate the feasibility of combining the two dielectric tests. One of the major apparatus repair complexes in the U.S. now uses only the D-1816 test on transformer oils in service. Additional data would be most beneficial.

2. Askarel (PCB)
   Though askarel, the synthetic "nonflammable" insulating fluid (PCBs), is no longer sold in the United States for environmental reasons, and askarel-filled transformers are being phased out as substitutes become available, its use as an insulating liquid may continue for some years to come.

   The major "enemy" of askarel is contamination by water. Therefore, for PCBs, a most important test is moisture content in parts per million in conjunction with the dielectric strength test.†

Crackle Test

A crude method of detecting water in oil apart from the dielectric tests has been the crackle test. Since this method only indicates the presence of free water in oil, it cannot determine the total water content of oil.

Chapter 3, Part 1.

†*Chapter 9, Part 3, Table 9.15.*

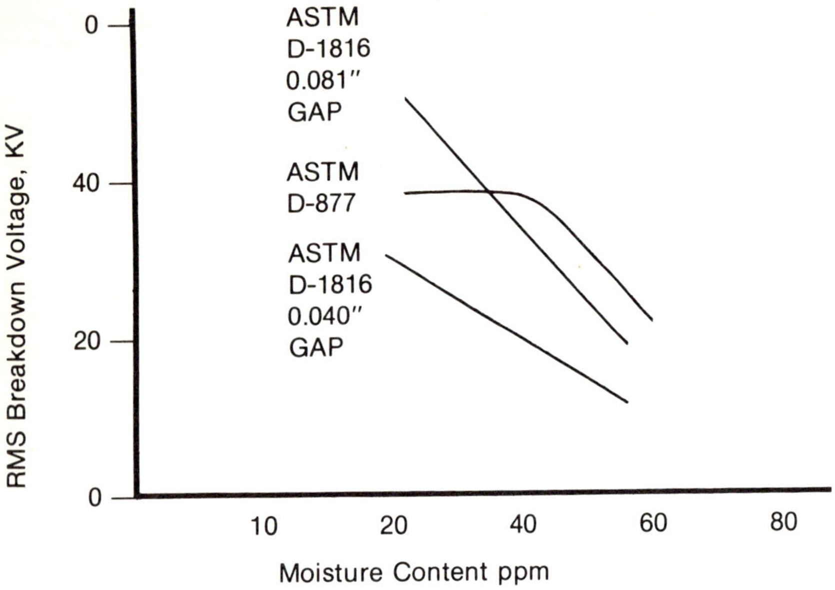

Figure 3.14 - Curve of dielectric test KV using the D-877 Flat Disc Method and the D-1816 Spherical Electrodes (VDE). ***ASTM Bulletin****, "A More Sensitive Method for Dielectric Strength of Insulating Oil," (1960).*

It may happen that testing apparatus is not available and an approximate indication must suffice. The test is then performed by heating rapidly over a silent flame a sample of oil in a test tube. If moisture is present a characteristic crackling will be detectable. A metal rod, heated to a dull red heat may also be lowered into the oil sample and used to stir it thoroughly. Crackling again would indicate free water.

Cloud Point Test

A practical test for estimating the degree of dryness of oil is the cloud test; recall Figure 2.56. It involves slowly cooling a test tube sample of the oil and recording the temperature at which cloudiness appears. If a cloud occurs above 0°C, then excessive moisture is indicated. Nonetheless further confirmation is still necessary.

Figure 3.15 shows a typical effect of moisture on the cloud point and the temperature at which the breakdown voltage is minimum. Forty-five ppm is the crucial level of moisture for new liquid.

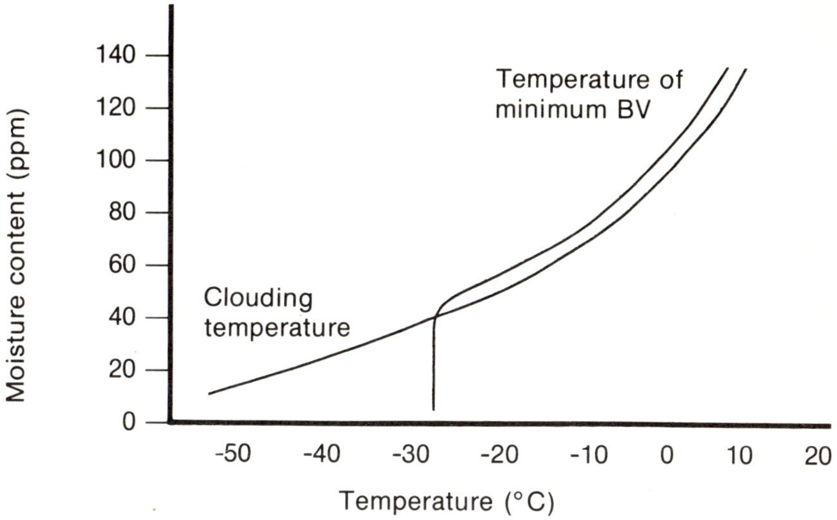

Figure 3.15 - Effect of moisture on cloud point and temperature at which breakdown voltage shows minimum value. T. Ishii, "Effect of Moisture on AC Breakdown Voltage..." (1971).

- Above 45 ppm—the cloud point and temperature of the minimum breakdown voltage are in agreement.
- Below 45 ppm—temperature of the minimum breakdown voltage is independent of the moisture content and is about 24° C. The cloud test should still be regarded as a *rough* indication only of the degree of moisture. It would not be applicable to badly contaminated oil.

Solid Insulation Qualitative Measurement

The de-energized qualitative testing of the solid insulation is often referred to as the "Doble Test". It requires a special piece of test equipment that can be purchased or leased.*

Mineral Oil-Filled Transformers

New transformers tested at the factory should have a power factor of 0.5 percent maximum. Units in-service can have power factor tests higher than 0.5 but any reading over two percent should be suspect. Typical relationships between the power factor in percent and moisture in percent dry weight are given in Figure 3.16.

Chapter 5, Figure 5.4.

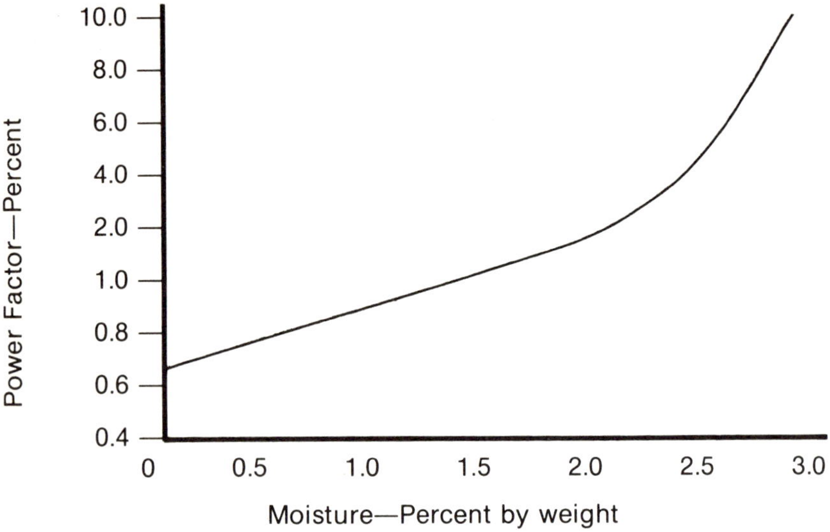

Figure 3.16 - Variation of power factor with the moisture content of oil-impregnated pressboard. W.L. Teague, "Dielectric Measurements on New Power Transformer Insulation," (1954).

Askarel-Filled Transformers

An increase in the power factor of the liquid will also affect the overall power factor of the transformer insulation system since the two are interrelated.

Recent test data for 50 typical service-aged askarel-filled and 50 oil-filled transformers revealed that the average power factor for the askarel-filled units was between four and five percent, while the average power factor for the oil-filled units was 1.1 percent or less. Tabulations were obtained with no distinction made for KVA or insulation ratings. All units were power and distribution transformers.

In time past variations up to 50 percent power factor values have been found acceptable for in-service askarel-filled transformers. However, the data also indicates that limits of five percent appear to be the norm. Power factors above five percent merit further investigation.

It is the *rapid departure* from normal values which is significant for both oil and askarel-filled units. The power factor value of transformers in service is a very important test number and takes on added value in comparison to the values previously recorded during the equipment's use. An unchanging value of power factor indicates insulation stability, while one slowly increasing during use signifies only normal insulation aging. Thus, a system showing rapid change between tests warrants definite investigation of its cause.

Liquid Quantitative Measurement

Moisture Content

A number of techniques have been investigated over the years to measure the quantity of moisture in a dielectric fluid, but the only method which has stood the test of time is that developed by Karl Fischer in the early 1930's. However, once again, it should be used in conjunction with other tests to determine evaluation on in-service oil or the insulation of the transformers.

The test, using what is described as "wet chemistry", has normally been conducted in a laboratory (Figure 3.17). The lab should be well ventilated as the chemicals are very tough on the olfactory nerves (i.e., it smells bad!). Fortunately, new equipment is available to perform field analysis, using an electronic test set (Figure 3.18).

The most obvious direct benefit is the elimination of even the possibility of further moisture contamination that might occur while a sample is being tested in the laboratory during analysis.

The traditional test method has depended on operator skill for its reliability and has tended to become inaccurate due to sensitivity of the wet-lab techniques. A thorough knowledge of the Fischer reagent and its reactions with other compounds that could form water is necessary, otherwise erroneously high results will be given. On the other hand, automatic titration capable of measuring moisture down to 10 ppm is now available and is recognized by ASTM. It can also be used as a guide to estimating the moisture content of insulation, since at equilibrium the degree of saturation of both the paper and oil are approximately the same.

Reliable estimates of moisture content of the cellulose, however, can only be made when the moisture equilibrium data for the oil in question are known or obtained. Therefore, the problem has been in trying to relate water content of the oil in ppm to the condition of the cellulosic insulation. Practical estimates can be made when the top oil temperature is known and the moisture content of the top oil sample in ppm is measured.

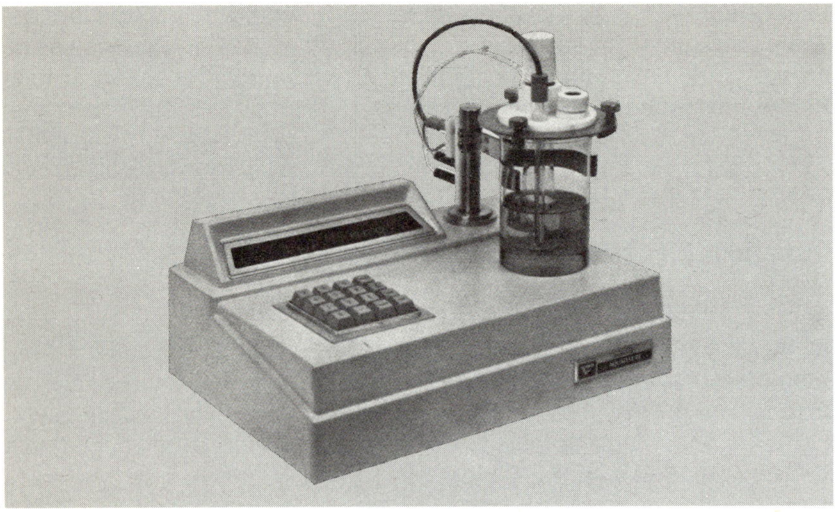

Figure 3.17 - Advanced laboratory instrument to measure moisture content in mineral oils using the Karl Fischer, water content method (D-1533). (Courtesy of Photovolt Corp.).

Figure 3.18 - Digital analyzer using the Karl Fischer, water content method (D-1533 and IEC) measures moisture in oil in the field. (Courtesy of the Micanite and Insulators Co., Ltd., Manchester, England).

Fifty parts per million (50 ppm) of water using a top oil sample is considered critical for oil-filled transformers. A water content of 40-50 ppm is the suggested maximum allowable for askarels (PCBs). It is almost automatic that under these conditions the paper insulation of the transformer is wet.

Relative Humidity

A recent approach to finding moisture content involves measurement of the relative humidity of the insulating oil. A number of important parameters of cellulosic insulation are functions of its relative humidity. These criteria include:

- Insulation resistance
- Percentage water content
- Thermal stability

Figure 3.19 shows how the humidity is virtually independent of temperature variation for a fixed amount of water in the solid insulation.

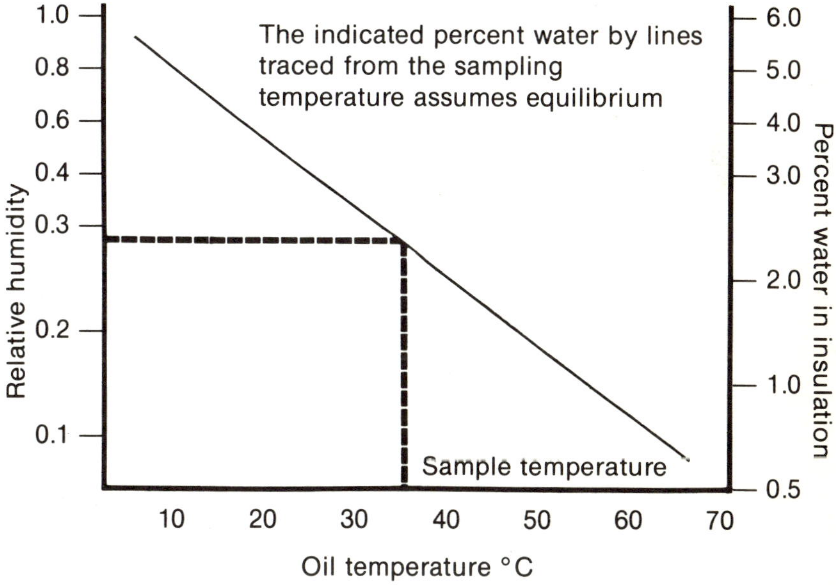

Figure 3.19 - Relative humidity versus temperature at fixed water content. C.P. Burns, "New Approaches to Testing Insulation Oils," (1976).

Since a small amount of water is in the oil compared with that in the cellulosic insulation, the oil acquires a relative humidity equal to that in the cellulose. Relative humidity may be more meaningful than parts per million when referring to the solid insulation which is independent of oil type, age, and temperature.

An oil dryness test set developed to measure relative humidity (RH) may be easily carried so it can be used on-site or in the laboratory (Figure 3.20).

Knowing the temperature of the oil sample when taken from the transformer, the equivalent RH can be obtained. By using a built-in heater to raise the oil temperature the relative humidity can be measured at different temperatures. From these results the RH/temperature curve may be plotted.

Solid Insulation Quantitative Measurement

It is imperative in preserving the life of a transformer to know the degree of moisture in a transformer's solid insulation to determine the need for dehydration or for evaluating the "end-point" of the drying process.

Foregoing sections have shown that the insulation life is (roughly) inversely proportional to the residual water content at a given temperature. *Ergo*, the lower the water content, the greater the possibility of prolonging the life of the transformer.

It has also been shown that the insulation resistance measurement is not sensitive enough to give a reliable means of judging the moisture condition of the solid insulation of a transformer in service or whether the drying process is needed or completed. The insulation power factor reading (Doble Test) is the only test sensitive enough to make these determinations.*

Dew Point Measurement

Prior to installation of a transformer, a dew point hygrometer (Figure 3.21) is used to establish the dryness in the tank on delivery. The dew point reading of the atmosphere inside the tank reveals the moisture locked into the insulation as a state of equilibrium will have been reached.

**Chapter 5.*

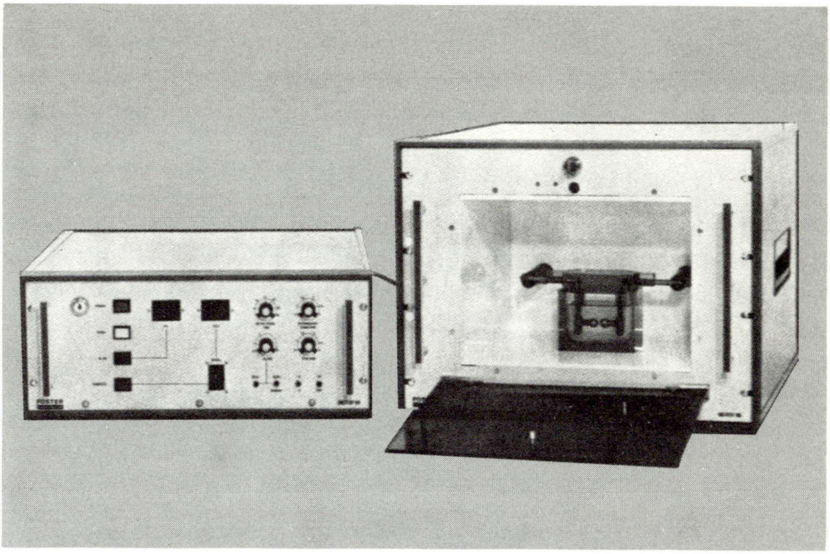

Figure 3.20 - A typical portable oil dryness tester. By using a built-in heater to raise the oil temperature the RH can be measured at different temperatures. From these results a RH/temperature curve can be obtained. (Courtesy of Foster Transformers Ltd., London, England).

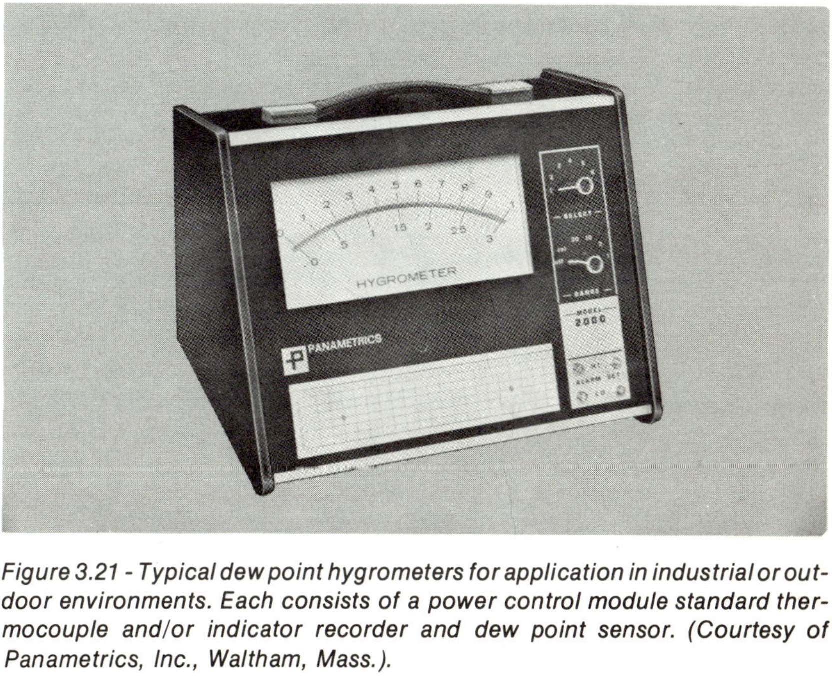

Figure 3.21 - Typical dew point hygrometers for application in industrial or outdoor environments. Each consists of a power control module standard thermocouple and/or indicator recorder and dew point sensor. (Courtesy of Panametrics, Inc., Waltham, Mass.).

Dew point determination within a closed unit, though responsive to the surface moisture on the insulation, actually measures the moisture in the gas space. Therefore, a state of equilibrium must be achieved within the unit for dew point readings to properly measure the average insulation surface moisture content.

An equalizing period of 12 to 24 hours duration is required to assure that equilibrium prevails. Equilibrium exists when the moisture content remains stabilized for at least 12 hours.

From the dew point temperature, the Piper moisture equilibrium chart (Figure 3.22)* can be used to obtain the vapor pressure in torr (or millimeters) equivalent to atmospheric pressure:

- 1 MM Mercury Absolute = 1 Torr = 1000 Microns Mercury Absolute.

The winding resistance and temperature are measured. The insulation temperature is assumed to be equal to the winding temperature. The average surface moisture content of the particular insulation system can then be found from the insulation temperature and the vapor pressure, using chart (Figure 3.23).

In time past, the measure of agreement between actual water and that determined from dew point has proved encouraging (Table 3.11). The dew point reading generally is used as a criterion of a drying "end point", or as a confirmation (Table 3.12) but not for continuous monitoring because of the time required in achieving equilibrium (at least 12 hours) and its interruption of the vacuum.†

Cold Trap Technique

Application of the cold trap—a periodic *monitoring* technique—has come into use, especially for EHV and other large transformers. This method of freezing out water vapor facilitates the measurement of progress in drying out the transformer.

The cold trap device is connected in the vacuum line between the transformer and a vacuum pump (Figure 3.24). Unless vapors are pumped with a cold trap, equilibrium pressures will quickly be reached in the system, preventing further reduction of water vapor pressure.

**Chapter 2, Part 5, Figure 2.64.*

†*Chapter 7, Part 2.*

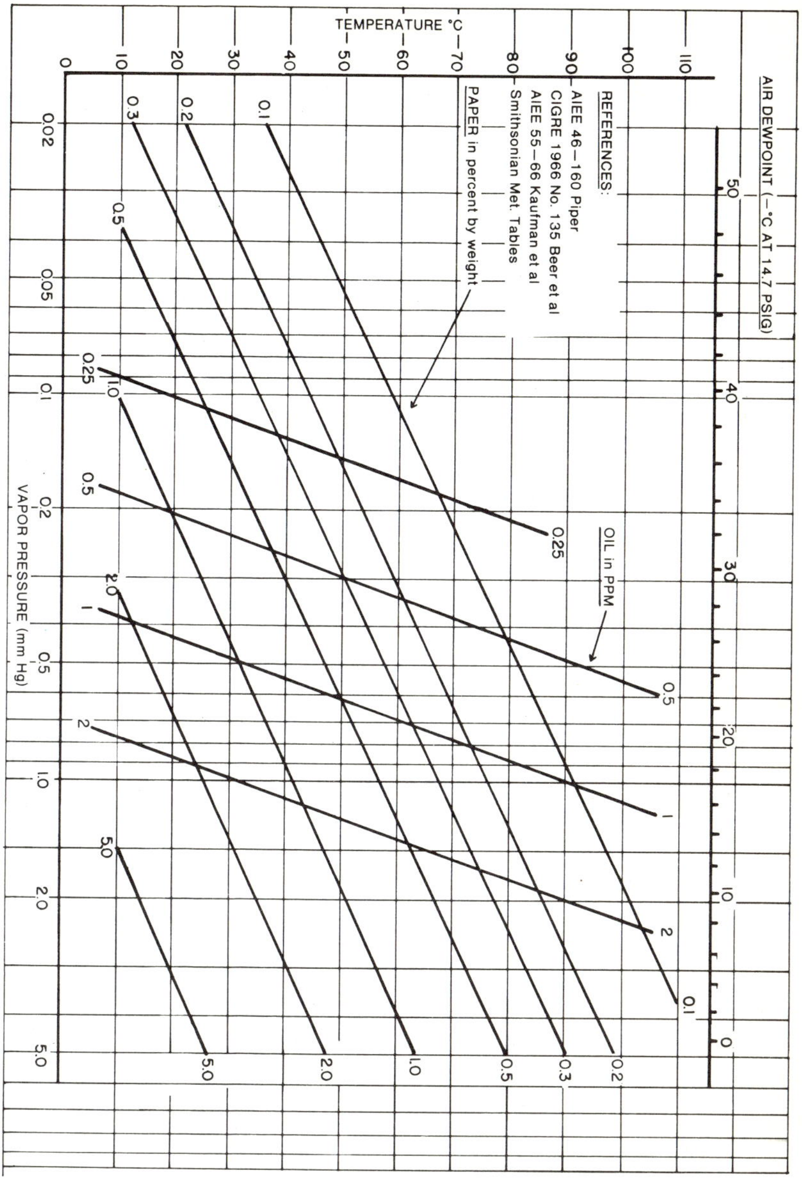

Figure 3.22 - Theoretical vapor pressure equilibria for cellulose, oil, and air "Piper Chart." J.D. Piper, "Moisture Equilibrium...(1946), and G.G. McCrae, "Transformer Insulation Drying Methods," (1975).

TABLE 3.11

COMPARISON OF CALCULATED AND ACTUAL WATER CONTENTS[1]				
Dew Point measured	Partial pressure of water vapor in Atmosphere	Residual water in insulation	Difference % water by weight from dew points	% by weight of water actually measured
+ 8 °C	105×10^{-4}	0.74	—	—
- 2.5 °C	47×10^{-4}	0.43	0.31	0.26
- 15 °C	17×10^{-4}	0.22	0.21	0.13
- 19 °C	12×10^{-4}	0.17	0.05	0.045

[1]E. B. Franklin, **Distribution of Water in Transformer Insulation** (1965).

TABLE 3.12

DEW POINT CONVERSION TO PARTS PER MILLION WATER VAPOR			
Dew Point	PPM	Dew Point	PPM
- 5°F	970	- 55°F	48.0
- 10°F	740	- 60°F	34.0
- 15°F	560	- 65°F	23.6
- 20°F	422	- 70°F	16.6
- 25°F	317	- 75°F	11.4
- 30°F	235	- 80°F	7.8
- 35°F	174	- 85°F	5.3
- 40°F	128	- 90°F	3.53
- 45°F	92	- 95°F	2.35
- 50°F	67	- 100°F	1.53

[1]U.S. NBS, Circular 564.

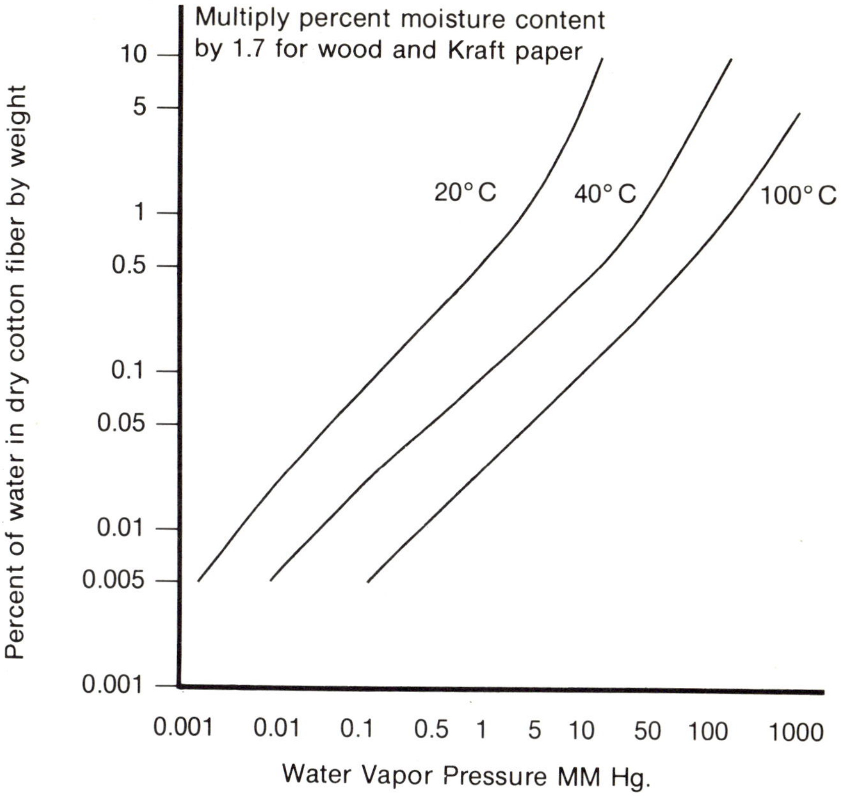

Figure 3.23 - Moisture equilibrium chart - J.D. Piper. F. Morin, "Vacuum Drying of Oil Transformers," (1967).

The dryness level is evaluated by measuring the quantity of the condensate removed from the transformer and its rate of removal. The total *quantity* of water in the insulation can be estimated by knowing the weight of insulation and its assumed moisture content. Comparing the accumulated weight of water removal against the estimated figure, permits a judgment of the condition of dryness. A comparison of the water collected in the form of ice by six hour periods will indicate the rate of water removed. One transformer manufacturer recommends continuation of the dry-out until no more than 1½ ounces of water is removed per six-hour period.*

**Chapter 7, Part 2.*

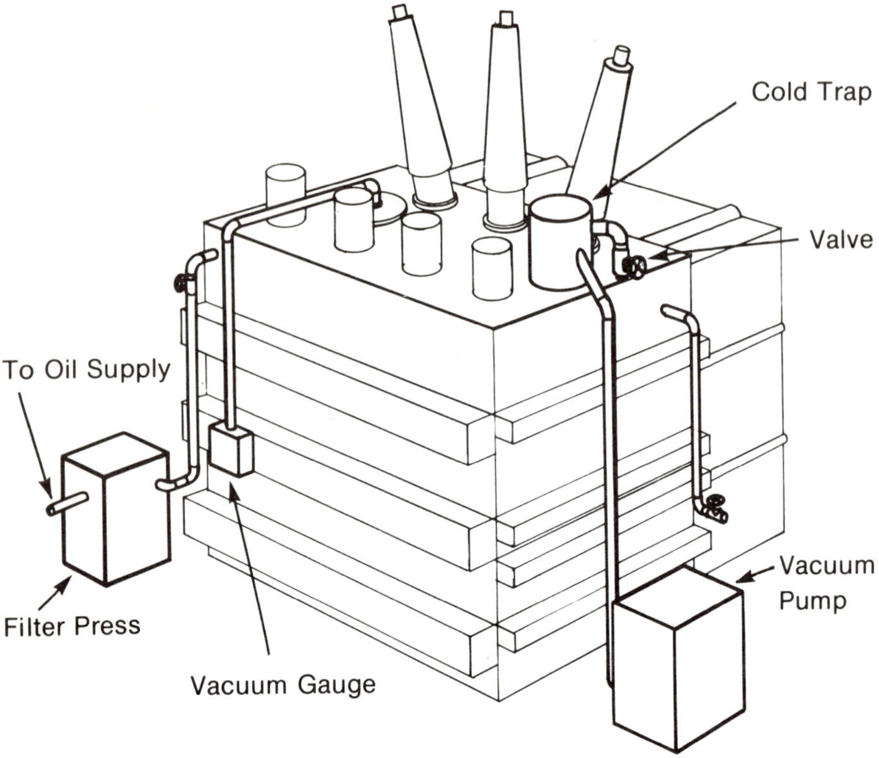

Figure 3.24 - Layout of transformer dry out, using cold trap device for periodic or "end-point" moisture measurement.

Cellulosic Test Probes for Excessive Moisture

When a transformer under repair has been exposed to the atmosphere for several weeks or months, the actual state of dryness of transformer insulation at the end of a drying process cannot be evaluated from the usual measurements, such as dew pont.

A suggested solution to the problem of deeply embedded moisture utilizes the insertion of a paper probe arrangement that is representative of the unit's insulation into a drying transformer. Monitoring of conditions takes place simultaneously. Further field studies with different insulation designs are necessary. Unfortunately, this method has not been extensively adoped because of fears that the probe may interfere with transformer operation.

How Much is "Too Much" Moisture?

During the service life of a transformer, the moisture content may increase by breathing damp air, natural aging of the cellulose insulation, oil oxidation, condensation from a workman inside of a unit during inspection and/or repair, or some accidental means, so that the water content of the paper may rise to five or even eight percent.

The present state of the art of transformer use—greater loading and high operating temperature—demands a unit with a dry insulation system at the factory, during the installation, and after every stage of maintenance requiring exposure of the system to the environment.

Measurements taken from units after a number of years in service have revealed residual moisture in the paper of approximately 4.5 percent and in some cases much more. This would include the initial moisture at installation, as well as that from oil and cellulosic deterioration. Typically the electrical breakdown of a transformer under normal operating conditions is not likely unless the water content of the cellulose exceeds 4.5 percent. Experience also indicates this percentage of moisture may be higher—for many older lightly loaded units may still operate with contents of five to six percent. On the other hand—with today's new transformer design (reduced BIL and oil/KVA) and increasinq demand—such high water content cannot be tolerated.

Laboratory data has shown that at a total moisture content of 4.5 percent by dry weight the insulation power factor puncture test (needle point to flat plane) is about 90 percent of that when the insulation is dry (Figure 3.25). Since the puncture test involves more stress than a megohm meter test, it means that the insulation resistance test, commonly relied upon to determine insulation resistance may not reveal a problem; on the contrary, it may give a false sense of security.

The presence of moisture—to whatever degree—does actual harm to insulation which is, in fact, permanent damage. Drying methods only reduce the rate of deterioration. Remember—previous loss of insulation is *never* regained even though consensus opinion agrees that moisture content of solid insulation should not exceed *two* percent to preserve dielectric strength margins; it is difficult to know *when* the moisture content reaches this critical level. Table 3.13 shows the results of insulation inspection of several Soviet 220-750 KV transformers under various conditions. Data indicates that insulation moisture may differ sharply. For a better evaluation of the state of insulation, consideration is given not only to the possible moisture content, but also to the nature of moisture distribution in the insulation.

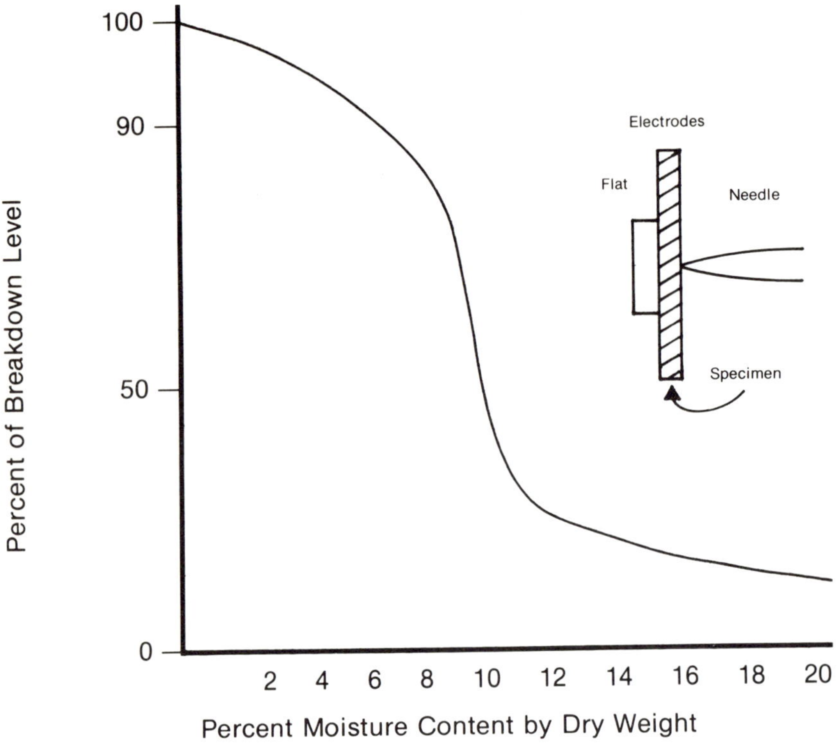

Figure 3.25 - Effect of water on the 60 Hz breakdown stress of oil-treated paper using a needle-to-plane electrode system. W.W. Guidi, "Mathematical Methods for Prediction of Moisture Take-Up...," (1974).

Unless the oil is maintained dry, the cellulose insulation, even if "dry" at the outset (say 0.5 percent residual), will inevitably pick up additional water from the oil. Therefore, it behooves the transformer user to minimize the presence of moisture through periodic preventive and predictive maintenance.

"But My Dielectric Was Good..."

...And your transformer failed anyhow! To add to this mystery, the reason for the failure turned out to be excessive moisture in the transformer. "How can this be?" you ask. The answer lies simply in what the standard dielectric test tells us.

TABLE 3.13

CELLULOSIC INSULATION MOISTURE CONTENT OF CERTAIN 220-750 KV TRANSFORMERS DURING INSTALLATION AND OPERATION[1]

Stages of Insulation Inspection	Variation in Insulation Moisture Content % by Dry Weight of Insulation	Nature of Moisture Distribution	Source of Information
Following storage without oil in transport state, less than 6 months	1% or loss increase	Non-uniform. Increase in moisture in layers to 0.5 mm deep by 0.4-10% of their mass	Inspection of 35 transformers
After long-term (more than a year) storage in transport state	Increase by 0.5-2.0%	Weakly non-uniform. Moisture of surface and inner layers differ slightly	Inspection of 9 transformers
Following desealing during installation and in operation under normal conditions	Increase by 0.1-0.2%	Sharply non-uniform. Possible increase in moisture surface levels at 0.5 mm deep by 1-1.5% and by 3-5% at the depth of 0.1 m	Inspection of insulation models
In operation. Protection of oil by means of silica gel air dryer	Average increase by 0.2% per year	Non-uniform. A difference has been noted in moisture of individual insulation zones of up to 1-1.5%	Inspection of 43 transformers operating up to 15 years
Nitrogen protection of oil	Moisture content of 1% or less of the average following 5.6 years of operation. Cases have been noted of an increase in moisture to 1.5-2%	Weakly non-uniform at moistures less than 1.0%. Non-uniform with difference in moisture in individual zones to 1.0% in transformers with high moisture	Inspection of 14 transformers operating up to 8 years
Film protection of oil	Average moisture following 2 and 6 years of operation 0.5% or less	Weakly non-uniform	Inspection of 2 transformers

[1]M.E. Ierusalimov, "Methods for Evaluating the Moisture Content of Large Power Transformers," (1978).

Setting the Record Straight

The classic dielectric test (ASTM D-877) can reveal conductive contaminants, such as the rust or other metallic cuttings, dirt, cellulosic fibers, filter dust, or free water but it will *not* pinpoint the presence of *dissolved* water, below 80 percent saturation, acids, or sludges. F.M. Clark found in early studies that the standard dielectric test does not give good protection against the insidious penetration of water and its absorption by the cellulosic insulation.

Keeping in mind today's higher operating temperature, experience has shown erroneous conclusions can be drawn (from the dielectric breakdown test) if the oil is tested at a temperature in excess of 40° C because of the phenomenon of water solubility.

Similarly, while the transformer oil contains an appreciable amount of dissolved water, this water would have no effect on the overall power factor of the transformer. However, if the unit was tested later after the oil had cooled off, the power factor may be higher due to the change of the excess water from a state of solution to a state of suspension, which does affect the power factor of the oil.

Confession was made in 1948 to what had already been known for several years, "The time-honored dielectric...was known to be not entirely reliable, even though it has been used many years in industry."*

Yet by 1962, Dr. Clark comments that many were still not convinced "that the dielectric strength of mineral oil containing moisture in *solution* is not affected by its presence." Moreover, ASTM states that a high dielectric strength is not a certain indication of the absence of all contaminants, especially moisture, that is, water in *all* its forms (D-117).† Still today, both Doble and IEEE surveys rank the D-877 dielectric test as the most used field screening test. The D-877 remains in this position because of tradition, availability of equipment, and of course, it's easy for maintenance people to use.

What Dielectric Tests Do Not Tell Us

Engineering calculations have found that well over 99 percent of the moisture in a transformer is locked up in its cellulosic insulation, while

**H.A. Cornelius, "Developing a Program to Keep Electrical Equipment Insulation High in Quality," (1948).*

†Chapter 2, Part 5.

the moisture in the insulating liquid is less than one percent. Why? Because the transformer manufacturer has built into each unit a "filter press" that has a fantastic affinity for water.* What is this "filter press"? The cellulose insulation system of the transformer.

Remember that "the seeds of destruction" inherent in the cellulosic paper—the weak secondary hydroxyls and its absorptive characteristic for oil decay products—perform these functions very early in the life of the transformer, including absorption of free water.† The dielectric test of the oil reveals *nothing about the condition of the solid insulation.*

Again in this connection, Figure 3.26 illustrates the equilibrium relationship that could exist between water in the transformer oil and water in the cellulosic insulation. Note that at 20° C a transformer oil having a measured water content of 10 ppm could be in a state of equilibrium with an insulating system that contains four percent of its own weight in water. For example, a 40 MVA unit having 10,000 pounds of insulation could have 400 pounds of water locked up in the cellulosic insulation. This is equivalent to nearly 48 gallons of water in the insulation in a state of equilibrium with oil containing only 10 ppm of water!

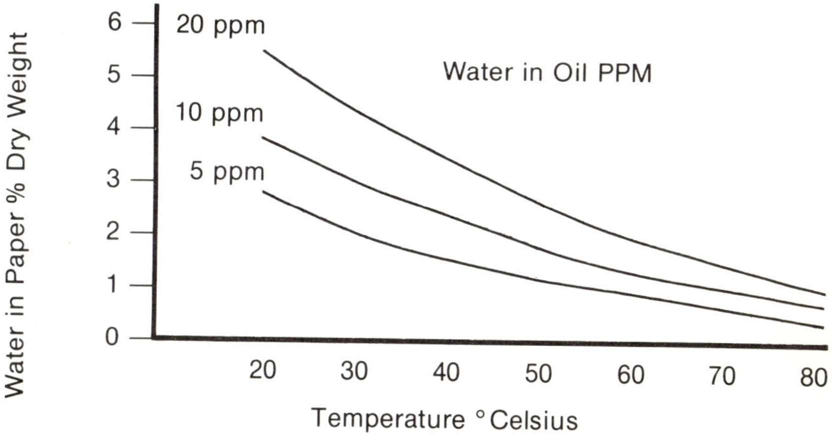

Figure 3.26 - Moisture equilibrium in oil paper system. E. Povazan, "Modern Techniques in Processing Transformer Oils," (1975).

* *Chapter 7, Part 3.*

† *Chapter 1, Part 1, Figure 1.45; Chapter 2, Part 5, Figure 2.32.*

As one example, consider a case history of three sister 500 KVA units (*typical* data for one of the units is given in Figure 3.27A). In the case of all three units, oil test data was good, particularly the dielectric! Yet electrical test data (Figures 3.27B and 3.27C) show how it is possible to have a good oil analysis (dielectric of 47) and yet have questionable windings because of high insulation power factor (6.8 percent) and low insulation resistance (16 M ohms).* Suffice it to say in this context that this abnormal data was positive indication of a moisture problem.

A second actual case history of a power transformer (approximately 4000 gallons of transformer oil with about 8000 pounds of cellulose insulation) also illustrates the facts presented. Multiple flat disc tests of the oil indicated that the dielectric varied between 38 KV and 41 KV. On this basis alone, the transformer was rated as good and dry. This unit was put back on the line and continued to operate for awhile without immediate problems. Other events, however, indicated problems. While the insulation resistance on the system was marginal, the power factor of the insulation was 45 percent! (Two percent at 20° C is cause for concern!) When the unit was untanked, approximately 130 gallons of water was found to be locked up in the insulating system! In fact, 1500 gallons of transformer oil were so loaded with water and sludges that the oil had to be thrown away as being beyond reclaiming.

Both theory and experience, therefore, point out dramatically that industry and utilities have for years placed unmerited confidence in the dielectric test. A good dielectric breakdown test (34 KV or higher) can mask a dissolved water problem in the winding of a transformer.

What Dielectric Tests Do Tell Us

It is not our purpose to say the ASTM dielectric tests D-877 and D-1816 do not serve any useful purpose. First, while they have their valid uses, they are subject to question because of the skill required in drawing the oil sample.

Sampling must be done in a relatively dry atmosphere which will not introduce the very contaminant for which later tests will be made. The *worst* possible environment is a sweaty electrician smoking a pipe or cigarette on a very humid day making a dielectric test!

Both dry and clean apparatus and hands, as well as a normal day (50 percent humidity or less), make for the most accurate sampling.

**Chapter 5.*

A

Customer City State

LOCATION OF TRANSFORMER

TC NUMBER

Substation Name or Number

Unit Or Tag Number

TRANSFORMER CONSULTANTS ®

☐ Indoor ☐ Outdoor ☐ Roof

☐ Ground ☐ Platform ___ Ft ☐ Pole

NAMEPLATE INFORMATION

Manufacturer West. Primary Volts 12000 Gal. 400 ☒ Oil ☐ Askarel

Serial Number 7375600 Secondary Volts 120/240 Impedence 6.7%

KVA Rating 500 Phase/Cycle 1/60 Other Insuldor

Radiators ☒ Yes ☐ No Conservator Tank ☐ Yes ☒ No Top Valve 1" Accessibility – Power on Power off

Fans ☐ Yes ☒ No Tap Changer Comp. ☐ Yes ☒ No Bottom Valve 1" Testing

H_2O Cooled ☐ Yes ☒ No Bushings: ☐ Top ☒ Side Other Access ReRef.

Other

ReRefining Parking ☐ Yes ☐ No Hose distance ___ Feet Power 3 ☐ Yes ☐ No Voltage ☐ 220 240 ☐ 440 480

TRANSFORMER LIQUID TEST DATA

	DATE	Service	LEVEL	TEMP	P V	DIEL	ACID	IFT	COLOR	SPECIFIC GRAVITY	VISUAL	LEAKS	PAINT	OIL CLASS	RECOMMEND
1	2/75	MTL				32	0.05	33.9	2.0	0.868	Clr	None	Fair	Good	Retest 1 yr
2	2/76	MTL				Not Tested - Per Customer Request									
3	7/76	MTL				47	0.05	31.2	2.5	0.855	Clr	Yes	Fair	Good	Retest 1 yr
4	7/76	ET				Electrical Tests Performed									
5															

B

DOBLE INSULATION TESTS

TWO-WINDING TRANSFORMERS

DOBLE ENGINEERING COMPANY
BELMONT, MASS
FORM MH-2W178

OIL ☒
ASKAREL ☐
AIR ☐
GAS ☐

COMPANY DIVISION DATE 7/26/76

LOCATION OF TESTS AIR TEMP. 80° TOP OIL TEMP. 25°C

TRANSFORMER Spare, Furnace WEATHER Hot, Clear % HUMIDITY

MFR. West. SERIAL NO. 7375600 AGE TYPE/CLASS SL/OA KVA 500

FREE BREATHING ☐ SEALED ☒ GAS BLANKETED ☐ CONSERVATOR ☐ GALLONS OF OIL 400

BUSHINGS: MFR. TYPE CLASS DWG. NO. CAT. NO. KV YEAR

HIGH SIDE KV 12K Y☐ Δ☐

LOW SIDE KV 120/240 Y☐ Δ☐ TC# 6

NEUTRAL DATE LAST TEST

COPIES TO LAST SHEET NO.

OVER-ALL TESTS CF .90

TEST	TEST CONNECTIONS: WINDING ENERGIZED	WINDING GROUNDED	WINDING GUARDED	TEST KV	EQUIVALENT 10 KV READINGS – MILLIAMPERES: METER READING	MULTIPLIER	MILLIAMPERES	WATTS: METER READING	MULTIPLIER	WATTS	% POWER FACTOR: MEASURED	COR. 20°C	KEY TO INSULATION RATING (G = GOOD, D = DETERIORATED, I = INVESTIGATE, B = BAD (REMOVE OR RECONDITION))	INSULATION RATING
1	HIGH	LOW		2.5	49	200	9000	20	10	200				B
2	HIGH		LOW	2.5	25	100	2500	19	10	190	7.6	6.84	C_H	B
3	LOW	HIGH									— —	— —		
4	LOW		HIGH										C_L	
CALCULATED RESULTS					— —	— —	7300	— —	— —	10	.14	.13	C_{HL} (TEST 1 MINUS TEST 2)	G
					— —	— —		— —	— —		— —	— —	(TEST 3 MINUS TEST 4)*	— —

Figure 3.27 - (A) Typical "Good" dielectric test data masking a moisture problem; (B) Electrical test (ET) data reveals a possible moisture problem.

INSULATION RESISTANCE - DIELECTRIC ABSORPTION TEST SHEET

POWER TRANSFORMER

TEST NO. ____________

COMPANY ____________ DATE 7/19/76

LOCATION ____________ TIME ____________

EQUIPMENT Spare, Furnace	RATING 500 KVA	VOLTAGE 12K/120-240	
TYPE SL/OA	MFR. West.	SERIAL # 7375600	
RECENT OPERATING HISTORY Being tested for future use.			WINDING AGE

STATE IF TRANSFORMER IS OIL FILLED AND OF THE CONSERVATOR, FREE-BREATHING OR GAS-FILLED TYPE Oil Filled, Sealed

IF OF THE FREE-BREATHING TYPE DESCRIBE THE BREATHER ARRANGEMENT AND DEHYDRATING MEDIUM USED

TC# 6

IF OF THE FREE-BREATHING TYPE NOTE INTERNAL CONDITIONS SUCH AS UNDERSIDES OF COVERS, TERMINAL BOARDS, WINDINGS ETC.

BUSHING DATA

OIL TESTS: ACIDITY ____ MG KOH/G. RESISTANCE IN STANDARD CUP ____ MEGOHMS. TOP OIL BREAKDOWN ____ KV.

STATE WHAT BUSHING PORCELAINS WERE GUARDED DURING TESTS

TEST DATA - MEGOHMS

PART TESTED				
GROUNDING TIME				
TEST VOLTAGE				
TEST CONNECTIONS	2500 TO LINE / PRI. TO EARTH / GND. TO GUARD	2500 TO LINE / PRI. TO EARTH / SEC. TO GUARD	TO LINE / TO EARTH / TO GUARD	TO LINE / TO EARTH / TO GUARD
¼ MIN.	23 M OHMS	1155 M OHMS		
½ "	22 M OHMS	1485 M OHMS		
¾ "	20 M OHMS	1650 M OHMS		
1 "	16 M OHMS	1980 M OHMS		
2 "	18 M OHMS	2046 M OHMS		
3 "	23 M OHMS	2310 M OHMS		
4 "	26 M OHMS	2475 M OHMS		
5 "	31 M OHMS	2475 M OHMS		
6 "	26 M OHMS	2475 M OHMS		
7 "	26 M OHMS	2475 M OHMS		
8 "	26 M OHMS	2475 M OHMS		
9 "	28 M OHMS	2640 M OHMS		
10 "	30 M OHMS	2640 M OHMS		
10/1 MIN. RATIO	1.87	1.33		

TEST MADE	HOURS DAYS AFTER SHUTDOWN
DRY-BULB TEMP.	75° °F
WET-BULB TEMP.	°F
DEW POINT	°F
RELATIVE HUMIDITY	%
ABSOLUTE HUMIDITY	GR./#
EQUIPMENT TEMP.	24°C °F(C)
HOW OBTAINED	
	C.F. 1.320 At
	24°C
"MEGGER" INST.:—	Biddle
SERIAL #	8638
RANGE	0-50 K M OHMS
VOLTAGE	500/1000/2500

REMARKS PRI. - GND. @ 500V - 55 M OHMS

TESTED BY E. Whittington

Figure 3.27C - ET data confirms a moisture problem in a transformer having a "good" dielectric.

Meaningful tests will not be attained if the oil sample is handled in dirty containers or subjected to contact with dirty laboratory equipment. It is especially important that the sample be protected from contact with light, air, and moisture. The sample should be drawn quickly, kept in a sealed container, and tested within a matter of a few hours. It is particularly important that the sample not be exposed to the ultraviolet rays of sunlight; exposure to sunlight for a matter of days will accelerate aging to the point where the sample is wholly unrepresentative of the condition of the transformer's oil.

Secondly, in the past too much has been read into the classical dielectric results. However, as a typical example, consider a case history of a 2500 KVA unit. Test results included a good dielectric strength of 35 KV and moisture content of 30 ppm. Yet later the unit was found to have a seven percent power factor (normal reading is below two percent!), and the cellulosic insulation was extremely wet! However, method D-877 does serve to indicate the presence of conductive contaminants such as free water, dirt, cellulosic fibers, filter dust, metallic cuttings, or splinters, but not dissolved water under 80 percent saturation, acids, or sludges!

Manufacturers acknowledge that a certain minimum amount of such fine contamination may remain on the internal components of large power transformers, even following the highest degree of flushing. A bad dielectric could thus mean significant solid contamination, (Figure 3.28) an arcing condition, the presence of combustible gas, or a tank or water leak. Yet carbon from arcing does not necessarily lower the dielectric strength.

On the other hand, a properly documented dielectric of 24 KV should require taking the unit off the line for a complete series of electrical insulation tests.* A reading of 22 KV or below necessitates an immediate drying out of the insulation system, because the cellulose paper is extremely wet.

The more sensitive ASTM D-1816 VDE method is recommended for the testing of filtered, degassed, and dehydrated oil prior to and during the filling of equipment rated above 230 KV, and for testing new or reclaimed oil after filling. Although this method may provide early warning of the presence of minute quantities of dissolved water in used oil, it *cannot* provide a complete picture of what's happening to the cellulosic paper insulating material.

**Chapter 5.*

One of the larger apparatus repair shops of a major transformer manufacturer will now only utilize the D-1816 test procedures in in-service oils of all types, contrary to ASTM D-1816 instructions.* Their purpose is to evaluate the D-1816 and build a data base. What's the purpose? To help do away with the classic D-877 flat disc test cup, which is recognized by nearly all authorities as almost useless.

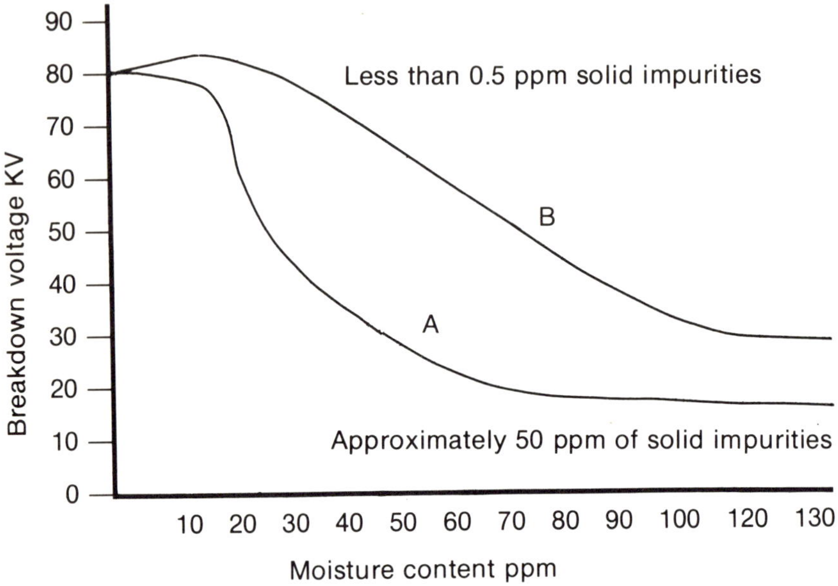

Figure 3.28 - Breakdown voltage of a transformer oil at 25° C between spheres of 12.5 MM diameter, 3.5 MM apart. Technically pure oil containing approximately 55 ppm of solid impurities (A); the same oil after filtering (B). R.W. Sillars, ***Electrical Insulation Materials and Their Application*** *(1973), p. 218.*

**The current ASTM Standards (D-877 and D-1816) are in the process of being revised (February, 1981).*

Chapter 4

Gas-In-Oil Analysis VS All Other Methods

Introduction

Two ways are available to the transformer owner/operator for monitoring units without power shutdown. The first is to make periodic oil tests in order to establish trends and to classify oils. The second, which is the focus of this chapter, is to analyze oil for *dissolved* combustible gases. Certain combustible gases are generated as a transformer undergoes abnormal thermal and electrical stresses. Both the oil and cellulosic insulating materials break down as a result of these stresses to yield gases.

The rate and amount of gas generation are important as normal aging produces gases but at an extremely slow rate; however, incipient or newly-formed fault conditions change these items drastically. More importantly, an overwhelming majority of these incipient faults give early evidence, and therefore can be detected when a transformer is subjected to proper analysis. The gas chromatograph (GC) is the most practical method available to identify the combustible gases. GC involves both a qualitative and a quantitative analysis of gases dissolved in transformer oil.

Current procedures involve the extraction and measurement of the gases that are dissolved in the oil, including the identification and measurement of the component gases. While mineral oil consists of some 2871 estimated liquid hydrocarbon compounds, it is remarkable that only nine gases are usually looked for in the insulating liquid. These gases are shown in Table 4.1. The measurement range is in parts per million (ppm).

These gases are usually the result of a stress acting upon organic insulating materials. The presence and quantity of these individual gases, extracted from the oil and analyzed, reveal the type and degree of the abnormality responsible for the gas generation.

Historical Background

The presence of these dissolved combustible gases in transformer oil has been known for over 60 years (since World War I). Laboratory work done details a characteristic of hydrocarbons little appreciated—that gases radically different from oil vapor, form on the passage of a disruptive discharge at or below the oil surface.

TABLE 4.1

TYPICAL GASES GENERATED BY TRANSFORMER FAULTS[1]	
Name	Symbol
Hydrogen[2]	H_2
Oxygen	O_2
Nitrogen	N_2
Methane[2]	CH_4
Carbon Monoxide[2]	CO
Ethane[2]	C_2H_6
Carbon Dioxide	CO_2
Ethylene[2]	C_2H_4
Acetylene[2]	C_2H_2

[1]Even though detectable and indentifiable, the presence of propane (C_3H_8), propylene (C_3H_6), and butane (C_4H_{10}) are not of significant concern in transformers.

[2]Denotes combustible gas.

Prior to 1920

M.S. Tswett (Tsvet), a Russian botanist, was having trouble analyzing a mixture of plant pigments. In analytical chemistry, the most difficult step is oftentimes separating the mixture prior to identification. Dr. Tswett first investigated this art of separation in 1903.

The Russian packed a glass column completely with a powdered calcium carbonate mixture. He then poured an unknown liquid through the column where it stratified in different colored bands. The tube was broken and each separate band was analyzed and found to be a fairly pure compound. Chroma is the Greek word for "color" and thus the designation "chromatography" (1906).

Actual realization of the gassing problem in energized transformers was discovered and then documented in the February 1919 issue of *The Electric Journal*:

> A further characteristic of hydrocarbons hitherto (known) but little appreciated is the formation, on the passage of a disruptive discharge, at or below the oil surface, of gases radically different from oil vapor. The vapors arising from the oil surface as a result of a gradual increase in oil temperature and the accumulation of which forms the basis of the flash point test, are still essentially oil and subsequently condense, on decreasing the temperature, to resume their original character. However, the gases resulting from the disintegration of the oil molecules are permanent over ordinary temperature ranges and are present in the following approximate percentages:
>
> | | |
> |---|---|
> | Carbon dioxide | 1.17 |
> | Heavy hydrocarbons | 4.86 |
> | Oxygen | 1.36 |
> | Carbon monoxide | 19.21 |
> | Hydrogen | 59.10 |
> | Nitrogen | 10.10 |
> | Methane | 4.20 |
> | Total | 100.00 |
>
> This analysis indicates the presence of a preponderating quantity of hydrogen and a relatively small amount of hydrocarbons. The nitrogen and oxygen in the reaction products show the presence of 13 percent occluded air. In view of the violence accompanying the reunion of oxygen and hydrogen, as well as the fact that hydrogen is explosive in air, within a range of 10 to 66 percent, it is evident that a destructive force can readily be obtained on the ignition of an atmosphere consisting of air and decomposed oil.
>
> With a proportion of hydrogen to oxygen of two to one existing in the gaseous mixture, thereby assuring complete combination of both gases and most rapid reaction, the speed of flame

propagation is about 11,000 feet per second, or approximately ten times that of sound waves in air.

With gases initially at atmospheric pressure, an explosion would produce a pressure of eight atmospheres or about 120 pounds. Assuming the evolution of gases from an arc below the oil sufficiently rapid to create a pressure above the oil, in case there is no ventilation, of 1.5 atmospheres, the maximum pressure of the explosion wave would be about 12 atmospheres or 180 pounds per square inch. As this wave, however, has a very steep front, due to its high velocity, the force applied to the structure is in the nature of an impact and the stress on the parts above the oil level may approximate 360 pounds per square inch, representing a loading sufficient to distort the container or shear the holding bolts.

Since the temperature required for the disintegration of the oil is necessarily very great, slight differences in flash or fire points of the oils used do not materially influence the ease with which such disruption is produced. No appreciable protection is, therefore, obtained by substituting a high flash for a low flash oil. In fact it has been observed that, due to the lower vapor pressure and consequent decreased dilution of hydrogen by the less active hydrocarbon vapors, explosions in the higher flash oils were occasionally more violent.

The Buchholz Relay (1928)

In response to the acknowledged problem of gas formation by faults, power engineers have developed methods of collecting the gases generated in the transformer with expectation of being able to diagnose the source of the problem. The classical example of such a device was the Buchholz relay introduced in 1928. This attachment would detect the passage of gas bubbles generated in the transformer as they passed through the pipe connecting the transformer to the conservator tank.

Other devices that could collect or detect gas formation were developed. Once operation of these attachments indicated evidence of gas, the gases were sampled and then transported to a laboratory for analysis.

Recent Developments

The inconvenience of depending on a laboratory for the analysis led H.H. Wagner in 1959 to develop a field combustible gas detector that could sample and identify the presence of combustibles in the head-space of a transformer. This fault gas detector would give a measure of the total combustible gas present, expressed as a percent.

Following the suggestion of A.J.P. Martin and R.L.M. Synge in a study for which they were later awarded the Nobel Prize, A.T. James and Martin introduced gas-liquid chromatography in 1952. Chromatography has since advanced to where it became possible to separate a mixture of gases. Therefore, today we have liquid chromatography and gas chromatography as useful tools. In the early 1960's, gas chromatography (GC) was first applied to the identification of fault gases dissolved in transformer oil.

Using gas chromatography for combustible gas identification has placed the problem back in the laboratory—where not only can the problem be quantified, but where we can often pinpoint the specific source.

Defining the Gas Problem

Sampling New Oil

If a barrel of new oil were sampled, the gases extracted and analyzed, a typical analysis would show approximately 69.8 volume percent nitrogen and 30.2 volume percent oxygen. Other atmospheric gases may be present in small amounts. Light hydrocarbon gases will usually not be present under normal conditions.

Sampling In-Service Oil

Now place the oil next to cellulose in an operating transformer and sample the oil for gases. If the unit is new and there is no fault existing, the analysis will show predominantly atmospheric gases (including the nitrogen from the head-pressure gas). If the unit is vacuum-filled, the majority of the gases present in the oil will be nitrogen, but at a greatly reduced percentage compared to a unit not so processed.

A transformer operating *normally* generates certain gases due to natural aging of the insulation. These gases appear in decreasing amounts: slight amount of carbon dioxide, carbon monoxide (very slight), and traces of hydrogen, methane, and ethane.

The magnitude of these normal concentrations depends largely on the unit's in-service age and history of loading; therefore, finding gas in an operating transformer is not a "go" or "no go" test for serviceability. This is why an interpretation of the test data is an art rather than an exact science.

When appreciable amounts of combustibles are present in the oil, something has happened in the transformer such as overheating, a loose connection, a break in the cellulose integrity, arcing, partial discharge, or corona.

Overheating by overloading is now being taken much more seriously, as there is hard evidence of gas bubble formation from cellulose breakdown at temperatures as low as 140° C. Yet this temperature is within the operating range covered by the current transformer Loading Guide. Caution must be exercised when considering transformer overloading.*

What Causes the Combustible Gases?

Components of the Insulation System

When we look at the transformer as a chemist does—that it is a chemical reactor that just happens to *transform* electricity—we can gain an insight into the causes of the gas generation. Recall that the major components of this "reactor" are the tank, the core and superstructure, the conductors, the bushings, the cellulosic insulation, and the insulating oil. At this point, our interest continues to be only the last two items as they are the Achilles' heel of the transformer, since they are organic and, as such, are subject to decomposition from various forms of stress.†

Primary and Secondary Causes of Transformer Faults

The *primary* causes of fault gases are the thermal, electrical, and mechanical stresses which result in the following conditions:

**Fortunately ANSI/IEEE C57.92 (1962) is undergoing revision (IEEE project Number 507).*

†Chapter 1, Part 3, p. 119; Chapter 2, Part 5, Figure 2.32, p. 212.

1. Corona (Partial Discharge) and Sparking

   a. Corona—electrical stressing resulting in ionization; first occurs around 12,000 volts over sharp edges of current-carrying conductors.

   b. Sparking—a single short electrical discharge lasting a microsecond or less.

2. Thermal Heating (Hot Spots and General Overheating)

   a. Hot spots—localized overheating. Incipient faults can reach 500° C without sufficient heat to char the cellulose.

   b. General overheating—without hot spots.

3. Arcing—a prolonged electrical discharge producing a *bright* flame-colored arc in contrast to *dim* corona-type glow.

These symptoms differ essentially in the intensity of energy that is dissipated by the fault.

The secondary causes of combustible gases accumulating in the oil include:

- Nitrogen blanket contaminated
- Old fault corrected—not degassed
- Atmospheric conditions (pollution)
- FOA transformer—motor burned out
- Hydrolysis (if free water present)
- Transformer shipped with CO_2 gas
- Transformer will retain ten percent of volume of oil

Combustible gases detected in the oil from these sources are indicative of abnormal conditions but *not* fault conditions.

> A classic incident occurred where the combustible gases in a transformer on the West Coast were so high that a telephone call was necessary to alert the customer to the danger of an imminent failure. When the transformer maintenance contractor told the customer to remove the transformer from the line at the earliest possible moment, if not immediately, his response was unexpected. He said, "What do you mean, ready to blow up? This is a brand *new* transformer and we haven't even energized it yet! As a matter of fact, we are just now building the substation where the transformer is to be installed."

Investigation revealed that the refiner had some problems at the plant and had shipped out several tankers of oil and that it was being recalled. This particular client ended up with some of this oil in his new transformer.

Laboratory Demonstration—Origin of Combustible Gases

If we were to take a transformer's oil and paper insulation system to the laboratory to conduct experiments on models that stress the components similar to those conditions found in an operating transformer, we would be able to collect the following data:

1. Cellulose
   a. Overheating of cellulose
   When cellulose is overheated in a closed system and the gases collected, analysis shows the primary products of decomposition at temperatures as low as 140° C to be:
      - Carbon monoxide
      - Carbon dioxide
      - Water

   b. Pyrolysis of cellulose
   When cellulose is heated to destruction (pyrolysis), Figure 4.1 illustrates the normal decomposition products. At temperatures above 250° C (in a sealed unit) more CO than CO_2 may be formed, perhaps as much as four times CO than CO_2 by volume.

2. Transformer Oil
   a. Overheating oil
   When insulating mineral oil is overheated at temperatures up to 500° C, these hydrocarbon vapors will be liberated:
      - Ethylene
      - Ethane
      - Methane

   Other products liberated below 500° C with some oxygen present are:
      - Carbon dioxide (400° C)
      - Water (200° C)

b. Pyrolysis of oil

When oil is subjected to extreme electrical stress (such as an electric arc), this quantity of gases will be liberated:

Hydrogen	60.0-80.0 percent
Acetylene	10.0-25.0 percent
Methane	1.5-3.5 percent
Ethylene	1.0-2.9 percent

Note the absence of carbon oxides in this case. This is surprising when we recall that mineral oil contains over 2800 hydrocarbons and only so few products are generated.

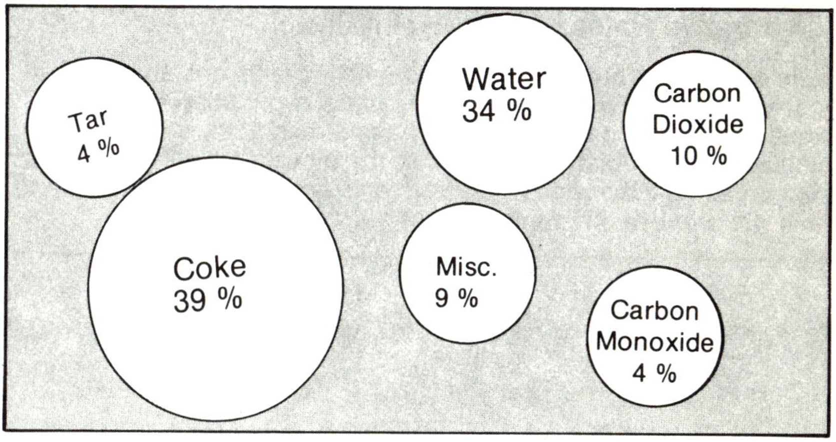

Figure 4.1 - Cellulose undergoing pyrolysis (heating to destruction).

Solubility of Gases in Mineral Oil

As we've already considered the basic sources of gas generation, add another piece of laboratory data—the oil solubility of different gases in one typical insulating oil. Table 4.2 shows how much the gases will dissolve in the oil.

Comparative Rates of Gas Evolution

As we continue to build our case of predicting incipient faults, look at Figure 4.2 which shows the relationship of decomposition temperature versus the type of gas generated in a fault condition. Thus, if one combines the solubility of the gas in the oil with the temperature it takes to generate the various gases, it can readily be seen that different faults will produce different evidences of their existence.

TABLE 4.2

SOLUBILITY OF GASES IN TRANSFORMER OIL		
Hydrogen[1]	H^2	7.0% by volume
Nitrogen	N_2	8.6% by volume
Carbon Monoxide[1]	CO	9.0% by volume
Oxygen	O_2	16.0% by volume
Methane[1]	CH_4	30.0% by volume
Carbon Dioxide	CO_2	120.0% by volume
Ethane[1]	C_2H_6	280.0% by volume
Ethylene[1]	C_2H_4	280.0% by volume
Acetylene[1]	C_2H_2	400.0% by volume
Static equilibrium at 760mm Hg and 25°C		
[1]Denotes combustible gas.		

Comparing Methods of Combustible Gas Detection

The earliest methods of detecting combustible gases involved direct flammability tests or chemical analysis of the gases over the oil. However, these methods were not sensitive enough by today's standards, nor, more importantly, timely enough to be of value.

Five methods of fault gas detection are used at the present time. We will consider both advantages and disadvantages, and document how dissolved gas analysis is rapidly gaining acceptance as the most informative method in today's computer-minded age. Table 4.3 summarizes the most pertinent facts.

Gas Collector Relay (Buchholz System)

This method has found limited application in the U.S. but extensive use in Europe. Recall that the generated combustible gases are highly soluble in oil. In a completely oil-filled transformer, small amounts of combustibles from an incipient fault may be totally absorbed in the oil before any appreciable free gas can accumulate in the collector relay. And when it does collect in the relay, you are now in possession of historical information; that is, your transformer *has had* a problem and you are now just discovering the evidence.

The Buchholz relay may respond to low energy partial discharges or local overheating. However, since the relay and collector relay are normally remote from the center of critical action—the windings and other key areas, it is impossible for the Buchholz system to respond to *early* stages of trouble in the heart of the transformer. In fact, as E. Dörnenburg reasoned, if the transformer is full of oil, oil must be in the immediate vicinity of the location of the fault. Thus, looking for the combustible gases in the oil is more efficient, considering their solubilities, than looking for them in the head-space or in a relay.

Fault Gas Detection

Fault gas detection, probably the most widely used field method in the U.S., measures *total* combustible gases (TCG). Its major advantages include its speed and adaptability to field use. In fact, it can be used to continuously monitor a transformer. However, it detects the combustible fault gases only, and is applicable only to those units having a gas-space above the oil. The major disadvantage of the TCG method is that it gives only a single value for the percentage of combustible gases, and therefore does not identify or quantify what gases are present. The results are subject to such factors as temperature and pressure, as well as solubility of the various gases of interest. Recall in Table 4.2 how widely the solubilities vary with temperature! Cold oil generally, but not always, dissolves more gas than hot oil.

William Henry's Law requires that subsequent analysis be taken at similar temperature and pressure; otherwise, any data would be meaningless: "The solubility of a gas in a liquid is directly proportional to the pressure of the gas above the liquid at a definite temperature." There may be little tendency for the combustibles to migrate into the gas-space if the concentration is low. Hence, a large volume buildup of combustibles in the oil may be required before the gases escape the oil and enter the gas-space above the oil to be sampled and analyzed.

Even a small amount (5 to 35 ppm) of acetylene, for example, portends great problems! Acetylene is released with great reluctance from the oil with the results that analysis of the free gas in the air-space may be quite misleading in terms of detectable combustible gases. Furthermore, keep in mind this part of Henry's Law: "The change in solubility with change in pressure follows no standard rules for the most soluble gases," such as acetylene.

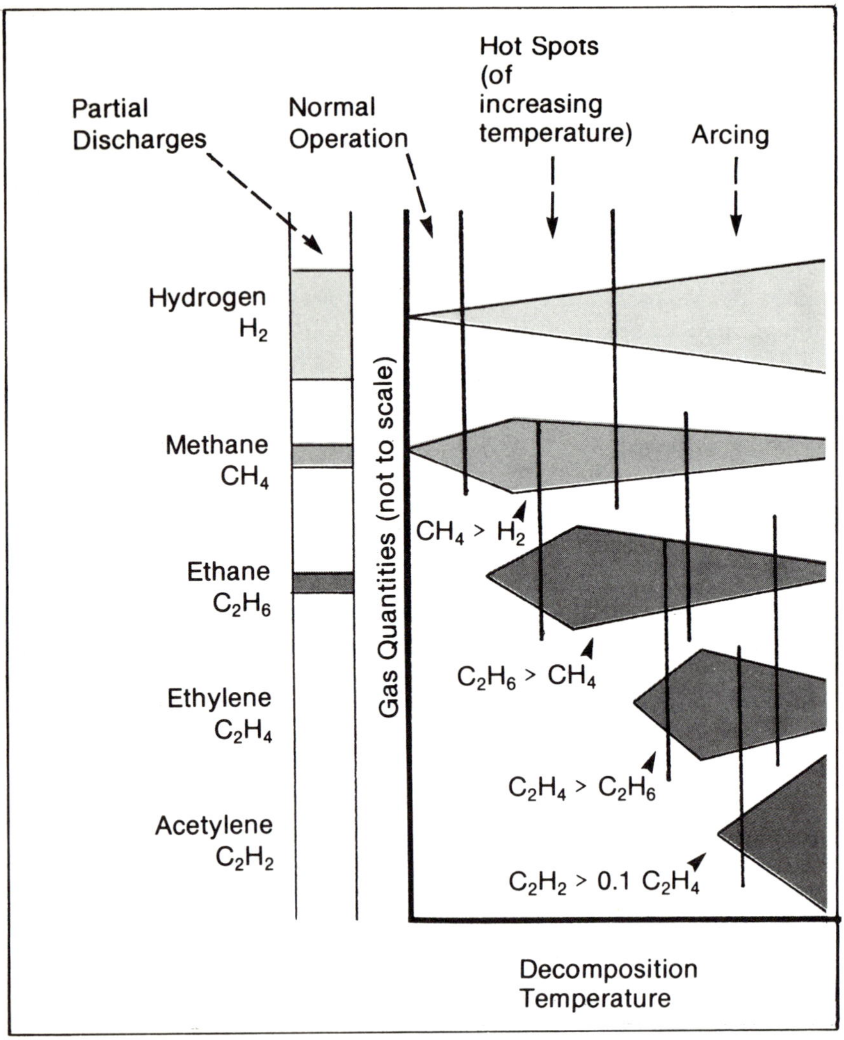

Figure 4.2 - Comparative rates of evolution of gases from oil as a function of decomposition energy. R.R. Rogers, "IEEE and IEC Codes to Interpret Incipient Faults on Transformers Using Gas-In-Oil Analysis" (1978).

Gas Blanket Analysis

A laboratory analysis of a gas sample in the transformer head-space identifies all the individual gas components—but only in the head-space. Thus, as with TCG, the gases must diffuse into the gas blanket

TABLE 4.3

COMPARING METHODS FOR DETECTION OF FAULT GASES[1]

Method	Location	Sampling Required	Gases Detected	Quantification	Transformer Application	Comments
1. Gas Collector Relay—Buchholz System (1928)	Field—on the transformer	Gas exceeding preset amount actuates "fault alarm"	Measure total gas accummulated	No—screen test only	Conservator type units only (between unit and conservator)	Primary use in Canada and Europe
2. Fault Gas Detection: Total Combustible Gases (TCG) 1959	Field	Gas sample; required head-space above oil	*Estimated total* combustible gases in head-space only	No—screen test only	Gas blanketed units	Possible continuous monitoring; results affected by gas solubilities and temperature
3. Gas Blanket Analysis ASTM D-2759	Laboratory	Gas sample	Identifies gases in head-space	Yes—parts per million (ppm)	Gas blanketed units	—
4. Dissolved Gas Analysis (DGA) 1963 ASTM D-3612	Laboratory	Oil sample	Identifies gas dissolved in oil where it is generated	Yes—parts per million (ppm)	All oil-filled transformers	Data of gas formation indicates the type of problem with repeated sampling and testing
5. Hydrogen in Oil 1979	Portable field analyzer	Oil sample	Identifies hydrogen dissolved in oil where generated	Yes—ppm screen test; if abnormal—lab DGA follow-up should be performed	All oil-filled transformers	Best single indicator of incipient fault; continuous monitoring; field sensitivity only

[1]ANSI/IEEE Standard C57.104-1978, pp. 7-13.

What you really need is a "prophetic view" of the transformer. Prophecy is history written before it happens. Dissolved gas-in-oil analysis is the "prophetic approach."

Dissolved Gas Analysis

The most informative method of fault gas detection is dissolved gas analysis (DGA). In this laboratory method an oil sample is taken from a transformer; the dissolved gases are then extracted, separated, identified, and quantitatively determined.

Various lab methods have been used, including infrared absorption and mass spectroscopy, but gas chromatography has emerged as the most popular technique. Primary criteria include sensitivity required, capital cost, facility of operation, and so forth.

Diffusion of gases between liquid and gaseous spaces takes time, during which serious equipment damage can occur undetected, if only gas samples from the transformer head-space are analyzed for combustibles (methods number 1, number 2, and number 3, Table 4.3).

Monitoring the oil for dissolved gases offers both the required sensitivity, while giving the earliest possible detection of a newly-formed fault. The only disadvantage for DGA lies in that it can't be done readily (as yet) in the field.* On the other hand, this method is not only applicable to all types of oil-filled equipment, it gives the information required to properly evaluate the transformer's ability to live up to its intended function.

Gathering the Data

Once it has been decided that dissolved gas-in-oil analysis will be used to evaluate a transformer's insulation system, these necessary steps are taken:

1. Representative Oil Sample
2. Gas Extraction From Oil
3. Gas Analysis by Gas Chromatography

**Attempts are being made to develop devices to attach to large transformers that will automatically analyze the gases in oil and signal for predetermined danger limits. These devices are quite expensive and as yet are not fully developed.*

- Gas separation
- Identification
- Individual gas quantitation

4. Interpreting the Data

Naturally, the first step concerns the oil sampling procedure.

Representative Oil Sample (ASTM Standard Test Method D-3613)

A transformer technician or an electrician can usually draw a sample in less than five minutes (Figure 4.3). However, extreme care must be taken in drawing samples to prevent contamination. Regardless of the sampling container type, certain precautions are required:

- Sample a unit only under positive pressure *
- Flush sampling valve—draw off approximately one quart of oil and discard to insure sample is not contaminated
- Protect sample from sunlight
- Forward to laboratory as soon as possible (prevent loss of dissolved gases)

Two sampling methods are almost exclusively used in the United States. A third method has been used in Europe but has not, so far, been generally accepted in the U.S. Table 4.4 summarizes pro and con views regarding each type used.

A known volume of oil (usually 40 cc) is drawn from the transformer and sent to a laboratory where the dissolved gases are extracted (Figure 4.4).

Typically, a 50-ml glass syringe fitted with a three-way stopcock is placed in a special shipping carton (Figure 4.5). A Tygon tube adapter is provided so that the syringe can be filled from the bottom filter press valve of most units.

Each syringe used for sampling should be checked periodically for very small gas leaks.

> The gas tightness of a syringe might be tested. . . by storing an oil sample with known amounts of dissolved gases in the syringe for two weeks, and analyzing the content of hydrogen and methane at the beginning and end of that period. A

**Don't try to draw a sample from a unit under a negative pressure. Danger—drawing an air bubble in the bottom valve could cause a failure.*

Figure 4.3 - A technician taking a syringe oil sample from a transformer's bottom filter press valve.

> syringe can be considered tight when the loss of hydrogen is less than 2.5 percent and loss of methane is less than 0.5 percent.

It follows then that such oil samples should be tested within a short period of time. Data has shown that storage for three additional weeks results in lower gas values. On the other hand, syringe samples do not appear to be affected significantly by ambient temperature cycline in transit.

In looking to the future, the flexible-sided "tin can" method, after additional work and subsequent modifications, *may* gain acceptance because it does offer the advantage of both the syringe and stainless steel bottle methods without their limitations.

Gas Extraction from Oil (ANSI/IEEE Standard C57.104-1978 and ASTM Test Method D-3612)

Once the oil sample containing the dissolved gas is taken to the laboratory, the first step is the extraction of the gases from the oil by

TABLE 4.4

COMPARING TYPICAL GC SAMPLING CONTAINERS ASTM STANDARD D-3613				
Type of Sample Container	Location Used	Pro	Con	Comments
Glass, Hypodermic Syringe ASTM D-923-73	U.S. utilities and industry (gas blanketed and open breather units)	Inexpensive; allow for volume change per temperature change.[1]	Breakage	Test within short time; periodic gas test suggested
Stainless Steel Bottle	U.S. industry and transformer manufac-turers	Contamin-ation-proof, leak-proof[1]	Rigid construc-tion; no visible inspection; can't see if bubble is in container	Use for extended storage
Flexible-sided Metal (tin) Can	Europe (transformer manufacturer) U.S.—some utilities	Allow for volume change per temperature change; larger container[1]	Limited use; insufficient data	Method under study in U.S. to determine its precision, future acceptance.

[1]Each container may be recycled for additional use.

some type of vacuum apparatus. Figure 4.6 shows a typical laboratory extraction set-up.

Direct injection of a small amount of oil into a gas chromatograph is not the best procedure for the detection of dissolved gases, because of insufficient amount of the gas sample, plus liquid hydrocarbons load up the column. Although this approach is being worked on and may be advantageous in the future, the best method has been *first* to extract the gases from the transformer oil, externally to the chromatograph, and to then inject the gas sample.

Three basic types of extraction apparatus are used. The equipment used depends on the volume of the test sample. Figure 4.7 gives the details of one type of vacuum apparatus. A typical sequence for gas extraction from insulating oil consists of these steps:

1. Injection of syringe oil sample into the extraction apparatus.

Figure 4.4 - A typical laboratory where gas-in-oil analysis by GC is performed.

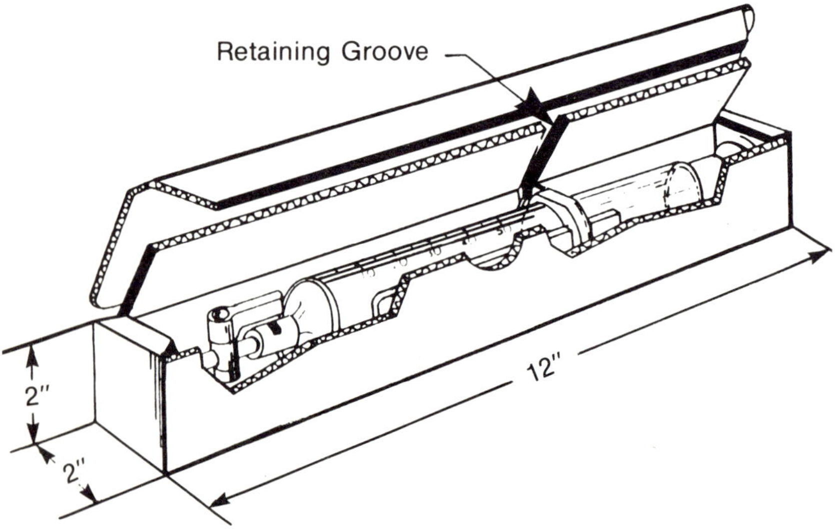

Figure 4.5 - A syringe, oil sample, and shipping carton.

Figure 4.6 - Typical laboratory glassware for extraction of gases from insulating oil.

Connect syringe to the degassing flask. In performing this step, make sure any air bubble is removed and watch for foaming of the gas. Then, gas only should pass into the boiling flask.

2. Boiling of the gas in the gassing flask, (Figure 4.8), should last for about 15 minutes. The gases are released from the oil by vigorous stirring under high vacuum.

3. Normalization of pressures for standard temperature (0° C) and pressure (760 Torr). A mercury "piston" is utilized to compress the gases to S.T.P.

4. Measurement (ml) of extracted gas in collection tube. The percentage of gas in the oil is calculated from the corrected volume of extracted gas. Figure 4.9 shows extraction of gas with a syringe for GC analysis.

Gas Analysis by Chromatography

A gas chromatograph is a means for "separating" gas components on a column and then "detecting" them (Figure 4.10).

Figure 4.11 illustrates how an extracted gas sample is first injected into the gas chromatographic instrument.

A typical gas chromatograph set-up (Figure 4.12) consists of a cylinder of carrier gas, a flow controller and pressure regulator, a sample injection port (sample inlet), adsorption column, detector (with necessary electronics), thermostats, a recorder, and a microprocessor for a data printout.

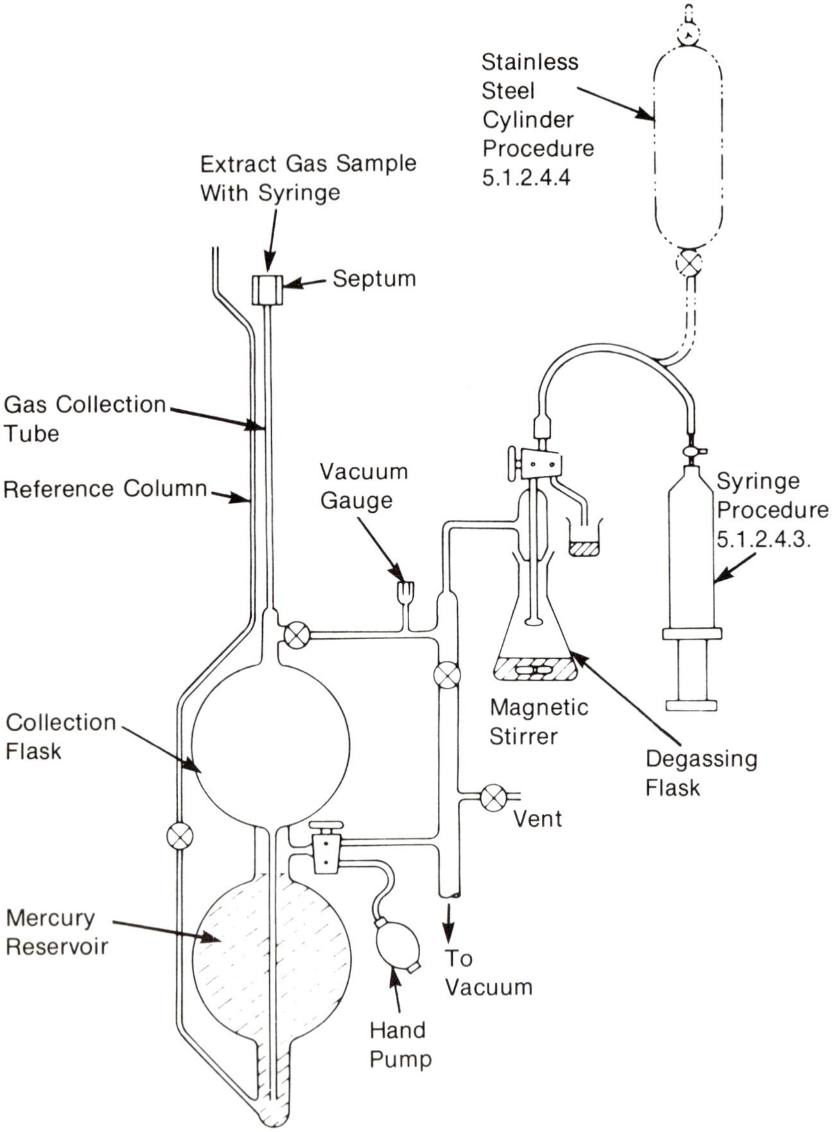

Figure 4.7 - Typical details of laboratory apparatus for extraction of gas from insulating oil. (Courtesy of ANSI/IEEE C 57.104-1978).

The entire analysis is completed in approximately 25 minutes. Accuracy attainable in the gas chromatograph depends upon the care exercised in obtaining the gas sample, the choice of detector, and daily calibration of the test instrument to assure reliable results.

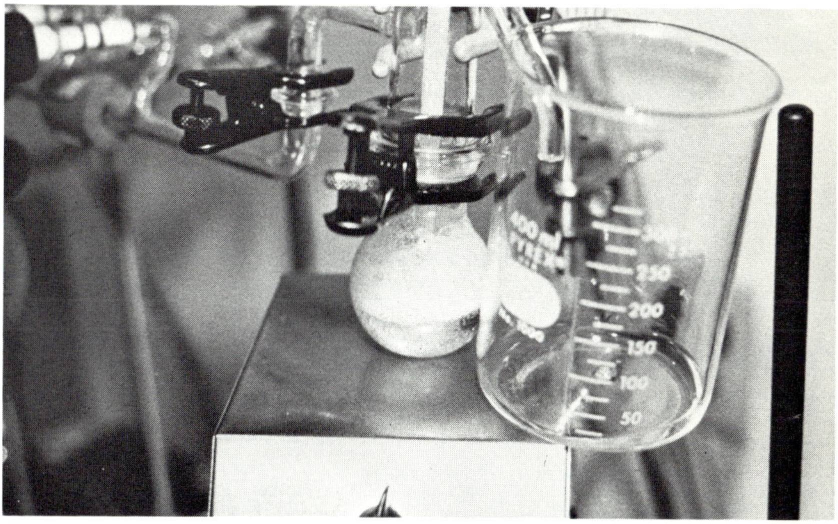

Figure 4.8 - Boiling of gas in the gassing flask. The gases are released from the insulating oil by vigorous stirring under high pressure.

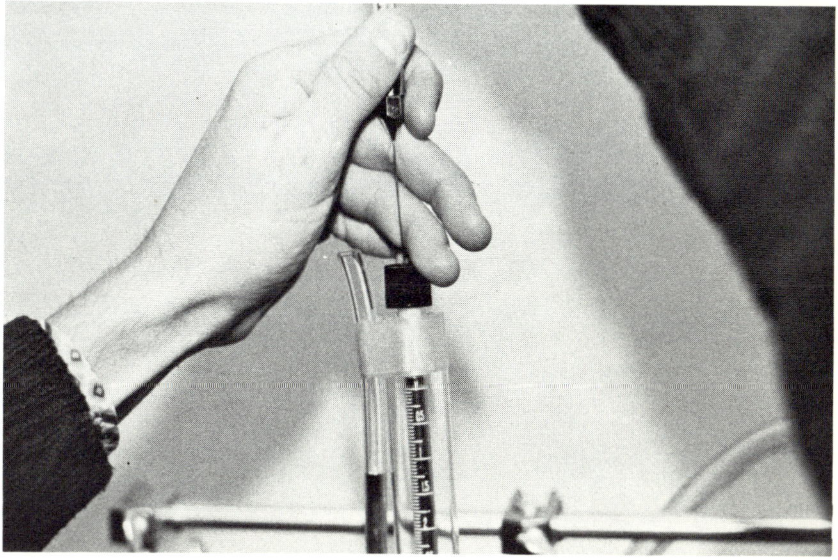

Figure 4.9 - The extraction of gas with a syringe for GC analysis.

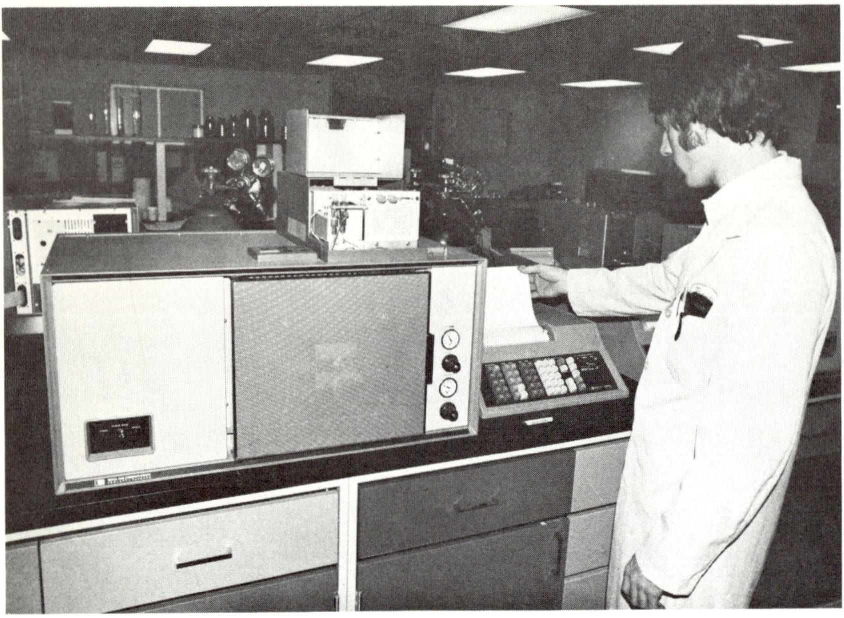

Figure 4.10 - A typical gas chromatographic system.

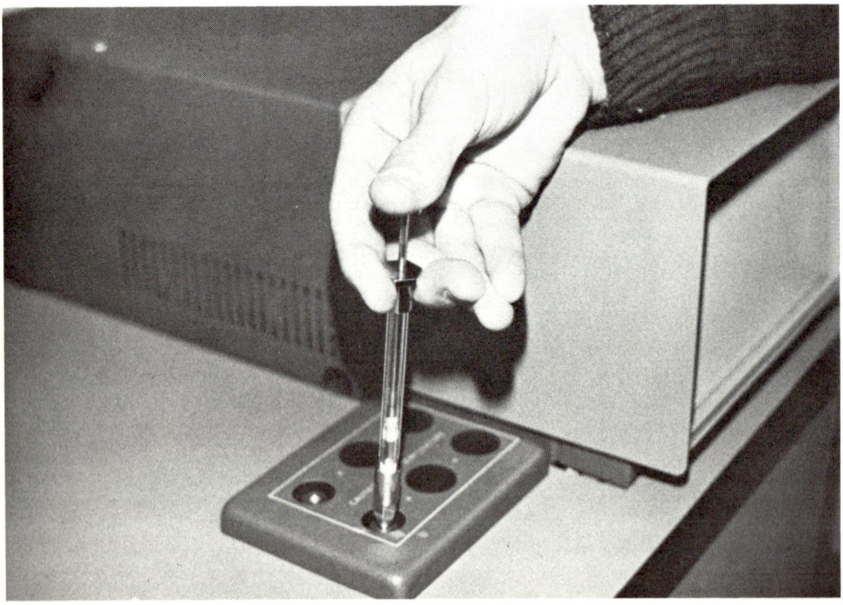

Figure 4.11 - Injection of the gas sample into the GC instrument.

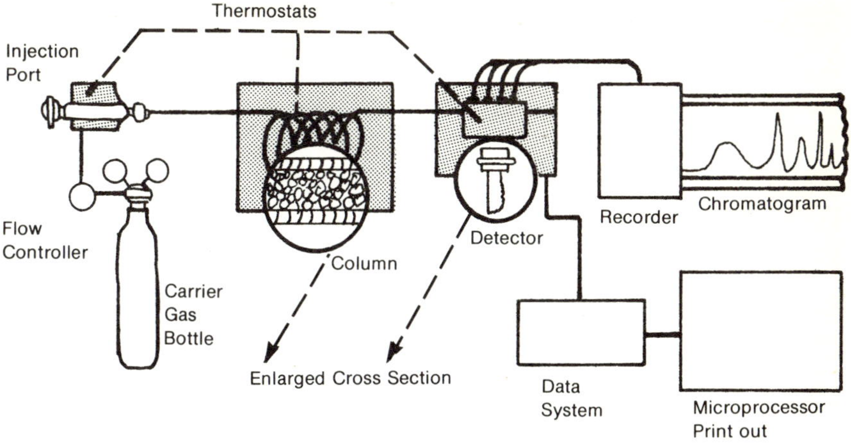

Figure 4.12 - Schematic drawing of a gas chromatographic system.

Major reasons why extensive application of GC has occurred include:

1. Its extreme sensitivity—down to the parts per million (ppm) range.
2. Only a few gases (nine) need to be analyzed in order to determine their cause.
3. Only a small sample size is required.

Other applications of GC besides gas analysis of transformer mineral oil include air and water quality, medical use, coatings, foods, pharmaceuticals, and drugs.

Today at least four different chromatographic procedures for dissolved gas analysis are in general use. Primary references include ANSI/IEEE Standard C57.104-1978, (p. 18), ASTM Standard Recommended Practice for General Gas Chromatography Procedures, Designation E-260, and IEC Publication 567 (1977), "Guide for the sampling of gases and of oil from oil-filled electrical equipment and for the analysis of free and dissolved gases."

Data Reproducibility

A question that inevitably arises concerning any laboratory data is, "What is the reproducibility of the dissolved gas analysis between various laboratories?" Table 4.5 shows ASTM desired results with a range difference from 0.016 to 0.171 percent, and averaging less than 0.1 percent difference. Thus, the data independently obtained by dissolved gas analysis from two qualified labs is scientifically and accurate (Figure 4.13). Nevertheless, interpreting the same data is clearly altogether another matter.

TABLE 4.5

DISSOLVED GAS-IN-OIL ANALYSIS BY GAS CHROMATOGRAPHY LABORATORY REPRODUCIBILITY					
Transformer Serial Number	MVA Rating	Gallons Oil (Thousand)	Combustible Gas (CGC) Content as percent of Total Gas Content		Percent Difference Combustible Gas Content (CGC)
			Transformer Consultants	Second Laboratory	
1. TT3-45M-2	45	13.5	0.409	0.425	0.016
2. TG3-45M-3	45	18.7	2.258	2.285	0.027
3. ZM3-45M-1	45	9.8	0.504	0.463	0.041
4. TZB-292M-1	292	24.0	0.520	0.685	0.165
5. TZ3-292M-2	292	24.0	0.903	0.732	0.171
6. TZ3-308M-4	308	23.7	0.477	0.395	0.082
Units 1 thru 6 = Average percent (CGC) 0.084 difference (approximately 0.1 percent)					

Figure 4.13 - Data printed out independently from different qualified laboratories, using the chromatographic method of combustible gas analysis in a transformer, demonstrates its reproducibility and therefore its scientific reliability.

Interpreting the Data

Once the laboratory has generated data, the obvious concern is correctly diagnosing a fault that is giving rise to the gases that have developed.

A number of methods are in use for the interpretation of the dissolved gas analysis (Table 4.6).

TABLE 4.6

COMPARING METHODS OF INTERPRETING DISSOLVED GAS-IN-OIL DATA

Interpretive Method	Value of Method	Source	Comment
Detected Gases	Types of probable faults	ANSI/IEEE C57.104-1978	Qualitative (Table 4.7)
Key Gas	Type of general problem	Doble Engineering	Qualitative (Figure 4.14)
Key Components	Type of general problem	E. Dörnenberg	Qualitative; all components must be identified
Dörnenberg Ratios	Type of general problem	E. Dörnenberg	Qualitative (Figure 4.15)
Amount of Key Gases	Severity of problem	California State University (CSUS)	(Table 4.8)
Total Combustible Gases	Severity of problem	Doble Engineering	500 or more ppm—retesting is critical—establish a trend
Rogers Ratios	Severity of problem	R. R. Rogers	No mention of CO, simultaneous faults, ambiguous analysis, concentrations important (Table 4.9)
Combustible Concentration Limits	Guidelines only	E. Dörnenberg	(Table 4.15)
90% Norms Concentration	Guidelines only	ANSI/IEEE C57.104-1978	Concentrations important; avoid legalistic application (Table 4.16)

Qualitative and Quantitative Interpretation

First of all, a generally accepted list of gases and associated conditions is found in Table 4.7.

TABLE 4.7

TYPES OF PROBABLE FAULTS	
Detected Gases	Interpretations
Nitrogen plus 5 percent or less oxygen	Normal operation of sealed transformer
Nitrogen plus more than 5 percent oxygen	Check for tightness of sealed transformer
Nitrogen, carbon dioxide, or carbon monoxide, or all	Transformer overloaded or operating hot, causing some cellulose breakdown. Check operating conditions
Nitrogen and hydrogen	Corona discharge, electrolysis of water, or rusting
Nitrogen, hydrogen, carbon dioxide, and carbon monoxide	Corona discharge involving cellulose or severe overloading of transformer
Nitrogen, hydrogen, methane with small amounts of ethane, and ethylene	Sparking or other minor fault causing some breakdown of oil
Nitrogen, hydrogen, methane with carbon dioxide, carbon monoxide, and small amounts of other hydrocarbons; acetylene is usually not present	Sparking or other minor fault in presence of cellulose
Nitrogen with high hydrogen and other hydrocarbons including acetylene	High energy arc causing rapid deterioration of oil
Nitrogen with high hydrogen, methane, high ethylene, and some acetylene	High temperature arcing of oil but in a confined area, poor connections or turn-to-turn shorts are examples
Same as above except carbon dioxide and carbon monoxide present	Same as above except arcing in combination with cellulose

1. Key Gases
Table 4.7 may be simplified by looking for "Key Gases" in the data and diagnosing the associated condition (Figure 4.14).

2. Key Components Technique
Presence of certain components and their relative concentration gave rise to this Dörnenburg technique.

- Thermal decomposition (hot spots)—mainly C_2H_2, CH_4; less C_2H_6, H_2, sometimes C_2H_2 (Reference Table 4.1).
- Electrical discharges (except corona)—mainly H_2, CH_4, then C_2H_2, C_2H_4, and C_2H_6.
- Internal corona—mainly H_2, smaller amounts CH_4, C_2H_6.
- Cellulosic insulation decomposition—CO and CO_2.

Once the *type* of fault is diagnosed, the severity of the problem becomes the next item of concern.

3. The Amounts of Key Gases
Since the detection limits of the GC instrument producing the data are in the parts per million (ppm) range, schemes such as Table 4.8 are being used to indicate the normal or abnormal levels of the individual gases.

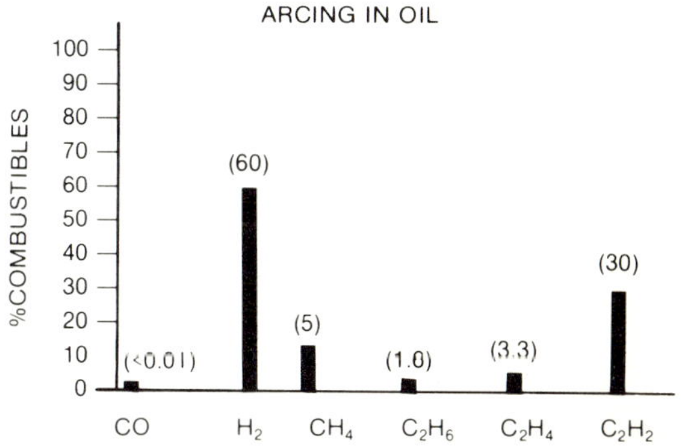

Key gas: Acetylene (C_2H_2). The least desirable evidence to be found. It nearly always indicates the symptom of an electrical arc in oil.

Figure 4.14 continued on next page

Figure 4.14 continued

Key gas: Hydrogen (H_2). Large quantities associated with corona (partial discharge) conditions.

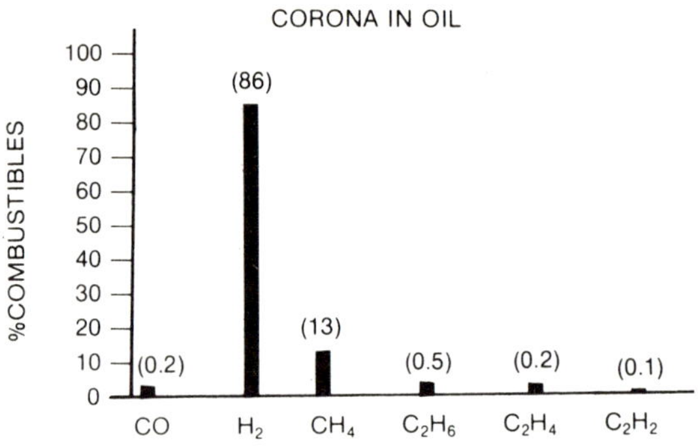

Key gas: Ethylene (C_2H_4). Thermal degradation of the oil will produce this gas.

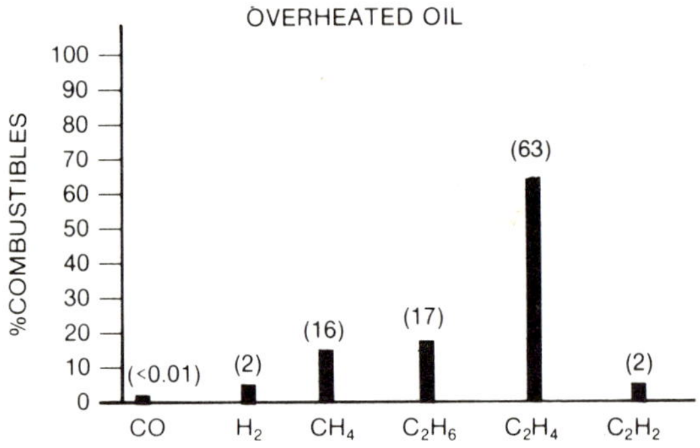

Key gas: Carbon Monoxide (CO). Thermal aging of the paper will produce this gas.

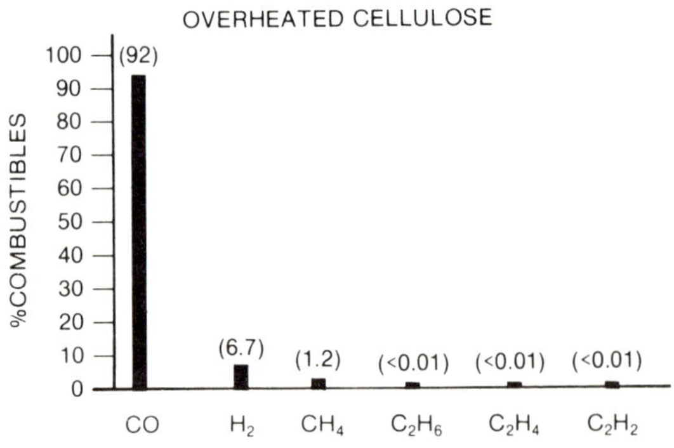

Figure 4.14 - Key gas diagnostic criteria.

TABLE 4.8

CALIFORNIA STATE UNIVERSITY, SACRAMENTO—GUIDELINES FOR COMBUSTIBLE GASES			
Gas[1]	Normal	Abnormal	Interpretation
H_2	< 150 ppm	> 1000 ppm	Arcing, Corona
CH_4	< 25	> 80	Sparking
C_2H_6	< 10	> 35	Local Overheating
C_2H_4	< 20	> 100	Severe Overheating
CO	< 500	> 1000	Severe Overloading
CO_2	< 10000	>15000	Severe Overloading
N_2	1-10%	N.A.	
O_2	0.2-3.5%	N.A.	
	0.03%	>0.5%	Combustibles

[1]Reference Table 4.1.

4. Total Combustible Gases
   Combustible gases in the range of 0-500 ppm are *normally* an indication of *satisfactory* operation of the transformer (for an exception, see Table 4.18).

   A range of 500-1000 ppm of combustible gas indicates that decomposition may be in excess of normal aging (suggesting a need for more frequent analysis).

   More than 1000 ppm of combustibles usually means that decomposition is *significant*. The next step here should be establishing a *trend.**

**If the amount of combustibles remains constant, then a possible self-healing effect may have taken place. If the amount increases, then the unit can be in the "danger" zone.*

Accumulation of more than 2500 ppm of combustibles indicates *substantial* decomposition. The integrity of the insulation system is in question. Monitoring is essential. The transformer should be inspected for fitness, continued operation, or for repair.

Ratio Analysis Methods

A second type of interpretation concerns the Dörnenburg and Rogers Ratio methods.

1. Dörnenburg Ratio Method
   One of the earliest methods is that of Dörnenburg in which two ratios of gases are plotted on log-log axes (Figure 4.15).

   The area in which the point falls indicates the type of fault that has developed.

2. Rogers Ratio Method
   A more recent method is the Rogers Ratio method (Table 4.9).

   This method of ratioing five different gases has probably received the most publicity.

   > Statistical study of nearly 10,000 gas analyses (United Kingdom utilities) suggested that certain types of fault conditions could be differentiated with more detailed ranges and combinations of gas ratios. This has been confirmed by internal examination of a number of suspect transformers together with post-mortems on faulty units...

Case Histories Illustrating Basic Transformer Faults

Even with a computer printout and various textbook schemes of data interpretation, the jury may still be out as to "When should a transformer be removed from service?" The decision is simply not always clear. The bottom line—the final decision—is therefore largely empirical, based on comparative field case histories that give data, diagnosis, and actual internal findings for a comparable situation. This data should also be compared with a similar *normal* operation's data.

With all of this input and an analysis of a transformer's history, an intelligent "guesstimate" can be made. The data bank of actual case histories continues to grow as both industry and utilities recognize the

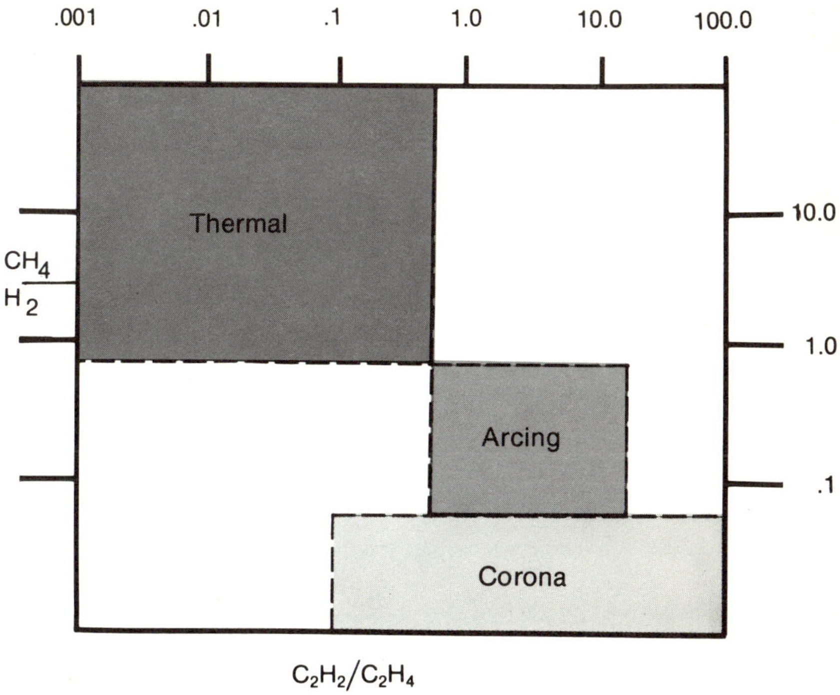

Figure 4.15 - Early Dörnenburg ratio plot of symptomatic indicators of transformer faults.

TABLE 4.9

SUGGESTED DIAGNOSIS FROM GAS RATIOS—ROGERS RATIO METHOD[1]

$\frac{CH_4}{H_2}$	$\frac{C_2H_6}{CH_4}$	$\frac{C_2H_4}{C_2H_6}$	$\frac{C_2H_2}{C_2H_4}$	Suggested Diagnosis
>0.1 <1.0	<1.0	<1.0	<0.5	Normal
<0.1	<1.0	<1.0	<0.5	Partial discharge—corona
<0.1	<1.0	1.0	≥0.5 or ≥3.0 <3.0	Partial discharge—corona with tracking

Table 4.9 continued on next page

Table 4.9 continued

>0.1 <1.0	<1.0	≥3.0	≥3.0	Continuous discharge
>0.1 <1.0	<1.0	≥1.0 or ≥3.0 <3.0	≥0.5 or ≥3.0 <3.0	Arc—with power follow through
>0.1 <1.0	<1.0	<1.0	≥0.5 <3.0	Arc—no power follow through
≥1.0 or ≥3.0 <3.0	<1.0	<1.0	<0.5	Slight overheating—to 150° C
≥1.0 or ≥3.0 <3.0	≥1.0	<1.0	<0.5	Overheating 150°-200° C
>0.1 <1.0	≥1.0	<1.0	<0.5	Overheating 200°-300° C
>0.1 <1.0	>1.0	≥1.0 <3.0	<0.5	General conductor overheating
≥1.0 <3.0	<1.0	≥1.0 <3.0	<0.5	Circulating currents in windings
≥1.0 <3.0	<1.0	≥3.0	<0.5	Circulating currents core and tank; overloaded joints

[1]Several simultaneously occurring faults can cause ambiguity in analysis.

importance of such documentation. Specific guidelines will eventually be developed after evaluation (possibly an EPRI study) of extensive numbers of case histories from various sources.

Needless to say, the first table (Table 4.10) illustrates typical data from normal operation (in this case, a 25 MVA furnace transformer).

Tables 4.11, 4.12, 4.13, and 4.14 illustrate the four basic transformer faults and should be self-explanatory.

When applying a method of interpretation in the diagnostic phase of analysis, *trends* are the first thing to look for. Note in Table 4.12 the numbers in themselves are relatively small, but when the trend and the rate of generation of carbon monoxide and acetylene have a 300 percent and 600 percent increase respectively in less than six months, an unplanned shutdown (failure) is inevitable. Consider then the *types* of gases and *volume* of oil in which they may dissolve.

Unless the fault is very obvious, these considerations will act as a guide for specific actions, such as shutdown or continued analysis to determine the progression of the fault.

TRANSFORMER CONSULTANTS

TRANSFORMER CONSULTANTS

DIVISION OF S.D. MYERS, INC.
P.O. BOX 3575
AKRON, OHIO 44310
PHONE (216) 929-2847
TOLL FREE (800) 321-9580

TC No. ______________________

Client ______________________

Address ______________________

City ______________ State __________ Zip ______

TABLE 4.10

A CASE HISTORY ILLUSTRATING NORMAL OPERATION
DISSOLVED GAS ANALYSIS OF TRANSFORMER OILS-GAS CHROMATOGRAPHY (GC)[1]

Location of Equipment: Furnace	Type of Equipment: Transformer
	KVA Rating: 25,000
Manufacturer: GE Serial No. G-851627-A	Gallons Oil:

GC Data - Values of Gases are expressed in PPM (Parts Per Million)

Date	Hydrogen	Oxygen	Nitrogen	Methane	Carbon Monoxide	Ethane	Carbon Dioxide	Ethylene	Acetylene	Total Gas Content	Combustible Gas Content	Recommendation
12/22/78	ND	12,408	68,570	Trace	101	ND	836	ND	ND	81,915	101	Retest 1 yr.

ND = none detected.

Diagnosis: Operation is *normal*. In 12 samplings over the last 12 months the combustibles have averaged from a low of 101 to a high of 174 ppm.

[1]While Table 4.10 is a complete data sheet used in practice, subsequent tables will present this data in a different form. This has been done to assist the reader.

TABLE 4.11

A CASE HISTORY ILLUSTRATING THERMAL OVERHEATING OF TRANSFORMER OIL

DISSOLVED GAS ANALYSIS OF TRANSFORMER OILS BY GAS CHROMATOGRAPHY (GC)

Location Substation | Type Transformer

Manufacturer Niagara | KVA Rating 1000

Serial Number 41686 | Gallons Oil 274

GC DATA-VALUES OF GASES ARE EXPRESSED IN PPM (PARTS PER MILLION)

Date	H_2	O_2	N_2	CH_4	CO	C_2H_6
11/6/78	2004	2799	66,637	9739	1737	2750

Date	CO_2	C_2H_4	C_2H_2	Total Gas Content	Combustible Gas Content
11/6/78	9596	5113	ND	100,375	21,343

RECOMMENDATION(S)

ND = none detected

Diagnosis: The Key Gas, *ethylene,* indicates the overheating of the transformer oil in the 200°-250°C range, possibly at a bad lead or loose connection in the terminal board or tap changer areas.

Findings: The iron bolts and nuts connecting the copper primary leads to the aluminum were carbonized and melted.

Solution: Repair and use brass nuts and bolts.

TABLE 4.12

A CASE HISTORY ILLUSTRATING THERMAL OVERHEATING OF TRANSFORMER CELLULOSIC INSULATION

DISSOLVED GAS ANALYSIS OF TRANSFORMER OILS BY GAS CHROMATOGRAPHY (GC)

Location City — Type Transformer

Manufacturer ASEA — KVA Rating 292,000

Serial Number 6598783 — Gallons Oil 23,670

GC DATA-VALUES OF GASES ARE EXPRESSED IN PPM (PARTS PER MILLION)

Date	H_2	O_2	N_2	CH_4	CO	C_2H_6
7/5/78	10	2950	15,100	5	48	2
12/11/78	57	2615	13,407	13	145	Trace
12/14/78	50	3540	15,300	10	133	2
12/29/78	ND	5901	22,134	ND	8	ND
1/8/79	ND	10,697	38,947	ND	Trace	ND
6/30/79	ND	9046	35,314	Trace	57	ND

Date	CO_2	C_2H_4	C_2H_2	Total Gas Content	**Combustible Gas Content**
7/5/78	163	2	2	18,282	69[1]
12/11/78	241	11	12	16,501	238[2]
12/14/78	290	10	12	19,347	217
12/29/78	96	ND	ND	28,139	8[3]
1/8/79	109	ND	ND	47,948	Trace[4]
6/30/79	270	ND	ND	44,687	57[5]

RECOMMENDATION(S)

ND = none detected.

[1]Analysis by a second laboratory.

[2]Diagnosis: Minor but significant increases in carbon monoxide and acetylene. The influx of these gases is normally indicative of an arc involving the cellulosic insulation.

Finding: A flashover from the end shield to the corona ring found on one bushing.

[3]Oil degassed.

[4]Too short of interval.

[5]Unit operating normally.

TABLE 4.13

A CASE HISTORY ILLUSTRATING TRANSFORMER ARCING

DISSOLVED GAS ANALYSIS OF TRANSFORMER OILS BY GAS CHROMATOGRAPHY (GC)

Location City Substation | Type Transformer

Manufacturer Standard | KVA Rating 20,000

Serial Number RIL-1717 | Gallons Oil 4095

GC DATA-VALUES OF GASES ARE EXPRESSED IN PPM (PARTS PER MILLION)

Date	H_2	O_2	N_2	CH_4	CO	C_2H_6
2/21/79	127	6215	71,620	107	174	11

Date	CO_2	C_2H_4	C_2H_2	Total Gas Content	Combustible Gas Content
2/21/79	973	154	224	79,605	797[1]

RECOMMENDATION(S)

ND = none detected.

[1]Diagnosis: The Key Gas, *acetylene*, indicates an arcing condition with *some cellulose* involved.

Findings: The unit had suffered a through-fault. When the secondary windings moved up there was an arc from the secondary to the core.

The unit is being repaired.

TABLE 4.14

A CASE HISTORY ILLUSTRATING CORONA IN TRANSFORMER OIL

DISSOLVED GAS ANALYSIS OF TRANSFORMER OILS BY GAS CHROMATOGRAPHY (GC)

Location: Main Plant Substation | Type: Transformer

Manufacturer: Delta Star | KVA Rating: 1500

Serial Number: W-208106 | Gallons Oil: 510

GC DATA-VALUES OF GASES ARE EXPRESSED IN PPM (PARTS PER MILLION)

Date	H_2	O_2	N_2	CH_4	CO	C_2H_6
4/19/78	35,882	1201	16,043	8470	97	2139
1/19/79	760	13,647	60,177	88	Trace	ND
9/21/79	1308	3105	56,338	335	116	112

Date	CO_2	C_2H_4	C_2H_2	Total Gas Content	Combustible Gas Content
4/19/78	1290	7	2	65,131	46,597[1]
1/19/79	189	ND	ND	74,861	848[2]
9/21/79	554	6	ND	61,874	1877[3]

RECOMMENDATION(S)

ND = none detected.

[1]Diagnosis: The Key Gas, *hydrogen,* indicates corona (low energy partial discharge) with *no cellulose* involved.

Findings: Design problem with unit. Unit will continue to operate until it can be replaced.

[2]Oil was degassed.

[3]Problem continues, but the condition is being monitored to determine the rate of gassing.

Cautions in Analyzing GC Data

Caution is necessary when you apply a method of analysis to your GC data. Concentration limits were published by Dörnenburg and others (Table 4.15). They suggested that even though these gases were present in the stated quantities, the transformer was thought to be operating perfectly.

These suggestions were refined and then incorporated into the ANSI/IEEE Guide (Table 4.16) as "90 percent probability norms." This means that if the concentrations are exceeded, the equipment merits further investigation.

Table 4.17 gives one example that was not predictable by any of the guidelines.

A second case history (Table 4.18) illustrates how every current man-made rule has an exception. In this case, this unit was not operating normally, even though the combustible gas content was substantially below 500 ppm (176 ppm). Fortunately, noting the *rapid* increase in acetylene indicated the possibility of a problem.

Further caution is necessary when you leave textbook cases and apply a method of analysis to your data. The ANSI/IEEE document C57.104-1978 was published as a guide. If a legalistic method of viewing the data is used, someone will demand that at the norm value plus one ppm the transformer must be shut down. This action is not usually necessary and is certainly not what the guide intended.

Comparing Current Methods of Interpretation of Dissolved Gas Analysis

As would be expected, differences of opinion exist concerning the preferred method of fault identification. The two most used methods are the Key Gas and Rogers Ratios techniques.

The best procedure appears to be the Key Gas method, with Rogers Ratios used as a back-up or a confirmation. It is clearly obvious at this time that interpretation of dissolved gas analysis is an art, not a science.

One study concludes in a preliminary evaluation that diagnosis of a fault by the gas ratio method, developed by the Central Electricity Generating Board of the United Kingdom, can identify a fault more specifically than diagnosis by the presence or absence of key gases.

TABLE 4.15

CONCENTRATION LIMITS	
Hydrogen	200 ppm(V/V)
Methane	50 ppm (V/V)
Ethane	15 ppm (V/V)
Ethylene	60 ppm (V/V)
Acetylene	15 ppm (V/V)
Carbon monoxide	1000 ppm (V/V)
Carbon dioxide	11000 ppm (V/V)

One ppm (V/V) of a gas denotes that 10^{-6} liters (or 1 mm^3) of this gas dissolved in 1 liter of insulating oil at a pressure of 1 kgf/cm^3.

TABLE 4.16

90 PERCENT PROBABILITY NORMS CONCENTRATIONS

	PPM V/V	
	Generating Equipment	Transmission Equipment
CH_4	160	120
H_2	240	100
C_2H_2	11	35
C_2H_4	190	30
C_2H_6	115	65
CO	580	350

TABLE 4.17

A CASE HISTORY ILLUSTRATING HOW THE NO-FAULT-DIAGNOSTIC METHOD CLEARLY INDICATES AN INTERNAL TRANSFORMER PROBLEM

DISSOLVED GAS ANALYSIS OF TRANSFORMER OILS BY GAS CHROMATOGRAPHY (GC)

Location City — Type Transformer

Manufacturer ASEA — KVA Rating 292,000

Serial Number 6598783 — Gallons Oil 23,670

GC DATA-VALUES OF GASES ARE EXPRESSED IN PPM (PARTS PER MILLION)

Date	H_2	O_2	N_2	CH_4	CO	C_2H_6
2/11/78	57	2615	13,407	13	145	Trace

Date	CO_2	C_2H_4	C_2H_2	Total Gas Content	Combustible Gas Content
2/11/78	241	11	12	16,501	238[1]

RECOMMENDATION(S)

[1]Diagnosis: Unit had failed (refer to Table 4.12).

The unit developed a flashover from the end shield to the corona ring. The problem is not clearly indicated by the "Key Gas" diagnosis method.

Dörnenburg's and the 90 percent probability norms (Tables 4.15 and 4.16) do not indicate a problem.

Rogers Ratios do suggest that the unit is not normal, but this method is still not definitive in its diagnosis.

TABLE 4.18

A CASE HISTORY ILLUSTRATING TRANSFORMER ARCING

DISSOLVED GAS ANALYSIS OF TRANSFORMER OILS BY GAS CHROMATOGRAPHY (GC)

Location: Substation | Type: Transformer
Manufacturer: Westinghouse | KVA Rating: 13,500
Serial Number: 6995019 | Gallons Oil: 6513

GC DATA-VALUES OF GASES ARE EXPRESSED IN PPM (PARTS PER MILLION)

Date	H_2	O_2	N_2	CH_4	CO	C_2H_6
10/3/78	ND	22,453	60,830	Trace	ND	9
11/13/78	35	9250	81,154	25	71	ND
11/16/78	ND	5162	22,301	1	1	ND
2/28/79	ND	12,059	68,122	2	4	2

Date	CO_2	C_2H_4	C_2H_2	Total Gas Content	Combustible Gas Content
10/3/78	1033	7	ND	84,332	16[1]
11/13/78	170	23	22	90,750	176[2]
11/16/78	35	Trace	Trace	27,500	2[3]
2/28/79	186	8	ND	80,383	16[4]

RECOMMENDATION(S)

ND = none detected.

[1]Oil was tested and unit was found to be operating satisfactorily.

[2]Diagnosis: Key Gases of *acetylene* and *carbon monoxide* indicate there is arcing involving *cellulose.*

Findings: Client reported the unit developed an arc from high voltage windings to the tank.

Special Note: A transformer is normally considered as operating satisfactorily below 500 ppm combustibles.

[3]The unit had its oil degassed after the repairs were completed on the transformer.

[4]Operation after three months appears to be normal.

However, consideration should be given to the transformer type (whether it is conservator or non-conservator). As an important bit of information, the ratio method was devised for use on conservator type units (relatively common overseas); whereas in the U.S., the majority of samples used in Transformer Consultants' evaluations are from either sealed or gas-blanketed transformers. In addition, look for a change in ppm if you have a bench-mark value.

In our own study, comparing the numerical values of the ratios with the suggested limits provides a suggested diagnosis of the source. We have found that correlation between the two methods ranges from poor to good, depending upon possibly the incipient fault and the ratio method used (Table 4.19).

Our lab experience continues to show an excellent record of predicting newly-formed faults in oil-filled transformers.

Case Histories Illustrating Priority Purpose for Dissolved Gas Analysis

Our three year laboratory experience (Table 4.20) illustrates that the possibility of a fault is not something remote. In fact, the data shows that one out of every four transformers tested by our laboratory using the GC technique has an operational fault.* This data considers units ranging in size from 25 KVA through 300 MVA.

In light of the foregoing data, the need is obvious to establish gas-in-oil analysis as a major facet of a protective maintenance plan. But what would be a possible standard for priority analysis?

We would suggest dividing analysis into three types of protective maintenance:

1. Predictive maintenance
2. Preventive maintenance
3. Corrective maintenance

Predictive Maintenance

Predictive maintenance involves critical units by location, by function, and by operating environment (Tables 4.21 and 4.22).

**This data does not mean that 25 percent of all units in-service are in trouble.*

Preventive Maintenance

Preventive maintenance of transformers is an integral part of an annual total maintenance program (Table 4.23).

Since GC can be done while the transformer is in operation, it is not unreasonable to suggest monitoring the operation on a yearly basis. Some units, because of their size and importance to your continued operation, may well justify dissolved-gas analysis on a semi-annual or even a quarterly schedule. It is also advantageous to perform GC on oil samples shortly after installation of new, large, and important units in order to determine if any shipping damage has occurred.

Corrective Maintenance

Corrective maintenance is concerned with units which have shown some definite warning signal, such as cloudy oil or unusual odor (Tables 4.24 and 4.25).

Practical Guidelines—
Transformer Serviceability and Appropriate Action

Although dissolved gas analysis is a major key to maintaining transformer health, it is still only one phase of a complete protective maintenance plan. Therefore, interpretation of combustible gas data will be enhanced by results of other basic maintenance tools—prior GC analysis, transformer oil screen tests, and electrical test data, as well as a loading profile of the transformer. This would include top oil and ambient temperatures. In addition, a report of all abnormal operating conditions should include periods of overload, faults, surges and malfunctions of equipment, such as cooling devices, switchgear, or other associated apparatus.

Practical guidelines developed over several years experience are given in Table 4.26. The frequency of analysis depends upon transformer size (power rating) and use conditions, and the critical importance of a given unit.

Three years of experience has revealed that normal accumulation takes place for the first nine years of the life of a transformer.* Only about three percent of these units tested revealed a fault condition. Yet for the third five-year period (10th thru the 14th year of in-service life),

**Non-random sampling of approximately 5000 units tested.*

the rate of critical faults had increased to 33 percent of the units tested. The per year rate of critical faults then progressively decreased as in-service age increased. Therefore, the analysis of dissolved fault gases represents a practical and effective method for the detection of incipient faults and the determination of their severity. Gassing only when a transformer is fully or heavily loaded likely indicates the problem is a loose connection rather than an insulation breakdown.

Gassing in transformer oil can cause other problems, such as deterioration of the electrical insulating properties of transformer oils and lowering of the flash point of the oil from 145° to somewhere between 50° to 80° C.

Since the transformer oil will hold the gases from old historical fault problems, it is imperative to degas the oil after repairs so that new problems won't be masked by the old. Long periods of very high vacuum and heat are required to remove the majority of the gases trapped in the paper insulation. Since residual gas should stabilize after a period of operation, follow-up analysis should be made.

The benefits of low gas content should not be underestimated since gas bubbles which might be formed in the transformer will then be quickly absorbed before they combine and cause a breakdown. Gas supersaturation, with the danger of the formation of free gas bubbles on sudden pressure drops, is thus completely eliminated.

State of the Art—
"Gas-in-Oil Analysis"

IEEE has concluded that the best method of early detection of *internal* transformer faults is an analysis of gas-in-oil. We have also confirmed through laboratory experience that dissolved gas analysis is the most valuable tool available to the power engineer. The GC technique detects newly-formed faults both accurately and consistently, and will oftentimes locate a fault that cannot be detected any other way. The growing importance of GC is demonstrated by increasing application.

1. Monthly GC analysis of large power transformers (>300 MVA) by Great Britain's Central Electrical Generating Board (CEGB)

2. Recommended monthly sampling by American Nuclear Insurers
3. Numerous other insurance carriers
4. Virtually every recognized transformer authority

Each proponent testifies to the reliability of gas-in-oil analysis. The present trend toward preventive maintenance has now been expanded to include SF_6 (sulfur hexafluoride) gas analysis.

Clearly then, if owner/operators of large equipment have grown to depend on GC analysis, it follows that the same dependence should occur for medium power units (1-100 MVA). The only difference would normally be less frequency of analysis.

Even though dissolved gas analysis has proven itself an excellent predictive maintenance tool, a word of caution is in order. In spite of the reliability of present GC instrumentation, considerable error can result from improper sampling, leaks in the vacuum system during extraction, and contamination of columns used in the separation of gases.

In addition, gas-in-oil analysis may not always be able to detect an extremely rapidly developing fault; for example, one initiated by a mechanical type of fault which in turn leads to an electrical fault which can occur within seconds. Of course, the oil could be sampled for combustible gas during the mechanical fault.

Justification for Dissolved Gas Analysis

The essential question for you as a transformer owner/operator (not already utilizing periodic gas-in-oil analysis) depends on what *total* benefits you can derive.

First of all, justification demands that early detection of trouble can be repaired for appreciably less than it would cost if the incipient fault went undetected and was left to develop until warning is given by a Buchholz relay or other similar device. This reduction in repair cost justification must be added to four other less tangible benefits of early detection.

1. Possible mitigation of oil fires and consequential damage.
2. Ability to schedule outages for units in the warning state (minimize down-time; plan for most convenient time).

3. Document first year in-warranty problems before out-of-warranty failure.* As one example, CEGB has reported, "A number of faults have been detected, although in each case the transformer had satisfactorily passed the routine tests."
4. Down-time cost for industrial units—emergency power or factory lay-offs, etc., and possibly increased insurance cost.

A case history of another utility illustrates how it proceeded from using only a gas-in-gas analyzer to gas-in-oil analysis.

> The early 1960's saw use of a portable gas analyzer (gas-in-gas sampling). The Buchholz protection system was relied upon to indicate incipient faults in completely oil-filled transformers. No record, however, has indicated that the gas detection relay ever successfully located a developing fault.
>
> Insufficient justification for added expense against expected benefits existed during the late 60's for use of gas-in-oil sampling.
>
> During the 1970's, however, a unit failed in-service; the result of internal inspection was the incentive to review diagnostic procedures.
>
> Re-evaluation then led to at least annual samplings for gas content of *all* completely oil-filled units.
>
> In the subsequent two years, dissolved gas analysis has prevented four known major failures.

Ongoing Research and Adoption of Universal Standards

Recognizing all the various parameters, it is evident that data interpretation is an art that will be continually improving. Because of its proven value, the growing use of the GC technique will in turn demand additional research and development, and the adoption of certain international standards. Naturally, this endeavor requires cooperation among all associated parties concerning the following areas:

1. Sampling—universal sampling container (instead of three, Table 4.4).

**A proposal has been made to include dissolved gas analysis as part of factory acceptance tests. Considerable support has already been given to this proposal. See Chapter 1, p. 112.*

2. Development of direct analysis of sample without need to extract gases first.
3. Gas analysis—universal detection limits for gases dissolved in oil (IEEE/IEC standard level of sensitivity). IEC already has such a standard (1977).
4. Interpretation of gas analysis—IEEE/IEC standard method of data interpretation (of those considered in Table 4.6); one primary and one established method as a back-up or confirmation.
5. Combustible gas limits for safe transformer operation (Fault Probability Guidelines)—concerning the approximate significance of total and individual combustible gases. This guideline requires a substantially increased analytical data base from all possible sources (both U.S. and overseas). As corollaries, consider also:
    a. Further understanding of the cellulosic insulation process and its effect upon transformer life.
    b. Specific guidelines on potential hazards of accumulated combustibles. With a growing data base, a *random* sampling will develop.
6. Metal-in-oil analysis—as a supplementary technique to gas-in-oil analysis to determine if field repairs are possible or if fault replacement is possible.

TABLE 4.19

ROGERS RATIO METHOD APPLIED TO CASE HISTORIES				
Case History Reference	$\frac{CH_4}{H_2}$	$\frac{C_2H_6}{CH_4}$	$\frac{C_2H_4}{C_2H_6}$	$\frac{C_2H_2}{C_2H_4}$
Case From Table 4.14[1]	$\frac{8470}{35882}$ .23	$\frac{2139}{8470}$ .25	$\frac{7}{2139}$ .003	$\frac{2}{7}$ .28
Case From Table 4.11[2]	$\frac{9739}{2004}$ 4.9	$\frac{2750}{9739}$ 0.3	$\frac{5113}{2750}$ 1.9	$\frac{ND}{5113}$ 0.0
Case From Table 4.18[3]	$\frac{25}{35}$ .7	$\frac{ND}{25}$ 0	$\frac{23}{0}$ 0	$\frac{22}{23}$ 1

[1]These ratios fit the "Normal" diagnosis row of Table 4.9. Since this is known not to be the case, it can be seen that it fits closely, but not exactly, the "Partial Discharge—corona" diagnosis row.

[2]These ratios fit in three of the four categories in the "Slight overheating—to 150° C" diagnosis row.

[3]These ratios fit the "Arc—no power follow-through" diagnosis in all four cases.

TABLE 4.20

LABORATORY EXPERIENCE[1]	
Units Tested	4899
Total Problems	1295
Overheating	1016
Corona	90
Arcing	189
PPM Combustibles	
Under 500 ppm	3543
500 - 1500	1063
1500 - 2500	91
2500 - 50,000	176
50,000+	26

[1]The data is not a random sample. In some cases there was evidence of a problem such as an odor in the oil, a color change in the oil, or a change in operational conditions, which prompted the analysis by our laboratory.

TABLE 4.21

A CASE HISTORY ILLUSTRATING PREDICTIVE MAINTENANCE

DISSOLVED GAS ANALYSIS OF TRANSFORMER OILS BY GAS CHROMATOGRAPHY (GC)

Location East Soaking Pits — Type Transformer

Manufacturer Wagner (1943) — KVA Rating 4000

Serial Number 366586 — Gallons Oil 1570

GC DATA-VALUES OF GASES ARE EXPRESSED IN PPM (PARTS PER MILLION)

Date	H_2	O_2	N_2	CH_4	CO	C_2H_6
5/13/75	ND	18,000	58,000	1100	5	1400

Date	CO_2	C_2H_4	C_2H_2	Total Gas Content	Combustible Gas Content
5/13/75	1500	4500	11	84,616	7016

RECOMMENDATION(S)

ND = none detected

Diagnosis: The Key Gas, *ethylene*, indicates severe overheating of the oil. Possible causes are poor core grounds or tap changer contact.

Findings: All three tap switches were severely burned.

TABLE 4.22

A CASE HISTORY ILLUSTRATING PREDICTIVE MAINTENANCE

DISSOLVED GAS ANALYSIS OF TRANSFORMER OILS BY GAS CHROMATOGRAPHY (GC)

Location East Lake Pump — Type Transformer

Manufacturer GE (1927) — KVA Rating 1000

Serial Number 3763068 — Gallons Oil 600

GC DATA-VALUES OF GASES ARE EXPRESSED IN PPM (PARTS PER MILLION)

Date	H_2	O_2	N_2	CH_4	CO	C_2H_6
10/76	770	11,069	59,497	1420	242	401

Date	CO_2	C_2H_4	C_2H_2	Total Gas Content	Combustible Gas Content
10/76	1985	1452	3	76,659	4108

RECOMMENDATION(S)

ND = none detected

Diagnosis: Severe overheating of oil. Key Gas, *ethylene,* possibly caused by loose connections or overheated joints and conductors.

Findings: Tap changer severely burned.

TABLE 4.23

A CASE HISTORY ILLUSTRATING PREVENTIVE MAINTENANCE

DISSOLVED GAS ANALYSIS OF TRANSFORMER OILS BY GAS CHROMATOGRAPHY (GC)

Location: City Substation — Type: Transformer

Manufacturer: GE (1967) — KVA Rating: 10,500

Serial Number: H-878597-C — Gallons Oil: 2030

GC DATA-VALUES OF GASES ARE EXPRESSED IN PPM (PARTS PER MILLION)

Date	H_2	O_2	N_2	CH_4	CO	C_2H_6
8/17/76	168	7217	55,884	1353	67	581

Date	CO_2	C_2H_4	C_2H_2	Total Gas Content	Combustible Gas Content
8/17/76	1386	3281	63	70,000	5513

RECOMMENDATION(S)

ND = none detected.

Diagnosis: Key Gas is *ethylene*, with methane and acetylene. This indicates a bad conductor joint or loose connection.

Findings: A loose connection was found in the terminal board. This was due to a cross-threaded nut that was not properly tightened at manufacturing.

TABLE 4.24						
A CASE HISTORY ILLUSTRATING CORRECTIVE MAINTENANCE						
DISSOLVED GAS ANALYSIS OF TRANSFORMER OILS BY GAS CHROMATOGRAPHY (GC)						
Location: Furnace			Type: Transformer			
Manufacturer: GE (1936)			KVA Rating: 4200			
Serial Number: 5442907			Gallons Oil: 1290			
GC DATA-VALUES OF GASES ARE EXPRESSED IN PPM (PARTS PER MILLION)						
Date	H_2	O_2	N_2	CH_4	CO	C_2H_6
8/17/76	354	20,887	60,779	4161	188	2476
9/27/77	39	33,878	88,690	426	10	1310
1/17/78	234	30,107	67,584	1867	Trace	1312
2/13/78	231	22,326	74,562	3233	Trace	2080
3/15/78	ND	6930	24,279	23	ND	39
6/20/78	Trace	24,473	70,962	1723	67	936
7/78	ND	22,297	52,323	1029	50	872
12/12/78	97	27,599	64,492	1607	ND	771

Date	CO_2	C_2H_4	C_2H_2	Total Gas Content	Combustible Gas Content
8/17/76	1896	6614	73	97,428	13,866[1]
9/27/77	1706	2275	Trace	128,334	4060
1/17/78	4213	3148	8	108,470	6569[2]
2/13/78	1937	5596	36	110,000	11,176[3]
3/15/78	64	95	ND	31,429	157[4]
6/20/78	1048	3066	49	102,324	5841[5]
7/78	1136	3139	ND	80,840	5093[6]
12/12/78	714	2019	8	97,307	4502[7]

RECOMMENDATION(S)

ND = none detected.

[1]Diagnosis: The Key Gas, *ethylene*, indicates extreme overheating of transformer oil. Low amount of carbon monoxide indicates cellulosic insulation not affected as yet. Possible cause would be the internal tap changer.

Findings: Physical inspection revealed the primary winding was almost completely burned off at one tap.

[2]Following repair and replacement of oil, unit still generating excessive amount of combustible gases.

[3]Unit still has excessive combustibles. Electrical Test (E.T.) on primary side—OK. Unable to disconnect secondary. Gases could be caused by a loose connection or a bad lead.

[4]Unit degassed.

[5]Substantial decrease in combustible gas content.

[6]Problem still seems to be overheating of oil. To check possibility of residual gases, retest one month. Long term: further E.T. secondary side.

[7]Level of combustible up again. The limit of degasification has been reached. Internal inspection and repair followed by degassing.

TABLE 4.25

A CASE HISTORY ILLUSTRATING CORRECTIVE MAINTENANCE

DISSOLVED GAS ANALYSIS OF TRANSFORMER OIL BY GAS CHROMATOGRAPHY (GC)

Location Substation — Type Transformer

Manufacturer GE — KVA Rating 13470

Serial Number D-583393 — Gallons Oil 1570

GC DATA-VALUES OF GASES ARE EXPRESSED IN PPM (PARTS PER MILLION)

Date	H_2	O_2	N_2	CH_4	CO	C_2H_6
8/5/76	9817	2488	76,772	36,962	837	11,608
8/30/77	23	28,365	90,420	506	Trace	2229

Date	CO_2	C_2H_4	C_2H_2	Total Gas Content	Combustible Gas Content
8/5/76	6649	62,815	2624	210,572	124,663[1]
8/30/77	2972	7836	213	132,564	10,807[2]

RECOMMENDATION(S)

ND = none detected

[1]Arcing odor detected in oil. Unit was tested and found to have extremely high combustible gas content in the oil.

Diagnosis: The Key Gas, *acetylene* with large amounts of *ethylene*, indicates extreme overheating of oil.

Findings: Problem found to be in tap changer. An inner-ring contact was badly burned, as was the stationary contact.

[2]Combustible gas problem (though considerably reduced) still exists. Retest one month to note how fast the gases are generating.

Recommendation: As combustible gas level remains high—repair and degas.

TABLE 4.26

TRANSFORMER SERVICEABILITY AND APPROPRIATE ACTION

Combustible Gas Content (Parts Per Million-PPM)	Why Gas-In-Oil Analysis	When Gas-In-Oil Analysis
New Unit or 0-500 PPM	*Normal* aging; transformer operating satisfactorily; very slow accumulation.	Establish a historical *data base;* include in operating history any unusual event concerning transformer.
500-1000 PPM	*Significant* decomposition of oil or cellulose insulation; developing incipient fault; unit could fail at any time. 10th thru 14th year are critical (non-random sampling of 5000 units tested).	Retest periodically to plot *trends*, noting *marked* changes. Suggested retesting time is critical: If retest *too soon*, insufficient time elapsed for meaningful gas concentration to build up. If retest *too late*, irreparable harm could be done to equipment.
1000 PPM or More	*Very substantial* decomposition of oil cellulose insulation; existing fault; transformer should be de-energized. The 14th in-service year has been the most critical year, with faults declining from this point.	Too late now!! Internal visual inspection (for loose connection); possible unit repair and degasification required.

Chapter 5

Field Electrical Testing of Transformer Insulation

Part 1 — Why Perform Electrical Tests in the Field?

Introduction

Preventive maintenance testing of in-service electrical equipment has the primary objective of monitoring conditions in the insulation and evaluating the useful life still available in the equipment tested. Whether or not the useful life, as revealed by the tests made, is sufficient to give a reasonable guarantee against a future operating failure is usually a matter of engineering judgment based on a background of experience. The experience of the plant engineer must be the guide to the selection of the tests to be used and to the establishment of criteria for gauging the reliability of the equipment for continued trouble-free service.

Although the Plant Engineer may have just had other tests run on energized transformers and received a clean bill of health, are the transformers really in good condition? Let us briefly go over several of these other tests to see what they do not reveal.

1. Oil Screening
   Having good oil specs does not always mean the transformer has good, dry windings. Does the cellulose have moisture in it? Has the insulation shrunk? Is it brittle? Does it have any holes in it? Oil screening tests are not Intended to tell you very much about the cellulosic insulation. *

2. Moisture-In-Oil
   This test shows if the transformer oil is dry — good — but what about the solid insulation? At 20°C, a measured water

**Chapter 3, Part 2.*

content of 10 parts per million in transformer oil can be in a state of equilibrium with a cellulosic insulating system that contains water equal to 4 percent of its own weight. A large transformer with 10,000 pounds of insulation and 10 ppm water-in-oil could have 400 pounds of water locked in the cellulosic insulation. Here good test results on oil could mask a serious moisture problem in the cellulose.*

3. Combustible Gas-In-Oil Analysis By Gas Chromatography
Even though dissolved gas analysis (DGA) has been proven reliable for detecting incipient faults, incidious (gradual and cumulative) faults† may not be detected even with on-line continuous DGA. When the rate of deterioration is insufficient to produce the heat and corresponding gases associated with DGA, electrical tests can expose the problem.

4. Thermographic Survey (IR)
A heat survey of transformers, bushings, and lightning arresters may show no problems. However, this is only an external view of your electrical system. What is really happening inside your transformer? Here again you are getting only a partial view of how good your electrical system is.‡ These are just a few suggestions why the data gained from the above tests are incomplete and why the electrical tests are necessary.

Other reasons for performing field electrical tests are as follows:

- Establish a test reference point — especially new transformers
- Determine the amount of insulation degradation since previous test
- Placing a unit in storage — test to see if the transformer is in good condition to justify storage
- Removing unit from storage — test to see extent of deterioration while unit was in storage

**Chapter 3, Part 3; Chapter 6, Part 2 (top and bottom oil or hot and cold oil samples). More than 50 ppm water content in the oil by ASTM D-1533 does indicate that windings are probably wet.*

†Chapter 4.

‡Chapter 8, Part 4.

- Unit moved to new location — has moving caused any internal damage to the unit?
- Smell/sound of arcing
- Rapid change in the color of transformer oil since last sample was drawn
- Cloudy oil
- Humming noise louder than normal
- Free water/cloudy oil in test sample
- Overheating/overloading contemplated
- Transformer has tripped off the line for any reason

A Complete Physical

For a clean bill of health, one needs a complete physical. The importance of the conditions of the oil and the tests used to monitor this condition have already been discussed.* The integrity of the solid insulation — the cellulose — is even more critical and can best be monitored with a series of electrical tests. As no single test procedure is adequate to supply all the information needed to properly evaluate a transformer, this basic EKG battery of tests should be conducted also (Figure 5.1).

Figure 5.1 - A variety of equipment is required for a complete battery of electrical insulation tests.

**Chapter 3, Parts 1 and 2.*

The frequency of these tests will be determined by many factors such as the type and class of the equipment, its importance, age, load, and history of operation. Scheduled outages should also be utilized as periods for testing.* If a fault appears to be developing, then more frequent electrical tests may be scheduled to monitor this condition.

These tests on transformers will fulfill three distinct but general functions:

- Prove the integrity of a piece of equipment at the time of acceptance
- Verify the continued integrity of the unit at periodic intervals of time
- Determine the nature of extent of the damage when a unit has failed

The specific tests required to fulfill each function may vary and will be tailored to meet the requirements of each piece of equipment and at the discretion of the technician. When a transformer has failed, then tests should also be made after the repair work is completed to verify the quality of the service done. Test results may be compared with factory test data.†

**Chapter 10, Table 10.2.*

†Available from the manufacturer for transformers built since 1960.

Part 2—A General Overview

Records

Be prepared to keep detailed records which include the following:

- Name of test, date, and results
- Scheduled date to retest
- Special requirements, recommendations etc. due to indications of problems detected

This data should become a part of the complete file kept on each transformer.

Safety

General Precautions

Some tests — DGA, Oil, IR — do not require that the unit be de-energized. This is not so with the electrical tests.* Special attention is called to the need for caution here. Personnel performing such tests **must** observe safety regulations at all times. Temporary safety barriers may be necessary to isolate the test area.

As the testing devices are very sophisticated, give special attention to all the manufacturer's instructions. *Restrict the use of this equipment to properly trained personnel* as mistakes may be hazardous to their health — and to the unit being tested.

Equipment to be tested must be disconnected from the power system using established operating procedures. Grounds should next be applied and left applied.† Personnel should be trained to treat all ungrounded apparatus as energized.

These same grounds must be removed to permit the application of test voltage or ungrounded low-voltage test leads. It is desirable to work and test in between grounds, and to provide within the grounds temporary disconnection of sufficient clearance for the test leads.

**Although not currently applicable, we foresee the day when certain tests may be done on energized transformers. A technique has been patented.*

†Even when grounded once if ground is removed, a residual charge can be built up again (much like a capacitor).

Precaution must be taken to prevent personnel from contacting all terminals of the apparatus under test. An observer should be stationed to warn approaching personnel and should be provided with a means of de-energizing the power source. There should be provision for grounding the circuit until all stored charges are dissipated.

When the test voltage exceeds the normal operating value of the apparatus insulation, there exists the possibility of failure under test. Before performing such a test, thought should be given to the time, material, labor, and cost that may be required to repair or replace equipment that may be destroyed. If the failure could cause a fire, then necessary fire extinguishing equipment should be available.

Entering Transformers (Adequate Oxygen)

1. New Units
   Large power transformers are usually shipped without oil to reduce weight. To prevent the entrance of moisture during transit, the tank is usually filled with dry air or nitrogen gas under pressure. If a unit is shipped under nitrogen, the gas must be replaced with dry air (at least 19% oxygen) before permitting anyone to enter the transformer.* Nitrogen gas will cause asphyxiation. Commercially available dry air should be used to prevent moisture from entering the transformer. (Dry air is air with an oxygen content of at least 19 percent and with less than 100 ppm moisture content.)

2. Undercover Inspection
   When oil is removed from a transformer prior to making an internal examination, dry air should be allowed to replace the oil as the oil is pumped out. When personnel are in the transformer, a dry air supply line should be connected to the bottom oil drain valve, and a dry air flow should be maintained at all times when manholes or bushing holes are open. The quantity of dry air being admitted should be such that its escape is noticeable when manholes or bushing holes are open. A plastic cover may be placed over open manholes to conserve air. The work should be performed in

**Two people should work together; one person staying outside the transformer but watching the person entering the unit. That person should have a safety rope around his waist.*

such a sequence that when accessories such as bushings are being installed, only one part other than the manhole will be open at any one time.*

Objectives

As indicated earlier, testing is performed to fulfill three functions.

Acceptance Tests

It is not uncommon to run acceptance tests on transformers immediately upon arrival and prior to removal from the carrier. Frequently, damage is done to the unit while in transit — many impact recorders have been deliberately removed in an attempt to conceal rough handling. This test data is then compared with the factory test data and any necessary remedial action taken. When the data is acceptable it is recorded for future reference.

Figure 5.2 is a plot of power factors recorded on approximately 1600 new transformers delivered over a four year period. It is evident that most are under 0.5%, the maximum acceptable for a properly dried transformer. Note, however, those with a power factor of ≥0.5%. A power factor of ≥ 2.0% on an "in-service" transformer is considered excessive and should have remedial attention.† In any case, the vendor should be prepared to supply reasonable justification for these higher values. If the cause is materials with inherently high power factor, their replacement should be encouraged because of their masking effect on an otherwise valuable tool. Under no circumstances should a transformer be energized if received with a power factor of ≥1.0%. Possible options would include internal inspection, consultation with the manufacturer, and drying the unit.

It is recommended that transformer purchase specifications stipulate that transformer manufacturer's power factor data be obtained in conformation with user's field practice (Method II of IEEE STD. No. 262).

**Courtesy of Doble Engineering Co., Watertown, Mass.*

†Askarel (PCB) units will normally be higher (4-10% at 20° C).

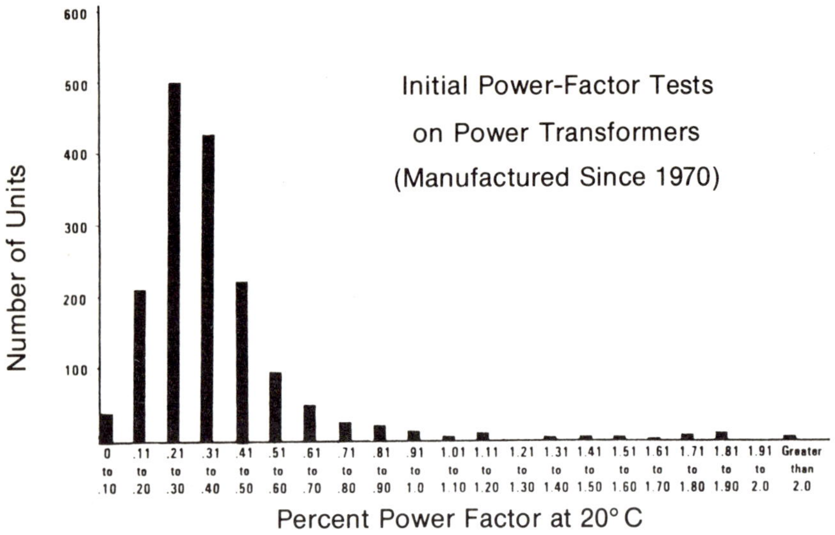

Figure 5.2 - Initial power-factor tests on power transformers manufactured since 1970. (Courtesy of Doble Engineering Co.).

Periodic Tests

All transformers undergo aging at all times. There is no such thing as new oil or cellulose in a transformer; the rate of aging never becomes static. It simply proceeds at its own inexorable rate — sometimes faster, sometimes slower but always there. This is true of transformers in all cases — units never energized, units energized, and stored transformers. Thus, it is prudent to generate test data to document the condition of the cellulose at each significant juncture of the unit's history. This will serve to monitor the rate of aging for a given block of time and will become the bench mark to use in evaluating the extent of later degradation. This, along with other test data will suggest on answer to the question, "What is the condition of this transformer's dielectric system?"

Test After Unit Failure

Caution — When a transformer has taken itself off the line, extreme care should be taken to determine the cause of the outage. Premature re-energizing may result in additional damage or complete failure. The possibility of serious injury to personnel also exists.* Fortu-

**If arcing has taken place the gases present may be explosive in nature.*

nately, troubleshooting guidelines provide major assistance in evaluating the problem and should be carefully followed. The manufacturer's assistance may also be procured.

Causes — These outages may be in response to primary breaker or fuse operation, pressure relay or mechanical relief operation, differential relay operation, etc., and may have been preceded by a voltage line problem. The inspection is intended to expose clues as to what triggered the operation of the protective device. This cause may range from failure of the transformer to a faulty protective device.

Clues — The initial visual inspection of the transformer's exterior may reveal such things as tank wall bulge, cracked or leaking seams, oil spills. Visual inspection through the manhole may reveal burned cellulose, melted copper, darkened oil, unnatural odor, and so forth.

Appropriate electrical tests, to be detailed later, would include Transformer Turns Ratio, Insulation Resistance, Winding Resistance, and Power Factor. High voltage tests should be avoided as they may excessively stress an already weakened insulation. This would be an appropriate time for dissolved gas-in-oil analysis.*

If determined that the outage was not due to a transformer fault, it may be re-energized with impunity.

Standards

Transformers that have been built according to American National Standard Institute (ANSI) Standards will comply with the following:

- ANSI C57.12.90-1980 (IEEE Std. 262-1980), "Test Code for Distribution, Power, and Regulation Transformers,"
- ANSI C57.12.00-1980 (IEEE Std. 462-1980), General Requirements for Distribution, Power, and Regulating Transformers" (Section 8)

Specific details on some of the tests listed in ANSI C57.12.90-1973 may be found in IEEE Standard 4-1978.

- ANSI C57.12.14/D6. Dielectric Test Requirements For Power Transformers for Operation at System Voltages From 115 KV through 230 KV.

**Chapter 4.*

- IEEE Std. 262B-1977. "Trial Use Standard Dielectric Test Requirements for Power Transformers for Operation in Effectively Grounded Systems, 345 KV and Above".

Application

The test requirement for any given repaired transformer may be unique. This is understandable, considering the complex variety of failure modes and repairs that may be involved. Consider the options available when a transformer fails.

- Repair in kind and risk the possibility of a repeat failure
- Repair employing updated materials and design philosophy — this may void original performance guarantees
- Scrap the unit and purchase a new one

If either of the first two options is chosen, acceptance tests must now be modified to accommodate these new specifications.

Part 3—Electrical Insulation Tests

Guidelines

In all cases, specific equipment manufacturer's recommendations supplied in the equipment instruction literature must be factored into testing procedures. The guidelines which follow are, of necessity, general in nature and must be modified to match specific equipment.

The following includes the more common tests used in the field to evaluate the condition of transformer insulation. When coordinated with data from oil tests, a reliable picture of the transformer interior evolves. From this information, decisions can be made about repairs and scheduling future tests.

- Insulation Power Factor
- Core Excitation Current
- Insulation Resistance
- Transformer Turns Ratio
- D.C. Winding Resistance
- D.C. Overpotential
- Core Ground
- Ground Resistance

Insulation Power Factor Series

Insulation power factor should not be confused with system power factor in an ac network. Insulation power factor provides an indication of the quality of the insulation, and may be understood from the following explanation. Any winding in a transformer is separated from all other windings and ground potential by solid insulation. Cellulosic insulation forms an effective capacitance network as indicated in Figure 5.3.

In each capacitance are dielectric losses which can be conveniently represented by a resistor in series with a capacitor. The insulation power factor is commonly defined as the ratio of the resistance to the impedance of this combination and can be measured by applying a voltage across this capacitance and measuring the amperes and watts loss and then calculating the power factor.

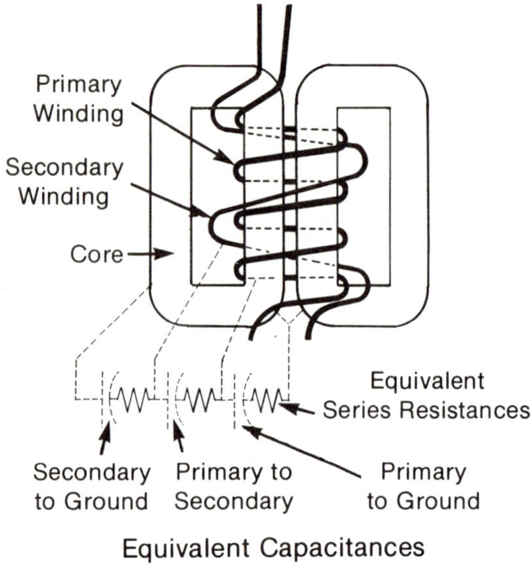

Figure 5.3 - Dielectric loss of each capacitor divided by capacitive volt-amperes is equal to power factor.

This measurable dielectric loss will develop heat in the insulation during transformer operation (in the equivalent resistor) and this heat, along with moisture and other factors, can cause deterioration of the insulation.* The interpretation of insulation condition depends primarily on comparison with previous test results from the same unit or from similar units that are considered in good condition.

The de-energized series of power factor tests, often referred to as "Doble Tests," use one of many types of bridges available (Figure 5.4) or can be calculated from readings taken with instruments that measure volts, current, and watts loss.

Winding Insulation Power Factor Test

As described above, this equipment measures the power loss through the insulating system to ground caused by dielectric loss or

**Primary heat deterioration losses come from copper or aluminum winding losses. If the dielectric losses from the insulation are high enough the combination of losses can lead to a run-away loss condition, resulting in failure. This is especially true under overload conditions.*

leakage current. All insulating systems will have minor leakage paths through the system that will permit a small current to flow. If the insulating system is perfect (a theoretical condition) there would be zero current leakage. Then we could calculate a power factor reading of zero percent. The greater the leakage path, the greater the power loss through the insulating system caused by the I^2R watts loss as determined by Ohm's law. If all of the input power were lost in heat in the insulating system then the power factor would be 100 percent.

Figure 5.4 - Typical winding power factor test in the field.

By definition, power factor is a measure of the power loss through the insulation system to ground caused by leakage current. It is equal to the circuit resistance (R) divided by the circuit impedance (Z). It may also be expressed as the cosine of the phase angle (Θ) between the current and the voltage.

A Power Factor reading of 0.03 percent is not unusual under ideal conditions. Data shown in Figure 5.5 gives a typical reading of 37.3 percent. The insulation is becoming an excellent conductor!

Tabulated in Figure 5.6B are the standard series of tests (1 through 4), two calculated values, and two supplementary tests, illustrating how data pertaining directly to C_H, C_L and C_{HL} are obtained without need for physical isolation of transformer parts or sections.

OVER-ALL TESTS

TEST	TEST CONNECTIONS: WINDING ENERGIZED	WINDING GROUNDED	WINDING GUARDED	TEST KV	EQUIVALENT 10 KV READINGS — MILLIAMPERES: METER READING	MULTIPLIER	MILLIAMPERES	WATTS: METER READING	MULTIPLIER	WATTS	% POWER FACTOR: MEASURED	COR 20°C	KEY TO INSULATION RATING (G = GOOD, D = DETERIORATED, I = INVESTIGATE, B = BAD (REMOVE OR RECONDITION))	INSULATION RATING
1	HIGH	LOW		4	59	1	59	88	2	176	— —	— —		
2	HIGH		LOW	4	15	1	15	28	2	56	37.3	37.3	C_H	B
3	LOW	HIGH		.5	21	1x4	84	28	2x4	224	— —	— —		
4	LOW		HIGH	.5	10	1x4	40	13	2x4	104	26.0	26.0	C_L	B
CALCULATED RESULTS					— —	— —	44	— —	— —	120	27.2	27.2	C_{HL} (TEST 1 MINUS TEST 2)	B
					— —	— —	44	— —	— —	120	— —	— —	(TEST 3 MINUS TEST 4)*	— —

*CURRENT AND WATTS SHOULD COMPARE WITH THOSE FOR C_{HL}

OVER-ALL TESTS

TEST	TEST CONNECTIONS: WINDING ENERGIZED	WINDING GROUNDED	WINDING GUARDED	TEST KV	EQUIVALENT 2.5 KV READINGS — MILLIVOLTAMPERES: METER READING	MULTIPLIER	MVA	MILLIWATTS: METER READING	MULTIPLIER	MW	% POWER FACTOR: MEASURED	COR. 20°C	KEY TO INSULATION RATING (G = GOOD, D = DETERIORATED, I = INVESTIGATE, B = BAD (REMOVE OR RECONDITION))	INSULATION RATING
1	HIGH	LOW		2.5	78	200	15600	4	20	80	— —	— —		
2	HIGH		LOW	2.5	51	100	5100	2	20	40	.78	.74	C_H	G
3	LOW	HIGH		2.5	21	1000	21000	4	20	80	— —	— —		
4	LOW		HIGH	2.5	51	200	10200	3	20	60	.59	.55	C_L	G
CALCULATED RESULTS					— —	— —	10500	— —	— —	40	.38	.36	C_{HL} (TEST 1 MINUS TEST 2)	G
					— —	— —	10800	— —	— —	20	— —	— —	(TEST 3 MINUS TEST 4)*	— —

*CURRENT AND WATTS SHOULD COMPARE WITH THOSE FOR C_{HL}

Figure 5.5 - Typical insulation power factor test data.

A

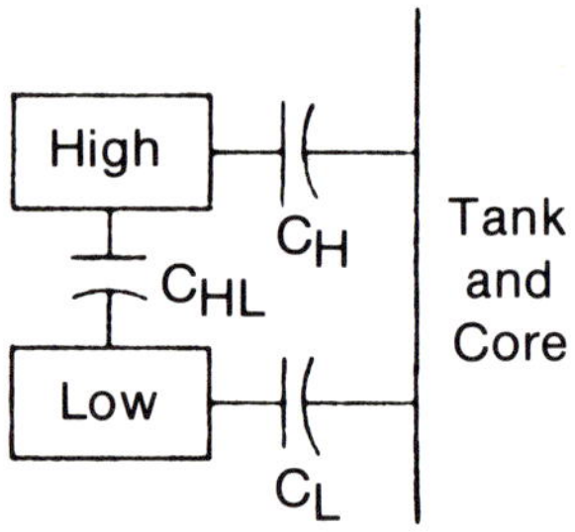

B

Test	Energized	Ground	Guard	Measure
1	H	L	—	$C_H + C_{HL}$
2	H	—	L	C_H
3	L	H	—	$C_L + C_{HL}$
4	L	—	H	C_L
5	Test 1 minus test 2			C_{HL}
6	Test 3 minus test 4			C_{LH}

Figure 5.6 - A simplified diagram of a typical two-winding transformer (A); Standard series of power factor tests applied to windings in-service. C_H refers to all insulation between the high voltage winding and grounded parts, including bushings, winding insulation, structural insulating members, and oil (B). C_L refers to the same parts and materials between the low voltage windings and grounded parts. C_{HL} refers to all winding insulation, barriers and oil between high and low voltage windings. (Courtesy of Doble Engineering Co.).

Figure 5.7 illustrates a system for lumped measurements on a two-winding transformer. Three measurements are made, the first two (A and B) include ground and interwinding insulation and the third (C) includes the combined ground insulations of the high and low voltage windings. The power factors recorded for the three measurements are the weighted averages of the two components involved in each.

Test	Energize	Ground	Measure
A	H	L	$C_H + C_{HL}$
B	L	H	$C_L + C_{HL}$
C	H_1L	—	$C_H + C_L$

Figure 5.7 - Alternative method for power factor testing: lumped measurements—more applicable for factory testing.

It is obvious that, if a condition of a high power factor exists in one component (C_H for example), it may be masked in Test A of Figure 5.6 but accentuated in Test 2 of Figure 5.6, which permits separate determination of C_H, C_L, and C_{HL}. The obvious solution is to make standard the method illustrated in Figure 5.6. It is consistent with Method II in IEEE Standard No. 262 (ANSI C57.12.90-1973) and is the test method used almost universally in the field. It increases the sensitivity of the power factor test to localized conditions. The method permits individual measurements of charging currents and dielectric loss for C_H and C_L, and power factors can be calculated from these values. Currents and losses for Tests 2 and 4 are subtracted from those recorded for tests 1 and 3 to obtain data for a power factor calculation for C_{HL}. If desired, C_{HL} may be measured directly by the UNGROUNDED SPECIMEN TEST (UST) method. Figure 5.8 gives typical test data and calculated power factor test results for a transformer in good condition.

From the data in Figure 5.8, it can be seen that the measured value of C_{HL} by the UST method closely agrees with the calculated value in Line 5 of Figure 5.8.

Oil-filled transformers that have been in-service can have power factors as high as two percent and still be deemed worthy of continued operation. Insulation power factors above two percent should be suspect and additional tests should be conducted at this point to further investigate the probable causes. The bushings and liquid insulation should be tested to see if they contribute to the overall high power factor.

OVER-ALL TESTS TEMP. CORRECTION FACTOR 0.49

TEST	TEST CONNECTIONS: WINDING ENERGIZED	WINDING GROUNDED	WINDING GUARDED	TEST KV	EQUIVALENT 10 KV READINGS — MILLIAMPERES: METER READING	MULTI-PLIER	MILLI-AMPERES	WATTS: METER READING	MULTI-PLIER	WATTS	% POWER FACTOR: MEASURED	COR. 20°C	KEY TO INSULATION RATING: G = GOOD, D = DETERIORATED, I = INVESTIGATE, B = BAD (REMOVE OR RECONDITION)	INSULATION RATING
1	HIGH	LOW		4	20	1	20	10	.2	2.0	— —	— —		
2	HIGH		LOW	4	33	.1	3.3	10	.02	0.2	.61	.30	C_H	G
3	LOW	HIGH		2.5	24	1	24	12	.2	2.4	— —	— —		
4	LOW		HIGH	2.5	39	.2	7.8	7	.1	0.7	.90	.44	C_L	G
5	CALCULATED RESULTS				— —	— —	16.7	— —	— —	1.8	1.1	.53	C_{HL} (TEST 1 MINUS TEST 2)	G
6					— —	— —	16.2	— —	— —	1.7	— —	— —	(TEST 3 MINUS TEST 4)*	— —

BUSHING TESTS

*CURRENT AND WATTS SHOULD COMPARE WITH THOSE FOR C_{HL}

LINE NO.		BUSH. NO.	PHASE	BUSHING SERIAL NO.	TEST KV	EQUIVALENT 10 KV READINGS — MICROAMPERES: METER READING	MULTI-PLIER	MICRO-AMPERES	WATTS: METER READING	MULTI-PLIER	WATTS	% POWER FACTOR: MEASURED	COR. 20°C	COLLAR TESTS (WATTS/CURRENT): TOP		INSULATION RATING
HIGH SIDE	1					CHL - UST MEASUREMENTS										
	2				4	82	.2	16.4	17	.1	1.7	1.0	.51	CHL		G
	3				2.5	82	.2	16.4	17	.1	1.7	1.0	.51	CHL		G
	4	N														

Figure 5.8 - Typical power factor test data and calculated test results for a transformer in good condition.

The effects of temperature on power factor is a most important consideration, since the magnitude of the power factor recorded on a given specimen varies directly with temperature. This must be taken into account when comparing data recorded for the same unit in the field with the factory, for the same unit on different occasions in the field, and in comparing similar units.

There are temperature correction charts published for the various types of transformers and insulating fluids. While the problem of temperature correction does exist, it is not complicated. A survey of a great number of tests indicates that a large majority of transformers tested in the field are found to be within 10 degrees of the 20°C base, thereby requiring little correction with minor error from available average curves.

When transformers must be tested at near freezing temperatures, or in excess of 20°C, or where some question exists as to the results recorded or to the accuracy of the temperature correction factor, the transformer should be retested if possible near 20°C.

The power factors recorded for routine overall tests provide interesting and valuable information regarding the general condition of ground and interwinding insulation of transformers. These values also provide a valuable index of dryness for a transformer. They are helpful in detecting undesirable operating conditions and failure hazards resulting from moisture, carbonization of insulation, defective bushings, contamination of oil by dissolved materials or conducting particles, improperly grounded or ungrounded cores, etc.

In an effort to improve the sensitivity of the overall tests by minimizing the amount of insulation included in the measurement, particularly in larger transformers, it is essential that separate tests also be performed on the bushings and liquid insulation in a transformer.

Winding Insulation Power Factor Tip-Up Test

On many occasions, insulation power factor tests for a transformer are higher than normal, showing that something is wrong in the insulating system. A short, easy to perform test called the Power Factor Tip-Up Test can provide more definitive data that may help pinpoint the trouble.

Using the power factor test set, the tip-up test can be performed by varying the applied test voltage from zero volts to the maximum in several equal steps. Calculate the percent power factor at each voltage. If the power factor does not vary as the test voltage is changed, moisture or other inherently high power factor/polar compounds are the probable cause of the high power factor. If the power factor increases as the test voltage increases, then ionization is occurring causing carbonization of the oil and transformer windings.

Figure 5.9 shows test data for a transformer that did have carbonization of the oil and windings taking place. It can be seen that the

Test KV	Power C_H	Factors C_T[1]	Corrected C_{HT}
10	0.60	4.52	0.81
2	0.40	3.05	0.55
1	0.17	2.56	0.45
[1]C_T - Tertiary Winding			

Figure 5.9 - Insulation power factor test results before untanking.

power factor values did increase as the test voltage increased.

When the transformer was untanked at a service shop, burned areas and shorted turns were found and fine carbon residue had been thoroughly deposited throughout the winding insulation.

Bushing Power Factor Tests

Bushings can be tested in several different ways without removal from the transformer. The most effective test method utilizes the potential taps, power factor taps, or split mounting flanges with which most modern bushings are equipped. This method as shown in Figure 5.10 is referred to as the UNGROUNDED SPECIMEN TEST (UST).

Figure 5.11 lists typical field test data. Care must be taken not to exceed the allowable tap test voltage. Consult the manufacturer's published information for the maximum permissible test voltage for the particular type of bushing being tested.

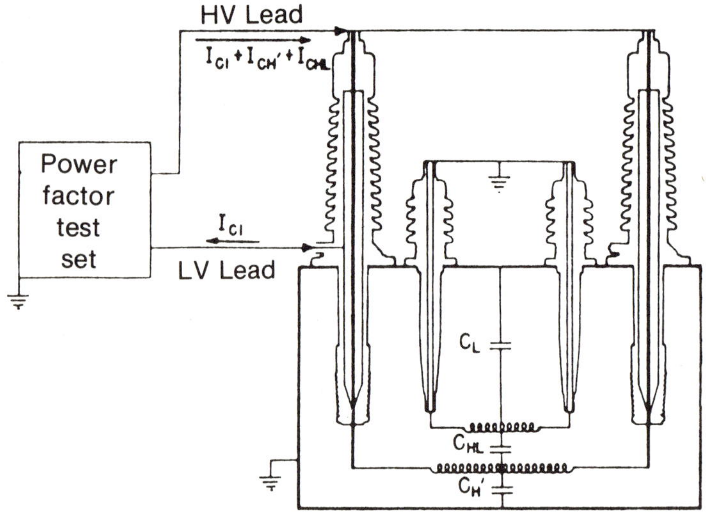

Figure 5.10 - Ungrounded specimen test on transformer bushings. (Courtesy of Doble Engineering Co.).

The Hot-Collar test was designed to detect localized defects (contamination or voids) in compound-filled bushings or cable potheads. This test is applied, using single or multiple collars on transformer bushings (Figure 5.12) if the UST method cannot be performed or to

BUSHING TESTS

*CURRENT AND WATTS SHOULD COMPARE WITH THOSE FOR C_{HL}

	LINE NO.	BUSH. NO.	PHASE	BUSHING SERIAL NO.	TEST KV	EQUIVALENT 10 KV READINGS: MICROAMPERES: METER READING	MULTIPLIER	MICRO-AMPERES	WATTS: METER READING	MULTIPLIER	WATTS	% POWER FACTOR: MEASURED	COR 20°C	COLLAR TESTS (WATTS/CURRENT): TOP		INSULATION RATING
HIGH SIDE	1					NAMEPLATE DATA										
	2					McGraw Edison - S/N 452575										
	3					Power Factor 0.54% UST at 22°C										
	4	N														
LOW SIDE	5					TEST RESULTS										
	6				2.0	53	.02	1.06	5	.01	.05	.47	.58	Temp.	30°C	G
	7					Temperature Correction Factor 1.24										
	8	N														
	9															
	10															
	11															
	12															
	13															
	14															
	15															
	16															
	17															
	18															
	19	OIL SAMPLE												OIL TEMP. °C		

Figure 5.11 - Typical field test data for a large transformer bushing. Ungrounded Specimen Test (UST).

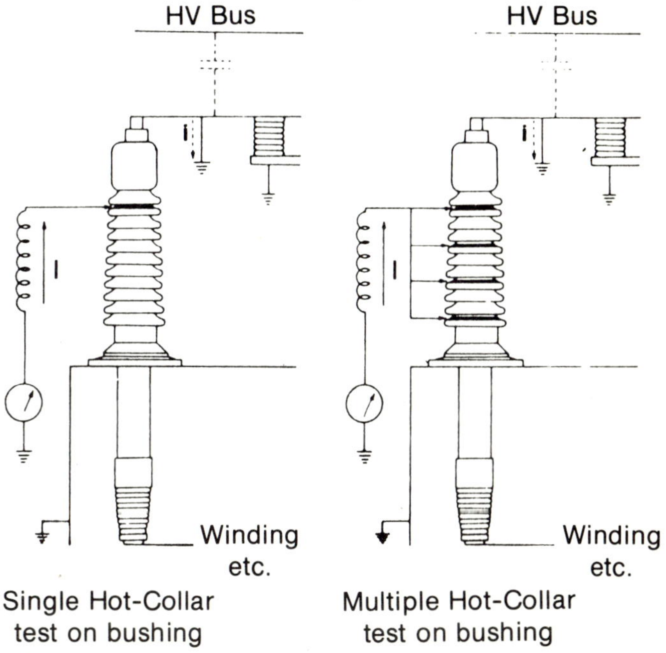

Figure 5.12 - Illustrating the Hot-Collar test method for testing of bushing insulators. (Courtesy of Doble Engineering Co.).

supplement the UST test when compound-filled bushings are involved. A single Hot-Collar test is also often applied to oil-filled bushings not equipped with liquid level gauges in an effort to detect low liquid levels.

Figure 5.13 shows some typical Hot-Collar test data.

BUSHING TESTS

*CURRENT AND WATTS SHOULD COMPARE WITH THOSE FOR C_{HL}

LINE NO.		BUSH. NO.	PHASE	BUSHING SERIAL NO	TEST KV	EQUIVALENT 10KV READINGS						% POWER FACTOR		COLLAR TESTS (WATTS/CURRENT)		INSULATION RATING
						MICROAMPERES			WATTS							
						METER READING	MULTIPLIER	MICROAMPERES	METER READING	MULTIPLIER	WATTS	MEASURED	COR 20°C	TOP		
HIGH SIDE	1							HOT COLLAR TESTS								
	2		A	452575	10	7.5	.01	.075	7.5	.002	.015	---	--	.015/.075		G
	3		A	452575	10	9	.01	.09	10	.002	.02	---	--	.02/.09		G
	4	N														
	5															

Figure 5.13 - Typical field test data—bushing hot-collar tests.

Since relatively low dielectric losses and currents are normally recorded for Hot-Collar tests, small changes in either value can result in misleading changes in calculated power factors. Because of this, it is recommended that Hot-Collar tests be evaluated by comparison of currents and losses obtained for similar tests on similar bushings and potheads under the same atmospheric conditions. Power factor values need not be calculated. As a general guideline, losses up to 0.10 watt at 10 KV and 6 milliwatts at 2.5 KV can be considered acceptable. Contamination and moisture will affect the measured losses, so be sure the bushings are clean and dry before testing.*

Liquid Insulation Power Factor Test

To complete the routine power factor testing of the transformer insulation system, a sample of oil must be drawn from the transformer and tested for power factor. Although this test by itself offers little information about the remaining life expectancy of the oil, when this information is properly applied along with other oil screening tests, it will assure you that the oil is in a safe and satisfactory condition for continued operation.† Oil samples are tested for power factor by the UST Method as shown in Figure 5.14.

Good, new oil should have a power factor of 0.05 percent or less at 20°C. Power factors can gradually increase in-service to a value as

**Chapter 8, Part 3.*

†Chapter 3, Part 1.

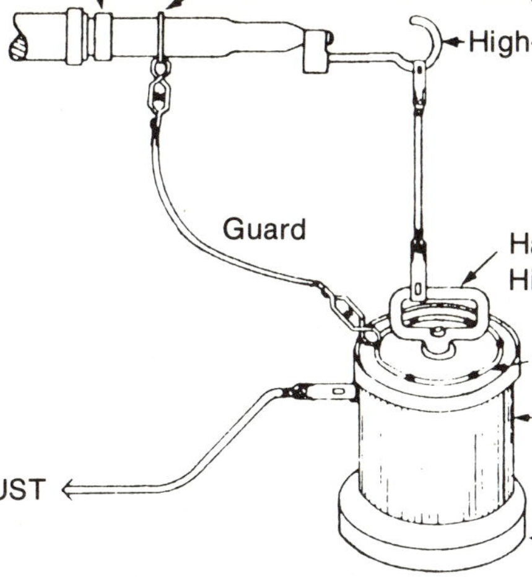

Figure 5.14 - Liquid-insulation cell connected for ungrounded specimen testing. (Courtesy of Doble Engineering Co.).

high as 0.5 percent at 20°C without, in most cases, indicating sufficient deterioration to warrant investigation. When the power factor exceeds 0.5 percent an investigation is indicated.* The question of what decision to make regarding disposition of the oil depends on what is causing the high power factor. Insulation Resistance or Karl Fischer moisture content tests may be made to see if moisture is present. The necessity for further tests will depend, to a large extent, on the magnitude of the power factor, the importance of the apparatus in which the oil is used, its rating, and the quantity of oil involved.

Insulation power factor tests can also be made on lightning arresters, air and oil circuit breakers, current and potential transformers, capacitors, cable (500 feet or less) and rotating machinery, such as motors and generators.

Core Excitation Current

The technique of using excitation current measurements for field detection of shorted turns and heavy core damage was developed in

**If a 100° C reading is more than 7 to 10 times the value at 20° C, then this usually indicates a soluble contaminant in the oil other than water.*

1967. This test method is a natural extension of the Power Factor Test and makes use of the same test equipment (Figure 5.3). Numerous cases of extensive core problems such as shorted laminations or core bolt insulation breakdown, have been detected by excitation current measurements.*

The following guidelines should be followed for routine excitation current tests.

1. All loads should be disconnected and the transformer de-energized.
2. Routine tests can be confined to the high voltage windings. Defects in the low voltage windings will still be detected and the excitation current required will be reduced.
3. Winding terminals normally grounded in-service should be grounded during tests, except for the particular winding energized for the test. For example, with a wye/wye transformer, the neutral of the high voltage winding would be connected to the UST (Ungrounded Specimen Test) circuit, while the neutral of the low voltage winding would be connected to ground.
4. Caution should be exercised in the vicinity of all transformer terminals because voltage will be induced in all windings during a test.
5. For routine tests the load tap changer (LTC) should be set to neutral; one step above neutral; one step below neutral; full raise or full lower; however, to ensure that the tap selector is functioning properly throughout the entire range of selection, it may be desirable sometimes to perform tests on all LTC positions.
6. Test voltages should not exceed the rated line-to-line voltage for delta-connected windings or rated line-to-line neutral voltage for wye-connected windings. Generally, these tests are made at 2.5, 5, or 10 KV as the capacity of the test equipment permits.
7. Test voltages should be the same for each phase and because of the nonlinear behavior of exciting current at low

**Other tests such as Transformer Turns Ratio (TTR) and Winding dc resistance tests must be used to detect tap changer contact problems, high resistance joints, and coil movement.*

test voltages, should be set accurately if results are to be compared.

8. For single-phase transformers excitation current test results are recorded with the high voltage winding energized alternately from opposite ends. This should also be done on the individual phases of three-phase transformers if the unit is suspect or if the initial exciting current measurements are questionable.

Figure 5.15 illustrates the excitation current test procedure for single-phase transformers. Figures 5.16 and 5.17 illustrate the excitation current test procedures for three-phase wye and delta connected units respectively. Note that all excitation current tests are performed by the Ungrounded Specimen Test (UST) method.

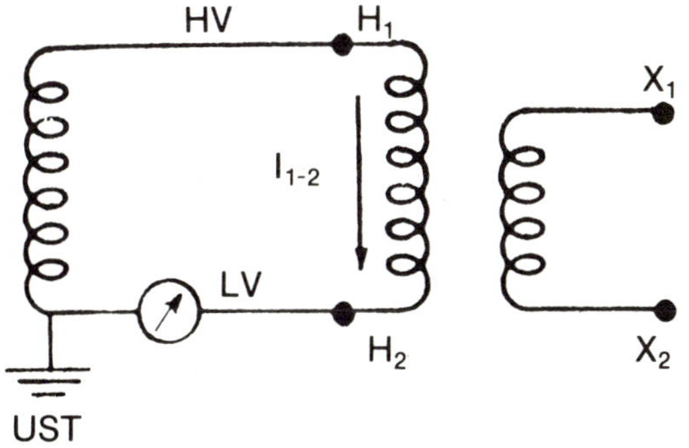

I_e	Energize	UST	Float
H_1-H_2 H_2-H_1	H_1 H_2	H_2 H_1	X_1X_2 X_1X_2

Figure 5.15 - Measurement of I_e in a single-phase transformer. (Courtesy of Doble Engineering Co.).

On single-phase transformers, the two currents obtained should be the same. Currents recorded for single phase transformers should be compared either with similar units or with data obtained from previous tests on the same unit.

The test results recorded on the individual phases of three-phase transformers are also compared. For a three-phase core-form transformer, a pattern of two similar currents and one low current is expected. The lower current for one of the phases, usually phase H_2

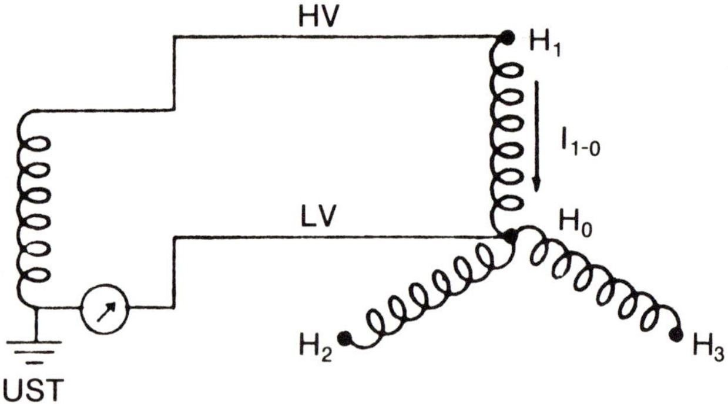

I_e	Energize	UST	Float	Ground
H_1-H_0	H_1	H_0	H_2H_3,$X_1X_2X_3$	*
H_2-H_0	H_2	H_0	H_1H_3,$X_1X_2X_3$	*
H_3-H_0	H_3	H_0	H_1H_2,$X_1X_2X_3$	*
*If X is Wye-Connected, X_0 is Grounded.				

Figure 5.16 - Measurement of I_e in a wye-connected transformer winding (routinė method). (Courtesy of Doble Engineering Co.).

H_0 for wye-connected windings and Phase H_1 H_2 for delta-connected windings, is associated with that winding which is wound on the center leg of a three-legged core. The magnetic reluctance of this phase is lower than the other two phases and, therefore, results in a lower excitation current value. Three-phase shell-form transformers may also yield a pattern of two similar currents and one low current but only if the secondary winding is delta-connected.

Sufficient experience has been gained with this low voltage excitation-current test method so that abnormal readings are easily recognized and numerous defects have been located.

When excitation current tests are included in factory specifications, these tests then could be performed in the field so that comparison of test data can show if any changes have taken place.

On repaired transformers the value of excitation current will generally increase along with but not necessarily proportional to the

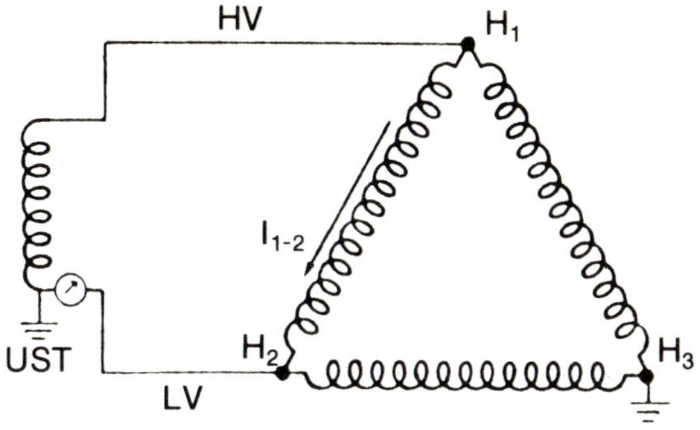

I_e	Energize	UST	Ground	Float
H_1-H_2 H_2-H_3 H_3-H_1	H_1 H_2 H_3	H_2 H_3 H_1	H_3,* H_1,* H_2,*	$X_1X_2X_3$ $X_1X_2X_3$ $X_1X_2X_3$
*If X is Wye-Connected, X_0 is Grounded.				

Figure 5.17 - Measurement of I_e in a delta-connected transformer winding (routine method). (Courtesy of Doble Engineering Co.).

increase in core loss. Joint construction severely affects the magnitude of the excitation current. Changes in the hysteresis and eddy current characteristics due to handling the steel also affect the excitation current.

High precision in excitation current measurements does not appear to be necessary. The serious faults found have increased excitation current magnitudes by greater than 10 percent over normal values.

Megohm Meter (Insulation Resistance) Series

The nondestructive insulation resistance measurement on equipment in service is perhaps one of the oldest techniques in testing. This resistance test and others in this series of electrical tests are usually measured by a specially constructed meter with both current and voltage coils and a meter calibrated to read directly in ohms or megohms. This megohm (or Megger)* meter has a built-in direct current generator that may be either hand cranked or electrically driven to develop a high dc voltage which causes a small current to flow through and over the surfaces of the insulation being tested (Figure 5.18).†

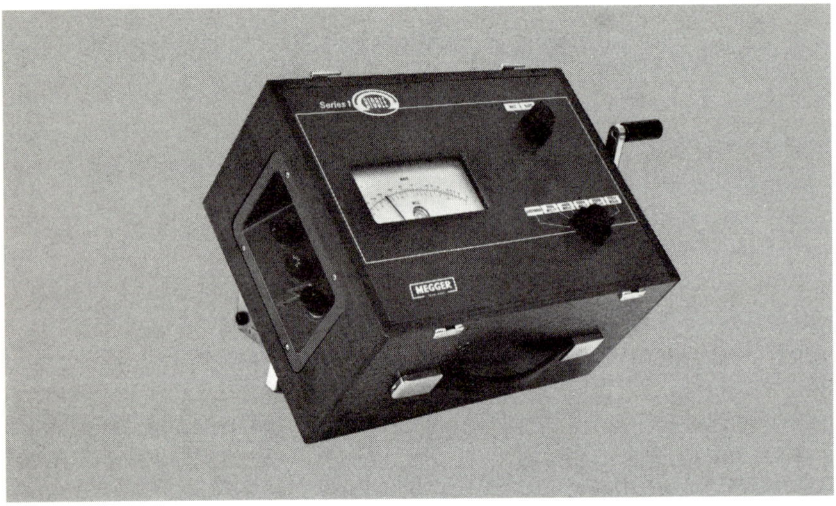

Figure 5.18 - This heavy-duty Megger® tester can be used for all measurements on large electrical equipment in the megohm meter (insulation resistance) series. (Courtesy of James G. Biddle Co.).

The total current through and along the insulation is made up of three components:

1. Capacitance Charging Current
   Current that starts at a high value and decreases after the insulation (capacitance) has been charged to full voltage.

2. Dielectric Absorption Current
   The current component resulting from absorption within imperfect dielectrics is caused by various polarizations, the most predominant being the interfacial type resulting from the barrier effect at the interfaces of materials within such composite structures. A dipole polarization resulting from

**Megger is the registered trademark of James G. Biddle Company, Plymouth Meeting, Penn.*

†Megohm meter testing of bushings is no longer practiced.

polar molecules and molecular chains will also be found at the dc end of the spectrum in some types of insulation. It is this energy that causes voltage to reappear at the electrodes or plates of a dielectric after the stored energy resulting from the capacitance has been dissipated by a short circuit and the short circuit removed. It is because of this fact that the short circuit should be held on the insulation under test for at least as long as the total test time or longer.

The absorption current starts at a high value and decreases rapidly with time.

3. Leakage Current
The leakage current is the most important component when attempting to evaluate the condition of insulation. The path of this current may be either through the volume of insulation or over leakage surfaces. This component, unlike the other two, represents a loss current. Theoretically, the leakage current should be constant with time for any one value of applied voltage. This constancy with time is a good indication that the insulation under test can withstand the voltage being applied. Any tendency for this current to steadily increase with time at a constant applied voltage is a warning that the insulation may be damaged if the test is continued at that voltage.

The insulation resistance is generally accepted as a reliable indication of the presence or absence of harmful contamination or degradation. However, its marked sensitivity to relatively small changes in the apparatus being tested can in itself result in confusing data. All possible variables must be given careful attention if a sound and safe interpretation of the test results is to be obtained. Such variables include the effects of temperature, humidity, and external leakage due to dirty insulators and bushings, duration of the test, etc.

The temperature of the equipment under test has a marked influence on the results obtained. Insulation resistance values decrease with increasing temperature. Therefore, if you wish to make a reliable comparison of readings taken at different time periods, the readings should be taken at approximately the same temperature. This may be difficult to do. If readings are taken at different temperatures, all readings should be corrected to a common base of 20°C. A short listing of temperature correction factors is shown in Table 5.1.

For unsealed transformers, testing in an atmosphere of high humidity should be avoided if possible. The cellulosic insulation is hygroscopic and moisture will appear on and in the fiber. This will affect the insulation resistance and cause erroneous test results. Attention should also be given to the dew point of the atmosphere before and during the time the actual test is made.* If the temperature of the equipment is for any reason, below atmospheric dew point, this will also encourage the migration of moisture into the cellulose. This moisture problem may sometimes be corrected with the circulation of a small amount of warm dry air through the exposed area.

TABLE 5.1

TEMPERATURE CORRECTION FACTORS FOR OIL-FILLED TRANSFORMERS		
Base Temperature 20° C		
°C	°F	Correction Factor
0	32	0.25
5	41	0.36
10	50	0.500
15	59	0.720
20	68	1.00
30	86	1.98
40	104	3.95
50	122	7.85

This test is relatively quick and easy to make, and is normally performed with a 500 V, or 1000 V, or 2500 V tester.

Insulation-resistance measurements are made from the individual windings to ground and also between windings. The tests consist of applying voltage to a winding for a period of ten (10) minutes. The generally accepted minimum value is 1000 megohms when corrected to 20°C temperature. Lower values may be acceptable for certain low-voltage windings which have a high cross-sectional area.

**Chapter 3, Part 3.*

It is advisable to watch the trend of successive readings as well as the minimum values. A downward trend over a period of time may reveal changes which justify investigation even though the readings are above an established minimum value. Conversely, allowances may be made for transformers in service that show periodic test values below the minimum value, providing such periodic values remain stable and consistent.*

Minimum Insulation Resistance

One way of determining the condition of the insulating system is to see if the measured resistance is above a minimum resistance value that can be calculated if certain facts about the system are known. This is done mainly for transformers and is a calculation made prior to the series of tests. This figure then becomes a reference point by which the one minute measured resistances may be compared.

$$R = \sqrt{\frac{CE}{KVA}}$$ Where

R = Insulation resistance of winding to ground with other winding grounded or between windings in megohms at 20°C.

C = 0.8 for oil-filled transformers at 20°C.

C = 16 for dry, compound filled, or untanked oil filled transformers.

E = Voltage rating of winding under test.

KVA = Rated capacity of winding under test.

If the insulation resistance tester is equipped with a guard terminal, then:

R = Insulation resistance of winding to ground with other winding guarded, or between windings with core guarded at 20°C.

*"*Proposed Transformer Test and Maintenance Guide," Doble Engineering Co., Watertown, Mass. (1980).*

C = 1.6 for oil filled transformers at 20°C.

C = 30.0 for dry, compound-filled or untanked oil filled transformers.

The above applies to single phase transformers or to separately tested windings of multiphase units. If the transformer is the three phase type and the three windings are being tested together, then:

E = Voltage rating of one of the single phase windings (phase to phase for delta connected units and phase to neutral for wye connected units).

KVA = Rated capacity of the complete three phase windings. Transformer with windings rated less than 100 KVA.

R = KV + 1, where KV equals voltage classification of the winding under test.

Insulation resistance values near or below these calculated minimum values are indicative of probable trouble and some type of preventive maintenance should be considered.

Figure 5.19 represents typical Insulation Resistance test data. Readings have been recorded at 15, 30, 45 and 60 seconds and then every minute thereafter for a total of 10 minutes. In Figure 5.19 the test voltage of 2500 V was connected Primary/Ground, Primary/Secondary, and Secondary/Ground. Now consider an evaluation of the test data.

Dielectric Absorption

These readings can be plotted on Log-Log Graph paper. The resulting curve for a good insulating system will have a regular rise with time (Figure 5.20).

Polarization Index (PI)

From the Dielectric Adsorption data, the Polarization Index can be determined. Polarization Index, the ratio of the 10 minute resistance to the 1 minute resistance value, is a dimensionless quantity often used in insulation evaluation. It is a pure numeric and offsets the

INSULATION RESISTANCE - DIELECTRIC ABSORPTION TEST SHEET

POWER TRANSFORMER

TEST NO. ______

DATE 10/2/74

TIME ______

______ COMPANY

______ LOCATION

EQUIPMENT	Transformer	RATING	18750 KVA	VOLTAGE	132KV/13.8KV
TYPE	OA/FA - TU	MFR.	G.E.	SERIAL #	C656340
RECENT OPERATING HISTORY	In Service			WINDING AGE	

STATE IF TRANSFORMER IS OIL FILLED AND OF THE CONSERVATOR, FREE-BREATHING OR GAS-FILLED TYPE: Oil - Sealed

IF OF THE FREE-BREATHING TYPE DESCRIBE THE BREATHER ARRANGEMENT AND DEHYDRATING MEDIUM USED.

IF OF THE FREE-BREATHING TYPE NOTE INTERNAL CONDITIONS SUCH AS UNDERSIDES OF COVERS, TERMINAL BOARDS, WINDINGS ETC.

BUSHING DATA

OIL TESTS: ACIDITY ______ MG KOH/G: RESISTANCE IN STANDARD CUP ______ MEGOHMS; TOP OIL BREAKDOWN ______ KV.

STATE WHAT BUSHING PORCELAINS WERE GUARDED DURING TESTS

TEST DATA - MEGOHMS

PART TESTED				
GROUNDING TIME				
TEST VOLTAGE	2500	2500	2500	
TEST CONNECTIONS	PRI. TO LINE	PRI. TO LINE	SEC. TO LINE	TO LINE
	GND. TO EARTH	SEC. TO EARTH	GND. TO EARTH	TO EARTH
	TO GUARD	TO GUARD	TO GUARD	TO GUARD
¼ MIN.	600 M OHMS	450 M OHMS	300 M OHMS	
½ "	950 M OHMS	850 M OHMS	650 M OHMS	
¾ "	1150 M OHMS	1300 M OHMS	1100 M OHMS	
1 "	1250 M OHMS	1550 M OHMS	1400 M OHMS	
2 "	1600 M OHMS	2250 M OHMS	2250 M OHMS	
3 "	1775 M OHMS	2750 M OHMS	2950 M OHMS	
4 "	1975 M OHMS	3250 M OHMS	3250 M OHMS	
5 "	2050 M OHMS	3500 M OHMS	3500 M OHMS	
6 "	2250 M OHMS	3750 M OHMS	3750 M OHMS	
7 "	2350 M OHMS	4000 M OHMS	4000 M OHMS	
8 "	2450 M OHMS	4000 M OHMS	4000 M OHMS	
9 "	2450 M OHMS	4250 M OHMS	4100 M OHMS	
10 "	2450 M OHMS	4250 M OHMS	4250 M OHMS	
10/1 MIN. RATIO	1.96	2.74	3.03	

TEST MADE	HOURS DAYS AFTER SHUTDOWN
DRY-BULB TEMP.	°F
WET-BULB TEMP.	°F
DEW POINT	°F
RELATIVE HUMIDITY	%
ABSOLUTE HUMIDITY	GR./#
EQUIPMENT TEMP.	°F(C)
HOW OBTAINED	
"MEGGER" INST.:—	
SERIAL #	1927133
RANGE	50 KM OHMS
VOLTAGE	500/1000/2500

REMARKS PRI-GND 500 V. D.C. Applied 1200 M OHMS

Transformer is in good condition.

TESTED BY K. Stutz

Figure 5.19 - Typical insulation resistance/dielectric absorption test data.

fact that previous test information may not be available. Since the leakage current increases at a faster rate with moisture present than does the absorption current, the Polarization Index is lowered for insulation in poorer condition. In as much as all the variables that can affect a single megger reading are essentially the same for both readings (that is, the one and ten minute reading) information of real value can be obtained using the PI. Table 5.2 lists the parameters whereby an insulating system can be categorized:

TABLE 5.2

POLARIZATION INDEX GUIDE FOR EVALUATION OF TRANSFORMER CONDITION	
Condition	Polarization Index (PI)
Dangerous	Less than 1.0
Poor	1.0 to 1.1
Questionable	1.1 to 1.25
Fair	1.25 to 2.0
Good	Above 2.0
Observe the Polarization Index calculations in Figure 5.19.	

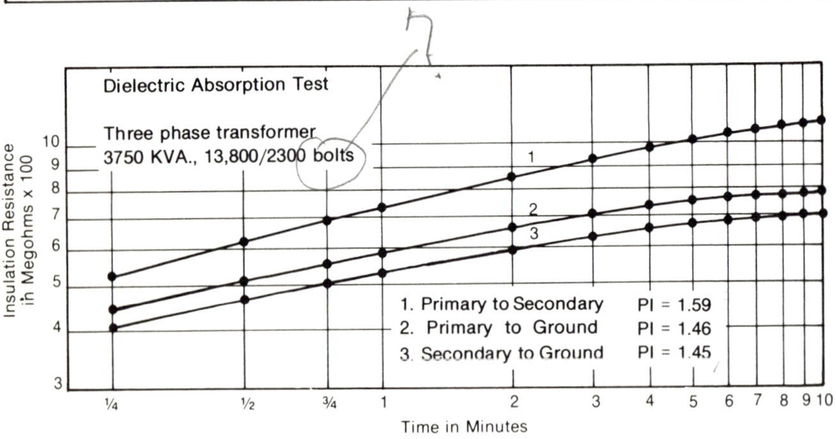

Figure 5.20 - Plot of dielectric absorption test from which can be calculated the PI (Polarization Index). (Courtesy of Factory Mutual).

Step Voltage Test

The Step Voltage Test is very useful to determine the presence of excessive moisture in the insulation of equipment that is rated at voltages equivalent to or greater than the highest voltage available from the insulation resistance tester. Even though the highest available test voltage does not stress the insulation beyond its rating, a two voltage test can often reveal the presence of harmful contaminants.

The applied voltages should preferably be in the ratio of 1 to 5 or greater (500 and 2500 V, for example) both applied for one minute.* Results to date show that a decrease in the insulation resistance of 25% or greater at the higher test voltage is usually due to the presence of excessive moisture.† Why is this? Water in an insulating system is polar positive and will be attracted to areas of high negative electrical intensity. Therefore, when a Megger® or similar equipment is used the negative lead is attached to the copper and the positive lead to the ground system. During the test, excessive moisture will be attracted to the area of high negative potential along converging field lines. In other words, a lower resistance will be measured if moisture is present by testing with center conductor at a negative potential. This phenomena is called electro-end osmosis and is often referred to as the Evershed Effect. If there is not sufficient moisture in the insulating system to create this problem, then the meter readings at the two voltages should be nearly the same.

Transformer Turns Ratio Test (TTR®)‡

The turns ratio test is primarily used as an acceptance test.It is also useful as a tool for investigating problems, as well as an integral part of a routine preventive maintenance program.

During the manufacture of new transformers, the turns ratio test should be performed on all tap positions to verify that the internal connections are correct and that there are no short circuited turns.

***Note**: *2500 V dc should never be applied to a winding insulated for less than 2300 V ac. It is advisable to apply a much lower voltage, say 250 V dc first, preferably with a hand-operated megohm meter to see if the winding is of sufficient strength to withstand the test voltages to be applied.*

†See application of the step voltage test, Chapter 7, Part 2, Figure 7.6B.

‡TTR is a registered trademark of the James G. Biddle Co.

During routine maintenance tests, the turns ratio test could be performed to identify short circuited turns, incorrect tap settings, errors in turn count, mislabeled terminals, and failure in tap changers. If the transformer has been modified or repaired in the field or if a fault has occurred dropping the unit off the line, a TTR test is a good way to check the condition of the transformer windings.

In a well constructed transformer, the voltage rating or ratio as you see it on the nameplate is determined by the number of turns of wire on the primary and secondary. Therefore, the turns ratio between the primary and secondary is equal to the voltage ratio of the primary to the secondary. The turns ratio does not tell how many turns of wire are on the primary or secondary coil, but only gives the ratio of the primary to secondary turns.

The ratio test can be made either of two ways. One method involves applying a known voltage on one winding and measuring the induced voltage in the other winding. On three phase transformers connected in wye with an external neutral, the ratio can be checked with a single phase potential. Other three phase transformers require three phase potential. The manufacturers recommend that a voltage of at least 10 percent of the rated voltage be used when making the ratio test by this method; however, this is not always a practicable method to use in the field.

The most well known method of checking turns ratios in the field utilizes the Transformer Turns Ratio test set which is available commercially (Figure 5.21). This test set has an internal alternator to supply the test potential. The ratio of a standard transformer in the test set is adjusted until its voltage is exactly equal to that of the transformer under test. The ratio of the standard transformer then indicates the ratio of the transformer under test. The measured ratio should compare with the nameplate ratio to within ±0.5 percent.

A turns ratio measurement can show that a fault exists but does not determine the exact location of the fault. An internal inspection or detanking may be needed to locate the problem.

A typical set of turns ratio test data would be as follows:

Transformer connected delta-delta
Primary voltage: Tap #1 4368 V
Secondary voltage: 2400 V
Calculated turns ratio = Voltage ratio $= \frac{4368}{2400} = 1.820$

Measured turns ratio

$H_1 - H_2 / X_1 - X_2 = 1.818$
$H_1 - H_3 / X_1 - X_3 = 1.817$
$H_2 - H_3 / X_2 - X_3 = 1.817$

These measured ratios fall within ± 0.5 percent of the calculated ratio.

Figure 5.21 - A transformer turns ratio (TTR) test set on the field.

DC Winding Resistance Test

The DC Winding Resistance test, made with a low resistance ohmmeter (or Ducter®),* will indicate a change in dc winding resistance when there are short circuited turns, poor joints or bad contacts. This reading should be compared to factory test information.

The low resistance is measured by the potential drop method. The "Ducter" has a high range scale of 0 to 5 ohms; the lowest range measures from 0 to 500 microhms. It has an accuracy within one percent of full scale deflection on all ranges. The idea is to apply current to the system. The resistance will be constant so the potential will show any voltage drop present. Readings should be in the milliohm range.† High readings could indicate loose or dirty connec-

**A trademark of James G. Biddle Company.*

†*Typical data is given in Chapter 6, Part 2, Figure 6.9A.*

tions. Once again, prior readings are necessary so that a comparison can be made.

On three-phase transformers, the measurements are made on the individual windings (phase-to-neutral) when possible. On delta connections, there will always be two windings (in series) which are in parallel with the winding under test. For this reason, on a delta winding three measurements should be made to obtain most accurate results.

Since the resistance of copper varies with temperature, all test readings must be converted to a common temperature to give meaningful results.

> Most factory test data are converted to 75°C and this has become the most commonly used temperature.* The formula for converting resistance readings on copper winding (to 75°C) is as follows:

$$R_{75^{\circ}C} = R_{test} \times \frac{234.5 + 75}{234.5 + \text{Wdg temp in } ^{\circ}C}$$

Due to the inability to obtain precise temperature measurements, the permissible deviation for this test in the field is two percent of the factory test values.

The DC Winding Resistance Test is also applicable to circuit breaker contacts, tap changer contacts, coils, and other electrical devices.

DC Overpotential Testing

Insulation failure even in properly designed equipment may result from such causes as aging and embrittlement due to evaporation of plasticizers, stabilizers, or antioxidants or electrolytic deterioration, chemical deterioration, lowered strength due to the absorption of moisture, physical damage, insulation cold or hot (polymers) flow or gradual tracking due to corona in internal voids or across the surface with eventual breakdown due to transients or abnormally high voltage stresses. While it is impractical to perform physical tests to pinpoint each of these possible defects, a properly performed Over Potential test can frequently point out weaknesses in the insulation regardless of cause (Figure 5.22).

*"*Proposed Transformer Test and Maintenance Guide,*" *Doble Engineering Co., (1980).*

DC Over Potential testing has been designated by some authorities as a potentially destructive insulation test as it stresses the solid insulation. Nonetheless, it has been rapidly gaining favor as an important tool in determining the condition of insulation in cables, transformers, rotating machinery, etc.

Figure 5.22 - Checking the insulation of a HV winding to ground using a Biddle 15 KV DC Dielectric Test set (Highpot). The measurement of the leakage current values through the windings insulation are the basis for determining if the insulation will be reliable for service. This test is just one of the ways to determine this reliability. (Courtesy of James G. Biddle Co.).

The use of dc for overpotential testing has several important advantages over an ac test potential.* DC overpotential test sets are usually much smaller, lighter in weight and lower in price. Properly used and interpreted, dc tests will give much more information than can be obtained with ac testing. There is less chance of damage to equipment under test. The high capacitance current associated with ac testing is not present during dc testing to mask the true leakage current, nor is it necessary to actually break down the material being tested to obtain information regarding its condition. Although dc testing does not simulate the operating conditions as closely as ac testing, the many other advantages of using dc makes it well worthwhile.

**The ac test voltage data should be given with the transformer's factory test data.*

The cable industry has published information that permits the determination of definite direct voltage values for making tests in installed cable. These values have been calculated from information contained in IPCEA*, NEMA*, AEIC EEI* and ANSI* Standards may be procured from EEI and ANSI Standards direct from that organization.

If published information for determining the direct-current voltage values needed during cable maintenance testing is not available, an alternate method of calculating the dc test voltage may be used. A figure commonly used in maintenance testing of in-service equipment is twice the rated voltage plus 1000 for ac tests. For dc testing, as discussed here, this figure is then multiplied by the ac to dc conversion factor of 1.6. This voltage is then derated by 65% for older equipment. Hence, the formula for maintenance testing of older equipment is:

$$E_T = (2\,E_N + 1000) \times 1.6 \times 0.65$$

ET = is maximum DC test voltage
EN = is nameplate voltage
1.6 = is ac to dc conversion factor
0.65 = derating factor for older equipment

The multipliers may vary depending on what organization is doing the testing.

The dc test voltage should be applied to the winding or cable in approximately 10 steps, recording the leakage current at each step. The test operator should plot the leakage current vs test volts as the test progresses. In this way, the condition of the equipment is under constant surveillance and the operator can stop the test if the current is rising too rapidly. As long as the leakage vs voltage curve is relatively flat (Figure 5.23, Point A to Point B), the item under test is considered to be in good condition. At some point the current will start rising at a more rapid rate (Point C). This will show up on the plot as a knee in the curve. It is very important that the voltage increments be small enough so that the starting point of this knee can be noted. If the test is carried beyond the start of the knee, the current will increase at a much more rapid rate and breakdown will soon occur if the test is not halted. It is the usual practice to stop the

IPCEA - Insulated Power Cable Engineers, Montclair, N.J. 07042. NEMA - National Electrical Manufacturers Association, Washington, D.C. 20037. EEI - Edison Electric Institute, New York, N.Y. 10017. ANSI - American National Standards Institute, New York, N.Y. 10018.

test as soon as the beginning of the knee is observed in the plot. With a little experience, the test operator can extrapolate the curve and estimate the point of breakdown voltage (D). With this procedure, the operator can anticipate the breakdown point and stop the test at the start of the knee of the curve and cause no damage to the insulation.

A comparison of leakage current from phase to phase or a comparison with the leakage currents obtained on tests made of the same equipment on prior occasions will give an indication of the condition of the insulation. In general, higher leakage current is an indication of poorer insulation; however, this is a matter difficult to evaluate from a single test. As long as the curve of "Leakage Current vs. Voltage" is of the general shape shown in Figure 5.23, and as long as the knee occurs at a sufficiently high voltage (above the maximum test voltage required) the equipment being tested may be considered satisfactory.

For transformers greater than 34.5, KV, the transformer must be detanked. There are those who do NOT recommend the over potential test because it can cause damage to the insulation. On the other hand, an imminent insulation failure is one of the problems this test is intended to reveal. Therefore, such damage caused by the test voltage should be considerably less than that which would develop

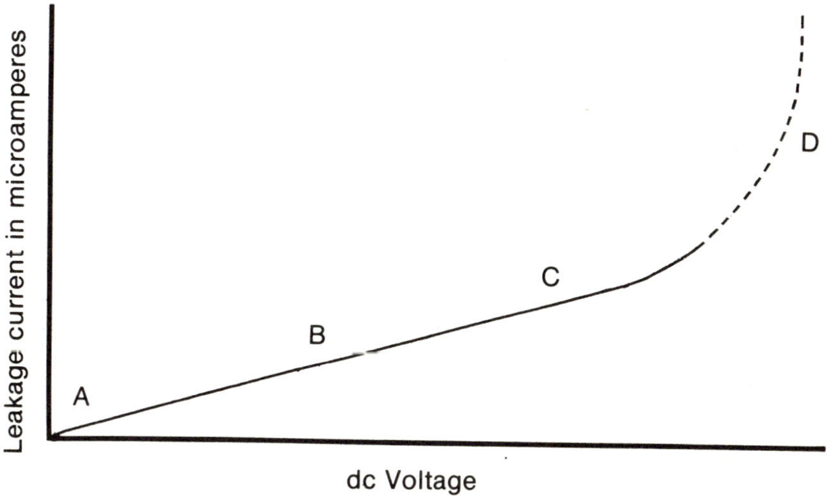

Figure 5.23 - Evaluating dc overpotential test results.

through in-service failure. Factory Mutual recommends that overpotential tests should not be run on machines rated 10,000 KVA or less unless either the short-time insulation resistance or PI values are satisfactory. On units rated above 10,000 KVA, it is advisable that both the insulation resistance and PI values be satisfactory before conducting overpotential tests.

Core Ground

Another test that may be feasible to perform in the field is the Core Ground check, done with the Megger which is also used for the Insulation Resistance tests. This test is most important as a check for shipping damage when the transformer arrives at the installation site, but consideration should be given to making the core ground check a routine test.

Heating will occur in a core form transformer due to circulating currents if the core laminations are grounded at more than one point. The laminations which make up the core are insulated from each other and from the end frames and other structural members. These structural members are normally connected to the main tank. The laminations are insulated from each other by a chemical treatment which leaves a very thin layer of insulation. This insulation between the laminations is only a few ohms and is sufficiently high to prevent damaging eddy currents within the core, and at the same time, is sufficiently low to permit the entire core to be effectively grounded by a connection to only one of the laminations.

The core ground connection is usually located at the top of the transformer. On some designs, this connection is not solid, but instead is made through a heavy-duty resistor in the 250 to 1000 ohm range. A resistor in this range still accomplishes the effective grounding of the core, and at the same time limits circulating currents. This ground connection is usually conveniently mounted under a manhole at the top of the transformer. Some designs bring the connection through the tank with a small low voltage bushing and the grounding strap is external. These accessible locations permit the ground to be easily removed and thus the core-to-ground insulation may be readily tested.

The core-to-ground insulation normally consists of a pressboard material which should withstand at least 2000 volts. Damage incurred in shipment can very likely affect this core-to-ground insulation. Therefore, it is recommended that it be tested before unloading the transformer from the railroad car or truck on which it was shipped. Using a 1000 volt Megger®, this insulation generally should read 200 megohms or more between ground strap and end frame.

If the Megger® shows a ground between ground strap and end frame switch to an ohm meter and read the resistance. If the resistance reading is only 1 or 2 ohms, it can be considered a solid ground and the unit should be repaired. If the resistance is 200 to 400 ohms, it is considered a high resistance ground and it is recommended that an effort be made to clear the ground before energizing the transformer. The following procedure is recommended:*

1. Apply 110 volts, single phase, 60 Hz, between the ground strap and the end frame. Be sure to use at least a 5 KVA power supply and test leads for 30 amps. It would be wise to use a knife switch and a 30 amp fuse in one side of the 110 V circuit. Make sure the grounded side of the 110 volt supply is connected to the end frame. If the fuse holds, apply 220 volts between ground strap and end frame.

2. Make sure the circuit for 220 volts is not grounded. It may be best to use an isolation transformer between the power supply and the 220 volt power applied to the ground strap and end frame. This will permit the grounding of the side of the 220 volt supply as was done for the 110 volt test circuit. Use a knife switch and 60 amp fuses in the 220 volt circuit. Listen for arcing in the transformer and see if it can be located.

3. Make another ohmmeter check. If the ohms have increased to approximately 1000, make another Megger check at 1000 volts. If the high resistance ground has definitely been removed, proceed to Step 4.

4. Apply 1000 volts, single phase, 60 Hz, between core ground strap and end frame. Be sure the grounded side of the power supply is connected to the end frame. Also make sure that the transformer and one side of the power supply are tied to the same ground. Use switch and fuses as in Step 2. Apply the voltage for one minute. Use a high voltage volt meter in the output circuit or use a clamp-on-ammeter to determine if

*__Caution:__ *The core ground procedures should first be compared with the manufacturer's instructions supplied with the particular megohm meter.*

there is a steady current flow in the power supply to the ground strap and end frame. If the ground has been completely removed, there should be no current flow.

5. Having now completed Step 4, the core to ground test has been performed the same as during the manufacturing process.

6. Repeat "Megger" insulation resistance tests and record for future reference.

   In some cases, dc welders can be used to remove unwanted core grounds. The following alternate method may be tried.

   a. Place the negative lead of the welder on the end frame and the positive lead on the ground strap. Apply approximately 40 amps, monitoring the current with a suitable ammeter. If the ground is removed under this condition, the current will decrease and the voltage will increase. It may be necessary to increase the current in 20 to 40 amp steps and repeat. Always check after each application with an ohmmeter or "Megger" tester to be assured the test is moving in the right direction.

   b. If the insulation resistance has improved, apply the procedure under Step 4 of the ac Method to prove the ground has been removed. If the applied 1000 volts ac holds for one minute, the core to ground insulation is considered to be good and the ground strap may be reconnected to the end frame.

Data should be recorded for the insulation value, including notes on the top oil temperature. If possible, all measurements should be made with the same instrument to avoid errors and to obtain comparative results.

If an unwanted core ground is found and proves extremely difficult to clear, it may be practical to install a resistor in series with the normal grounding strap so as to make a high-resistance loop. This resistor should be from 250 to 1000 ohms.

In shell form transformers, it is not important that the laminations be grounded in only one spot, since the flux distribution in this type of unit differs from core-form transformers.

Ground Resistance Measurements

What is ground resistance? Ground resistance is the ohmic resistance between a grounding electrode and a remote or reference electrode.

You may wonder why ground resistance is of much concern in a substation or any power system network. After all, one can see that the transformers all have copper conductors attached to their respective metal bases and the conductors extend down into the earth. It would appear that all the units are properly grounded, right? NOT NECESSARILY SO!!! What has happened to the copper conductors beneath the soil? The grounding electrode or ground grid may have corroded sufficiently to increase the ground resistance to an unsafe level. If the equipment (transformer, tower, etc.) is not properly grounded, under certain conditions dangerous potentials could exist between the equipment and ground. A person touching the equipment would then be subjected to this potential. Therefore, ground resistance in substations and power system networks is a vital concern to plant maintenance personnel and should be measured periodically to see that it remains at a safe, low value.

How is this ground resistance measured? The basic technique which is universally used for the measurement of a grounding system resistance is known generally as the "fall-of-potential" method. The basic diagram is shown in Figure 5.24.

Several manufacturers supply equipment utilizing this method of measurement. Present day instruments provide a readout directly in ohms, thus eliminating any calculation. Although the newer equipment is more sophisticated, it still performs the basic functions for fall-of-potential testing.

The soil condition, temperature, and moisture content of the soil have a great bearing on the value of resistance that is measured. A single ground rod does not always provide a sufficiently low resistance for large power systems. A ground grid consisting of a system of grounding electrodes interconnected by bare cables buried in the earth may have to be constructed to obtain the safe, low ground resistance. If the ground electrode or grid is known to be in good condition but the measured resistance is high, the soil may need to be conditioned with water or chemicals.

A good grounding system keeps the ground resistance to an acceptable low value. This condition serves to protect personnel and equipment. The *National Electrical Code* contains detailed rules for grounding. These rules specify that the grounding electrodes have a resistance to ground of not more than 25 ohms. However, it is recommended that the resistance be kept at 5 ohms maximum.

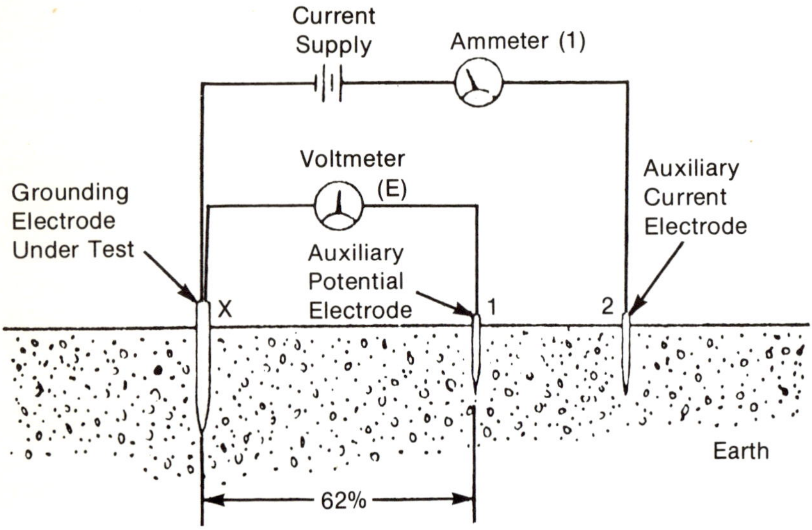

Figure 5.24 - Fall-of potential method of ground resistance testing.

Chapter 6

Basic Protective Maintenance Practice for Oil-Insulated Transformers

Part 1—Prevention of Severe Oil Deterioration

Introduction

Transformer mineral oil and cellulosic insulating materials — just like other organic materials — are changed chemically while in-service under the influence of moisture, oxygen, heat, and catalytized by copper and iron building materials (Figure 2.32). This process is termed "aging", and the end result is to limit the useful life of a transformer.

Over the years, the emphasis on maintenance of oil-filled transformers has been on the words — "prevent", "severe", and "deterioration." The very early units had little maintenance attention, as everyone considered the transformer to be a "miracle of modern industry." When the need for maintenance became apparent, the unit usually was on its last legs.

This experience led the manufacturers to try to design deterioration out of the units. Oil was improved, units were sealed to keep the atmosphere (oxygen and moisture) out, and windings were varnished.* Yet the deterioration of oil and paper insulation still persisted. Why? Because oil, if not vacuum processed into the equipment, can dissolve 10 to 12 percent gas by volume. If this gas had oxygen in it, you were on the road to oxidation. Copper was found to be a catalyst to oil oxidation so the transformer was varnished to prevent contact with oil.† But, water and iron are also catalysts in oil deterioration. Cellulose when heated also provides a source of moisture and oxygen.

**Chapter 1, Part 2, p. 81.*

†Varnish did not prevent or hinder oil oxidation. As a matter of fact, it compounded the problem.

When it became evident that oil and cellulosic deterioration could not be eliminiated, then, the next step was to control the severity of the deterioration. Tests were run to monitor the insulation system. When the oil reached the sludging state, the school solution was to schedule the transformer for an overhaul to remove the oil and its decay products. No consideration was given to the paper insulation. Nobody even considered taking a transformer out of service to change oil or clean the unit. It was only an academic solution — not at all practical.

In some respects, older units were better made. You probably have heard someone say, "They don't make them like they use to." This is true, they don't. The amount of oil per KVA is now much less than in the earlier units.* While oil has continued to perform the same three functions, in time past, there was more oil to soak up the abuses. Unfortunately, as the units were modernized the abuses became more severe. Still another problem with reduced physical size and thus less paper insulation has only recently been recognized. The smaller the amount of paper, the faster the oil will oxidize and degrade.

The actuality now gaining favor is that the **prevention** of sludge formation is much easier than the **removal** of sludges. Transformer oils should be serviced before they reach the critical acid number (sludging stage) in order to assure long life.† No economic or operating advantage is gained by permitting the oils to degrade beyond this point (Figure 6.1).

Knowing what this point is (the sludging stage) and knowing when you have reached that point presumes an oil testing program.

Types of Maintenance

Three kinds of maintenance are normally recognized:

- Unscheduled maintenance — inevitable breakdown
- Ordinary maintenance — repairs, adjustment or replacement of parts shown to be necessary by visual inspections made at irregular intervals

**Chapter 1, Part 3 (Table 1.14).*

†Chapter 3, Part 1 (Figure 3.6).

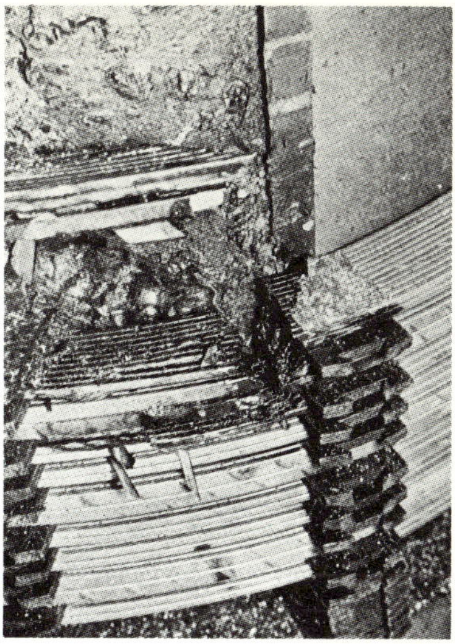

Figure 6.1 - Deteriorated oil caused the failure of this secondary coil of a rectifier transformer. Through inspection and periodic oil testing, the costly damage could have been prevented. (Courtesy of Kemper Insurance Companies).

- Preventive maintenance — regularly scheduled inspections and periodic dismantling or testing of equipment to check every detail likely to cause trouble.

In contrast to this traditional approach, we believe that a "protective maintenance" program should consist of three major phases:

1. Preventive Maintenance
   Monthly "hands off" inspection*, annual energized testing of equipment (oil testing, gas-in-oil analysis, infrared inspection), and de-energized biennial or triennial dismantling or testing of equipment to check every detail likely to cause trouble (electrical insulation tests, switchgear, and so forth).

2. Predictive Maintenance
   More frequent monitoring (inspection and testing) of critical equipment.

"Hands-off" inspection consists of what a person can see, hear or smell without touching energized equipment.

3. Corrective Maintenance
   When equipment performance begins to tail off, the deterioration is recognized and probable causes are pinpointed. Maintenance needed then can be scheduled when convenient.

Operations Maintenance Guidelines for Controlling Heat

It is a fact that most insulation (both oil and paper) ages more rapidly, indeed a good deal more rapidly, as the temperature rises.* Therefore, while prevention that we are discussing obviously presumes a testing program to monitor the condition of the oil, our real goal is to protect the paper and oil and thus presumes also a compatible operations program. The first consideration should be to control the heat within reasonable limits. Heat is a function of load carried by the transformer. Basic features of such a program include the following:

- Records of what loads are added to a given line.
- Recording devices to see that loads are kept within designated limits.
- Installations of alarms and protective devices. Liquid temperature alarms should be set to operate at a temperature not higher than 80°C.

Such load monitoring will predict oil overheating sooner than just monitoring the top oil temperature.

Next in this operations scheme, make sure that in forced-air and water-cooled units that monitors provided are working. Many units are constructed so that the additional cooling is available as the load rises and the sensors call for it. All too many units have these sensors hanging outside the transformer monitoring the atmosphere. Remember Murphy's Law — "If everything seems to be going well, you've overlooked something."

For indoor units adequate ventilation of the transformer room or vault should also be provided to prevent overheating of a unit. Outdoors, fire barriers and sound baffles can restrict air flow.

**Chapter 2, Part 3, Figure 2.13; Chapter 3, Part 1, Figure 3.6.*

Separate self-cooled distribution and small power units by a space of at least 24 to 30 inches (610-760 cm.) to permit free air circulation.

Ventilating air should be dry, free of dust, corrosive gases, or other contaminants.

In vaults ventilated to an outdoor area without ducts or flues, the combined net area of all ventilating openings after deducting the area occupied by screens, gratings, or louvers should not be less than one square foot. (0.09 m^2) for any capacity under 50 KVA. If forced ventilation is required approximately 100 cubic feet per minute of air per KW of transformer loss is needed for adequate cooling.

An air intake line cannot double as an exhaust.

For water-cooled units the difference in temperature between inlet and outlet water should not be more than 10°C. Approximately ¼ gal. (0.001 m^3) of water per minute at 72°F (23°C) is required per KW of transformer loss.*

Defining Load Limits for Transformer Oil

Since heat is directly proportional to the I^2R heat losses, what can we define as limits to this loading? Heat rise is one indicator we can look to for guidance. Transformers prior to 1963 were rated 55°C rise as the maximum allowable top oil temperature rise (105°C hottest spot temperature).† Here is the problem. Modern thermally upgraded cellulose insulation immersed in good clean oil will take 120°C indefinitely with very little harmful effects from the heat. Since we are dealing with an insulation system and the oil is part of that system, it will not readily tolerate these high temperatures. Since 1963, transformers have gone to a 65°C rise insulation, but this system has the same oil limiting factor.‡

If the oil deteriorated under thermal stress were a straight line function, we could load transformers to their insulation temperature,

**Factory Mutual Data Sheet 5-4/14-8 (1978).*

†Chapter 1, Part 3 (Table 1.15).

‡Chapter 1, Part 3 (Table 1.15).

and, while monitoring the oil tests, could predict long range when the oil no longer was in the sludge-free life of the oil. Unfortunately, oil does not act this way (Figure 6.2).* We believe that a 60°C temperature is the criterion for oil. For each 10°C, the oil exceeds this temperature, useful oil life is halved. For example, assume a 20-year useful life for oil. Running a transformer at 70°C would reduce the life expectancy to ten years. Running the unit at 80°C would shorten the life to five years. An oil temperature of 90°C would give only a few years life expectancy to the oil.†

The ANSI/IEEE standard is the second guideline.‡ This standard gives several capability tables for both 55°C and 65°C rise self-cooled and water-cooled transformers with moderate sacrifice of life expectancy for various equivalent loadings, before application of the overload. Table 6.1 is a condensation of one of these tables.

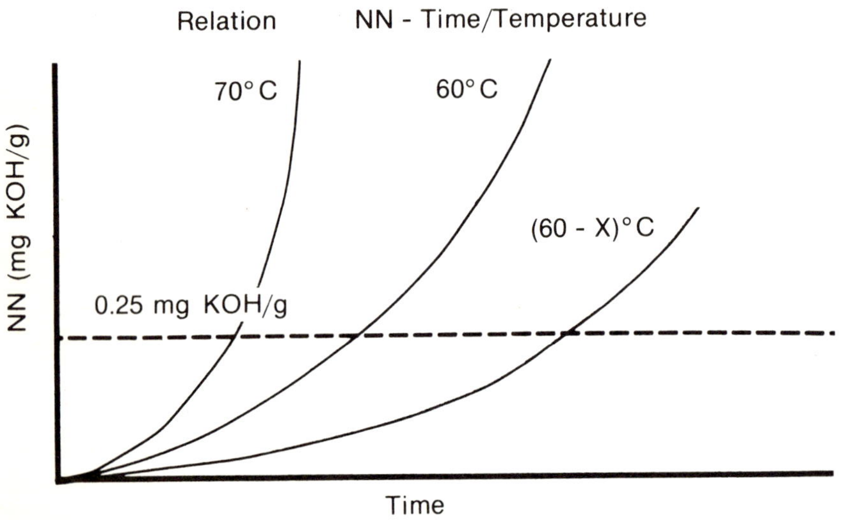

Figure 6.2 - Heat is the greatest accelerator of oil deterioration. The critical top oil temperature is 60° C. At 70° C oil life is one-half that at 60° C, and oil in nitrogen-blanketed transformers will deteriorate rapidly at that temperature.†

*Chapter 2, Part 3, Figure 2.13.

†*Chapter 3, Part 1, Figure 3.6; Chapter 7, Part 4, Table 7.13.*

‡*Chapter 2, Part 4, Loading Guides for Oil-Immersed Transformers, p. 206.*

TABLE 6.1

CAPABILITY TABLE FOR 55° C RISE
SELF-COOLED (OA) AND WATER-COOLED (OW)
TRANSFORMERS WITH 0.25 PERCENT
SACRIFICE OF LIFE EXPECTANCY[1-3]
(Equivalent Load Before Peak Load =
100 Percent of Rating)

Peak Load in Hours	Hottest Spot Temperature Reached Degrees C	Allowable Peak Load Times Name-plate Rating in 30° C Ambient Temperature
½	146	2.00
1	142	1.92
2	142[2]	1.69
4	135	1.50
8	127	1.36
24	115	1.22

[1]IEEE Project 507, May 1979 to replace C57.92-1962 for power transformers.

[2]140° C maximum recommended by IEC and recent IEEE reports.

[3]Keep in mind that this table assumes clean and dry oil.

Unfortunately, people today are accustomed to dealing in terms of built-in overprotection. This leads them to believe that if a 10 or 20 percent safety factor has been built into a product, it should be used. If a transformer is rated 100 percent at certain conditions, "it is built to take more than that."* For this reason, ANSI also has an emergency overload rating table. We keep building transformers so they operate better but keep working them harder thinking they will last longer. This is not true. †

Admittedly, these guide factors affect total life expectancy and some companies amortize this loss of life into the price of their product. But in days of oil shortages and required energy conservation, a long hard look must be taken at operating conditions (Figure 6.3).

Chapter 1, Part 3.

†Chapter 7, Part 4, Use of a Transformer Cooling System.

Figure 6.3 - Proper maintenance would have prevented this high voltage coil damage caused by short ciruit coil fault in a 25 MVA transformer. (Courtesy of Hartford Steam Boiler Inspection and Insurance Co.).

Preventive maintenance instructions are available from many sources — manufacturer, electrical codes and guides, insurance companies and experience. Unfortunately, none of these really speak adequately to the problem or solution.

Physical Inspection and Basic Equipment Check Points

Since a transformer does not have an actual "window" through which to evaluate unit operation, we must find other places to monitor what is happening inside. A good place to start is with the records available. What should the unit be doing? What loads are connected to it and in what fashion? How is it acting now in comparison to the day it was energized? Too often these items are left to memory. This should be the beginning of a log book for each unit. Any change can be significant, but benchmarks must be known before change is recognized.*

A regular inspection of the transformer and its auxiliary devices should be the next source of information as to how the unit is

**Chapter 10, Part 2.*

operating. Each of the items of Table 6.2 should be inspected and recorded.

A frequency of inspection should be established and maintained. Some items should be checked daily; others weekly; still others monthly or semi-annually (Chapter 10, Tables 10.1 and 10.2) Whatever the schedule, it should be followed religiously and the results recorded.

These inspection items are quite simple, but may reveal a potential problem in its early stages. On the other hand, insulation system checks are more technical in nature. These tests consist of chemical and electrical tests of the oil and electrical tests of the solid insulation.* The external inspections will give an indication of the interior of the unit but not to the degree that the oil and insulation tests can. The frequency of these tests is often much less than that of the external inspection. However, in both cases, it is important to record the results and look for any developing trends. One complete set of test data on a 50 year old unit, no matter how elaborate, can compare to the information gathered over a regular period of time. Thus, we recognize the vital need to run inexpensive, fast, and easy to perform oil screening tests to monitor, among other things, the effects of transformer loading.† Fifteen years of field experience is the basis for the suggested oil maintenance schedules (Tables 6.3 and 6.4).

As originally stated in 1952, the following quote is even more applicable to the energy-dependent modern world.

> It appears that proper servicing can result in large financial returns to operators of power systems through the saving and improvement of many millions of gallons of insulating oil now in-service and deteriorating by oxidation. Still greater savings can accrue from the resulting life extension of the apparatus with which the oil is associated. ‡

**Chapter 3, Part 1 (oil tests); Chapter 5 (insulation tests).*

†Chapter 6, Part 2.

‡ASTM Special Technical Publication #152, ***Symposium on Insulating Oils,*** *June 23, 1952.*

Only in very recent years have we fully recognized the importance of applying this prophetic remark to the 1.73 billion gallons of mineral oil now in-service.*

TABLE 6.2

TRANSFORMER AND AUXILIARY DEVICES CHECKLIST	
Transformer	Auxiliary Devices
Load Voltage Liquid Level Temperatures Ambient Liquid Windings Normal hum Gasket or other leaks External surfaces for rust Air—in and out (dry type only) Water—in and out (water-cooled) Oil—in and out (oil-cooled) Pressure/Vacuum-forced (oil-cooled) Nitrogen pressure (N_2 pressure sealed)	Pumps and fans Breathers Pressure relief devices Gas detectors—Pressure relays Alarms Tap changers Grounds Ground resistance Bushings Valves Gaskets Fuses Lightning arresters

Chapter 10, Parts 1 and 3, Justification for a Protective Maintenance Program.

TABLE 6.3*

PREVENTIVE MAINTENANCE GUIDELINES NEWER TRANSFORMERS		
Unit Years of Operation[1]	Oil Properties	
	Neutralization Number (NN), Maximum	Interfacial Tension (IFT), Minimum
1 - 5 Years	0.05 mg KOH/g	35 dynes/cm
6 - 10 Years	0.06 - 0.10 mg KOH/g	30-35 dynes/cm
	> 0.10 mg KOH/g	< 30 dynes/cm

[1]The transformer and transformer oil should be treated as soon as the NN exceeds 0.10 mg KOH/g or the IFT drops below 30 dynes/cm. The oil is the atmosphere in which the transformer lives. Letting the oil degrade below this standard is like people living in an atmosphere laced with carbon monoxide.

TABLE 6.4*

RECOMMENDED TRANSFORMER TESTING SCHEDULE				
Test Method	Top Oil Temperature Range (Continuous)[1]			
	60°-70° C	70°-80° C	80°-90° C	90°-100° C
Oil Testing (Table 6.5)	Annual	Semi-Annual	Quarterly	Monthly
Dissolved Gas-In-Oil Analysis (Table 4.26)	Annual	Quarterly	Monthly	Weekly
Moisture Content PPM (Karl Fischer)	Annual	Semi-Annual	Quarterly	Monthly

[1]Hottest spot estimate should be no more than 15° C additional temperature, with 140° C the absolute *maximum* allowable. See Chapter 7, Part 4 (Auxiliary cooling) in order to reduce the frequency of testing.

**Tables 6.3 and 6.4 and Tables 10.1 and 10.2 are considered as the basis for suggested transformer maintenance scheduling. (Chapter 10, Part 2).*

Part 2—Preventive and Predictive Maintenance For Oil-Insulated Transformers

Introduction

Preventive Transformer Maintenance begins with recognizing the inevitable factors that contribute to insulation degradation. The presence of free and dissolved moisture, heat, oxidation of the oil and cellulosic insulation through normal usage and aging, and dissolved combustible gases in the insulating oil are significant factors in the total insulation deterioration problem. These areas include acids and other chemical impurities whose end result is sludge, as well as the moisture in both the oil and the paper.

Continuous production is today's way of life. Therefore, preventive maintenance should be a critical concern.

Preventing Premature Transformer Faults

Significant preventive maintenance areas are highlighted in Figure 6.4, a flow chart of basic power transformer preventive maintenance procedures.

The history of transformer testing dates back to the 1890's. At that time, insulation resistance (of solid insulation) was considered the criterion. Then, shortly thereafter, the dielectric breakdown strength was first conducted for oil. From 1921, when first standardized, until the present, the flat disc dielectric strength (ASTM Method D-877) has remained the classic test. Introduction of the last of the traditional tests — the insulation power factor — took place in the mid-1930's.

In the early 1940's, practical attention was first directed toward the sensitive Interfacial tension test (D-971). Nevertheless, it has been only in the last 20 years that users recognize the importance of periodically conducting multiple transformer tests.

Periodic analysis of the oil and dissolved gas in the oil are considered today the most practical methods of early detection of insulation problems.

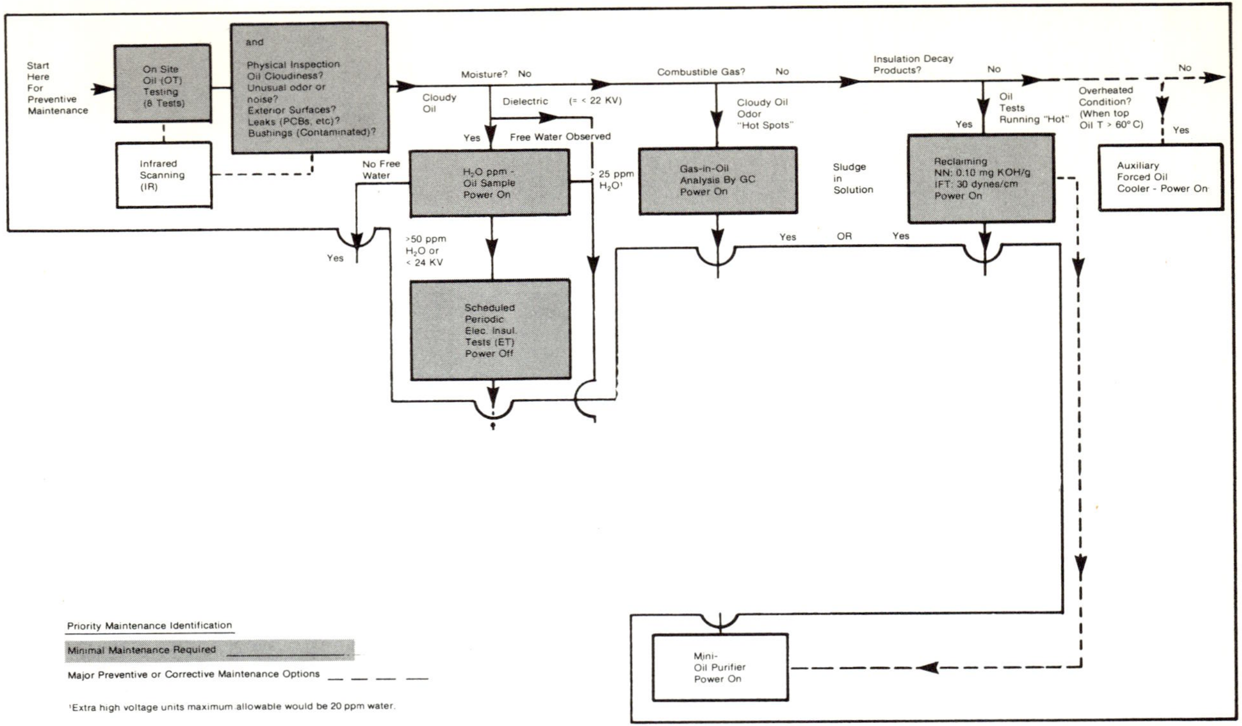

Figure 6.4 - Chart shows basic power transformer maintenance procedures. Preventive Maintenance starts in the upper left hand box ("on-site testing") and works to the right and downward. Sequence follows lines and arrows.

Periodic Analysis of Mineral Oil

Purpose

Consensus opinion holds that the first key to preventive maintenance of transformers is annual oil testing. Its primary purpose is to pinpoint the condition of a transformer's insulation system so that it may operate in the sludge-free range.

The American Society for Testing and Materials (ASTM) lists 33 tests in its publication D-117, "Standard Methods of Testing Electrical Insulating Oils." However, the nine field tests of Table 6.5 are the most useful for in-service transformer oil.* No one test is sufficient to judge oil condition or reveal information regarding transformer insulation. An acceptable degree of accuracy may be expected if the combined results of all these laboratory tests are considered.

What Causes the Problem? — Oxidation

On several occasions mention has been made of oxidation. Nothing can stop this gradual process.† The relatively few unstable hydrocarbon compounds of the oil plus the oxygen contained in the oil and generated by degrading cellulosic paper plus the catalytic effect of the copper winding yield various unwanted polar contaminants. These oxidation by-products, including peroxides, acids, alcohols and ketones, are formed as the initial step in the deterioration process. Sludge is the final step of oil deterioration. In everyday language, sludge comprises everything solid which collects in a transformer. This process is accelerated in a heavily loaded, abused, or neglected apparatus.

Detecting an Oil Problem

For too long, however, preventive maintenance has been inadequate, limited primarily to the standard ASTM D-877 flat disc dielectric test.

Though the dielectric test is an important test, the neutralization number (NN) and interfacial tension (IFT) tests speak more directly to the oil quality and in turn the tensile strength of cellulosic insulation.

**Chapter 3, Part 1, Table 3.1.*

†Chapter 2, Part 3.

TABLE 6.5

THE NINE MOST IMPORTANT ASTM FIELD TESTS FOR IN-SERVICE TRANSFORMER MINERAL OIL[1]			
ASTM Test Method[1]	Criteria for Evaluating Test Results	Average Limits For Continued Use	Information Provided By Test
D-877/ D-1816 Dielectric Breakdown Strength	New oil should not break down at 30 KV or below.	25 KV (minimum) 20 KV (minimum)	Conductive contaminants present in the oil such as metallic cuttings, fibers or free water.
D-974-Neutralization (or acid number NN)	Milligrams of potassium hydroxide required to neutralize 1 gram of oil (0.03 or less for new oil).	0.10 mg KOH/g (maximum)	Acid present in oil.
D-971-Interfacial Tension (IFT)	Dynes per centimeter (40 or higher for new oil).	27 dynes/cm. (minimum)	Sludge present in oil.
D-1524-Color	Compared against color index scale of 0.5 (new oil) to 8.0 (worst case)	2.7 (maximum)	*Marked* change from one year to next indicates a problem.
D-1533-Moisture Content Parts Per Million (PPM)	Below 25 PPM—New Medium Power equipment[2]	35 ppm, 69 KV and below; 25 ppm, 69 KV-288 KV; 20 ppm; more than 345 KV (all maximum)[3]	Reveals total water content. Leak or cellulosic deterioration.
D-1298-Specific Gravity	Specific gravity of new oil is approximately 0.875	—	Provides a quick check for presence of contaminants.
D-1524-Visual evaluation of transparency/opacity	Good oil is clear and sparkling, not cloudy	Clear	Cloudiness indicates presence of moisture or other contaminants.
D-1698-Sediment	None/slight/moderate heavy	—	Indicates deterioration and/or contamination of oil.
D-924-Power Factor	Power factor of new oil is 0.05% or less.	0.70% (maximum)	Reveals presence of moisture, resins, varnishes, or other products of oxidation in oil or of foreign contaminants such as motor oil or fuel oil.

[1]For comparative purposes, specifications of new insulating oil can be obtained from ASTM publication D-3487, Standard Specifications for Mineral Insulating Oil in Electrical Apparatus, available from American Society for Testing and Materials, 1916 Race Street, Philadelphia, PA 19103.

[2]EHV units are much more critical - below 20 ppm.

[3]Proposed IEEE Project 637, April, 1980.

Figure 6.5 illustrates a typical case history of a unit having a "good" dielectric, but with sludged oil. Notice the *instability* of the dielectric test results compared to acid and IFT values.

When annual oil testing is practiced, the problem of polar contaminants will be detected before the sludge enters into the insulation. If testing is not performed, sludges form undetected and permanent damage is done to the transformer's insulation system; for example, clogged oil ducts in windings, cooling coils, etc.)

Though Neutralization Number (NN) and Interfacial Tension (IFT) are the most important oil evaluation tests, a complete oil testing program recognizes the importance of the seven OTHER tests of Table 6.5, especially moisture in parts per million for units more than five years old.

NAMEPLATE INFORMATION

Manufacturer		Primary Volts	63500	Gal. 4020	☒ Oil ☐ Askarel
Serial Number	1217835 (1925)	Secondary Volts	13800	Impedence	6.2
KVA Rating	10000	Phase/Cycle	1/60	Other	

Radiators ☐ Yes ☐ No — Conservator Tank ☐ Yes ☐ No — Top Valve ____ — Accessibility – Power on / Power off
Fans ☐ Yes ☐ No — Tap Changer Comp. ☐ Yes ☐ No — Bottom Valve ____ — Testing ____
H_2O Cooled ☐ Yes ☐ No — Bushings: ☐ Top ☐ Side — Other Access ____ — ReRef. ____ — Other ____

ReRefining Parking ☐ Yes ☐ No — Hose distance ____ Feet — Power 3 ☐ Yes ☐ No — Voltage ☐ 220/240 ☐ 440/480

TRANSFORMER LIQUID TEST DATA

	DATE	Service	LEVEL	TEMP.	P V	DIEL	ACID	IFT	COLOR	SPECIFIC GRAVITY	VISUAL	LEAKS	PAINT	OIL CLASS	RECOMMEND
1	4/71	MTL				35+	0.42	20.8	4.0	0.861	Clr	None	Good	V. Bad	Hot Oil Cln
2	10/72	MTL				39	0.49	17.0	4.0	0.870	Clr	None	Good	V. Bad	SLP
3	10/73	MTL				28	0.48	17.3	4.0	0.870	Clr	None	Good	V. Bad	SLP
4	10/75	MTL				Not Tested - Per Customer Request									
5	11/76	MTL				36	0.54	16.2	4.0	0.870	Clr	None	Good	V. Bad	SLP
6	10/77	MTL				34	0.59	17.4	4.0	0.870	Clr	None	Good	V. Bad	SLP
7	10/77	MTL				Conservator - Dielectric 36									
8	8/78	* SLP				39	0.02	36.7	1.0	0.860	Clr			Good	Retest 1 yr
9															
10															
11															
12															
13	*Oil Reclaimed Energized; sludgpurged 20 recirculations of hot oil.														
14															
15															

Figure 6.5 - Typical transformer case history: good dielectric with sludged oil.

Detecting a Moisture Problem

Dissolved moisture, not free water, is the most critical factor for cellulosic insulation. Since the classic dielectric test only detects free water above 80 percent saturation, the amount of dissolved water is thus often overlooked.

Karl Fischer test equipment (ASTM Method D-1533) may be used to determine the total moisture content in parts per million (ppm). Above 50 ppm total moisture in transformer oil should be reason for concern, especially units more than five years old. Why? Wet cellulosic insulation! Inasmuch as the paper insulation is greedy for water, if there is sufficient water available to share 50 ppm with the oil, then the paper is definitely loaded with water. Here is what you should do if free water is visible to the naked eye. While carefully drawing an oil sample from the bottom filter press valve, with the transformer operating, check these items:

- Load in percent of nameplate rating.
- Top oil temperature (a thermometer is usually located in the vicinity of the top filter press valve).
- Estimate hot spot temperature.
- Do not increase load.
- Do not shut down transformer. Why? The oil will immediately start to cool down, give up its moisture to the paper insulation and thus compound the problem. When moisture is locked up in the paper, the key to the "lock" is hot transformer oil.
- Carefully draw a top oil sample and record the top oil temperature. Hot oil extracts moisture from the paper because of the oil's capacity to hold it.
- Estimate ppm content of the oil sample.

Now refer to Tables 6.6 through 6.8 and/or Figure 6.6 as guidelines to determine the extent of a problem. Laboratory studies have shown the significance of this ppm measurement. Table 6.6, for example, shows at 80°C top oil temperature with 25 ppm water in new oil indicates a dissolved water content in the cellulose of about 2.4 percent. When the transformer had operated for a longer period at 60°C, the same oil condition indicates a cellulose water content of 3.3 percent, at 40°C, five percent and so on. It is a fact that most transformers today are operating at temperatures much higher than 60°C*; therefore, the water in the paper has truly become insulation's

**Chapter 7, Part 4, Table 7.12.*

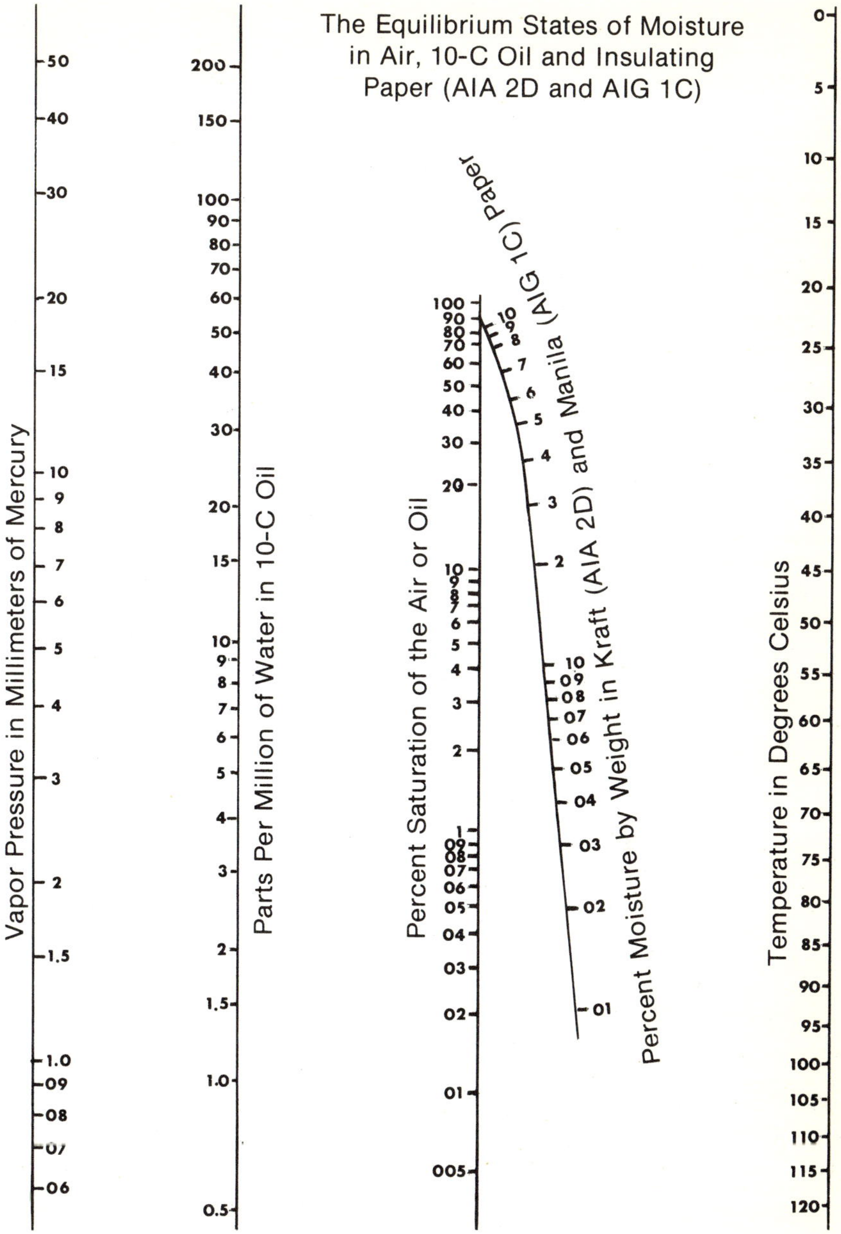

Figure 6.6 - The equilibrium states of moisture in air, new 10-C oil and insulating paper (A1A 2D and A1G 1C). (Courtesy of the Westinghouse Electric Co.).

TABLE 6.6[1]

DISSOLVED MOISTURE IN CELLULOSIC INSULATION				
Top Oil Temperature	Oil Condition	Service Time	H_2O PPM In Oil	Percent H_2O By Dry Weight In Cellulose
80° C	Good	6 Months	25	2.4%
60° C	Marginal	6 Months	25	3.3%
40° C	Bad	? Years	25	5.0%
[1]A. Stannett. "The Problems of Water in Transformers" (1965).				

hidden enemy number one! Next, look at Tables 6.7 and 6.8 to determine the seriousness of a given water problem; first recall the aging rate of insulation, and second, how much water (percent dry weight) is safe.

In order for dielectric and power tests to have any real significance, insofar as moisture problems are concerned, it is necessary to study them at the transformer's operating temperature. Since both top and bottom oil samples are not always possible, the next best step is to convert the data so that the oil has the same percent saturation as that of the operating temperature.* Figure 6.6 may be used to obtain an approximate percent saturation. A straight edge placed diagonally connects the temperature of the oil (degrees Celsius) with the water content (ppm) in the sample.

> As an example, a sample of new oil taken at 60°C with 40 ppm of water gives a value of 16 percent saturation. Then connect this percent saturation with room temperature (25°C) to find the ppm of water which corresponds at this lower temperature to the percent saturation found in the operating transformer. In the example cited, this 16 percent saturation value at 25°C is found to be 11 ppm of water (Table 6.9).

Finally, consider Figure 6.7 a realistic guide (in terms of moisture content and top oil temperature) as to whether a given unit should

**Or compare the moisture content of a hot and cold oil sample, remembering that hot oil has greater solubility for water than cold.*

be immediately shut down. Therefore, we can again see that it is very possible for a transformer to have a "good" dielectric strength and yet still have wet paper!

TABLE 6.7[1]

THE EFFECT OF MOISTURE ON CELLULOSIC INSULATION	
Transformer H_2O Content By Percent Dry Weight in Cellulose	Aging Rate (Reduction in Tensile Strength)
0.3%	1.0
2.0%	6-16x
4.0%	12-45x

[1]A. Stannett (1965).

TABLE 6.8[1]

THE DANGER OF FLASHOVER		
Percent H_2O By Dry Weight in Cellulose	Temperature Top Oil	Condition
0.5%	95° C	Good
1.5%	95° C	Good
3.3%	95° C	"Roulette"
4.5%	90° C	Flashover
7.0%	50° C	Flashover
8.0%	20° C	Flashover

[1]A. Stannett (1965).

TABLE 6.9

TABULATION OF OIL DATA — ESTIMATING MOISTURE CONTENT			
Oil Sample	Transformer Operating Temperature Degrees Celsius	Water Content (Parts Per Million)	Percent Saturation In Oil
1	60° C	40	(Unknown) 16%
2	25° C	(Unknown) 11	16%

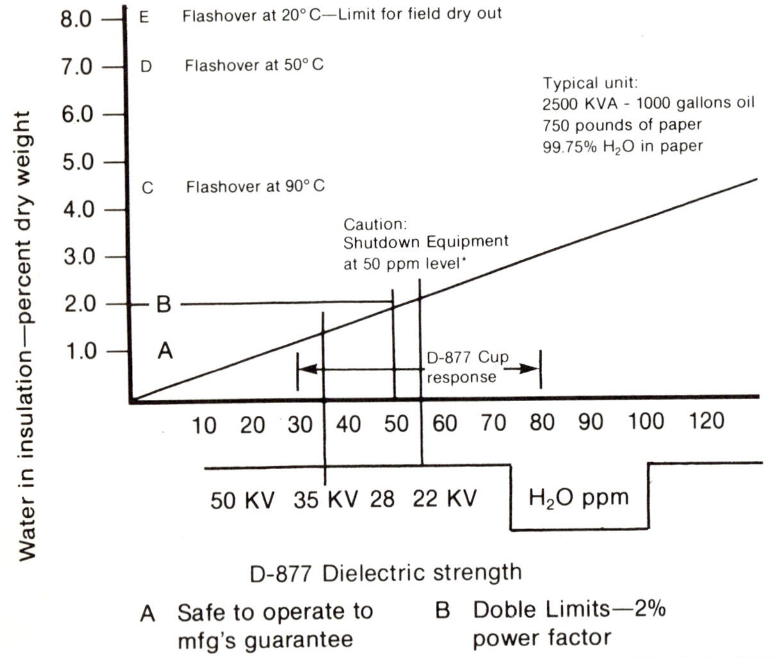

*For EHV shutdown equipment at > 20 ppm.

Figure 6.7 - A practical guide for dehydration maintenance of medium power transformers. H_2O in oil versus H_2O in paper percent dry weight for a typical transformer.

Consider one case history — free water was detected in a 750 KVA unit. Successive analysis over a three month interval indicated a constant moisture content of 71 ppm; therefore, de-energized electrical insulation testing was recommended. The tests confirmed extremely wet insulation to be the problem. With a large power transformer, consideration should be given to applying an on-site oil purification system. Let this continuous vacuum dehydration of the oil dry out the insulating system over an extended period of time without transformer shutdown. This is a possibility that merits a good sound engineering approach.*

The alternative would be to immediately take the unit off the line for a series of electrical insulation tests.†

What the Tests Tell Us About the Oil

Cleanliness and quality control are absolutely essential prerequisites for reliable oil testing. No test will mean anything if dirty containers or laboratory equipment are used, or if air, moisture, or sunlight is allowed to enter the oil sample.

Once a good sample has been taken and properly tested, the oil may now be classified so that the proper service can be recommended.

What Oil Tests Tell Us About Transformers

Several independent studies have shown that an increased neutralization or acid number (NN) should normally be followed by a characteristic drop in interfacial tension (IFT), as in Figure 3.11.‡ Knowing the NN and IFT tells us not only how naphthenic insulating oil has performed but, more importantly, about the insulation and whether or not the transformer is operating in the sludge-free range.

Combining this with the ASTM field sludge probability study of 500 in-service transformers, Table 6.10 permits oils to be classified.§

**Chapter 3, Part 3; Chapter 7, Part 2.*

†*Chapter 5 and Chapter 7, Part 1.*

‡*Figure 3.11 is found in the special Color Section following Chapter 10.*

§*Table 6.10 is found in the Color Section; also see Chapter 3, Part 2, Table 3.8.*

Using the Oil Classification Color Guide

These seven oil classifications in Table 6.10 are realistic. This table accurately describes the effect of each oil category on the transformer. The key to keeping transformers operating over a long period of time is to recognize the value of this transformer oil classification system and apply it religiously.

Physical Inspection

While drawing transformer oil samples for the foregoing laboratory tests, key observations also serve as early warning signals of internal transformer conditions (Figure 6.8).*

1. Oil Cloudiness
   An oil sample should be free of any particles in suspension or cloudiness. Good oil is clean and sparkling. Cloudiness reveals moisture, sludge, or other undesirable suspended material.

2. Strange Odor or Noise
   In the drawing and testing of a transformer oil any unusual odor should initiate further testing. Oil with a strong acid content will sometimes mask the "sweet' smell of transformer arcing. This situation calls for dissolved gas analysis and may also require a series of electrical insulation checks.

   The presence of aromatic gases emanating from the oil is a sensitive indication of problems inside the transformer —hot spots, sustained partial discharges, or arc overs.

3. Critical Unit Monitoring Points
   Check transformer liquid level, pressure/vacuum and unit top oil temperature, and compare with prior results. Check the pressure relief device to see if it has tripped (indicator in a vertical position) and a sudden pressure fault has occurred. Whenever an oil filled unit has a tripped pressure device, a dissolved gas-in-oil analysis should be recommended.†

**Chapter 3, Part 1; Chapter 6, Part 1 (Table 6.2).*

†*Chapter 4.*

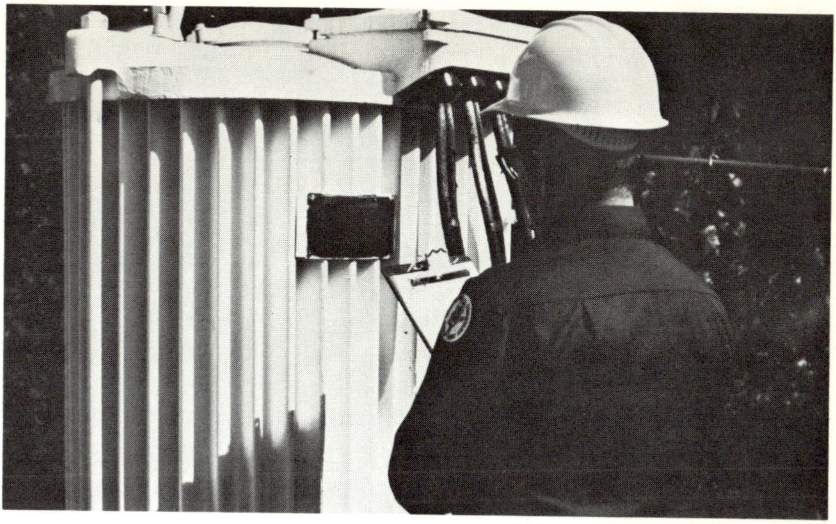

Figure 6.8 - An on-site engineering survey of your transformers should be made periodically.

Some transformers have a "rapid pressure rise relay" to distinguish between rapid and slow pressure rate. The relay will trip a switch if pressure changes faster than a specified rate.

4. Cooling System
   Are the valves to the fins opened or closed? If they are cold, they are closed. Do the cooling fans operate properly (or at all)?*

5. Unit Leakage
   Another observation involves at least monthly inspection of transformers for leaks (continuous dripping) or seepage (gradual leakage). This should include PCBs (polychlorinated biphenyls) in oil-filled transformers (i.e., fluids containing more than 50 parts per million of PCBs) as defined by U.S. Environmental Protection Agency regulations.† The regular practice of labeling, monitoring, and proper storage or disposal has become law, subject to

*Chapter 7, Part 4.

†Chapter 9, Part 2.

penalty under the U.S. Toxic Substance Control Act (Public Law Number 94-469, *Federal Register,* May 31, 1979, Part VI, pp. 31514-31558).

Inspect units carefully to insure that they are structually sound. Check for leaks around gasket areas, such as cover, manhole, or bushings. Special attention should be given to fittings (such as sampling valves) and cooling radiators. Any evidence of leaking could mean that a "sealed transformer" has become a "free breather," or perhaps even worse, a uni-directional breather. Therefore, analytical testing of all oil-filled units, beginning with the most critical units (those near water ways), and those near the "end-of-life" expectancy, should be scheduled to minimize failures and to assure the transformer's maximum design lightning strike capability.

Testing for PCB content in oil by electron-capture gas chromatography (GC) is the most prevalent laboratory method. It has the sensitivity to detect the PCBs down to parts per billion.

> Caution — When oil testing and visual inspection do not indicate a problem, do not automatically rule out further tests, for experience has shown that incomplete preventive maintenance may be risky. Figure 6.4 diagrams a suggested complete preventive maintenance program.

First consideration should go to an analysis of the dissolved gas-in-oil. Even though a 1977 Electric Power Research Institute (EPRI) survey revealed that only eight percent of utilities were using this test, an increasing number of utilities have begun to recognize the value of some of the things we have considered up to this time.*

Dissolved Gas-in-Oil Analysis by Gas Chromatography

As a transformer ages, incipient (or newly-formed) faults may develop. Certain combustible gases in turn give early evidence of this internal condition. These combustible gases may lower the flash point of oil and thus could cause premature failure.

Dissolved gas-in-oil analysis is the most informative test to reveal either existing or past faults in a transformer's insulation system.

Note the trend in the Bibliography for Chapter 4.

The most popular analytical method is gas chromatography (GC), whose results are not influenced by the differing solubility characteristics for the various gases. GC also provides data indicating both the progressive rate of total combustible gas formation and the type of problem so that corrective steps can be taken. Keep in mind normal thermal aging slowly produces carbon monoxide and carbon dioxide.

As the state of the art increases, X-ray and infrared scanners, have been used successfully to locate or confirm the source of combustible gas.* Remember the four major gases of concern and their effect on a transformer (Table 6.11). By analyzing the KEY combustible gas, it is possible to determine where harmful effects may be occurring.

Practical guidelines developed over several years' experience have already been given. The frequency of analysis depends upon transformer loading, as well as the critical importance of a given unit.†

TABLE 6.11

KEY COMBUSTIBLE GAS GENERATION IN TRANSFORMER OIL	
Condition	Key Gas
1. Overheating	Ethylene (C_2H_4)
2. Corona	Hydrogen (H_2)
3. Arcing	Acetylene (C_2H_2)
Key Combustible Gas Generation in Cellulose	
4. Overheated Cellulose	Carbon Monoxide (CO)

**Chapter 8, Part 4.*

†*Chapter 4, Table 4.26.*

Scheduled Electrical Insulation Testing

The final phase of preventive maintenance involves a scheduled shutdown of equipment for a series of electrical tests (ET).* These nondestructive tests are performed without any risk of insulation damage.

1. Power Factor Series
   - Insulation power factor
   - Insulation power factor tip-up test
   - Excitation current
2. Megohm Meter Series
   - Insulation resistance
   - Dielectric absorption
   - Polarization Index
   - Step voltage ratio†
3. Other Associated Tests
   - Transformer turns ratio
   - DC winding resistance

Both the power factor and megohm meter series of tests are only the starting point. As experience has shown, the full battery of eight tests should be scheduled for the next regular shutdown to establish a data base, comparing actual field data with original factory test results. No known **single test** is adequate to supply all the data required to properly evaluate a given transformer's insulation.

In addition, some disagreement concerns how often these tests should be subsequently performed, whether annually or less often. We believe if all the aforementioned energized tests (oil, gas-in-oil, etc.) are performed consistently, the ET monitoring need be scheduled only as test data dictates, or every five years, even if the "energized tests" do not demand it.

ET should be done as soon as possible for such warning signals as excessive overheating, unusual odor, rapid change in oil color cloudy oil, dielectric strength below 24 KV, or excessive free water in an oil sample.

**For test details, see Chapter 5, and field applications upcoming in Chapter 7, Part 1.*

†Step voltage ratio test should be used only for at least a 2300 Vac insulating system. (See p. 421).

Full Assurance Against Power Outages

Each of the foregoing tests or guidelines has proven an effective tool in averting "premature transformer failure," that is, units that fail long before their anticipated life expectancy.

Nearly everything considered thus far has focused primarily on the *internal* transformer condition. However, a complete preventive maintenance program does not overlook the all-too-obvious. A transformer whose internal insulation condition may be rated "excellent" could still fail prematurely. Table 6.12 lists some other areas that demand attention. Two areas stand out for special mention —relays and corrosion control.

An incorrectly set protective relay was the basic cause of the 1965 "Great Northeastern U.S. Blackout"! Therefore, preventive maintenance of switchgear is fundamental.*

TABLE 6.12

ASSURANCE AGAINST POWER OUTAGES
A. Basic External Areas of Concern 1. The reliability of *switchgear* and associated surge protection devices. 2. Corrosion control on *external surfaces* of electrical apparatus. 3. The condition of transformer bushings/standoff *insulators*.
B. *Infrared* camera—Detects any problem where *abnormal* heat variation is a factor, such as loose electrical connections or cracked bushings.
C. Field Monitoring of Transformer Performance 1. Mini-oil purification—Control the operating climate of a transformer, including oxygen content, *contamination*, and *moisture*. 2. Auxiliary forced oil cooling—Dissipate *excessive* heat so that continuous operation would not exceed 60° C average top oil temperature.

**Chapter 8, Part 1.*

Consider also the unique characteristics of a transformer: internally generated heat, 60 Hz vibration, surge voltages, expansion or contraction and uneven surface temperature — hot at the top, cold at the bottom, hot inside and cold outside, etc.

Since transformers are located in at least four seasonal environments and manifold industrial settings, it is well to remember that many kinds of contaminants attract moisture, initiate rust in the radiator fins, tubes, or tank structure, and cause inevitable flashover on the transformer bushings. Look for initial signs of rust, peeling, or cracking on the cooling fins and tubes. In addition, are the bushing skirts free of cracks? Are they clean? Are there signs of tracking? If oil-filled, are the levels normal?*

When "audible frying" is heard or blue-violet streamers are seen, then the unit may already be beyond preventive maintenance because flashover could be imminent.†

Finally, infrared scanning can detect any problem where abnormal heat variation is a factor‡, such as defective electrical connections or cracked bushings/insulators. A power factor reading of dirty bushings may not be accurate. Therefore, this test should only be done after thorough cleaning.§

The economic results of a major failure in service of a large power transformer justifies an integrated attack on both the insulation maintenance and supervision problem. This approach involves field monitoring of transformer performance. Applying this principle in two additional areas (Table 6.12 C) could, under well-defined conditions, add years of uninterrupted service life, especially to medium power transformers (1-100 MVA).**

Keep in mind continually the never ceasing Murphy's Law — "If it's possible for one more thing to go wrong, that thing will go wrong." This could involve the failure of a transformer oil pump.

In most large transformers the insulating oil is circulated through a heat exchanger. This forced movement of oil is accomplished by

**Chapter 8, Parts 2 and 3.*

†Chapter 8, Part 3.

‡Chapter 8, Part 4.

§Chapter 5.

*** Chapter 7, Parts 2 and 4 respectively.*

use of an oil immersed pump. Most often this unit is designed in such a way that all or part of the oil flow passes through the motor. In both configurations the bearings are lubricated and most motor losses are removed by the oil.

Unfortunately, these pumps and motors have failure modes which may introduce metallic debris and other foreign substances directly into the transformer oil. On record are accounts of combustible gases detected by Gas Chromatography — acetylene from arcing in motors.*

A significant number of oil pumps have failed and thus introduced debris (bearings, impellers, etc.) into the transformer. Even if this does not cause transformer failure, downtime and removal of debris are expensive.

Predictive Maintenance

Definition

Preventive maintenance procedures are normally performed on an annual basis on all transformers. However, both industry and utilities are recognizing the need for more frequent maintenance of critical equipment in order to continue uninterrupted service. Predictive maintenance has thus become the most recent phase of protective maintenance.

Detection of impending breakdowns in the early stages allows lead time to schedule corrective action. In contrast, the typical preventive maintenance program consists of an annual evaluation. Projections are then made only on a theoretical breakdown based on historical data.

A predictive maintenance program combines the data base with diagnostic tools that establish an automatic warning of imminent failure.

The first step in the predictive program involves the classification of all equipment. The two highest priorities should go to transformers that:

1. Can shut down the entire plant (or substation) or:

Chapter 4.

2. An entire department; this would include units operating continuously with temperatures greater than 60°C*, transformers with odd voltages, and units constituting main substations.

Arc furnace, rectifier, and water-cooled transformers should also require more frequent attention, such as semi-annual or quarterly evaluation and on-site inspection.

When the classification of transformers has been completed, diagnostic tools should be selected for monitoring of transformer performance. Some may be done by in-plant personnel, while others will require an outside service contractor, utilizing special equipment not ordinarily found in most industrial plants or smaller utilities.

The critical areas of predictive maintenance should include: oil testing, infrared inspection, "hands off" external inspection, and particularly dissolved gas-in-oil analysis by gas chromatography (GC).

Two case histories illustrate the value of predictive maintenance, especially detecting possible manufacturing or design faults in critical transformers.†

Why Multi-Testing Is Required

Electrical insulation testing alone, for example, may be unable to pinpoint a particular problem. Yet a fault could still exist somewhere in the insulation system.

As a case in point, none of the insulation tests indicated **why** a particular oil was cloudy and rapidly changing color. Typical test results are given in Figure 6.9.

A gas-in-oil analysis by gas chromatography was then made. The results are given in Table 6.13. Since the key combustible gas was carbon monoxide, overheating of the cellulose insulation was taking place. Repair and degasification were performed. Follow-up oil and GC analyses confirmed the overheating problem had been solved. Today, the unit continues to operate satisfactorily. The field data of Figure 6.10 show why multi-test procedures are necessary for reliable protective maintenance.

**This inventory may cover nearly all units. Chapter 7, Part 4, Table 7.12.*

†Chapter 4, Tables 4.21 and 4.22.

DOBLE INSULATION TESTS

TWO-WINDING TRANSFORMERS

OIL ☒
ASKAREL ☐
AIR ☐
GAS ☐

DOBLE ENGINEERING COMPANY
BELMONT, MASS
FORM MH - 2W178

COMPANY ____ DIVISION ____ DATE 3/14/75

LOCATION OF TESTS ____ AIR TEMP. ____ TOP OIL TEMP. 25°C

TRANSFORMER Furnace WEATHER Inside % HUMIDITY ____

MFR. Penn. SERIAL NO. A223681 AGE 1952 TYPE/CLASS OW KVA 1250

FREE BREATHING ☒ SEALED ☐ GAS BLANKETED ☐ CONSERVATOR ☐ GALLONS OF OIL 1265

HIGH SIDE KV 4.16 Y☐ Δ☒

LOW SIDE KV taps Y☐ Δ☒

NEUTRAL

BUSHINGS: MFR. | TYPE | CLASS | DWG. NO. | CAT. NO. | KV | YEAR

DATE LAST TEST

COPIES TO

LAST SHEET NO.

OVER-ALL TESTS X.79

TEST	Winding Energized	Winding Grounded	Winding Guarded	TEST KV	Milliamperes: Meter Reading	Milliamperes: Multiplier	Milliamperes	Watts: Meter Reading	Watts: Multiplier	Watts	% Power Factor: Measured	% Power Factor: Cor. 20°C	Key to Insulation Rating	Insulation Rating
1	HIGH	LOW		4	25	1	25	25	.1	5.0	— —	— —		
2	HIGH		LOW	4	54	.2	10.8	21	.2	2.1	1.94	1.5	C_H	I
3	LOW	HIGH									— —	— —		
4	LOW		HIGH										C_L	
CALCULATED RESULTS					— —	— —	14.2	— —	— —	2.9	2.04	1.6	C_{HL} (TEST 1 MINUS TEST 2)	I
					— —	— —		— —	— —		— —	— —	(TEST 3 MINUS TEST 4)*	— —

KEY TO INSULATION RATING: G = GOOD; D = DETERIORATED; I = INVESTIGATE; B = BAD (REMOVE OR RECONDITION)

*CURRENT AND WATTS SHOULD COMPARE WITH THOSE FOR C_{HL}

BUSHING TESTS X.79

Side	LINE NO.	BUSH. NO.	PHASE	BUSHING SERIAL NO.	TEST KV	Microamperes: Meter Reading	Microamperes: Multiplier	Microamperes	Watts: Meter Reading	Watts: Multiplier	Watts	% Power Factor: Measured	% Power Factor: Cor 20°C	Collar Tests (Watts/Current): Top	Collar Tests	Insulation Rating
HIGH SIDE	1								UST							
	2				4	71	.2	14.2	29	.1	2.9	2.04	1.6			I
	3															
	4	N						EXCITATION CURRENT								
LOW SIDE	5			A-B		Could not measure										
	6			A-C		Could not measure										
	7			B-C		Could not measure										
	8	N														
	9															
	10							PRIMARY WINDING RESISTANCE - TAP A								
	11			A-B		.150	OHMS									
	12			A-C		.150	OHMS									
	13			B-C		.150	OHMS									
	14															
	15							SECONDARY WINDING RESISTANCE								
	16			A-B		350	U OHMS									
	17			A-C		400	U OHMS									
	18			B-C		550	U OHMS							X.79		
	19	OIL SAMPLE			10	40	20 U	800	1	.01	.01	.125	.10	OIL TEMP 25 °C		G

N = NEUTRAL

DIAGRAM

Tap A
Could not measure Turns Ratio
because of the Delta Primary.

REMARKS:

TEST SET NO. 1938 TEST BY K.S. CHECKED BY ____ SHEET NO. ____

doble

Figure 6.9A - Typical electrical test (ET) data (winding power factor) suggests further investigation.

INSULATION RESISTANCE - DIELECTRIC ABSORPTION TEST SHEET

POWER TRANSFORMER

TEST NO ______

COMPANY ______ DATE 3/14/75

LOCATION ______ TIME ______

EQUIPMENT	Furnace Transformer	RATING	1250 KVA	VOLTAGE	4160/Taps
TYPE	OW	MFR.	Penn.	SERIAL #	A-22868-1
RECENT OPERATING HISTORY	Overheated with rapid color change in oil.			WINDING AGE	1952

STATE IF TRANSFORMER IS OIL FILLED AND OF THE CONSERVATOR, FREE-BREATHING OR GAS-FILLED TYPE: Oil-Free Breather

IF OF THE FREE-BREATHING TYPE DESCRIBE THE BREATHER ARRANGEMENT AND DEHYDRATING MEDIUM USED: ______

IF OF THE FREE-BREATHING TYPE NOTE INTERNAL CONDITIONS SUCH AS UNDERSIDES OF COVERS, TERMINAL BOARDS, WINDINGS ETC: ______

BUSHING DATA: ______

OIL TESTS: ACIDITY ______ MG KOH/G. RESISTANCE IN STANDARD CUP ______ MEGOHMS. TOP OIL BREAKDOWN ______ KV.

STATE WHAT BUSHING PORCELAINS WERE GUARDED DURING TESTS: ______

TEST DATA - MEGOHMS

PART TESTED				
GROUNDING TIME				
TEST VOLTAGE	2500	2500		
TEST CONNECTIONS	PRI. TO LINE	PRI. TO LINE	TO LINE	TO LINE
	GND. TO EARTH	SEC. TO EARTH	TO EARTH	TO EARTH
	TO GUARD	TO GUARD	TO GUARD	TO GUARD
¼ MIN	1300 M OHMS	1550 M OHMS		
½ "	1475 M OHMS	1875 M OHMS		
¾ "	1550 M OHMS	2000 M OHMS		
1 "	1625 M OHMS	2125 M OHMS		
2 "	1750 M OHMS	2375 M OHMS		
3 "	1800 M OHMS	2500 M OHMS		
4 "	1875 M OHMS	2525 M OHMS		
5 "	1925 M OHMS	2550 M OHMS		
6 "	1950 M OHMS	2550 M OHMS		
7 "	2000 M OHMS	2600 M OHMS		
8 "	2000 M OHMS	2625 M OHMS		
9 "	2000 M OHMS	2625 M OHMS		
10 "	2000 M OHMS	2625 M OHMS		
10/1 MIN RATIO	1.23	1.23		

TEST MADE	HOURS DAYS	AFTER SHUTDOWN
DRY-BULB TEMP.	25°C	°F
WET-BULB TEMP.		°F
DEW POINT		°F
RELATIVE HUMIDITY		%
ABSOLUTE HUMIDITY		GR./#
EQUIPMENT TEMP.		°F(C)
HOW OBTAINED		
"MEGGER" INST.:—		
SERIAL #	1927133	
RANGE	0-50 K M OHMS	
VOLTAGE	500/1000/2500 V	

REMARKS: PRI - GND at 500 V 1900 M OHMS

Tap A

TESTED BY: K. Stutz

Figure 6.9B - An example where "Good" insulation data (insulation resistance/dielectric absorption) may not be what it appears. It fails to indicate why a particular transformer oil is "cloudy" and rapidly changing in color.

NAMEPLATE INFORMATION

Manufacturer Lectromelt | Primary Volts 4160 | Gal. 1265 | [x] Oil [ ] Askarel

Serial Number A-22368-1 (1952) | Secondary Volts 3389 | Impedence

KVA Rating 1250 | Phase/Cycle 3/60 | Other

Radiators [ ] Yes [x] No | Conservator Tank [ ] Yes [x] No | Top Valve 1" | Accessibility – Power on Power off

Fans [ ] Yes [x] No | Tap Changer Comp. [ ] Yes [x] No | Bottom Valve 1" | Testing

H$_2$O Cooled [x] Yes [ ] No | Bushings: [x] Top [ ] Side | Other Access None | ReRef.

Other

ReRefining Parking [ ] Yes [ ] No | Hose distance ______ Feet | Power 3 [ ] Yes [ ] No | Voltage [ ] 220/240 [ ] 440/480

TRANSFORMER LIQUID TEST DATA

		DATE	Service	LEVEL	TEMP	P·V	DIEL	ACID	IFT	COLOR	SPECIFIC GRAVITY	VISUAL	LEAKS	PAINT	OIL CLASS	RECOMMEND
	1	9/72	MTL				42	0.04	32.4	2.0	0.880	S/Cld	None	Good	Good	Retest 1 yr
1	2	1/74	MTL				28	0.02	33.7	1.0	0.875	S/Cld	None	Good	Good	Retest 1 yr
	3	3/75	ET				Electrical Test Performed (All Tests Normal)									
2	4	4/75	GC				Gas Chromatography Analysis									
3	5	5/75	HOC				45	0.01	36.2	1.5	0.875	Clr			Good	Retest 1 yr
	6	12/75	MTL				43	0.02	37.6	(4) 2.5	0.880	Clr	None	Good	Good	Retest 1 yr
	7	3/77	MTL				42	0.02	34.7	2.0	0.880	Clr	None	Good	Good	Retest 1 yr
	8	4/78	MTL				31	0.03	36.2	2.0	0.880	Clr	None	Good	Good	Retest 1 yr
	9	6/79	MTL				30	0.04	31.7	2.5	0.880	Clr	None	Good	Good	Retest 1 yr
	10	6/80	MTL	N	48		Not Tested - Per Cusomter Request									
	11															
	12															
	13	1 - Oil changed in January 1973 when top oil temperature reached 100°C.														
	14															
	15	2 - The new oil, which was not subjected to the high temperature was shown by the														
	16	small amount of ethylene gas evolved (31 ppm) compared to carbon monoxide (840 ppm).														
	17															
	18	3 - Oil degassed energized (HOC = Hot Oil Cleaned). Both load and operating														
	19	temperature reduced.														
	20															
	21	4 - Color also is a very subjective criterion.														
	22															
	23															
	24															
	25															

Figure 6.10 - Field data illustrates why various test procedures are necessary.

TABLE 6.13

CASE HISTORY—GAS IN OIL ANALYSIS	
Gas	Parts Per Million
Hydrogen	25
Oxygen	12,000
Nitrogen	54,000
Methane	140
Carbon Monoxide	840
Ethane	19
Carbon Dioxide	7600
Ethylene	31
Acetylene	4
Total Gas Content	74,659
Combustible Gas Content	1059

Practical Maintenance Scheduling

A transformer insulation system has four potent enemies—water, oxygen, excessive heat, and contamination. Likewise, multi-testing by the various methods (oil testing, dissolved gas-in-oil analysis, electrical insulation testing) and periodic inspection provide the most complete protective maintenance of power transformers.

Keep in mind that transformer authorities as well as plant insurers tell us that the first ten years of a transformer are the most critical, especially the first three! In addition, recall that the ANSI/IEEE Loading Guide is at best only a "rough assessment" of the effect of transformer overload.*

While ANSI has established criteria for the cellulose insulation, no mention in the guide has been made concerning the oil condition. Yet before the transformer's solid insulation can properly perform its function, it is assumed that it operates in a good, clean, dry, oxygen-free environment. This means the transformer oil must run in the "good" range throughout the unit's life in order to obtain the

**Chapter 2, Part 4 (Loading Guides for Oil-Immersed Transformers).*

maximum life for the paper. We also believe that since oil is the limiting factor preventing severe and rapid oil deterioration is a must!

Therefore, especially for newer units (ten years old), a mobile mini-oil purifier (INSUL-AIDER^{TM}), applied for no more than four months, is recommended for controlling the operating climate in which the insulating system finds itself.* This will assure longer life for any given transformer.

When actual oil test data exceeds defined values in the preventive maintenance program, however, then corrective action is obviously required (Figure 7.1). While the suggested "corrective action" will stop any excessive transformer deterioration, it can never bring back the equipment life, already sacrificed because of failure to initiate consistent preventive maintenance.

More than 15 years of field experience have become the basis for a series of suggested maintenance schedules.†

Conclusions

Field data tells us that most transformer failures can be attributed to the breakdown of the oil and paper insulating system. These so-called "weak links" can still provide indefinite unit life. Although no one test by itself can fully indicate the inner condition of a transformer, a complete preventive maintenance program can detect with near certainty any unknown internal problem.

Preventive maintenance is essential to both older transformers (before 1963) and the newer thermally upgraded units (since 1963). A trade-off has occurred between those older units having untreated Kraft paper with more oil per KVA, while the newer transformers using thermally upgraded paper do not appreciably compensate for the increased thermal effect on reduced oil per KVA.‡

Regardless of test methods chosen, we repeat that the maintaining of accurate, consistent records is fundamental to everything else that takes place.

** Chapter 7, Part 2 (Figure 7.18).*

†Chapter 10, Tables 10.1 and 10.2.

‡Chapter 1, Part 3, Table 1.14.

Chapter 7

Corrective Maintenance Procedures for Oil-Insulated Transformers

Part 1—General Guidelines

Introduction

When preventive or predictive maintenance has disclosed a serious problem of moisture, combustible gas, and/or oil decay products (especially acids and sludge), certain corrective maintenance procedures should be followed. These are highlighted in Figure 7.1, a flow chart of basic transformer field maintenance procedures.

Water trapped in oil-impregnated cellulosic materials, indicated by a low dielectric breakdown strength and confirmed by increased water content (and/or de-energized winding power factor and other electrical tests), requires some form of dehydration of the transformer's insulation system. Keep in mind that drying procedures can only reduce the rate of loss of mechanical strength but **never** regain that which has already been lost through neglect.

The removal of dissolved combustible gases, determined by the technique of gas-in-oil anaylsis by gas chromatography (GC), generally requires unit repair and degasification.*

Finally, oil decay products deposited in the form of sludge in and on vital internal transformer parts require some degree of desludging.

Later, we will consider how these three groups of impurities can be effectively dealt with by one solution — Hot Oil Cleaning — while transformers remain in service.† When performed in sufficient time, all three problems — moisture, combustible gases and sludges — have been under certain conditions simultaneously eliminated, as the SLUDGPURG® desludging process meets both ASTM dielectric and chemical criteria.

**Chapter 4. A few instances of self-healing faults are known but are extremely rare.*

†Chapter 7, Part 3.

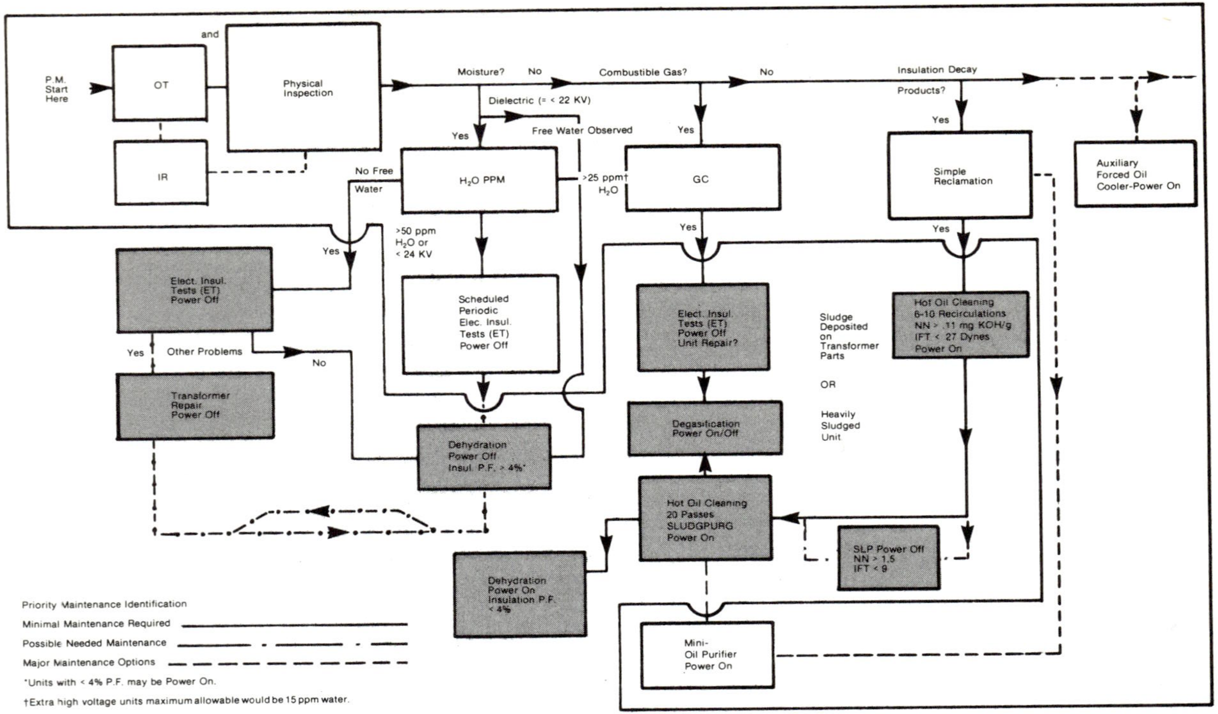

Figure 7.1 - Chart shows basic power transformer maintenance procedures. Preventive maintenance starts in the upper left hand box ("on-site oil testing") and works to the right and downward. Sequence follows lines and arrows. Shaded areas concern facets of corrective maintenance. (See Figure 6.4).

Pinpointing a Moisture Problem

When preventive maintenance has found free water (water visible to the naked eye), certain steps should immediately be taken.

First, consider the possibility of dehydrating the transformer while energized.*

Second, before ET (electrical tests), the transformer needs to be de-energized (remove connections from the unit's primary and secondary loads at the bushings).†

Yet, there need **not** be panic when free water does appear. Next, these specific steps should be taken:

1. Apply positive pressure using dry nitrogen, but do not exceed two to three pounds per square inch.
2. Determine where the leak is by applying a soap solution around all the joints and fittings.
3. Make necessary repairs.
4. Make an above the core inspection of the transformer's super-structure and tap changer to locate the sources of the free water, (when the problem takes on major complications of massive invasions of water). Entrance of this water could be from: bad gasketing, cracked insulation, a rust hole, a loose manhole cover, a ruptured explosion diaphragm, etc.

From these steps we should know that the oil is wet, and where the water came from. The next bit of information needed is whether or not the windings are wet and to what extent. The answer to these questions will determine the action to be taken.

Recall that cellulose in the windings is very hygroscopic (that is, adsorbs and absorbs moisture), while the oil is hydrophobic (that is, tends to give up water).

Any water that enters the system will initially dissolve in the oil up to a few hundred ppm or it will settle to the bottom of the oil-filled unit. If the windings are hot, the water that is in the oil will go readily into

Chapter 7, Part 2 on how to arrive at the right decision.

†*Chapter 5.*

solution. When conditions of free water are found, it may sometimes be better to leave the unit energized so the dissolved water does not phase out free water and get soaked up by the cellulose.

Large amounts of free water entering the unit may or may not affect the windings depending on the location of the leak. If the water enters the transformer from the top of the unit, it will probably go on and over the windings and be absorbed like a sponge. However if the water enters from the side, it may simply run down the inside of the tank and settle at the bottom.

A typical example was given by a major midwestern utility. One of their units was very low on oil so that the oil level was below the cooling fins and the transformer was running extremely hot. As "Murphy's Law" would have it, no oil was available. So water was pumped into the bottom of the tank (carefully) to raise the oil level. The oil level was raised and the unit ran without incident until oil was available. The water was drained out and replaced by oil, and the unit never gave any trouble for its remaining life. While this is not recommended, it serves to prove the point that when free water appears, it may not necessarily be reason for panic.

The applications of winding insulation tests that immediately follow are the absolute minimum necessary as they dictate the solution to the problem of moisture and/or sludge. Remember these tests are nondestructive since they are performed without any risk of insulation damage. Once a leak is repaired, it is quite easy to dry out the oil, but it is altogether another matter to dehydrate the windings.

Field Application of Electrical Insulation Tests

The Power Factor Series

The first insulation check should be the "Doble" winding power factor test, using the appropriate equipment.*

Once again, in-service units can have power factors as high as two percent and still be deemed worthy of continuous operation. Insulation winding power factor tests over and above two percent should be suspect, especially for newer transformers. Older units may deviate from this general rule. When the winding power factor is HIGH, proceed to the "tip-up test".

**Chapter 5, Figure 5.4.*

If the tip-up test does not vary as the test voltage is varied, then moisture in the insulating system is the probable cause. Look at a typical data sheet that illustrates the tip-up test situation (Figure 7.2). The tip-up test results indicate that the transformer probably has excessive moisture in it, causing the insulation power factor to be very high. The insulation power factors of older transformers should not exceed four percent. At this point we would suggest wet insulation is the cause of the problem. Nevertheless, further tests are required to confirm this suspicion.

If the "tip-up" is high and varies with voltage, then ionization is probably occurring in the unit, causing carbonizing of the oil.

The Megohm Meter Series

The megohm meter series consists of four tests; the most basic is the minimum insulation resistance.

Using our example of Figure 7.2 look at Figure 7.3 and compare this value with a calculated recommended minimum insulation resistance.

Typical example — small transformer:

2400 Volt Primary
150 KVA Rated

$$R = \left(\frac{CE}{KVA}\right) = \left(\frac{0.8 \times 2400}{150}\right) = 157 \text{ megohms}$$

A definite problem was confirmed by comparing the extremely low secondary to ground resistance (99.2 megohms) to the calculated value (157 megohms). Continuing the "megger test" for a full ten minutes results in the dielectric absorption test.

The dielectric absorption test is a refinement of the standard insulation resistance test method. Temperature is not a factor and conclusive data can be had without records of past tests.

Using a motor driven Megger® (or solid state megohm meter), dc voltage is applied over a period of time, usually ten minutes, to the equipment under test. Insulation resistance readings are recorded at specific times (15, 30, and 45 seconds and 1 minute; and successive minutes until 10 minutes) and the results plotted on log-log paper. Now look at the data of Figure 7.3. A trend will be observed.

DOBLE INSULATION TESTS

DOBLE ENGINEERING COMPANY
BELMONT, MASS
FORM MH 2W17B

TWO-WINDING TRANSFORMERS

OIL XX
ASKAREL
AIR
GAS

COMPANY		DIVISION	DATE 3/21/76
LOCATION OF TESTS		AIR TEMP.	TOP OIL TEMP. 68
TRANSFORMER	Distribution	WEATHER Clear	% HUMIDITY
MFR. West.	SERIAL NO. 3160325	...E 1942	TYPE/CLASS S	KVA 150
FREE BREATHING ☐	SEALED ☐	GAS BLANKETED ☐	CONSERVATOR ☐	GALLONS OF OIL

		BUSHINGS: MFR.	TYPE	CLASS	DWG. NO.	CAT. NO.	KV	YEAR
HIGH SIDE KV 2.4	Y☐ Δ☐							
LOW SIDE KV .12	Y☐ Δ☐	TC# 61, Sub 16, Tag #6551						
Single Phase	NEUTRAL					DATE LAST TEST		
COPIES TO						LAST SHEET NO.		

OVER-ALL TESTS

TEST	WINDING ENERGIZED	WINDING GROUNDED	WINDING GUARDED	TEST KV	MILLIAMPERES: METER READING	MULTIPLIER	MILLIAMPERES	WATTS: METER READING	MULTIPLIER	WATTS	% POWER FACTOR: MEASURED	COR 20°C	KEY TO INSULATION RATING: G = GOOD, D = DETERIORATED, I = INVESTIGATE, B = BAD (REMOVE OR RECONDITION)	INSULATION RATING
1	HIGH	LOW		2	71	.1	7.1	57	.2	11.4	— —	— —		
2	HIGH		LOW	2	23	.1	2.3	48	.1	4.8	20.9	20.9	C_H	B
3	LOW	HIGH									— —	— —		
4	LOW		HIGH										C_L	
CALCULATED RESULTS					— —	— —	4.8	— —	— —	6.6	13.7	13.7	C_{HL} (TEST 1 MINUS TEST 2)	B
					— —	— —		— —	— —		— —	— —	(TEST 3 MINUS TEST 4)*	— —

*CURRENT AND WATTS SHOULD COMPARE WITH THOSE FOR C_{HL}

BUSHING TESTS

LINE NO.	BUSH NO.	PHASE	BUSHING SERIAL NO.	TEST KV	MICROAMPERES: METER READING	MULTIPLIER	MICROAMPERES	WATTS: METER READING	MULTIPLIER	WATTS	% POWER FACTOR: MEASURED	COR 20°C	COLLAR TESTS (WATTS/CURRENT): TOP		INSULATION RATING
HIGH SIDE 1							UST								
2				2	53	.1	5.3	62	.1	6.2	11.7	11.7			B
3															
4	N						TIP-UP								
LOW SIDE 5				1	38	.1x2	7.6	54	.1x2	10.8	--	--			B
6				1	11	.1x2	2.2	24	.1x2	4.8	21.8	21.8			B
7							5.4			6.0	11.1	11.1			B
8	N			1	26	.1x2	5.2	31	.1x2	6.2	11.9	11.9			B
9															
10							EXCITATION CURRENT								
11															
12			A-B	1	60	10x2	1200								G
13			B-A	1	60	10x2	1200								G
14															
15															
16															
17															
18															
19	OIL SAMPLE			10	Free Water								OIL TEMP °C		B

N = NEUTRAL

DIAGRAM

REMARKS:

TURNS RATIO	CALCULATED
0.050	2400/120 = 20
0.050	1/20 = 0.050

TEST SET NO. 124 TEST BY J.C. CHECKED BY SHEET NO.

doble

Figure 7.2 - Typical data using the power factor tip-up test.

INSULATION RESISTANCE - DIELECTRIC ABSORPTION TEST SHEET

POWER TRANSFORMER

TEST NO. ______

______ COMPANY   DATE 8/10/76

______ LOCATION   TIME 10:15 a.m.

EQUIPMENT: Distribution Transformer	RATING: 150 KVA	VOLTAGE: 2400/120-240	
TYPE: S	MFR. West.	SERIAL #: 3160325	
RECENT OPERATING HISTORY: In Use		WINDING AGE: 1942	
STATE IF TRANSFORMER IS OIL FILLED AND OF THE CONSERVATOR, FREE-BREATHING OR GAS-FILLED TYPE: Oil Filled Sealed			
IF OF THE FREE-BREATHING TYPE DESCRIBE THE BREATHER ARRANGEMENT AND DEHYDRATING MEDIUM USED.			
	Sub #16		
IF OF THE FREE-BREATHING TYPE NOTE INTERNAL CONDITIONS SUCH AS UNDERSIDES OF COVERS, TERMINAL BOARDS, WINDINGS ETC.	TC# 61		
	Tag #6551		
BUSHING DATA	First test taken before Dehydration		
	process began.		
OIL TESTS: ACIDITY ______ MG KOH/G:	RESISTANCE IN STANDARD CUP ______ MEGOHMS:	TOP OIL BREAKDOWN ______ KV.	
STATE WHAT BUSHING PORCELAINS WERE GUARDED DURING TESTS			

TEST DATA - MEGOHMS 23°C CF 1.24

PART TESTED					TEST MADE	HOURS DAYS	AFTER SHUTDOWN
GROUNDING TIME					DRY-BULB TEMP		°F
TEST VOLTAGE	2500	2500			WET-BULB TEMP.		°F
TEST CONNECTIONS	PRI. TO LINE	PRI. TO LINE	TO LINE	TO LINE	DEW POINT		°F
	GND. TO EARTH	SEC. TO EARTH	TO EARTH	TO EARTH	RELATIVE HUMIDITY		%
	TO GUARD	TO GUARD	TO GUARD	TO GUARD	ABSOLUTE HUMIDITY		GR./#
¼ MIN.	99.2 M OHMS	93 M OHMS			EQUIPMENT TEMP.		°F(C)
½ "	99.2 M OHMS	91.5 M OHMS			HOW OBTAINED		
¾ "	99.2 M OHMS	91.5 M OHMS					
1 "	99.2 M OHMS	91.5 M OHMS					
2 "	93 M OHMS	90 M OHMS			"MEGGER" INST.:—		
3 "	93 M OHMS	90 M OHMS			SERIAL #		
4 "	93 M OHMS	90 M OHMS			RANGE		
5 "	93 M OHMS	88 M OHMS			VOLTAGE		
6 "	93 M OHMS	88 M OHMS					
7 "	91.5 M OHMS	87 M OHMS					
8 "	91.5 M OHMS	87 M OHMS					
9 "	90 M OHMS	87 M OHMS					
10 "	90 M OHMS	87 M OHMS					
10/1 MIN RATIO	.90	.95					

REMARKS

TESTED BY

Figure 7.3 - Data using the insulation resistance test reveals excessive moisture.

Insulation in good condition will have a dielectric absorption curve that is a straight line, increasing appreciably with testing time. Insulation with moisture or contamination will result in a line that rises slowly, if at all, and flattens out with time.*

A Polarization Index of less than 1.0 may indicate excessive moisture (our example of Figure 7.3). The Polarization Index ratio (ten minutes divided by 1 minute reading) for the primary to ground and primary to secondary was less than 1.0, indicating that moisture had been absorbed in the windings. The Polarization Index for a good dry unit should be well above 1.0.

The recommendation made for our particular unit case history was that the transformer oil and windings be dehydrated to remove the moisture that was in the unit.

Figure 7.4 test sheet number 8, shows the end result of the dehydration process with a measurement of 1466 megohms, primary to ground. Figure 7.5 shows both the beginning and the final results of dehydration pumping in graph form. Insulation resistance readings in megohms were recorded to determine when the transformer was sufficiently dried to return to service.

Another test to firm up our conviction that the insulation is wet is the step-voltage test using a megohm meter. Typical test data that confirms moisture in another winding insulation is given in Figure 7.6B. The conclusion is that when the difference between the high and low voltage readings (2500 vs. 500V) exceeds 25 percent, the winding is wet.

The results of these insulation tests — power factor, dielectric absorption, step-voltage — will answer conclusively the question — Is the winding wet or not?

Completing the Battery of Tests

These additional electrical tests should also be made to fully evaluate the insulation system; excitation current, transformer turns ratio, and dc resistance of the windings. Illustrations of typical normal test results are also given in Figures 7.2, 6.9A, 7.6A and 7.8.

**Chapter 5, Figure 5.20.*

INSULATION RESISTANCE - DIELECTRIC ABSORPTION TEST SHEET

POWER TRANSFORMER

TEST NO. 8

COMPANY ______ DATE 8/12/76

LOCATION ______ TIME 4:00 p.m.

EQUIPMENT	Distribution Transformer	RATING	150 KVA	VOLTAGE	2400/120-240
TYPE	S	MFR.	West.	SERIAL #	3160325
RECENT OPERATING HISTORY	In Use			WINDING AGE	1942

STATE IF TRANSFORMER IS OIL FILLED AND OF THE CONSERVATOR, FREE-BREATHING OR GAS FILLED TYPE: Oil Filled, Sealed

IF OF THE FREE-BREATHING TYPE DESCRIBE THE BREATHER ARRANGEMENT AND DEHYDRATING MEDIUM USED

IF OF THE FREE-BREATHING TYPE NOTE INTERNAL CONDITIONS SUCH AS UNDERSIDES OF COVERS, TERMINAL BOARDS, WINDINGS ETC: TC# 61, After 48 hours on Keene Bowser Dehydration Unit, and 1 hour of cooling off. 140°F (60°C) CF 15.85

BUSHING DATA

OIL TESTS: ACIDITY ______ MG KOH/G. RESISTANCE IN STANDARD CUP ______ MEGOHMS. TOP OIL BREAKDOWN ______ KV.

STATE WHAT BUSHING PORCELAINS WERE GUARDED DURING TESTS

TEST DATA - MEGOHMS

PART TESTED				
GROUNDING TIME			Hand Megger	
TEST VOLTAGE	2500	2500	250V	
TEST CONNECTIONS	PRI. TO LINE	PRI. TO LINE	SEC. TO LINE	TO LINE
	GND. TO EARTH	SEC. TO EARTH	GND. TO EARTH	TO EARTH
	TO GUARD	TO GUARD	TO GUARD	TO GUARD
¼ MIN.	1347 M OHMS	1585 M OHMS		
½ "	1347 M OHMS	1664 M OHMS		
¾ "	1347 M OHMS	1585 M OHMS		
1 "	1347 M OHMS	1664 M OHMS		
2 "	1426 M OHMS	1664 M OHMS		
3 "	1426 M OHMS	1585 M OHMS		
4 "	1426 M OHMS	1744 M OHMS		
5 "	1426 M OHMS	1744 M OHMS		
6 "	1426 M OHMS	Began having trouble with inner		
7 "	1426 M OHMS	scale on Megger. Next test 5:00 p.m.		
8 "	1426 M OHMS			
9 "	1466 M OHMS			
10 "	1466 M OHMS	Ratio based on 1 minute to 5 minute readings.		
10/1 MIN. RATIO	1.089	1.048		

TEST MADE	HOURS DAYS	AFTER SHUTDOWN
DRY-BULB TEMP.		°F
WET-BULB TEMP.		°F
DEW POINT		°F
RELATIVE HUMIDITY		%
ABSOLUTE HUMIDITY		GR./#
EQUIPMENT TEMP.		°F(C)
HOW OBTAINED		
"MEGGER" INST.:—		
SERIAL #		
RANGE		
VOLTAGE		

REMARKS

TESTED BY E. Whittington

Figure 7.4 - Typical data: insulation test results following dehydration.

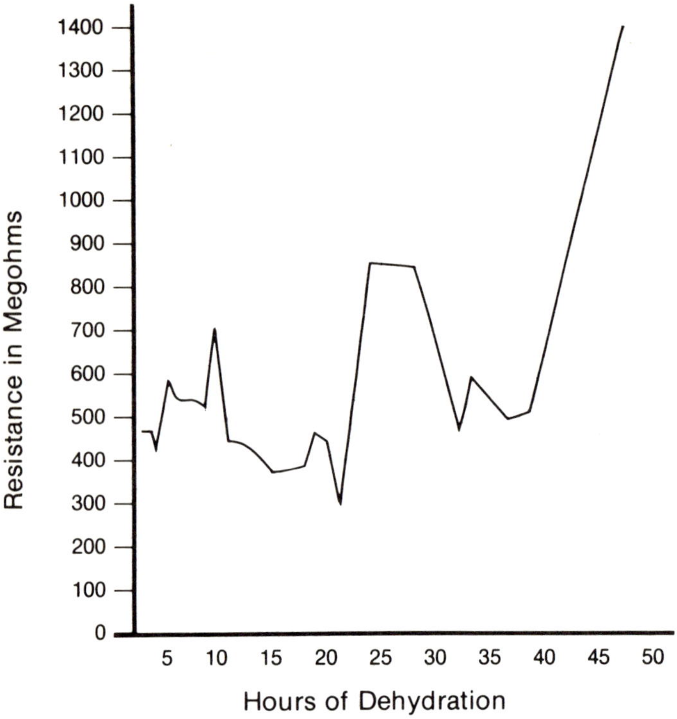

Figure 7.5 - Typical data: progressive effect on insulation resistance during dehydration process.

The excitation current test uses the same test equipment as the insulation power factor test equipment. The excitation current is the current flow into the primary when the secondary is open-circuited. Excitation current can detect shorted turns and heavy core damage. Figure 7.7 gives a typical example of the test data. Results of the standard Doble tests on three transformers, corrected to 20°C are as follows:

Transformer Unit Serial No.	C_H	C_L	C_{HL}
2714172	.68%	.39%	.33%
2714173	1.27%	.84%	.94%
2714174	.69%	.41%	.37%

DOBLE INSULATION TESTS

TWO-WINDING TRANSFORMERS

DOBLE ENGINEERING COMPANY
BELMONT, MASS
FORM MH-2W178

OIL [X]
ASKAREL [ ]
AIR [ ]
GAS [ ]

COMPANY	DIVISION		DATE 3/1/76
LOCATION OF TESTS		AIR TEMP.	TOP OIL TEMP. 30 C
TRANSFORMER Distribution		WEATHER Cloudy	% HUMIDITY 100
MFR. G.E.	SERIAL NO. 6910670-014 AGE 1942	TYPE/CLASS HS	KVA 167
FREE BREATHING [ ]	SEALED [X]	GAS BLANKETED [ ]	CONSERVATOR [ ] GALLONS OF OIL

	BUSHINGS	MFR.	TYPE	CLASS	DWG. NO.	CAT. NO.	KV	YEAR
HIGH SIDE KV 4.16 Y[ ] Δ[ ]								
LOW SIDE KV .48/.277 Y[ ] Δ[ ]		TC# 10						
Single Phase NEUTRAL						DATE LAST TEST		
COPIES TO						LAST SHEET NO.		

OVER-ALL TESTS CF .80

TEST	WINDING ENERGIZED	WINDING GROUNDED	WINDING GUARDED	TEST KV	MILLIAMPERES: METER READING	MULTIPLIER	MILLIAMPERES	WATTS: METER READING	MULTIPLIER	WATTS	% POWER FACTOR: MEASURED	COR. 20°C	KEY TO INSULATION RATING: G = GOOD, D = DETERIORATED, I = INVESTIGATE, B = BAD (REMOVE OR RECONDITION)	INSULATION RATING
1	HIGH	LOW		4	50	.2	10.0	61	.2	12.2	— —	— —		
2	HIGH		LOW	4	46	.02	.92	71	.02	1.42	15.4	12.3	C_H	B
3	LOW	HIGH		.5	17	.2x4	13.6	36	.1x4	14.4	— —	— —		
4	LOW		HIGH	.5	11	.1x4	4.4	43	.02x4	3.44	7.8	6.2	C_L	B
CALCULATED RESULTS					— —	— —	9.08	— —	— —	10.78	11.9	9.5	C_{HL} (TEST 1 MINUS TEST 2)	B
					— —	— —	9.2	— —	— —	10.96	— —	— —	(TEST 3 MINUS TEST 4)*	— —

BUSHING TESTS CF .80

*CURRENT AND WATTS SHOULD COMPARE WITH THOSE FOR C_{HL}

	LINE NO.	BUSH. NO.	PHASE	BUSHING SERIAL NO.	TEST KV	MICROAMPERES: METER READING	MULTIPLIER	MICROAMPERES	WATTS: METER READING	MULTIPLIER	WATTS	% POWER FACTOR: MEASURED	COR. 20°C	COLLAR TESTS (WATTS/CURRENT): TOP		INSULATION RATING
HIGH SIDE	1							UST								
	2				4	45	.2	9.0	56	.2	11.2	12.4	9.9	CHL		B
	3				.5	22	.1x4	8.8	27	.1x4	10.8	12.3	9.8	CHL		B
	4	N														
LOW SIDE	5							EXCITATION CURRENT								
	6				.5	49	10x4	1960								G
	7				.5	40	10x4	1960								G
	8	N														
	9				1	25	.2x2	10.0	61	.1x2	12.2					
	10				1	47	.01x2	.94	71	.01x2	1.42	15.1	12.1	CH		B
	11							9.06			10.78	11.9	9.5	CHL		B
	12															
	13				2	50	.2	10.0	60	.2	12.0					
	14				2	46	.02	.92	71	.02	1.42	15.4	12.3	CH		B
	15							9.08			10.58	11.6	9.3	CHL		B
	16															
	17															
	18															
	19	OIL SAMPLE				Water in oil - Could not Test								OIL TEMP. °C		

N = NEUTRAL

DIAGRAM

Turns Ratio - 4.997

REMARKS:

TEST SET NO. ______ TEST BY K.S. CHECKED BY ______ SHEET NO. ______

doble

Figure 7.6A - Typical data: using the Power Factor Test Series.

INSULATION RESISTANCE - DIELECTRIC ABSORPTION TEST SHEET

POWER TRANSFORMER

TEST NO. ____

COMPANY ____ DATE 3/1/76

LOCATION ____ TIME ____

EQUIPMENT	Distribution Transformer	RATING	167 KVA	VOLTAGE	2400x4160/480x240
TYPE	HS	MFR.	G.E.	SERIAL #	6910670-014
RECENT OPERATING HISTORY	In use but unit has water in it.			WINDING AGE	
STATE IF TRANSFORMER IS OIL FILLED AND OF THE CONSERVATOR, FREE-BREATHING OR GAS-FILLED TYPE	Oil - Sealed				
IF OF THE FREE-BREATHING TYPE DESCRIBE THE BREATHER ARRANGEMENT AND DEHYDRATING MEDIUM USED					
IF OF THE FREE-BREATHING TYPE NOTE INTERNAL CONDITIONS SUCH AS UNDERSIDES OF COVERS, TERMINAL BOARDS, WINDINGS ETC					
BUSHING DATA					
OIL TESTS: ACIDITY	MG KOH/G	RESISTANCE IN STANDARD CUP	MEGOHMS	TOP OIL BREAKDOWN	KV.
STATE WHAT BUSHING PORCELAINS WERE GUARDED DURING TESTS					

TEST DATA - MEGOHMS

PART TESTED					TEST MADE	HOURS DAYS AFTER SHUTDOWN
GROUNDING TIME					DRY-BULB TEMP	°F
TEST VOLTAGE	2500	2500	500		WET-BULB TEMP	°F
TEST CONNECTIONS	PRI. TO LINE	PRI. TO LINE	SEC. TO LINE	TO LINE	DEW POINT	°F
	GND. TO EARTH	SEC. TO EARTH	GND. TO EARTH	TO EARTH	RELATIVE HUMIDITY	%
	TO GUARD	TO GUARD	TO GUARD	TO GUARD	ABSOLUTE HUMIDITY	GR./#
¼ MIN.	1138 M OHMS	1090 M OHMS	693 M OHMS		EQUIPMENT TEMP.	°F(C)
½ "	1247 M OHMS	1385 M OHMS	732 M OHMS		HOW OBTAINED	
¾ "	1287 M OHMS	1585 M OHMS	772 M OHMS		Temp. Correction	
1 "	1305 M OHMS	1683 M OHMS	792 M OHMS		Factor 1.98	
2 "	1345 M OHMS	1782 M OHMS	840 M OHMS		"MEGGER" INST.:—	
3 "	1365 M OHMS	1782 M OHMS	890 M OHMS		SERIAL # 1927133	
4 "	1385 M OHMS	1830 M OHMS	890 M OHMS		RANGE 0-50 K M OHMS	
5 "	1385 M OHMS	1830 M OHMS	910 M OHMS		VOLTAGE 500/1000/2500	
6 "	1385 M OHMS	1830 M OHMS	930 M OHMS			
7 "	1385 M OHMS	1880 M OHMS	940 M OHMS			
8 "	1385 M OHMS	1880 M OHMS	950 M OHMS			
9 "	1385 M OHMS	1880 M OHMS	970 M OHMS			
10 "	1385 M OHMS	1930 M OHMS	970 M OHMS			
10/1 MIN. RATIO	1.06	1.15	1.22			

REMARKS PRI. - GND at 500V 1782 M OHMS

A change of more than 25% in resistance at 2500 V as compared to the resistance at 500V shows that there is moisture present.

TESTED BY K. Stutz

Figure 7.6B - Typical data: using the step-voltage test.

Excitation current tests, all at 10 KV, gave the following data:

Serial Number 2714172

Charging Current Milliamperes	Watts	Power Factor	Energize	UST	Ground
14.2	80.5	56.6	1	2	3
28.0	150	53.5	2	2	1
28.5	154	54.1	3	1	2
42	230	54.8	1	2 & 3	—
42	230	54.8	2	1 & 3	—
55	300	54.8	3	1 & 2	—

Serial Number 2714173

11.6	67	57.7	1	2	3
28.5	148	51.9	2	3	1
50	240	48.0	3	1	2
63	310	49.2	1	2 & 3	—
41.5	215	47.9	2	1 & 3	—
79	385	48.8	3	1 & 2	—

Serial Number 2174174

15	82	54.6	1	2	3
30	160	53.3	2	3	1
30	160	53.3	3	1	2
46	245	53.3	1	2 & 3	—
45	240	53.4	2	1 & 3	—
60	320	53.4	3	1 & 2	—

Figure 7.7 - Typical data example: using the excitation current test. R.I. Tudor, "Detection of Transformer Faults Using Exciting Current Test," (1969).

In the results obtained during the standard Doble test, transformer Serial No. 2714173 had power factors appreciably higher than the other two units, but NOT sufficiently high to warrant condemnation without other confirming evidence. Insulation resistance tests conducted at the same time gave no indication of a problem. The exciting current tests, however, showed unmistakable indications of a problem in the transformer, especially when compared to the data obtained on the other two units.

Two of the transformers gave a normal pattern of results, while the third transformer had charging currents which did not fit the pattern. In addition, the power factors of the three windings on the third unit varied considerably from each other, while on the other two units the power factors were closely related. Next, we make use of the TTR® (transformer turns ratio) test.

In our example (Figure 7.7) TTR tests were conducted on three units. The Tap B (68.8 KV) ratios were as follows:

Unit Serial Number	H1 H3-X1X0	H1 H2-X0X2	H2 H3-X0X3
2714172	28.60	38.62	28.61
2714173	26.41	28.62	28.61
2714174	28.60	28.62	28.60

The turn-to-turn test clearly confirmed that short-circuited turns existed, but it could not be used to pinpoint the problem. In our example, based on all of the data, transformer Serial Number 2714173 was taken to a repair shop for untanking and inspection. Turn-to-turn shorting was found on the H1-H3 winding as expected. After repair and return a new series of tests were performed on the transformer with good results. A second example of the TTR test, except where the excitation current test was "good", is given in Figure 7.8. The actual values were high when compared to the calculated values, indicating shorted turns on the secondary.

Today, an infrared scanner could possibly locate the faults of Figures 7.7 and 7.8.*

Finally, a Ductor® will indicate a change in dc winding resistance

**Chapter 8, Part 4.*

when there are short-circuited turns, poor joints or bad contacts. This reading should be compared to factory test information.

DOBLE INSULATION TESTS

DOBLE ENGINEERING COMPANY
BELMONT, MASS
FORM MH - 2W178

TWO-WINDING TRANSFORMERS

OIL [X]
ASKAREL [ ]
AIR [ ]
GAS [ ]

COMPANY ____ DIVISION ____ DATE 5/25/75

LOCATION OF TESTS ____ AIR TEMP. ____ TOP OIL TEMP. 35°C

TRANSFORMER Distribution WEATHER ____ % HUMIDITY ____

MFR. A.C. SERIAL NO. 2-6342-01145-1 AGE 1959 TYPE/CLASS OA/FA KVA 3750

FREE BREATHING [ ] SEALED [X] GAS BLANKETED [ ] CONSERVATOR [ ] GALLONS OF OIL 1838

		BUSHINGS: MFR.	TYPE	CLASS	DWG. NO.	CAT. NO.	KV	YEAR
HIGH SIDE KV	34.4 Y[ ] △[X]							
LOW SIDE KV	5.04 Y[ ] △[X]	LAPP	POC					
NEUTRAL								

DATE LAST TEST ____

COPIES TO ____ LAST SHEET NO. ____

OVER-ALL TESTS CF .71

TEST	WINDING ENERGIZED	WINDING GROUNDED	WINDING GUARDED	TEST KV	METER READING (MILLIAMPERES)	MULTIPLIER	MILLIAMPERES	METER READING (WATTS)	MULTIPLIER	WATTS	% POWER FACTOR MEASURED	COR. 20°C	KEY TO INSULATION RATING: G = GOOD, D = DETERIORATED, I = INVESTIGATE, B = BAD (REMOVE OR RECONDITION)	INSULATION RATING
1	HIGH	LOW		10	31	1	31	29	.2	5.8	——	——		
2	HIGH		LOW	10	52	.2	10.4	4	.1	.4	.375	.27	C_H PRI - GND	G
3	LOW	HIGH		5	60	1	60	33	1	33	——	——		
4	LOW		HIGH	3	40	1	40	28	1	28	7.0	4.97	C_L SEC - GND	B
CALCULATED RESULTS					——	——	20.6	——	——	5.4	2.6	1.86	C_{HL} (TEST 1 MINUS TEST 2)	I
					——	——	20	——	——	5.0	——	——	(TEST 3 MINUS TEST 4)*	——

BUSHING TESTS CF .71

*CURRENT AND WATTS SHOULD COMPARE WITH THOSE FOR C_{HL}

LINE NO.	BUSH. NO.	PHASE	BUSHING SERIAL NO.	TEST KV	METER READING (MICROAMPERES)	MULTIPLIER	MICROAMPERES	METER READING (WATTS)	MULTIPLIER	WATTS	% POWER FACTOR MEASURED	COR. 20°C	COLLAR TESTS (WATTS/CURRENT) TOP		INSULATION RATING
HIGH SIDE 1							UST								
2				10	20	1	20	28.5	.2	5.7	2.85	2.0			I
3				3	20	1	20	5	1	5.0	2.5	1.8			I
4	N														
LOW SIDE 5							EXCITATION CURRENT								
6			A-B	8	72	1	72								G
7			A-C	8	72	1	72								G
8	N		B-C	8	58	1	58								G
9															
10							PRIMARY BUSHING - POWER FACTOR TAP								
11			593038	2	18	.1x2	3.6	1	.1x2	.2	.55	.55			
12			672200	2	15	.1x2	3.0	2	.1x2	.4	1.33	1.33			
13			671358	2	13	.1x2	2.6	2	.1x2	.4	1.54	1.54			
14															
15															
16															
17															
18			T.Changer	10	40	20u	.80	1	.01	.01	.125	.09			G
19	OIL SAMPLE			10	40	20u	.80	1	.01	.01	.125	.09	OIL TEMP. 35°C		G

N = NEUTRAL

DIAGRAM

Turns Ratio

7.232
7.238
7.232

Calculated

$$\frac{34400}{5050} = 6.825Y$$

Tap C

REMARKS:

Measured turns ratio too high - shorted turns on secondary. Secondary power factor also too high.

TEST SET NO. 1938 TEST BY K.S. CHECKED BY ____ SHEET NO. ____

Figure 7.8 - Typical data: normal excitation but abnormal TTR test results.

Part 2—Dehydration of Transformers

Introduction

The continued reliability of an insulation system requires the highest measure of dryness. For high voltage transformers (220 KV primary and above), it is particularly important to minimize moisture contamination in-service and, when contamination does occur, to remove it as completely as possible. The time for moisture buildup in a transformer should not be underestimated. Consider the example of a 220 MVA unit not more than three years old and operating at only 60 percent of its capacity. It was found after only three months of continuous monitoring that the unit started to yield water at a rate of several grams per day. This constitutes proof-positive that drying of insulation cannot begin too soon. All drying methods share a common objective — the removal of free and dissolved moisture from the cellulosic paper, as well as the insulating fluid.

Significant factors concerning the choice of drying method include the type of transformer, ambient and operating temperatures, time limitations, thickness of the insulation pieces, and the vapor pressure of moisture in the surrounding atmosphere. A drying method can be considered as any process which reduces and maintains at a low level the material's water vapor pressure.

Moisture removal can be accomplished by several methods using various combinations of these factors, Table 7.1. The two general categories of drying practice involves use of heat and/or vacuum, performed using factory repair shop or field methods — either power off or power on. Even with all its benefits, improper drying can result in damage to the transformer insulation!

> Caution
>
> Concern for static charges when transformer oil flows through pipes or tanks must be emphasized. The explosive mixture of oil vapors and air that may be present is also a cause for concern. Proper grounding of all devices is essential. No smoking or open flame should be permitted. Fire extinguishers, preferably the carbon dioxide (CO_2) type should be available nearby before beginning any dry out procedure.

TABLE 7.1

SUMMARY OF INSULATION DRYING PRACTICES[1-2]

Processing Site	Heat (Removal of Major Portion of Water)		Heat and Vacuum (Final Removal of Water and Degasification)	
	Power Off	Power On	Power Off	Power On
Factory and Repair Shop Methods	Oven Elevated temperature Dry air (80° C-100° C)	— — —	Traditional heat (90° C) and vacuum Combinations with heat, such as hot oil and short circuit Vapor phase Hot oil spray and vacuum	— —
Field Practice	Short circuit (95° C) (core type units) Forced hot air (90° C) (shell type units) Circulating hot oil through external heater (85° C) Oil-immersed resistor (85° C)	Natural draft Dry air (6 months-1 year) Portable dry air supply Dry air circulation Use of silica gel (1-2 years)	Dry ice cold trap and high vacuum Water vapor trap and high vacuum Oil cycling and high vacuum Hot oil spray (80° C) and high vacuum Vapor phase	Oil conditioner (Purivac) Oil degasification (Keene Bowser) External mounted dehydrator Stationary installation Hot oil reclaimer Mini-oil purifier

[1]Heat only for older units without full vacuum bracing capacity.

[2]Vacuum only for units with high ambient temperature.

Each of the following methods requires regular attention and monitoring, especially the temperature. A transformer must not be left unguarded or unattended during any part of the drying process.

Insulation Drying Practice

Drying Through Heat Only

A number of HEATING methods are used; any method should take into account the relatively large quantities of heat required, the poor thermal conductivity of the oil-soaked transformer insulation, and the necessity for control of temperature to avoid overheating beyond 90°C. Older units not suitably braced for full vacuum require heat only, particularly the forced hot air or circulating hot oil methods.

1. Factory and Repair Shop Methods

   a. Oven Drying
   If the corrective work is being done in a service shop, the core and windings can be removed from the tank and dried as a unit in an oven. Larger units are dried in their tanks or may be placed in an insulated temporary enclosure.

   b. Elevated Temperature
   Heated shop air has proved successful for older insulation designs if the dew point is less than 0°C or less than 25 percent relative humidity at 20°C.

   c. Dry Air
   Keeping the insulated transformer in a confined enclosure and by removing the moisture from the air continually, drying will occur but over a long period of time (limited vapor pressure differential).

2. Field Methods-Power Off

   a. Heating by Circulating Current in the Windings (Short Circuit)
   A source of power to heat the transformer is connected on one winding, with the other winding short-circuited. Currents up to full rated current can be used continuously to heat forced oil cooled transformers.

The temperature should be monitored so that heating is controlled — a temperature high enough to drive out the moisture but not so high as to damage the insulation. For "self-cooled" units, up to full rated current may be used while the transformer is cold, then reduced as the transformer approaches rated temperature (maximum winding temperature — 95°C).

For other than extra-high-voltage (EHV) units, this method may be the most rapid. The chief disadvantage concerns the necessity for the on-site availability of a proper power supply.*

Drying of circular coil, core type, transformers can best be done by the short-circuit method. However, the forced air method is recommended for rectangular coil, shell type, oil-immersed transformers because of the mass and location of fibrous cellulosic insulation.

b. Heating By Circulating Forced Hot Air
With the oil removed, the transformer tank is blanketed in order to minimize heat loss and to keep it at uniform temperature to prevent condensation in the tank interior.

Suitable ducts and a blower or fan are assembled to blow clean dry air (90°C maximum) through the core and windings either at the bottom or top.

Although forced hot air is the simplest method, it is not very effective and is somewhat hazardous.

c. Heating by Circulating Hot Oil
This method requires an external heating system, including a suitable oil filter with a vacuum-type drier or filter press plus a heater to provide 85°C hot oil to be circulated until a satisfactory dryout has taken place.

The method by itself is slow. In addition, use of the traditional filter press or cartridge filter is of no practical value because even the driest paper cannot remove soluble water from the oil. The use of paper filters to dry

**This method requires about 1/3 of nameplate KVA and a source of voltage equal to primary volts times the impedance of the transformer, known as impedance voltage.*

out hot oil is absolutely contrary to all known engineering principles. Therefore, the hot oil method is most effective with use of the vacuum drier, and thus modern practice uses a combination of heat and vacuum.*

d. Heating by Oil-Immersed Resistor
This method consists of drying out the complete insulation system in the unit's tank by means of a specially constructed resistor unit. It is lowered between the windings at the bottom of the tank. Resistor units are spaced symmetrically as possible around the inside of the tank in order to distribute the heat. The resistors can be connected to single phase ac or dc low voltage supply.

This method has particular use for sites where mobile systems are not available and provides the temperature (not exceeding 85°C), where it is most normally required—in the oil and on the outside layers of insulation.

> Unless a transformer has had a thorough soaking, any moisture present will be partly in the oil and partly deposited on the outside layers of insulation. It will not have penetrated much to the inside layers.

3. Field Methods — Power On

a. Heat by Dry Air (Silica Gel)
Keeping the insulation dry by use of dry air circulation has been practiced for many years. However, time necessary to dry transformers with such special attached equipment has varied from six months to two years. In one example, the oil is removed and passed through a filter of silica gel or other drying agent. The dried gas is then returned to the top of the transformer tank where it will absorb more moisture from wet oil. The process is then repeated until the desired level of dryness is reached.

b. Heat by Natural Draft
This traditional method can also be done while the transformer is in-service. It requires a radiated heat

See Table 7.3.

source and air enclosure for the transformer. Above all, this method requires careful control and is the slowest of any of the other alternatives.

These two examples alone, with their need for extended time periods illustrate why the use of heat and vacuum is so much in demand for modern dehydration.

Drying Through Heat and Vacuum

1. Definitions
   Vacuum — the ultrapure environment found throughout the universe — remains difficult to simulate here on earth. The classic definition of vacuum concerns, "A space absolutely devoid of all air or gas or in practice, a space that has been exhausted to a high degree by a pump." A more useful definition is that, "Vacuum is any pressure less than atmospheric pressure (14.7 pounds per square inch absolute or 29.9 inches of mercury absolute) measured at sea level."

   The use of low pressure (high vacuum) is by no means a new technology. Vacuum freeze-dried foods and pharmaceuticals have become familiar household items. In the electrical field, vacuum has been used for drying and impregnation for many years.

   As higher vacuum practice has become necessary for EHV transformers, new terminology has come with it. The units "Torr" and "micron" have come into common use and now are considered the basic units of measurement for low pressure/high vacuum. Figure 7.9 illustrates the relationship between readings in the Metric and English systems of units. Table 7.2 summarizes the various relationships.

2. Equilibrium and Vapor Pressure (Use of Piper Chart)
   Keep in mind equilibrium for a paper and oil insulation system can be expressed as a condition in which the water vapor pressure in the paper is equal to that in the oil at a given temperature. In using the Piper chart*, keep in mind that the vapor pressure is not the same as the pressure of the gas, but is only the partial pressure of the water vapor in the gas.

**Chapter 3, Part 3, Figures 3.22 or 3.23.*

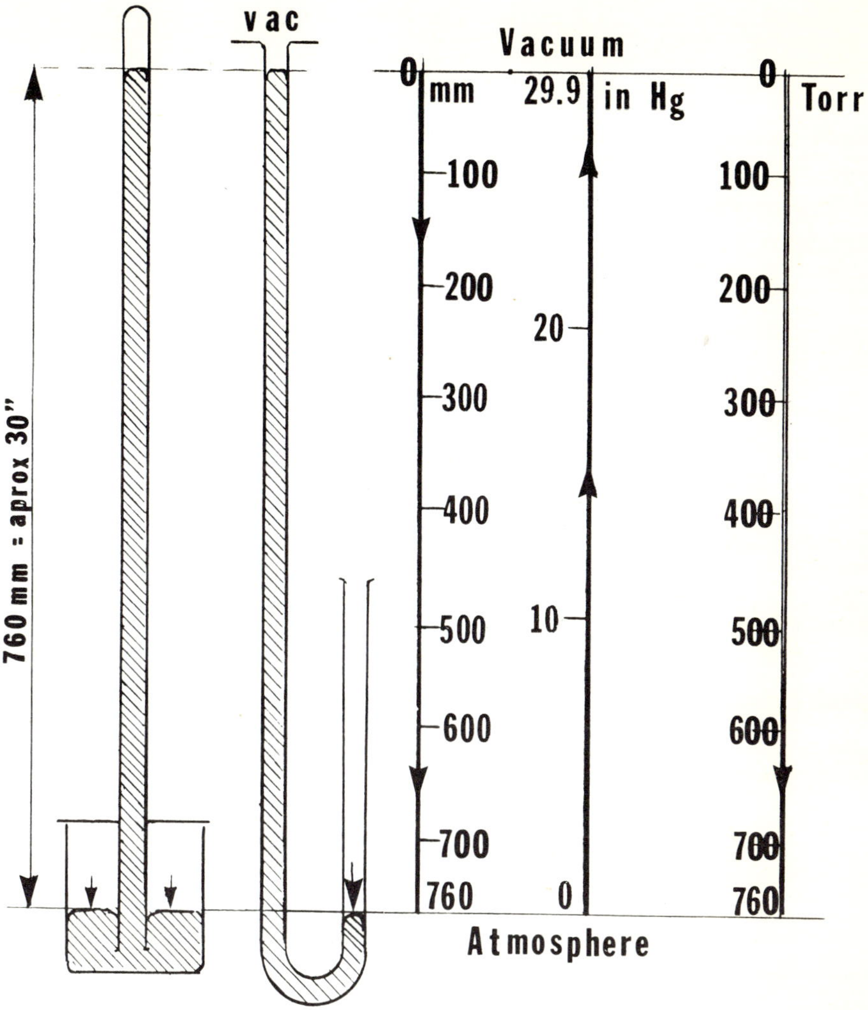

Figure 7.9 - Units of vacuum.

3. Determination of the "End Point" of the Drying Process
   The degree of dryness in a transformer's solid insulation depends upon the efficiency of the dehydration equipment as well as the length of time the drying procedure is continued. The most common methods of monitoring transformer dryness consist of:

TABLE 7.2

HIGH VACUUM UNITS OF MEASUREMENT				
Absolute Pressure				
1 Standard Atmosphere = 29.92 inches = 760 MM				
			=	1.013 KPa (KiloPascals)
1 Millimeter	=	1 Torr	=	1000 Microns
1 Micron	=	10^{-3} Torr	=	.001 Millimeter

- Dew Point
- Use of "Piper" vapor pressure equilibrium chart
- Cold Trap
- Winding power factor
- Polarization Index

a. Use of Dew Point and Piper Chart
While dew point is an effective means for determining the drying "end point" and the Piper chart is a valuable aid, both require an equalizing period of 24 hours to assure a reasonable equilibrium condition prevails.*

b. Cold Trap
As a good indicator of the water removed from a transformer, most transformer owner/operators discontinue drying when the rate of moisture collected is no more than 1½ ounces of water per six-hour period. The dehydration equipment and technique can no longer economically justify the removal of very small amounts of moisture. Yet, we know additional moisture remains in the windings, but the question is how much.

c. Use of Electrical Insulation Testing
The winding power factor test does not require a moisture equilibrium among the various insulating numbers, but measures the overall average dryness. Therefore, two of the Doble Power Factor tests can give an indication of deeply embedded (internal) moisture.

Chapter 3, Figure 3.22.

Caution

Never apply voltage to a transformer under vacuum. When checking the unit's power factor, the proper procedure is to first break the vacuum with dry air or nitrogen. Then connect the high voltage test cable to the high voltage winding and the low voltage test lead to the low voltage winding.

Use one of the two Doble tests to check the interwinding insulation — either the Ungrounded-Specimen Test (UST) or the Grounded-Specimen Test (GST) technique. The test voltage generally should not exceed three to ten percent of the normal voltage rating of the high voltage winding.

The procedure consists of measuring the power factor after periodic intervals of drying. Terminate the dehydration process when the power factor readings begin to level off.

Polarization Index (PI) and dc insulation resistance are other tools for checking transformer dryness.*

4. Factory and Repair Shop Practice
The tanks for **most** power transformers built since 1965 are braced for full vacuum, and therefore, the addition of vacuum to heat is the best way to go. At the same time vacuum alone without additional heat is sufficient only at the higher ambient temperatures (over 90°F).

Traditionally, therefore, manufacturers have used a combination cycle which uses heating by dry air to drive the major portion of water in the form of water vapor outward from the windings; simultaneously, vacuum is applied to the transformer in order to evacuate the water vapor for the larger units. However, this method requires larger vacuum tanks and considerable time (the more efficient the vacuum, the more rapid the extraction of moisture). The oil impregnated in the insulation, such as found in in-service transformers, increases the time by a factor of at least five.

**Chapter 5.*

Here are two modern drying methods:

a. Vapor Phase (Vapotherm)
The vapor phase method of transformer drying was developed by the General Electric Company (Figure 7.10). Almost all manufacturers now use some form of "vapor phase" dehydration to satisfy the uniform and exacting demands made by designers, particularly for EHV units. The drying process' basic principle does not differ from the traditional heat and vacuum. Nevertheless, the method of heating is different.

1) Heat Cycle
The interior of a transformer is placed in a vacuum tank, which is evacuated to approximately 35 mm Hg (35 Torr) pressure level.

The fluid, a special grade of kerosene that has a boiling point much higher than that of water, is heated to 100°C (and approximately 100 mm Hg absolute) and introduced into the vacuum chamber holding the core and coil to be dried out.

The hot solvent vaporizes and recondenses on cooler surfaces whether metal or paper and penetrates into the cold interior parts of the cellulosic insulation. The fluid then gives up its heat of vaporization to the object on which it condenses, causing a rise in temperature.

Heat transfer is uniform and of a high rate. The latent heat of the solvent rapidly increases the temperature of the core and coil assembly when its hot vapors condense. These solvent vapors are extracted and recovered at the condensers to be reused to complete the process.* The pressure is reduced to less than one mm Hg absolute. With the rapid increase of temperature, the vacuum strips the moisture from the core and coil.

Where time is the utmost concern, this method is capable of removing moisture rapidly and efficiently.

**The condensate may also remove some particulate matter from the windings such as carbon.*

Figure 7.10 - Factory medium power transformer vapor phase treatment equipment. Windings and pressboard insulation are dried in an oxygen free atmosphere. (Courtesy of Westinghouse Electric Corp.).

In addition, advantages include near perfect control of the process temperature, and "cleaning action" for an oil-impregnated structure; thus, less residual moisture remains than in any other method. As expected, special equipment is required to utilize this system. Its widespread use for medium and large power transformers, nonetheless, is limited by its high capital investment cost and the complexity of the system. As of this writing (circa 1981) this service is only available in large service shop and manufacturing facilities.

Research and development is currently going on to improve past field techniques that would be appropriate for critical units and where time for drying is strictly limited.

2) Vacuum Drying
By continuing vacuum treatment, the final condition of dry insulation can be quickly achieved. Typical ultimate vacuum is 0.5 Torr at 100°C insulation

temperature. A moisture content as low as 0.1 percent can be obtained. Drying time can be as low as 24 to 96 hours depending on initial moisture content load and the ultimate dryness required. Cold trap moisture measurement is normally used.*

Three common instruments are used for required checking of the pressure that exists in a transformer as it is being evacuated. The Bourdon gauge is used only for reading rough vacuum from atmosphere to about 20 Torr; the thermocouple gauge to read from one micron to 20 Torr, and the McLeod gauge as a reference instrument to check vacuum pumps and equipment. Both heat and vacuum, being in excess of the equilibrium level, force water rapidly out of the cellulose.

3) Oil Impregnation
After the dryout, regardless of method used, the tank is flooded with dry, degasified, and filtered oil to seal the insulation dry condition before reblocking and retanking. Then using torque clamping devices, the unit is structurally tightened to compensate for the shrinkage caused by drying.

b. Hot Oil Spray and Vacuum
The hot oil supply is directed over the core and coil assembly. Heat is applied (using a special manifold) to spray evenly over the entire assembly. Heat (80°C) slowly penetrates the insulation, while the unit is under vacuum. The process is somewhat similar to vapor phase. It is not as efficient as vapor phase for obtaining insulation dryness. Equipment cost is lower but the drying time is longer, and with some types of units, part of the insulation may never be reached by the oil spray. The hot oil spray and vacuum method has been adapted for use in the field.

5. Field Practice — Power Off

a. Cold Trap Drying
This field procedure uses a vacuum pump, capable of

Chapter 3, Part 3, Figure 3.24.

handling large volumes of vapor and of pumping down to an absolute pressure of 50 microns or lower, utilizing a cold trap to collect the water.* It may be done without the high temperature required for all the other alternatives (Figure 7.11).

Oil is drained from the unit, where the tank is then filled with dry nitrogen, while all external connections are removed and all leaks are eliminated. The cold trap and vacuum pump are connected to a suitable pipe connection on the tank. Water removal from the insulation begins when the residual vapor pressure in the tank is reduced below the vapor pressure of water in the insulation. The residual water content of the insulation may be estimated from a moisture equilibrium chart.† Air laden with vapor is pumped out of the unit, which is completely sealed by a high vacuum pump, and passed over a super cold trap that condenses and freezes the moisture. The dry out is continued as long as water is removed or may be terminated when dew point measurement indicates that residual moisture has been reduced to the desired amount. Supplementary heat may be needed to achieve the desired dehydration for EHV transformers. Even so, cold trap drying is an efficient and relatively fast method of drying.

Two types of cold traps have been used. Traditionally, dry ice has been the vehicle (Figure 7.12). Recently, a mechanically refrigerated water vapor trap has eliminated dependency on a supply of dry ice and provides an efficient method of defrost for the unit. This permits more accurate determination of the quantity of trapped water.

b. Oil Cycling

In this method, heat is applied by one or several total immersions of the assembly in hot oil, followed by periods of vacuum. It is not as efficient as cold-trap drying, but it is both a less complex and less costly drying practice.

**Chapter 3, Part 3, Figure 3.24.*

†Chapter 3, Part 3, Figure 3.22 or 3.23.

Figure 7.11 - Before dehydration can be performed the transformer tank and all fittings must be tightly sealed to prevent oil leakage, entrance of moisture, and loss of inert gas protection. (Courtesy of Ohio Edison Company).

Figure 7.12 - The traditional method of cold trap dehydration requires periodic changing out of dry ice. (Courtesy of Ohio Edison Company).

6. Field Practice — Power On

a. Automated Oil Purifiers

Many transformers operate in places and under circumstances that do not permit de-energizing from service. In time past, there was hesitancy toward oil purification (and reclamation) in energized equipment. Numerous experiences over the last 30 years have shown that with proper equipment this can be an economical and safe procedure. A typical energized pumping run of 48 hours was given in Figure 7.5. Yet even today some authorities do not recommend use of in-service drying. This method could introduce bubbles into the unit or particulate matter stirred up may move into high stress areas, thereby causing possible flashover.*

The concept had its origin in the 1940's with a vacuum chamber using spray nozzles followed by a filter press for removal of solid particles. This system is now obsolete.

Another type of oil conditioner (Figure 7.13) was developed shortly thereafter in which deteriorated oil was picked up by an oil pump and passed through two bulk-type clay (fuller's earth) filters. From the filters it passed through an oil heater which brought up the temperature to 140°F. The oil is then picked up by a second pump and delivered to a spray-type dehydrator under high pressure.

In the middle 1960's, another major system advance was made; a vacuum degasifier was used in combination with a high capacity vacuum pump package for transformer dryout and filling (Figure 7.14). Capacities for mobile units run up to 60 gallons per minute. Units are housed in 25 to 40 foot trailers, frequently incorporating a laboratory facility.

This version includes an interstage chilled trap between vacuum pump and booster, filter/water separator for interception of free water, and a monitoring system for the moisture.

Chapter 7, Part 3 (on-site energized complete desludging of transformers).

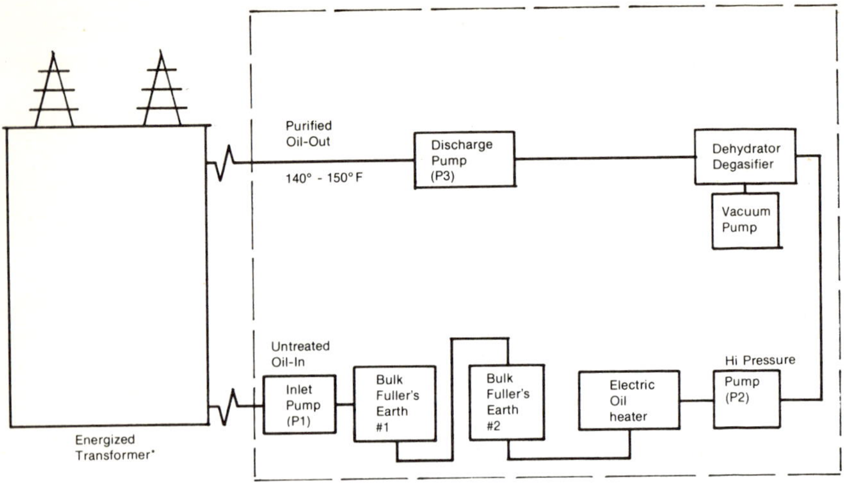

Figure 7.13 - Flow diagram of CFC Honan-Crane Purivac oil purifier.

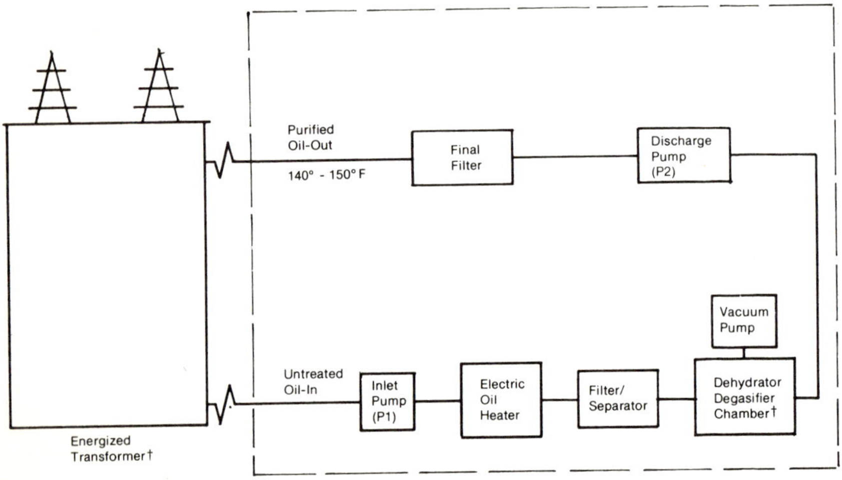

Figure 7.14 - A typical flow diagram of Keene high vacuum oil purifier.

**Fail safe controls - minimum temperature and correct vacuum required.*

†If preset pressure of vacuum chamber is exceeded, the system shuts down and/or sounds an alarm.

The trend toward faster and better dryout continues. Since then a number of mobile portable systems with variations based on this pattern have become available. These units may use deep bed clay filters (removal of chemical impurities) and additional stages of filtration (improved quality control).*

b. Externally Mounted Oil Dehydrators
Other means of oil maintenance, where moisture is the only problem, are the externally mounted oil purifiers removing water continuously. They are based on the principle of coalescence and have the advantage of not tying up on-site process equipment.

A most commonly used device to minimize water in oil is the conservator with the silica gel dryer. Though simple and relatively cheap it has the major drawback of recurring need for maintenance. This system is simply of no practical value for EHV units, even when additional cost is required.

One new method of automatic insulation drying involves freezing the moisture out of the air above the oil in the conservator in units greater than 275 KV.

This permanent accessory establishes low values of water vapor pressure in the air contacting the oil. It thus achieves the highest degree of dryness.

The dryer has been primarily designed to keep essentially dry units that way. Tests have also confirmed that besides drying air "breathed in", the unit slowly removes water from the internal insulation. The unit is automatic; therefore, it doesn't require the periodic replacement of a desiccant charge.

A second method — oil circulation through a molecular sieve material — has been used to dry units in which the insulation had attained a high water content in normal service.

The molecular sieve unit uses the principle of double filtration — a paper filter plus molecular sieve pellets

**Chapter 7, Part 3, Table 7.1.*

(synthetic zeolites). A probe is inserted to monitor the water content of the pellets. A typical dry-out takes up to two months, the process depending largely on the operating temperature.

c. Stationary Installations
A complete permanent on-site oil reclamation system is illustrated by the flow diagram of Figure 7.36.*

Results of the Drying Process

The Bonus of Degasification

The primary advantage for dehydration of oil with high vacuum in contrast to heat alone lies in a side effect —degasification — which simultaneously occurs whether needed or not. This does not occur with other methods of water removal.

In recent years U.S. transformer manufacturers, industry and many utilities have begun to recognize the value of both a high degree of dehydration and degasification of oil — both at the factory prior to impregnation and filling and secondly, in the field. A high degree of degasification brings these benefits:

- Minimizes partial discharge (corona) effect (from inadequate oil impregnation)
- Removal of unwanted oxygen
- Removal of combustible gases
- Extremely dry oil

The development of automated portable, mobile, and stationary high vacuum degasifier units has thus resulted from this almost universal demand for higher quality control. Here are the standard "specs" for the finished product, a transformer oil reading:

Total water content — maximum 10 ppm by ASTM Method D-1533
Total gas content — maximum 0.25 percent by ASTM Method D-2945
Dielectric breakdown — minimum 40 KV by ASTM Method D-877

Complete degasification is accomplished by exposing the oil in a thin layer to a high vacuum.

**Chapter 7, Part 3.*

Drying Efficiency

The efficiency of dehydration regardless of method depends on a number of factors, including the oil temperature and the degree of vacuum pulled. The challenge is to use a system that avoids use of excessive temperature, which promotes degradation of the cellulose, yet brings about the removal of the maximum degree of moisture from the insulation system.

Factory residual moisture content (0.3-0.5 percent) shows equilibrium with vacuum (Figure 7.15) as follows:

- At 40°C (105°F) vacuum required minimum 70-200 microns.
- At 60°C (142°F) vacuum required minimum 300-1200 microns.
- At 80°C (178°F) vacuum required minimum 1200-4000 microns.

Note: Other studies have further confirmed and added strength to the validity of the vapor pressure equilibrium chart for use in vacuum drying and processing of high voltage transformer insulation both at the factory and in the field.

For efficient dryout, vacuums in excess of the equilibrium values are required. The drying operation occurs when the oil-cellulose equilibrium condition is disturbed favorably.

Other factors that determine the efficiency of water removal include: initial water content in the oil and in the cellulose, oil film thickness, time of exposure to vacuum, and accelerators, such as agitation, sharp points, or rough surfaces.

Table 7.3 summarizes and compares the various field practices of oil purification.

Figure 7.16 likewise summarizes the various free water situations along with suggested troubleshooting procedures.

Dehydration and Degasification Only?

In the case of just wet oil, or wet oil and wet windings, a couple of recirculations of the oil through a vacuum dehydration system will dry the oil. The dry oil will be contaminated by the wet windings as they seek equilibrium condition. Repeating this cycle many times can correct the situation.

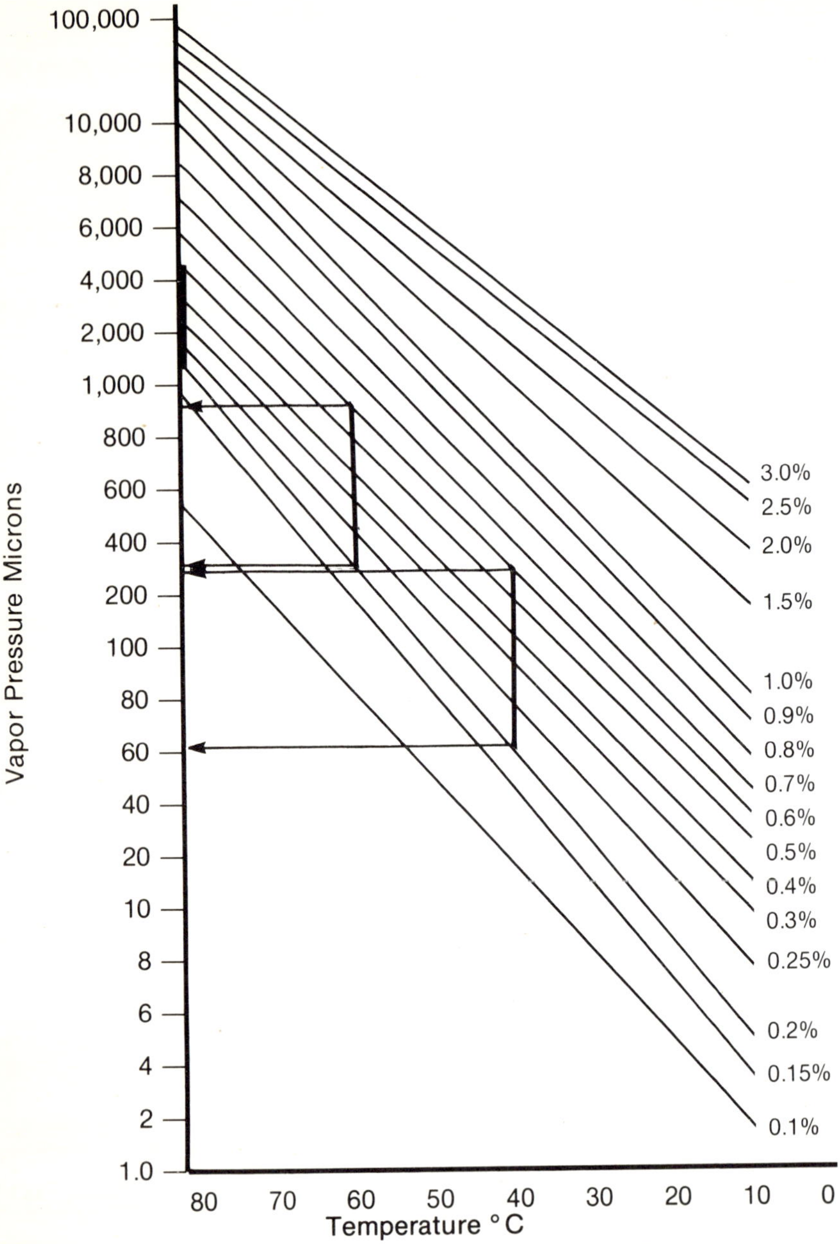

Figure 7.15 - 'Piper Chart' showing minimum vacuum required at typical temperatures in degrees Celsius. L.B. Barnowski, "Latest Trends in Transformer Drying and Filling Practices" (1971).

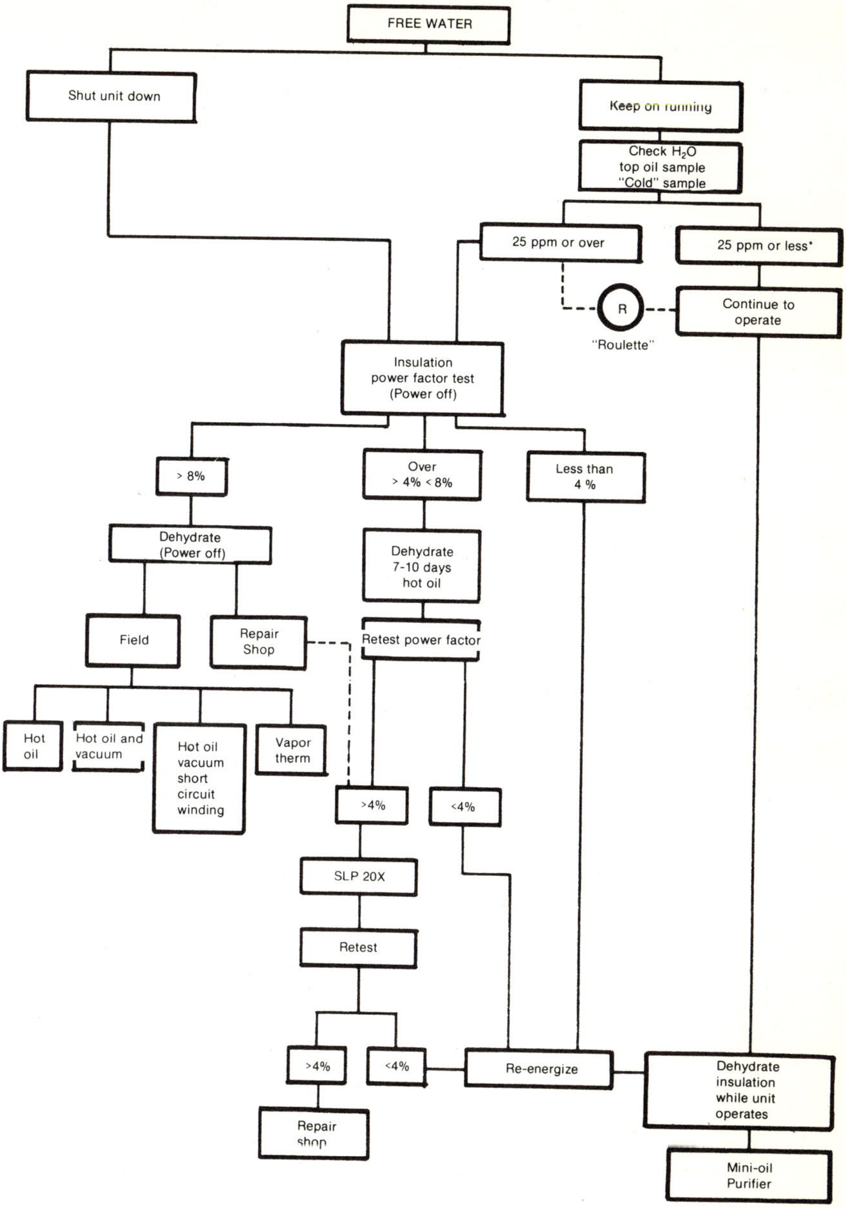

Figure 7.16 - Troubleshooting sequence for checking various free water situations.

**Extra high voltage (EHV) units maximum allowable would be 15 ppm water.*

TABLE 7.3

OIL PURIFICATION
FIELD PRACTICE—POWER OFF

Method	Heat/ High Vacuum	Time Duration For Drying	Cost	Efficiency	Special Notes	Contamination Removed					
							Water		Air and Gas	Acids & Sludges	
						Solid	Free	Dissolved		Volatile	Other
Cold Trap	Heat/ High Vacuum	60 = in 4½ days	2	1	IEEE preferred; EHV units; may not need heat	No	Yes	Yes	Yes Partial	No	No
Hot Oil Spray	High Vacuum	—	—	2	80°C heat	No	Yes	Yes	No	No	No
Hot Oil	High Vacuum	8 = 3 days at 50°C	3	3	85°C heat use for older units; not EHV	No	Yes	Yes	No	No	No
Short-Circuit	High Vacuum	40 hours	4	4	95°C heat use for old core types; not EHV; somewhat hazardous	No	Yes	Yes	No	No	No
Hot Air	Heat	1-2 weeks	1	5	95°C heat use for shell type units; somewhat hazardous	No	Yes	Yes	No	No	No
Oil-Immersed Resistor	Heat	40 hours	—	6	85°C heat; use where mobile plant not available (Europe)	No	Yes	Yes	No	No	No
Vapor Phase (Vapor Therm)	High Vacuum	40-60 hours	High	7	Not practical for the field; best for the factory EVH unit	No	Yes	Yes	No	No	No

(Table 7.3 continued on next page)

(Table 7.3 continued)

FIELD PRACTICE—POWER ON

Method	Heat/ High Vacuum	Time Duration For Drying	Cost	Historical Application	Special Notes	Contamination Removed					
							Water		Air and Gas	Acids & Sludges	
						Solid	Free	Dissolved		Volatile	Other
Oil Conditioner (Purivac)	Heat/ High Vacuum	—	Buy	1	150° F heat; portable; also for gas filled units	Yes	Yes	Yes Partial	Yes Partial	Yes Partial	Yes Limited
Centrifugal oil purifier (DeLaval)	Gravitation setting	—	Buy	2	Stationary	Yes	Yes	No	No	No	No
HiVac Degasifier (Keene Bowser)	Heat/ High Vacuum	Days to Weeks	Lease	3	Mobile	Yes	Yes	Yes Partial	Yes	Yes	No
Oil Purifier (Vacu-Dyne/ Baron)	Heat/ High Vacuum	—	Buy	4	110° F heat; Mobile Portable	Yes	Yes	Yes Partial	Yes	Yes	Yes Limited
Stationary Installation	Heat/ High Vacuum	—	High Capital Purchase	5	Stationary	Yes	Yes	Yes	Yes	Yes	No
External Mtd Dehydrator (Dryco)	Freeze Drying of air	Slow for older units	Buy	6	Permanent mounted; primarily for new units	No	Yes	Yes	No	Yes	No
Fuller's Earth (or Activated AL treatment)	Adsorption	Hours	Optional	—	Mobile unit; 150° C heat	No Some Colloids	Yes Limited	No	No	Yes	Yes
Hot Oil Reclaimer	Heat/ High Vacuum	Hours	Lease	7	Mobile; 180° F heat	Yes	Yes	Yes	Yes	Yes	Yes
Mini oil Purifier	Heat/ High Vacuum	Up to 4 months	Lease/ Buy	8	Portable Modular Options; Follow up Hot Oil Reclaiming	Yes	Yes	Yes	Yes Partial	Yes	Yes Limited
DEOX™	Heat/ High Vacuum	Varies	High Capital Purchase	9	Monitors continuously gas rate and moisture content	Yes	Yes	Yes	Yes	Yes	No

The recirculation of the oil through an energized transformer utilizing a "dehydrator—degasifier" will ultimately reduce the insulation moisture content. By keeping the oil at the safe temperature of 70°C for instance, the oil can absorb ten ppm of water from insulation containing one percent moisture content. An insulation weight of 20,000 pounds, (not unusual in EHV units) with a dehydrator differential pickup of five ppm may require up to 800 days at three GPM to reduce insulation content from 1.0 percent to 0.5 percent. That is slow!

It should be clearly understood that EHV equipment is not the only candidate for a high degree of dryout. Most of today's transformers, including distribution units of the pole, pad type, and power transformer varieties, are processed in the factory to the highest standard of vacuum dryout and impregnated with oil under vacuum. Anything less than such a high standard will reflect in shortened life of the apparatus once it is filled or shop refilled, without similar care.

You'll note in Tables 7.1 and 7.3 reference is made to reclaiming and Hot Oil Cleaning. Very shortly, we'll consider this phase of corrective maintenance of the insulation system.*

Ongoing Oil Purification

Limitations of Dehydration Systems

Modern oil purification systems are most effective in removing water from transformer oil. Regardless of how efficient any method of dehydration of oil might be, the reality of the integrity of the insulation system confronts the transformer owner.

In a dryout, eventually, the rate of water removal is so small that continued servicing becomes economically prohibitive. While up to seven to ten days water removed can be measured at one to four quarts daily, ultimately the amount of water removed may decrease to no more than a teaspoon per day. Although the oil is dry at this point, the process may or may not dry the transformer. Why? Because the rate of drying depends on the rate of diffusion of water through the paper which contains initially over 99 percent of the moisture in the unit. Thus, after several weeks of normal operation following the initial treatment the transformer appears to be nearly as wet as its initial condition.

**Chapter 7, Part 3.*

The cause of this problem is based in large measure on the insulation design. Consider Figure 7.17, a detailed cross-section of an insulated conductor. Older units were constructed with thick-solid insulation structures.

These thick vapor barriers in the insulation require a much longer drying time.

Newer transformers consist of a built-up structure in which relatively thin solid insulation barriers (usually less than 1/8 inch thick) are alternated with oil ducts of similar thickness. Nevertheless, even in this newer, more efficient insulation, a great deal of time is still necessary to attain the degree of dryness desired. For example, one case history illustrates this point.

> A 150 MVA, 400 KV transformer with about 15,000 pounds of paper may require well over two weeks for drying out and processing at the factory. After ten days, at an absolute pressure of 100 microns the dielectric characteristics

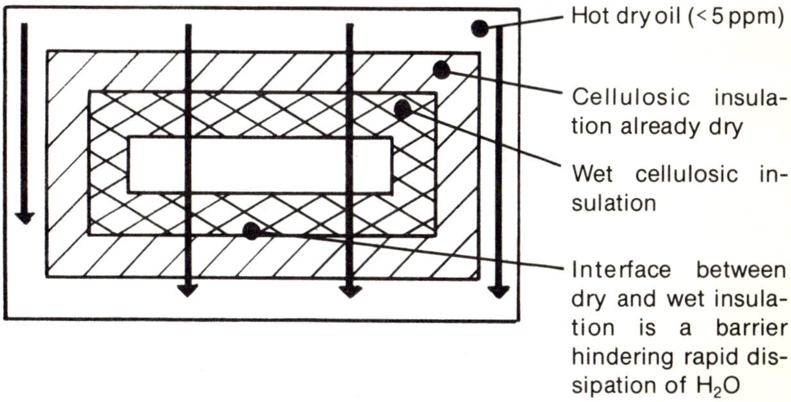

1. Under pictured condition insulation resistance and power factor will improve.
2. As water redistributes through insulation readings degrade rapidly.

Figure 7.17 - Cross-section of an insulated conductor.

(power factor, etc.) had reached a constant value and 453 pounds of water had been removed. In the ten subsequent days these characteristics remained the same, even as an additional 37 pounds of water was removed.

The need therefore becomes obvious for a method of follow-up that would not require repeat processing to substantially reduce the total water content of the transformer.

One solution is the application of a mini-oil purifier (INSUL-AIDER™)* that could be operating on either medium or large power transformers for several months at a time while the unit remains Power On.

A Possible Solution — Mini-Mobile Oil Purification Systems

A mini-oil purification system could be "Power on" operated during a three or four month interval (or what is necessary) so that "the black box" would dehydrate the entire insulation system. It could be made available on either a lease or purchase basis.

A typical mini-unit (Figure 7.18) could include: electric controls-vacuum, pressure, temperature, and flow gauges; flow switches; alarm system; water separator, vacuum chamber, prefilter, magnet, and final filter, pumps, and electric heater(s).

This application usually would require that the mini-dehydrator be plumbed permanently onto the transformer via the filter press valves. The oil is then recirculated continuously through the vacuum chamber and back into the transformer. A slight heating action may be applied to enhance the moisture exchange.

This mini-unit (Figure 7.19) is recommended for use with transformers three to ten years old as the "prime time" servicing and transformers older than ten years where the oil is still in the "Good" range (i.e., an acid number below 0.10 mg KOH/gram of oil and an Interfacial Tension of 30 dynes or more). While the incoming oil contains from 40 to 55 ppm of moisture, the outgoing oil to the transformer will be less than 20 ppm. The initial effect will be an

**INSUL-AIDER™ is a registered trademark of S.D. Myers, Inc., Manufacturing Div., for oil handling equipment, specifically designed to operate continuously with energized transformers.*

increase of the dielectric strength of the oil. But the sought after long term effect will be continuous slow drying of the cellulosic insulation.

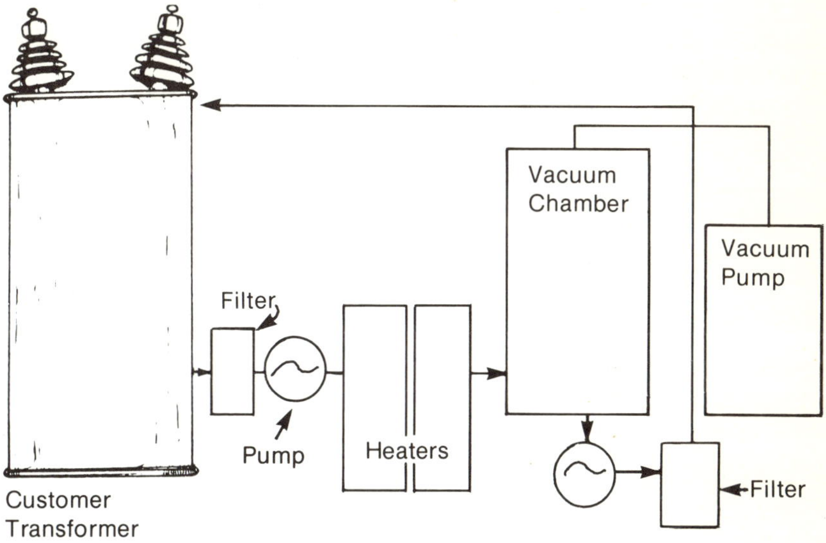

Figure 7.18 - Mini-oil purifier—typical flow diagram.

Figure 7.19 - Mini-oil purifier—basic system.

The goals should be for a situation where the cellulose contains less than 0.5 percent by weight of water (as in a new transformer). Yet more than likely we might find the moisture content as high as 4.5 percent by weight of water in the cellulose of an operating transformer that has oil with a good dielectric strength. Removal of this water will increase the dielectric strength of the transformer, minimize further loss of cellulosic mechanical strength, and slow down the rate of oxidation — all other items remaining the same.

A typical application could be in conjunction with a conservator type transformer. If the cellulose has a low moisture content only a short period of maintenance is necessary. If the moisture content is high, several months of operation may be necessary to effectively remove the water from the cellulose.

As a complete preventive maintenance tool, the mini-oil purifier can also provide several options.

In new transformers, for example, the insulating system (oil and paper) would have liberated sufficient oxygen to start the initial mechanism of oxidation. By applying a vacuum to the oil for a continuous period of time the oxygen content can be lowered significantly. Keep in mind that an "oxygen-free" insulation system may operate at 20°C higher temperature without sacrificing insulation life.*

A third use of the mini-purifier or INSUL-AIDER™ system is for removing various types of contamination. This would include the lint fibers and debris not removed by filtration at the factory. To remove other types of contamination such as sludges, a portable tank containing a minimum of 550 pounds of fuller's earth absorbent material should be added into the system (Figure 7.20). A few recirculations of the oil through this system will improve the oil by increasing the dielectric strength, lowering the acidity, and raising the IFT in addition to clarifying the oil and partially degassing it.

For badly deteriorated oil (greater than 0.15 mg KOH/gram) a second fuller's earth tank would be necessary. If the electric heaters that are part of the basic system cannot supply sufficient heat to raise the temperature of the oil to the aniline point at the desired flow rate,† or if not enough electrical power is available at the job

Chapter 2, Part 5, Figure 2.36.

†*Chapter 7, Part 3, (Desludging of Transformers).*

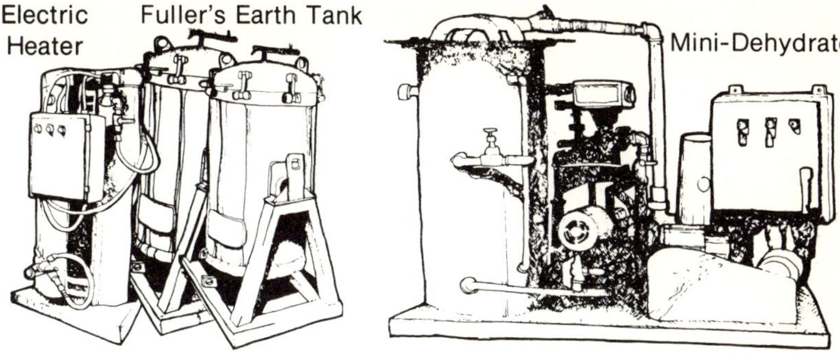

Figure 7.20 - Mini-oil purifier—additional modules.

site, then an oil-fired heater should also be added as a module to the basic system. Such mini-oil purification systems have proven feasible for an underground mine situation, as well as industrial and utility applications.

The modules of the mini-system can be further modified to suit other jobs. If oil circuit breakers (OCBs) or tap changers have carbonized oil in them, they can have the oil purified and degassed by the proper selection of filters, adsorbents, and processing time.

Most of the foregoing situations can be done while the transformer is energized. Nevertheless, certain cases and situations might dictate that the work be done de-energized. In ALL situations safety of personnel and the equipment must be a paramount concern.*

- The equipment must be adequately grounded to station ground.
- Flow rates, pressures and temperatures must be kept in prescribed limits.
- No vacuum should be directly applied to an energized transformer.
- Installation of proper fail-safe devices in the system when the equipment is plumbed for continued operation.

Automatic continuous monitoring of the combustible gas rate and moisture accumulation may be justified for very large critical transformers. Such a system which simultaneously dehydrates and degasifies a very large power transformer (200 MVA) has been tested

**Chapter 7, Part 3 (Reliability of Energized Transformer Desludging—Safety Precautions).*

both in the laboratory and in the field by ASEA — a unit called DEOX™. Other similar type systems are also operational. The cost of this DEOX™ system, however, would naturally limit its extensive use. The DEOX™ system has been so named to indicate the importance of oxygen removal to slow down the insulation aging process. Nevertheless, the DEOX™ type system certainly bears witness to the established need for an integrated attack on the insulation evaluation and maintenance problem.*

**Use of a mini-oil purifier is one phase of a Transformer Insulation Maintenance Service (TIMS). Chapter 10, Part 2.*

Part 3—Transformer Oil Treatment
vs
Transformer Desludging

Introduction

Most electrical power transformers require an insulating oil that serves two primary functions:

- Provides a dielectric medium
- Provides a cooling medium

The first purpose requires that the oil be free of water and suspended organic or inorganic matter, thus acting as a good electrical insulator. The second function requires an oil of relatively low viscosity, volatility, and good heat transfer capability. Deterioration of the oil, through either oxidation and/or contamination, can cause the transformer to overheat and prematurely fail.

American industry and utilities no longer can afford the luxury or option of installing a new transformer and forgetting about it. Many transformers operate in places and under circumstances that do not permit de-energizing and removing from service; therefore, they receive little, if any, maintenance.

Unfortunately, many transformers have been ignored for so long that only *corrective* maintenance—reclamation of the oil and desludging of the transformer—can rectify the situation. Here is the story on oil:

> When a heavily sludged used oil, completely unfit for further use in the transformer, is examined, it may be found that at least 80 percent of the hydrocarbons present in the used oil are unchanged and can be reused if the oxidized products are completely eliminated. It is correct to state that the oxidation of an oil is that of the impurities present in the oil and not that of the hydrocarbons. Pure hydrocarbons are not easily oxidized under normal conditions.

With 1.73 *billion* gallons of mineral oil now in thousands of in-service transformers, it behooves us to give more thought to proper maintenance. The classic statement, "We've always done it this way," is not a viable approach. Conservation and cost effectiveness are the battle cry.

The predicted world shortage of good grade naphthenic crudes (suitable for production of transformer oil) and the escalating prices of oil not only justifies but also demands the need for a practical approach to proper use, maintenance, and recycling of electrical insulating oils. The alternative and parallel approach is through research to develop new dielectric liquids or perhaps gases to solve this problem. The cost of such fluids is expected to be many times greater than the natural petroleum product; therefore, it also behooves industry and utilities to conserve this limited resource.

Viewing Oil Reclamation In Retrospect

The use of mineral oil in transformers dates back to 1887. Those early oils were paraffin-based crudes. As equipment was made larger and placed outdoors, it soon became evident that the use of paraffin-based crudes was not adequate for northern American winters because of the inherently high pour point temperature. The increase in viscosity and loss of heat transfer resulted in numerous transformer failures during the mid-1920's.

Paraffin crudes were successfully replaced with naphthenic oils made from crudes having an asphaltic base. They have performed admirably for more than 50 years. But naphthenics introduced a different problem. Sludge or oil decay products formed much sooner in naphthenics than in paraffinics. Paraffinic oils are more stable than the naphthenics; however, when the paraffinic oils oxidize, their oxidation products are more objectionable.

The first attempt at reclaiming naphthenic oils lost favor when it was discovered that the reclaimed or recycled oil deteriorated at a much faster rate than new oil in the same transformer. "Changing out" oils remained popular because the new oils gave a longer life than the reclaimed transformer oils. This was attributed to new oil having a "natural inhibitor" that prolonged its useful life. Thirty years of research resulted in the identification of a suitable antioxidant called 2,6-Ditertiary-Butyl-Para-Cresol (DBPC®)* which when added to a reclaimed oil, would prevent rapid oxidation.

Oil reclamation then regained some favor when the American Society for Testing and Materials (ASTM) D-9 committee announced a new

**DBPC is a registered trademark of the Koppers Company, Pittsburgh, Pa.*

concept in 1952: "Insulating oils, when properly serviced, can be given practically unlimited extension of life, free from the formation of sludge...due to oxidation."

But despite this authoritative pronouncement, much of industry and the utilities chose to replace old oils (at 15 cents per gallon!) supposedly in the interest of economy. To their credit, they chose to do something. Nonetheless, this study has revealed their choice to be a poor one.

It took the latest chapter in the history of transformer oil to awaken Americans to what must be done—a chapter summed up in the word "shortage". Overnight, people who once threw away their old oil now frantically sought to preserve it. By 1973, there was a shortage of both transformer oil and the inhibitor DBPC®. Prices shot up. And because only three percent of crudes are suitable for refining to transformer grade oil, and these now in critical supply, transformer oils are now a blend. This blend may include more than 50 percent paraffinics and aromatics.

Reviewing the Problem of Sludge

Confronting the transformer owner/operator are not only the problems of oil scarcity, cost, and deterioration, but ultimately—sludge. The key to preventive maintenance of transformers is thus annual oil testing. Its primary purpose is to pinpoint the condition of a transformer's insulating system so that it may operate in the "sludge-free range".

The normal aging of a transformer brings sludge—all the solid crud that collects in a transformer. It will appear faster in a heavily loaded, hot running, abused, or neglected unit. This process can be hastened by moisture and free access to oxygen. The sludge builds up in progressive layers with varying degrees of hardness, depending on how the unit has been operated and how long the equipment has been ignored (Figure 7.21).

Several independent studies have shown that an increase in neutralization or acid number (NN) should normally be followed by a characteristic drop in interfacial tension (IFT) and a deepening of color (Figure 7.22). Knowing the NN and IFT gives an insight on how much usable life may be left in the oil, the probability of sludge accumulation, and what action is required to prevent or remove sludge deposits.*

**Chapter 3, Part 2, p. 284.*

Data from an ASTM field sludge probability study of 500 in-service transformers (Tables 7.4 and 7.5) permits oils to be classified, also shown in Figure 7.22. A marked change in color between test periods has some significance. As a rule of thumb, once transformer oil changes from the yellows into the ambers and browns, the oil has degraded to the point where the vital parts of the transformer are being seriously affected.

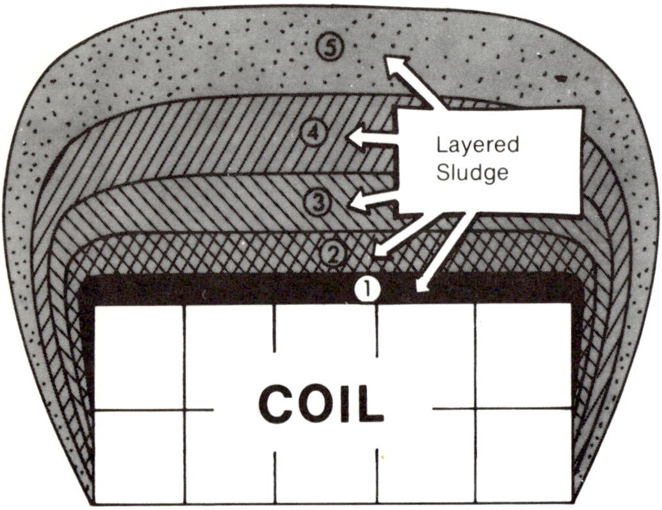

*Figure 7.21 - Shows cross section of transformer coil with 5 successive layers of sludge buildup. Layer 1 has already solidified and is now a "permanent part" of the transformer.**

TABLE 7.4

CORRELATION BETWEEN NEUTRALIZATION NUMBER AND SLUDGE FORMATION IN OIL-FILLED EQUIPMENT	
Neutralization Number, mg KOH/gm of Oil	Percent of Units in Which Sludge was Found
0.00 to 0.10	0
0.11 to 0.20	38
0.21 to 0.60	72
Above 0.60	100

*Chapter 2, Part 3, p. 186.

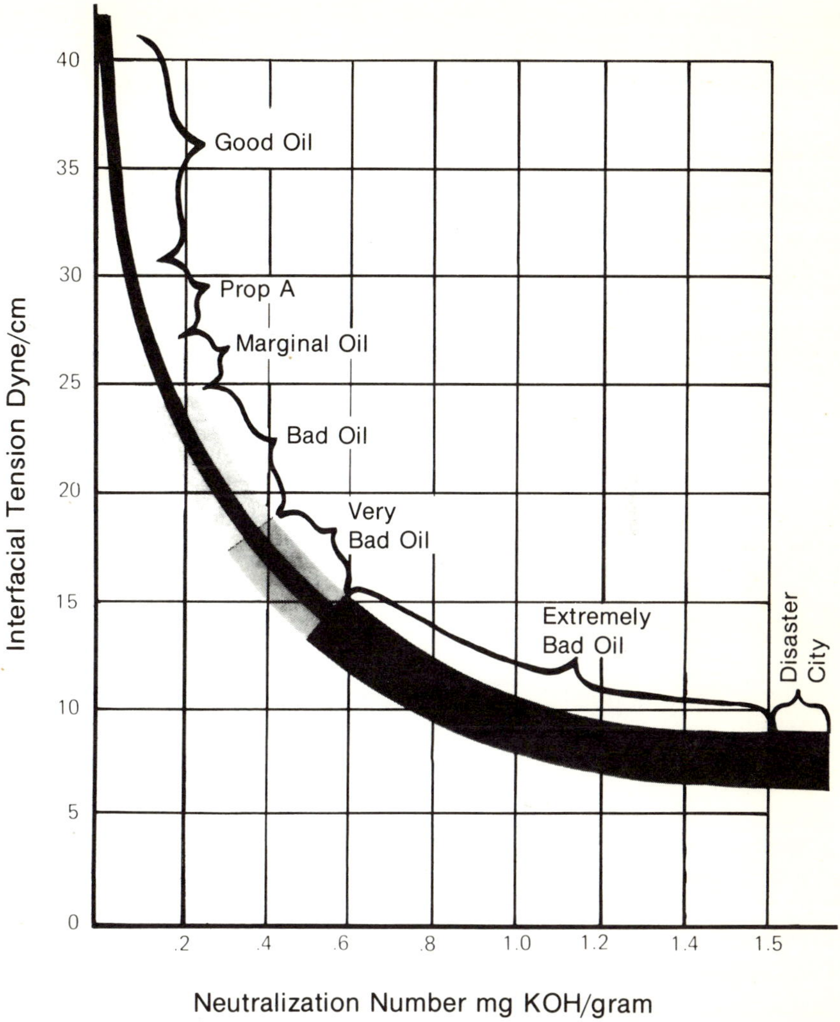

Figure 7.22 - A precise classification of transformer insulating oils involves the unique relationship between interfacial tension (IFT) and neutralization number (NN). Several 1940 studies show that an increase in NN should normally be followed by a typical drop in IFT.*

If the values of IFT versus NN for a given oil sample do not fall within a reasonable range on either side of the median line, further investigation is in order. The results of ***other tests*** *of Table 6.5 should be evaluated in combination to get a true picture of the condition of both the oil and the transformer.*

**Chapter 3, Part 2, p. 284.*

TABLE 7.5

CORRELATION BETWEEN INTERFACIAL TENSION AND SLUDGE FORMATION IN OIL-FILLED EQUIPMENT	
Interfacial Tension, dynes per cm (millineutons per meter)	Percent of Units in Which Sludge was Found
Below 14	100
14 to 16	85
16 to 18	69
18 to 20	35
20 to 22	33
22 to 24	30
Above 24	0

Definitions

Now that we've identified the problem, let's consider certain definitions before investigating various solutions.

1. Reclamation
   "Any method or process which results in a beneficial change in the oil composition" (restoration of transformer oil that has deteriorated through oxidation). Typical methods employed are treatment with adsorption agents, acid and alkali treatment, and so forth.

2. Reconditioning
   "Removal of moisture and solid materials and dissolved gases by mechanical means." This actually should read "*physical*" not "*mechanical*" means, as the processes covered by this definition include physical phenomena of absorption and vacuum distillation.

3. Reclaiming
   "Removal of acidic and colloidal contaminants and products of oxidation by chemical and adsorbent means."

   The term "rerefining" is being dropped from the vocabulary to avoid confusion. Rerefining describes processes that are used in a refinery, and are not rightly processes available for "in-the-field" work.

In addition to the preceding basic terms, the following are frequently used in reference to electrical oil treatment:

1. Inhibiting
   "Adding an inhibitor." An inhibitor is any substance which, when added to an electrical insulating fluid, retards or prevents undesirable oxidation (ASTM D-2864).

2. "Adding an Oxidation Inhibitor" (or "antioxidant")
   An oxidation inhibitor is any substance added to an insulating fluid to improve its resistance to deleterious attack in an oxidizing environment. For example, 2,6-ditertiary-butyl-para-cresol and/or 2,6-ditertiary-butyl-phenol are sometimes added to petroleum insulating oil to improve its oxidation stability (ASTM D-2864).

3. Upgrading
   "Improvement of new oil by removal of trace contaminants." This treatment by means of vacuum and adsorbents is frequently practiced at the manufacturer's level for removal of water and trace contaminants to improve power factor, dielectric breakdown voltage, and interfacial tension over and above the values of the new oil as described in New Oil Specifications (ASTM D-3487).

4. Degasifying
   "Removal of dissolved air and gases." This treatment by means of high vacuum may or may not be coincidental with removal of water and other volatile liquids. It can be applied to processing new oil or reclaiming used oil.

5. Transformer Hot Oil Cleaning
   Hot naphthenic oil, once it has been thoroughly reclaimed and elevated in temperature to the oil's aniline point (160° F-180° F), acts as a solvent for its own decay products. Undoubtedly, the transformer oils in 99 percent of U.S. operating transformers have been refined from naphthenic-based crudes. In the near future, the trend will be toward a paraffinic base (whose aniline point is higher), as the supply of asphalt crudes becomes depleted.*

**In contrast, most non-U.S. transformer oils have been refined from paraffinic-based crudes.*

6. SLUDGPURG® Processing (Complete Desludging)
The unique process of removing sludge already deposited *on* and *in* transformer windings and insulation, etc., by multiple passes of hot naphthenic oil through the transformer until the sludge is redissolved and removed from the transformer.

7. Oil Quality Index Number*
The ratio of interfacial tension (IFT) to acid or neutralization number (NN), indicates a relative measure of reclamation required.

8. Adsorption
A process in which one substance attracts and holds tenaciously another substance to its surface area (e.g., a magnet attracting metal filings or some methods of oil reclamation).

9. Absorption
A process in which one substance is soaked in or up in another (e.g., the soaking of water in a sponge or some methods of reconditioning).

10. Aniline point
The temperature at which one substance (oil) dissolves aniline, an aromatic substance (ASTM D-611).

General Guidelines Concerning Oil Reclamation

Preventive maintenance is the best approach for any oil-filled equipment. By far the greatest benefit is the additional life gained for the equipment, transformers in particular. A secondary benefit is the extended life of the transformer oil.

Reclamation is at the middle of the spectrum, wherein the transformer oil has deteriorated to the point where only corrective maintenance can fully solve the transformer sludge problem.

In addition, between new oil and badly oxidized oil are *special* cases to be considered. For example:

1. An oil company had 250,000 gallons of new transformer oil with power factor (over 3.5 percent) brought about by contamination with motor oil during transfer from a tanker to the tank farm. The oil was returned to new oil specifications in the

**Oil Quality Index Number is also known as the Myers Index Number™. See Chapter 3, Part 2, p. 288.*

field using a mobile reclamation system at a considerable savings to the oil company. The alternative was to send the barges from the U.S. East Coast back to the Texas refinery.

2. Good oil from faulted equipment may contain an excessive content of combustible gas in solution, as well as carbon particles. This oil can be returned to new oil conditions with modern reclamation equipment containing a high-vacuum degasifier.

3. Badly oxidized oil, to be disposed of, will probably have to be evaluated in the new light of the higher cost of oil, better reclamation techniques, and government regulations regarding disposal. The criteria for discarding used transformer oil should be reviewed.

Economic Justification

The IEEE Transformer Committee asserts that the greatest benefit derived from widespread reclamation may be in the substitution of suitable reclaimed oil for large quantities of expensive new oil which are normally required in-service for maintenance or make up oil. In turn, this will free a considerable volume of new oil for use in new critical apparatus.

The cost of reconditioning or reclaiming oils by various users will vary significantly and is dependent upon the type of system, field inspection operations, and laboratory and shop facilities. In determining whether or not oil reconditioning and/or reclaiming are economically justifiable, IEEE suggests the following factors:

- Cost of materials
- Disposition of service-aged and/or contaminated materials
- Total cost of process versus quality of end product
- Equipment maintenance and amortization
- Cost of collection and storage of oil
- Labor and transportation costs
- Laboratory costs
- Cost availability of new oil versus cost of reprocessed oil
- Loss of oil during reprocessing
- Cost and availability of oxidation inhibitors and cost of blending process
- Value of service-aged oil when used for some other purpose

In addition, IEEE believes that no general guideline can determine whether reconditioning and/or reclaiming should be performed by company personnel or contractors. The final decision must be resolved on the basis of factors applicable in each particular case.

Precautions Concerning Sources of Oil to be Reclaimed

Care must be taken in the selection of oil for reclaiming so that it is not contaminated with any of the following:

1. Askarel or Polychlorinated Biphenyl (PCB)
   Oil containing askarel with PCBs should be handled in accordance with federal (U.S. EPA 40 CFR 761), state/provincial, or local regulations and should not be reclaimed, but be disposed of in an approved manner.

2. Silicone Fluids
   Oil containing traces of silicone fluid will foam excessively, and therefore should not be reclaimed.

3. Other fluids with properties not conforming to ANSI/ASTM standard method D-3487 should not be pooled.*

4. Suspended Carbon in Oil From Failed Transformers
   Oil containing significant amounts of suspended carbon should be generally processed separately.

Quality of Reclaimed Oil

Service-aged insulating oils generally contain contaminants as the result of oxidation, or from contact with structural or insulating materials of the apparatus.

> The reclamation process may not adequately remove these contaminants and may also remove some of the desirable natural components found in new oil. Furthermore, any adverse effects due to the contaminants may not be demonstrated by standard laboratory tests.

Therefore, IEEE recommends that reclaimed oil not be used in new equipment under warranty unless agreed to by manufacturer and user.† On the other hand:

**American National Standards Institute (ANSI).*

†At least two major manufacturers now have made such an agreement, (February 1981).

> Most significant...is the increasing awareness that reclamation does more than restore the oil to a "like new" condition. The reclamation process also has a rejuvenating effect on the transformer.

During *some types* of reclamation, the oil is continuously recirculated through the transformer, purging the apparatus of contaminants that could not be removed by oil replacement, or by any other means currently available to industry or utilities.

Confronting the Sludge Problem

Introduction

Many transformer experts realize the problem of sludge in operating units. This happens when the acids attack the iron, copper, varnishes, paints, etc., and these materials, coming into solution, combine together to form sludge. This sludge eventually precipitates out of solution and forms a heavy tarry substance which adheres to the insulation, the side walls of the tank, lodges in ventilating ducts, cooling fins, etc. Sludges are also formed in the cellulose fibers of the insulating system itself, even before the sludge precipitates onto the interior surface. These sludges, once deposited *on* the insulation and formed *in* the insulating system, immediately go to work to "leech" out of the insulation, varnishes, and cellulose materials, resulting in severe shrinkage of the insulation.

Since this sludge problem is one of the prime causes of premature transformer failure, it is imperative that a solution be found to confront the problem.

Seven Alternative Approaches

If the goal is to keep transformers operating in the SLUDGE FREE RANGE during the life of the transformer, then positive action is required early in the game. If however, the transformer is already in the sludging stage, there are seven approaches to the problem:

- Ignore the situation
- Replace with new oil
- Service shop flushing
- Recondition the oil
- Reclaiming the oil

- Hot Oil Cleaning of transformer
- Complete desluding of transformer

1. Ignore the Situation
   Procrastination is an answer, but not viable in today's economy. Historically, it was made because of ignorance of methods available. This decision is simply playing roulette where the only sure thing is a premature transformer failure!

2. Replace with New Oil
   Another approach is to simply drain out the deteriorated oil, flush the core and coils down with oil, and refill with new oil.

   Flushing down a dirty transformer reaches only 10 percent of the interior surface. Twenty-four hours after new oil is added, the contamination level will probably return to its initial state.

   Why is that? Because up to 10 percent of the volume of oil in a transformer is entrapped in the cellulose insulation. It is a known and stated fact that a "small amount of badly oxidized oil can ruin very large quantities of new oil." Putting new oil into an operating transformer that has not entered into the sludging stage may actually cause a sludge fallout! It remains then a calculated risk when topping off a transformer with new or used oil that the mechanism of sludge fallout will be one of the results.

   Changing oil does not remove the deposited sludge on the core and coils and therefore is not a solution to the problem. Disposal of the "changed out" oil is now also a problem!

3. Service Shop Flushing
   Detanking the transformer in a nearby service shop to flush it down with oil will remove the outer layers of sludge. Neither flushing nor hand cleaning can remove the sludges that have permeated the voids of the insulation. The best estimates state that only 25 percent of the surface of the transformer winding can be reached by this method. Thus, the service shop can only partially clean the insulation. Moreover, the new oil will not hold up long as the residual sludges will dissolve in the oil and trigger the oxidation process. Finally, as the result of shop service, subsequent insulation damage may result.

   a. Embrittled insulation due to age—badly deteriorated oil may be damaged beyond repair by mechanical disturbance during the moving process (Figure 7.23).

b. Cracking of insulation—removal of oil buoyancy. When oil is drained, the additional strain may cause breaking of the insulation so that the unit may be subject to failure shortly after being put back in service.

The first three approaches are not methods to even begin solving the sludge problem. Reclamation, by definition, involves the use of a process which results in *definite beneficial change* in the oil composition. Table 7.6 lists three basic oil reclamation practices and pinpoints the types of contaminants removed by each process.

Figure 7.23 - In older transformers, mechanical vibration embrittles the cellulosic insulation, possibly causing damage beyond repair. Thus, "a sick transformer may not only get sicker, it may also die!"

4. Reconditioning the Oil

a. Filter Press

Traditionally, the answer to the first known problem in oil maintenance—moisture—led to the development of a mechanical filter originally known as a filter press. It has been used for all oil problems, most of the time misapplied. The porous filter medium used consists of untreated Kraft or bleached paper made of cellulose and cotton fibers. Free water and, to a limited degree, some dissolved water and carbon are removed from oil by absorption, depending on the dryness of the filter paper and process temperature. Blotter presses and other similar devices will not remove air, but in fact, tend to aerate the oil.

Drying the paper in an oven is very awkward and cumbersome, particularly in "field" operations. If the paper is not dry, the moisture migrates from the paper into the oil, seeking an equilibrium which fully defeats the purpose of the filtering. If the oil is hot, the water will dissolve in the oil. Furthermore, when the processed oil contains much contamination, it is necessary to frequently change the filter medium.

An interesting question also arises. How do you know when the filter material is too wet to continue to use? The answer is—you don't! Even though means are now available that permit a continuous indication of the water content of the outgoing oil, this doesn't guarantee the end result the user expects to accomplish. When the filter medium has picked up enough moisture to increase its combined weight by three percent, the same filter medium begins to disintegrate and fibers are released to the oil. This could be a bad situation.

Even though a large percentage of utilities still use filter presses, some authorities do not recommend use of this method to remove acid or dissolved sludge because the results can be so misleading, and the process dangerous, if done on energized equipment. Nonetheless, the filter press can successfully remove carbon from circuit breaker oil.

The manufacturers of filter presses also recommend against using these units on energized equipment. In fact, they suggest you leave the transformer de-energized for 12 hours after completion of filter pressing to permit bubbles to escape from the windings.

b. Centrifuge

The centrifuge is a mechanical means for separating gross amounts of water and carbon. The device is generally used for bulk cleaning of contaminated oil and tends to aerate the oil. This process usually leaves a residue of one percent water.

TABLE 7.6

COMPARING OIL TREATMENT PRACTICES						
	Types of Contamination Removed					
Oil Reclamation Method	Solid	Free Water	Soluble Water	Air and Gas	Acids, Sludge, etc. Volatile	Other
Reconditioning						
Mechanical Filter (Blotter or Filter Press)	Yes	Yes	Partial	No	No	No
Centrifuge	Yes	Yes	No	No	No	No
Coalescing Filter	Yes	Yes	No	No	No	No
Precipitation/Settling	Yes	Yes	No	No	No	No
Vacuum Dehydration	No	Yes	Yes	Yes	Most	No
Reclaiming						
Contact Process	Yes	Yes	Yes	No	Yes	Yes
Percolation by Pressure	Yes	Yes	Partial	No	Yes	Yes
Percolation by Gravity	Yes	Yes	Partial	No	Yes	Yes
Activated Carbon-Silicate Process	Yes	No	No	No	Yes	Yes
Trisodium Phosphate	Yes	No	No	No	Yes	Yes
Oil Reclamation Systems						
Stationary	Yes	Yes	Yes	Yes	Yes	No
Mobile On-Site	Yes	Yes	Yes	Yes	Yes	Yes
Mini-Oil Reclaimer	Yes	Yes	Yes	Yes partial	Yes	Yes limited

Since this is completely unsatisfactory for electrical grade insulating oil, further processing would be required for use in transformers.

c. Coalescing Filter

A coalescing filter acts somewhat like a centrifuge but has no moving parts. Figure 7.24 shows a single element coalescing filter consisting of two stages: filter and separator. Oil with dispersed water enters the coalescing fiber glass element, which traps small water droplets. The resultant increasing pressure differential across the filter medium forces the droplets of water together, releasing water in the form of large droplets at the outer surface of the filter element. The water is retained, collected by gravity at the bottom of the filter, while the dry oil passes through the separator screen.

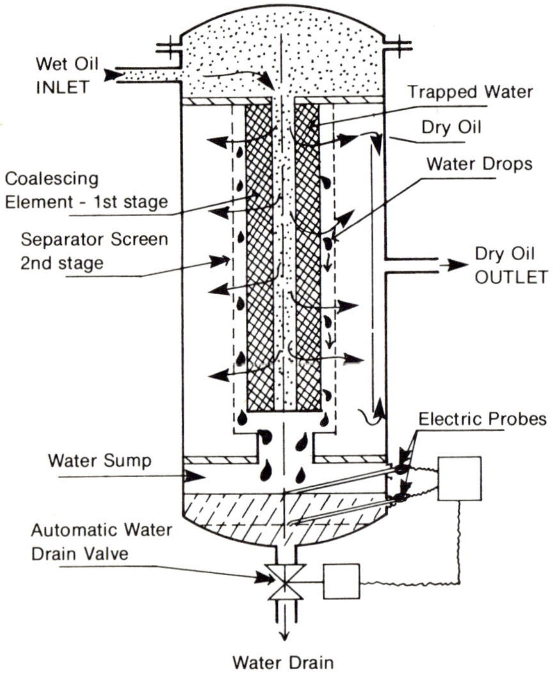

Figure 7.24 - Coalescing filter/water separator.

In contrast to centrifuging, this filter removes all free and dispersed air. Because the effluent oil contains less than 5 ppm free water, it has been used as a prefilter for vacuum or clay treatment to remove free water.

d. Precipitation Settling
Precipitation is a very useful and the least expensive way to remove free water and solid contamination heavier than oil from bulk oil storage. Over a period of time, water settles to the bottom of the tank and is drained periodically, while oil is decanted from the top. The process is temperature sensitive and the degree of water saturation may or may not be satisfactory for immediate use.

e. Vacuum Dehydration
The vacuum dehydrator is a highly efficient means of reducing the gas and moisture content of transformer grade insulating oil, as well as the few volatile acids present in the oil. Hence, by its name, the chief function is water removal and because such devices don't remove solid contamination, further consideration of dehydration is beyond the immediate context.*

Suffice it to say at this point that if an oil contains solid matter, the oil should pass through some kind of mechanical filtration before it is pumped into the vacuum dehydrator.

It is very evident none of the reconditioning devices mentioned thus far are designed to treat the oil chemically. Figure 5 illustrates this difference between reconditioning by the filter press and reclaiming by fuller's earth adsorption.

5. Reclaiming the Oil
The process of reclaiming is simply the passing of badly deteriorated transformer oil through an *adsorbent* bed such as fuller's earth to remove most oil oxidation decay products, polish the oil, and restore (with sufficient antioxidant added) to "like new" operating characteristics.

What is adsorption? Adsorption is the tendency of a liquid, a gas, or a small particle to cling to the surface of another substance by physical rather than chemical means.

Most of the contaminants in oil, including water, are polar in nature and are therefore easily adsorbed. Several types of materials are readily available. The most widely used is fuller's

**Chapter 7, Part 2.*

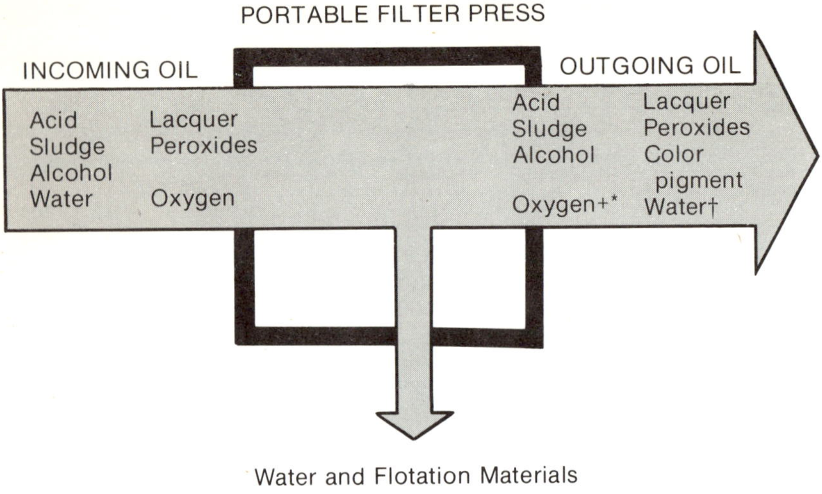

These decay products are the result of hydrogen and carbon combining with oxygen in the presence of copper.

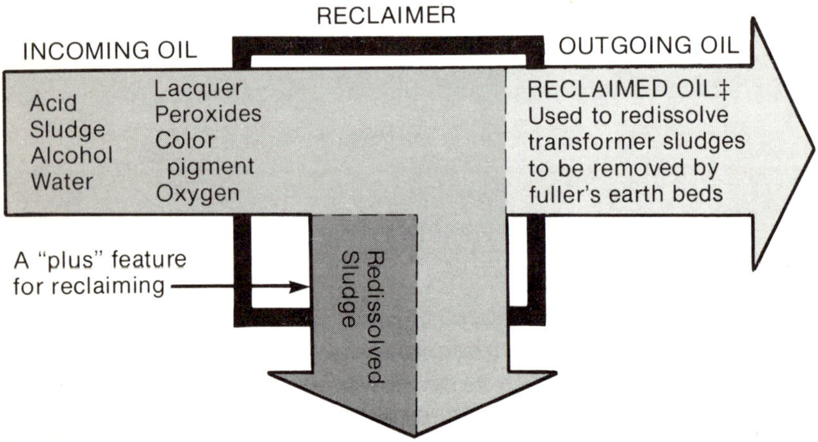

Figure 7.25 - Showing contrast between Reclaiming and Portable Filter Press for transformer oil treatment.

**A portable filter press adds fresh oxygen to the oil (undesirable).*

† Unless properly done water can be added to the oil.

‡It is impossible to remove all of the color pigmentation by Fuller's Earth Treatment alone. However, the vast improvement in color is remarkable to witness, and extremely impressive.

earth, while activated alumina and molecular sieves are much more expensive.

a. Materials

1) Fuller's Earth

The term fuller's earth refers to a class of naturally occurring adsorbent clays rather than to a specific mineralogical species. The main constituent in this class is attapulgite clay, mined principally in southern Georgia and northern Florida. Specifically, attapulgite clay has been found most satisfactory for purifying transformer oils because it has the abilities to adsorb polar compounds and decolorize to a clear oil.

> Fuller's Earth easily adsorbs impurities from oil and fat. Its most important use is as "bleaching clay" in the refining of mineral, vegetable and animal oils. Its color ranges from white to brown and it resembles potter's clay. It was originally used in a finely powdered form to remove grease from cloth and wool. It gets its name from this process, which is called fulling.

The Bible in Mark 9:3 also refers to the use of fuller's earth. The Apostle said, speaking of the appearance of Jesus' garments on the Mount of Transfiguration: "And his raiment became shining exceeding white as snow; so as no *fuller* on earth can whiten them."

Figure 7.26 represents a grain of fuller's earth greatly enlarged with contaminated oil passing by. The adsorption principle is pictured where the groups of molecules, identified by the letter "B", adhere to the surface of the fuller's earth particle, and then clean, or reclaimed, oil continues past the fuller's earth.

What makes attapulgite so unique is its crystalline structure. As mined, the clay is a hydrated magnesium aluminum silicate. During processing, the clay is crushed, heat activated, ground, screened, and bagged. The temperature of the heat activation or drying stage determines the degree of internal porosity. This porosity contributes to the clay's high surface area and hence adsorption capacity.

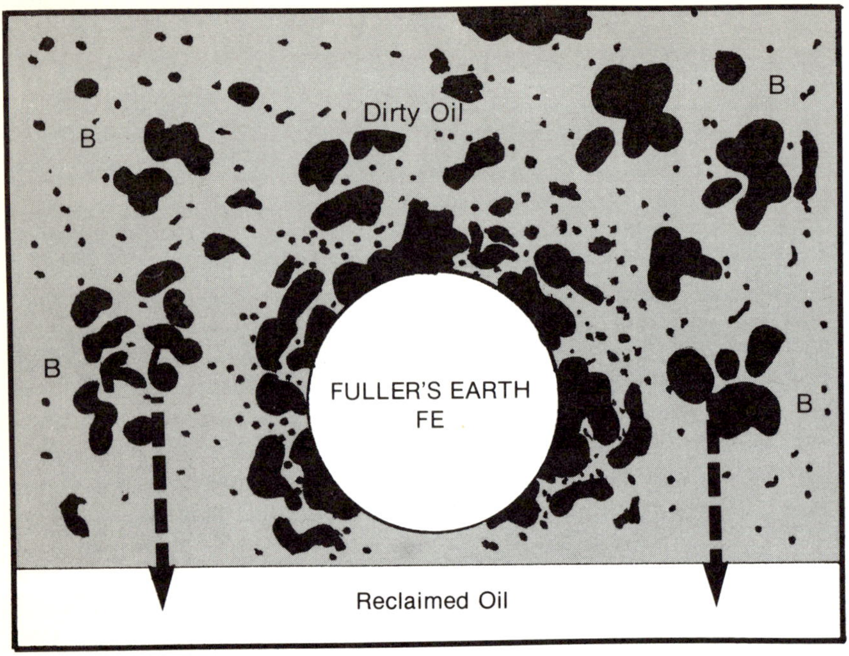

Figure 7.26 - Shows principle of ADSORPTION. FE is fuller's earth particle to which oxidized oil clings by mechanical processes.

The earth is activated by burning or calcifying at the temperatures of 800° to 1100°F. This treatment develops porosities in clay as high as 13 acres per pound of clay (125 m^2/gram).

Choice of clay particle size or mesh size is a function of the type of reclamation equipment in use. Thus, fuller's earth is marketed in various grades, representing screen meshes of approximate particle sizes: 15-30, 30-60, 50-80 in the granular products, and (100+ and 200+) in fine powder products.

Commercial granular products and fine products are sold as RVM (Regular Volatile Matter) with six percent free moisture or LVM (Low Volatile Matter) with one percent free moisture measured as percent of weight at 220°F.

The material can be purchased in bulk or in 50 pound bags with a vapor barrier. Care must be taken not to expose the fuller's earth to humid air or moisture, as it can pick up to 25 percent water by weight. Fuller's earth can be dried using elevated temperatures, particularly under high vacuum.

The grades used for the reclamation of insulating oils are LVM 30-60 mesh for filtering and percolation, and fine grade for the batch contact process. The specific gravity is approximately 2.45 and the bulk density is 34 pounds per cubic foot.

In order to maximize speed of adsorption, the smallest granular size should be used. In the 30-60 size, there are 30 holes/inch and every grain will pass through this screen, whereby at the upper unit of 60, none of the grains will pass through the screen in the 20 minute screen analysis. On the other hand, the larger the grain size, the slower the adsorptive capacity.

Activated attapulgite clay can adsorb large quantities of acid as shown in Figure 7.27. The amount of acid removed, correlated as the Neutralization Number, depends on many factors since adsorption is a dynamic equilibrium process. Temperature, flow rates, viscosity of oil, residence time, and initial level of acid, all affect the rate and capacity of adsorption. Hence, this chart may be used as a guideline for typical expected values only. It represents the amount of fuller's earth required under less than ideal or efficient methods utilized in modern mobile oil reclaimers.

2) Activated Alumina or Bauxite
Activated alumina consists essentially of aluminum oxide in porous form and is activated by thermal treatment alone. Although it has an initial high cost, it can be reused 20 to 25 times without damage; however, regeneration is an awkward procedure.

3) Molecular Sieves
These adsorbents belong to a class of materials known as zeolites. The zeolites are principally alumino-silicates with the unusual characteristic of being able to undergo dehydration with essentially no change in crystal structure. The type 4A is used with transformer oil; it can likewise be regenerated, but it is significantly more expensive than other adsorbents. Nonetheless, it is better used in the laboratory than on production work for large transformers.

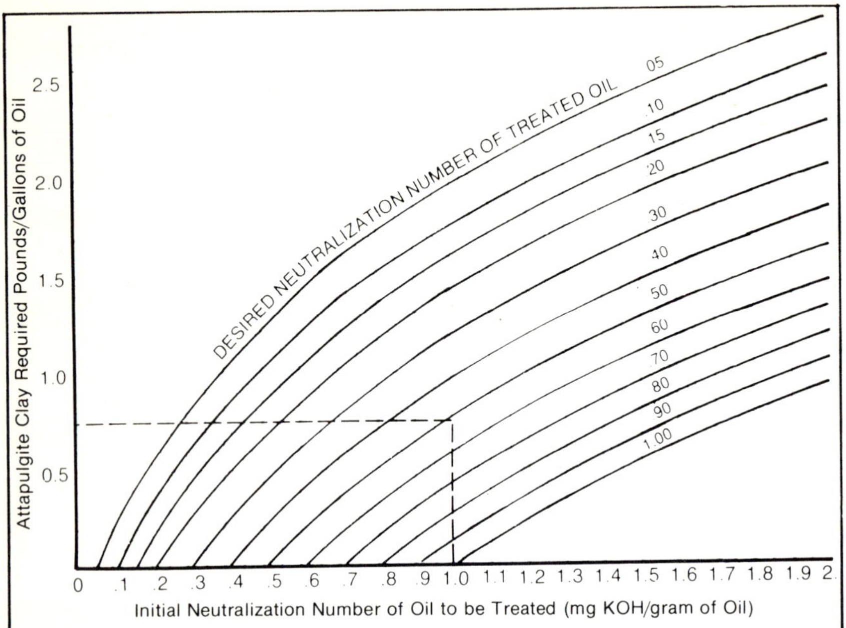

Example: Assume that a spent oil has a Neutralization Number of 1.0 mg KOH/grams of oil. In order to reduce the Neutralization Number to 0.5, approximately 3/4 pound of attapulgite clay will be required for each gallon of oil treated.

Figure 7.27 - Adsorption by attapulgite clay. (Courtesy of Engelhard Minerals and Chemical Corporation).

Reclaiming, using fuller's earth, is accomplished using one of two methods:

a) Contact at an elevated temperature with powdered clay.

b) Percolation through granular clay, using either pressure or gravity to force the oil through the clay.

b. Contact Process

Fine grades of fuller's earth are mixed with hot oil in a batch-type process. At the end of the contact period, the oil is settled and decanted through a filter press to separate the spent clay from the oil.

The contact method is relatively slow. When properly done, this method is most efficient but most costly. In addition, the contact process produces a uniform product and it is suitable for *small* quantities of oil. Other contact materials can also be used such as activated carbon (but costly) and trisodium phosphate.

c. Percolation by Pressure

In pressure percolation the oil to be reclaimed is forced through the adsorbent by a pump. Commercially available cartridge equipment has a chamber to hold a container such as a bag or cartridge filled with the adsorbent. The chamber is designed so that oil is admitted around the periphery of the adsorbent pack and must pass through the adsorbent before leaving the chamber.

These devices are capable of processing large volumes of oil in a relatively short time. Since the amount of adsorbent is relatively small with respect to the amount of oil, and the "dwell time" of oil to the adsorbent is short, frequent changes of the adsorbent are required.

An advantage of such machines is that they are portable and may be brought to the job and used directly on apparatus whose oil is to be reclaimed. Several types of pressure percolators are available. Fuller's earth is normally used as the adsorbent.

1) Bulk Filters

Each of the various types functions in a similar way. Large pressure tanks have fine mesh screen across the bottom and are filled with granular clay through an open top cover. In operation, the hot oil flows down through the clay slowly by pressure from an inlet oil pump (Figure 7.28).

Clay may be shoveled out through a side opening, or in some instances the entire pressure tank can be tilted for dumping the spent cake. Cost of operation of bulk filters is lower than other percolation methods.

2) Deep Bed Filtration

Today's trend is toward a series of two or three cylindrical columns packed with loose clay by means of vacuum or conveyors. Spent cake is discharged through a bottom opening cover (Figure 7.29).

(Deep bed filtration) assures *long contact time* of oil

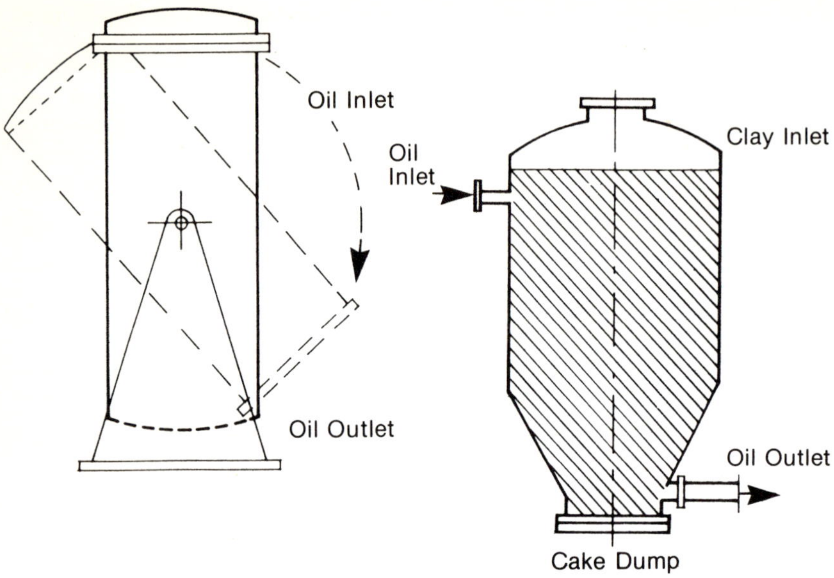

Figure 7.28 - Bulk filters using percolation processing.

flowing from top to bottom, minimizes channeling and provides the greatest improvement of oil conditions in a single pass. Two or three such columns or 'towers' piped in a series provide for a better utilization of clay. Only the first tower in a series is replaced with new clay and switched into the final tower position. In this manner, a continuous run with a consistent quality of effluent can be obtained.

The towers may contain from 500 to 3000 pounds of clay each. Even in mobile operation the tanks contain 200 to 1000 pounds each.

The spent cake is dried by pushing the oil slowly into storage with compressed air, or by pulling it with a pump suction. Fuller's earth retains from ½ to 1 pound of oil per pound of clay and this oil shrinkage must be taken into account.

The flow rate of oil through the clay towers is 10 to 20 gallons/minute in mobile installations and up to 50 gallons/minute in stationary plants. One of the largest installations in a utility is one consisting of three towers with 3000 pounds of clay each.

Most of the fuller's earth installations include electric or steam heaters to heat the oil to the optimum temperature of 140° to 160° F (60° to 70° C) to activate the clay. At lower temperatures, adsorption is less efficient.

3) Throwaway and Repackable Cartridges
Granular types of clay are packaged in throwaway canisters, holding 10 to 30 pounds of material, which are placed inside pressure filter tanks. The oil flow is from the outside in toward a center tube.

Throwaway cartridges and canisters are relatively expensive and can be justified only in emergencies when there is no alternative.

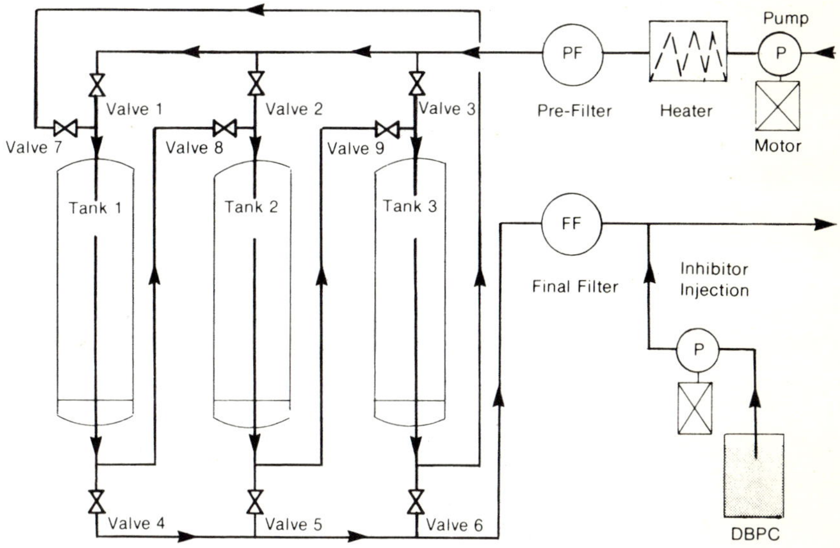

Figure 7.29 - Schematic of modern three clay tower system.

Repackable cartridges are usually of larger size, holding as much as 50 pounds or more of clay each. In most cases, this type is more economical than the throwaway cartridges, but 30 pounds of material will not process much oil.

d. Percolation by Gravity

Gravity percolation makes use of gravity (no pump) to force the oil through a column of adsorbent (Figure 7.30). By this method, the oil can be treated to any desired degree.

A slow flow rate results in long contact time with the filter medium permitting efficient use of the adsorbent. This process is used by major utilities at their service and repair facilities.

e. Activated Carbon-Sodium Silicate Process

Although today this method is not in use to any great degree, it is capable of variation to meet the requirement of different grades of service-aged oil. The oil is heated to 85° C and is maintained at this temperature until the final filtering operation is reached. The method consists of the following basic treatments:

1) An activated-carbon treatment in which one to two percent by weight of activated carbon is used (only for oils of high acidity).
2) A treatment involving 30 percent by volume of a two percent sodium silicate solution (followed by centrifuging).
3) A clay treatment in which one to two percent by weight of activated fuller's earth is used (followed by centrifuging).

The first step consists of a treatment by agitation with activated carbon in cases where the acid rating of the oil is 0.5 mg KOH/gram or over. This step is necessary to prevent subsequent emulsification of acid oil with the sodium silicate solution. Where the acid value is low, this process may be omitted. The oil is then run into a receiving tank and allowed to cool. Finally, it is filtered and run into storage tanks.

When *used without activated carbon*, the process is continuous with an output of 150 gallons per hour (570 liters per hour). When the activated-carbon treatment is necessary, the process becomes a batch process with an output of 500 gallons per day (1900 liters per day).

f. Reclamation by Trisodium Phosphate

The trisodium phosphate method consists in agitating a mixture of oil and trisodium phosphate solution maintained at

80° C for an hour and then allowing the mixture to separate. Most of the spent phosphate solution is drained from the tank, while the balance is washed from the oil with a water spray. The oil is then decanted through a centrifuge and a heater into another tank where 200 mesh (77 mesh per centimeter) clay is added and the mixture agitated. This agitation with clay is maintained for 15 minutes and the clay allowed to settle out overnight. The oil is again washed with hot water, decanted through a centrifuge, and then dehydrated by passage through a vacuum dehydrator or filter.

The process is economical and capable of yielding a uniform product by varying the amounts of reclaiming agents as determined by the analysis of the deteriorated oil. The method may be more economical than reclaiming with clay alone for large quantities of badly deteriorated oil, but phosphate disposal may be a problem. Only one known company uses this process, and only because the equipment, chemicals, and "in-plant know-how" related to their company's product makes it feasible.

6. Desludging of Transformers (Hot Oil Cleaning)
Reclaiming of transformer oil is energy efficient and less expensive today than refilling an older transformer with new oil. Nevertheless, the complete solution is not just simple reclaiming of transformer oil. The sludges must be dissolved and removed from the transformer itself.

It goes without saying that treating the oil by various mechanical methods of reconditioning doesn't even begin to solve a *chemical* sludge problem. Likewise, the shipping of equipment to a service shop fails to sufficiently deal with the central issue of sludges permeating the entire fabric of the insulating system.

a. The Sludge Problem Pinpointed
As Figure 7.21 has shown, multiple layers of sludge, deposited periodically over a long period of time and in various degrees of advanced oxidation, can be the problem. It is apparent that each of the foregoing solutions (including oil reclaiming) fails to deal with the primary issue of *soluble* sludge *in* and *on* the transformer cellulosic winding materials.

What is really needed to clean the insulation of sludges is a

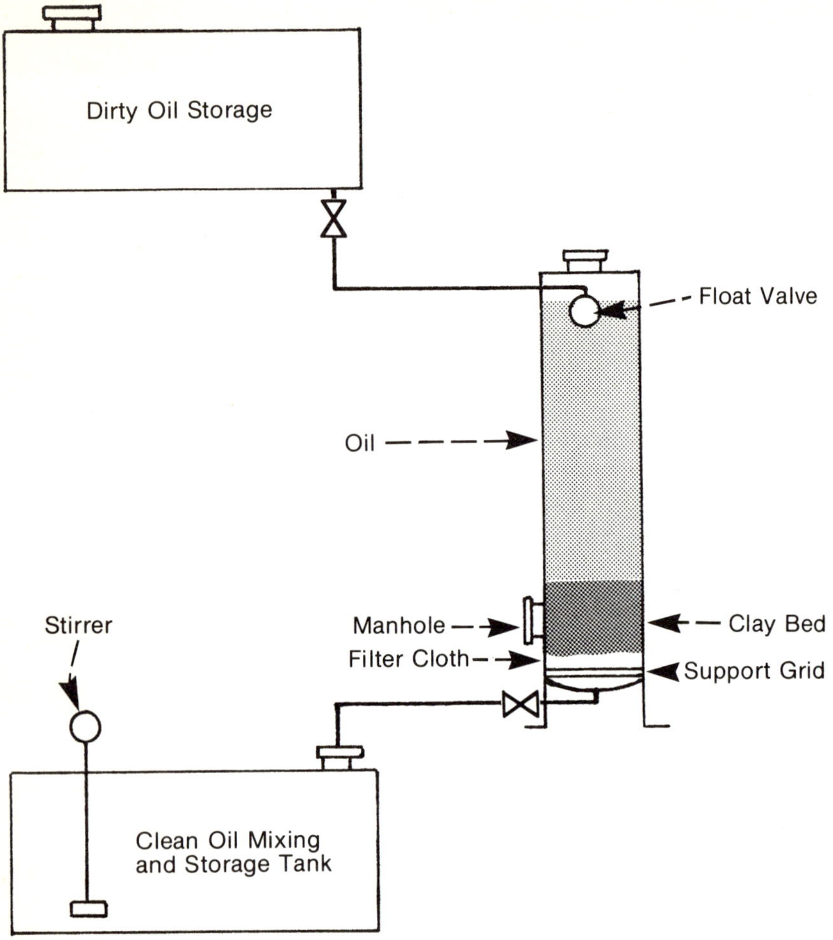

Figure 7.30 - Schematic diagram of gravity-percolation reclaiming apparatus.

solvent. Unfortunately, most solvents that will dissolve the sludges can also dissolve some of the winding materials and lower drastically the flash point of the unit. It is fortunate, however, that *hot* transformer oil is a solvent for its own decay products. The degree of heat necessary is indicated by the ASTM Aniline Point Test (D-611) which measures the temperature that the oil dissolves aniline, an aromatic substance. The degree of mutual solubility increases with temperature. Paraffinic hydrocarbons usually have the *least* solvency for aniline and therefore have a high aniline point.

Aromatics have the greatest solvency power and the lowest aniline point. Naphthenic oils are intermediate.

Transformer oil specifications list the aniline point at 160° to 180°F (72° to 82°C). Coincidentally, when heated to this temperature, transformer oil becomes an effective solvent for its own decay products.

Hot Oil Cleaning (as applied to transformers) combines the features of reclaiming and shop cleaning plus the solvency power of the reclaimed oil to attack the sludges within the transformer itself. This method thus consists of three steps—heat, adsorption, and vacuum (water removal and degassing). While reclaiming cleans the oil, Hot Oil Cleaning gradually restores the interior to a clean condition. Hot Oil Cleaning of a transformer thus is the only method that fulfills all *three* goals of a positive oil maintenance program.

- Operate transformers in the "sludge free range".
- Prevent premature transformer failure due to transformer oil oxidation products.
- Conserve two valuable resources—oil and transformers.

Hot transformer oil (75° to 85°C) is thus an excellent solvent for sludge. Carried sufficiently far, the Hot Oil Cleaning process becomes complete desludging using SLUDGPURG® procedures (option number 7).

b. Energized or De-energized Approach

Sludge can be removed by draining the old oil from the transformer, circulating new hot oil through the transformer, disposing properly of the contaminated oil, and filling the unit with new oil. With this archaic approach, sludge will be removed only up to the point at which the oil becomes saturated.

In a second approach, the reclamation equipment is arranged in a closed-loop system at the transformer. Transformer oil is heated as it flows from the transformer. The oil in turn is passed through a filter and/or centrifuge to remove gross contamination and water, and finally the oil is treated with fuller's earth and is returned through the transformer again. The number of recirculations (or passes) through the transformer (usually 6 to 20) depends on the amount of sludge. This amount is determined by oil screening tests.

Treatment continues until certain oil quality criteria are exceeded.

Using the second approach, transformers can be serviced either energized or de-energized. When de-energized, the transformer can be desludged without removing the oil.

Desludging an energized transformer requires no downtime, reduces labor costs, and eliminates transportation or rigging expenses. Nevertheless, most transformer manufacturers and a few large users of equipment are still hesitant to keep a transformer energized throughout the reclaiming process.

Yet the fact remains that experience over the years has proven that with proper equipment this can be both a safe and economical procedure.

Historically, a number of utilities have desludged energized transformers. A literature review shows these companies have been successful in working on energized transformers:*

- 1947 Narragansett Electric Company
- 1948 New York State Gas and Electric Company
- 1950 Central Maine Power Company
- 1952 Georgia Power Company
- 1953 Ten companies out of 53 report reclaiming oil in energized transformers
- 1955 Fifteen companies including Ohio Power Company
- 1965 Consumers Public Power District, Nebraska, reports work on 466 energized transformers in 3½ year period
- 1969 United States Bureau of Reclamation
- 1964-1980 Independent transformer oil reclamation contractors
- 1980 Niagara Mohawk Power Corporation

Properly designed and operated equipment can be used on energized transformers up through 230 KV.

c. Energized Transformer Hot Oil Cleaning
When oil testing reveals a neutralization number greater than

**Doble Engineering Company, Watertown, Mass., continues to publish periodic surveys of energized units so treated. A complete chronological bibliography is included in the IEEE Committee Report No. 637.*

0.15 mg KOH/gram of oil and an interfacial tension of less than 24 dynes/cm (Figure 7.22), Hot Oil Cleaning procedures are required.

While transformer oil can be reclaimed in just one pass of the oil through appropriately designed equipment, it takes up to ten recirculations of hot clean naphthenic* transformer oil to properly service a transformer with "Prop A" (Class 2) or Marginal (Class 3) oil (Figure 7.22).

The basic Hot Oil Cleaning process consists of a closed-loop mobile system (Figure 7.31) using connecting hoses between the energized transformer and uniquely designed semi-trailer reclaimer unit (Figure 7.32). This system uses a combination of continuous heat (150° F minimum), deep bed fuller's earth treatment, vacuum dehydration and fine final filtration.

d. How Sludge is Removed From Oil and Transformers

The entire system including hoses is filled with oil before the processing is initiated. Then, as the old oil is drawn from the bottom of the transformer, purified oil enters the top; therefore, an adequate oil supply always remains in the transformer.

The deteriorated oil is pumped into the Reclaimer. Here, the oil first enters the transformer oil heater which will raise the incoming oil temperature 100° F above ambient and to the oil's aniline point temperature. The outgoing oil will rise beyond the aniline point (160°-180° F) to 200° F. The system will maintain this 200° F within a few degrees throughout the processing of the oil and the desludging of the transformer.

How long it takes for the incoming oil to reach 180° F depends on the ambient temperature, the type of transformer, and its load.

After the transformer oil leaves the oil heater, it is pumped into a valve control center (Figure 7.33). On the first several recirculations or passes, the oil will go through two series-connected tanks of fuller's earth. Later, the oil can be redirected first to tank two as a new one is substituted for tank one, and then from the bottom of the second one back to the top of the refilled tank. Thus, the oil is passed through both

**Keep in mind that the oils used in transformers have been refined from naphthenic crude oil for the last 55 years.*

tanks in series to realize the maximum oil cleaning effort. The cleanest earth is always closest to the transformer. Pressure gauges for the entire system are located in this area so that the operator has complete knowledge of the equipment while it is in operation.

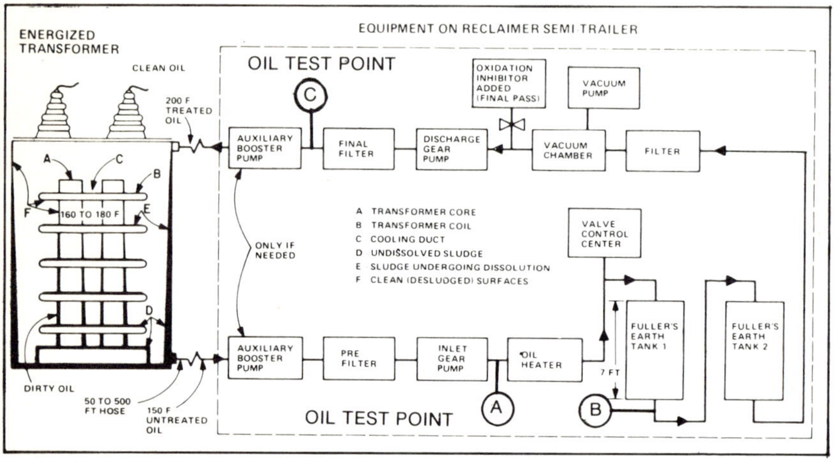

Figure 7.31 - Flow diagram for the energized desludging process. A pictorial side view of the transformer is shown. (Courtesy of Plant Engineering®).

Figure 7.32 - Transformer desludging semi-trailer reclaimer.

Figure 7.33 - Interior of semi-trailer reclaimer unit. Valve control center for alternating two fuller's earth tanks. This allows the cleanest earth to always be closest to the transformer.

Each of the two tanks holds 1000 pounds of fuller's earth, giving a total of 2000 pounds arranged so that the oil will pass through a 14 foot depth of bed of earth during the process. One pound of fuller's earth presents 13 acres of adsorption surface to the oil. It then follows that 2000 pounds of fuller's earth would present to the oil 1,118,000,000 square feet of adsorption surface to the oil being processed.

After leaving the fuller's earth beds, the oil is filtered through a special screen and then introduced into the vacuum chamber (Figure 7.34). Here the oil is held to the temperature of 200° F and vacuum from 25 to 29 inches of mercury is held continuously to completely dehydrate and to degasify. Temperature and vacuum must be closely controlled to prevent cracking of the hydrocarbons.

When the final degree of Hot Oil Cleaning of the transformer has been obtained, as determined by laboratory tests made on the Reclaimer (at testpoints A, B, and C of Figure 7.31), then oxidation inhibitor is added to the hot oil on the final pass to assure long life characteristics. It is added to restore the inhibitor lost by the aging process, contact with fuller's earth, and the effect of heat and vacuum. The inhibitor, purchased in

Figure 7.34 - Interior of semi-trailer reclaimer unit. In the foreground is the electric heater and in the background are matching interchangeable micron filtration units and a piggyback double-vacuum chamber that dehydrates and degases deteriorated insulating oil.

flake form and mixed to the proper concentration in a tank, is metered into the transformer through a solenoid valve. The tank is designed so that a one inch drop in oil level represents a known quantity of the concentrate and the level of inhibition, 0.3 percent of DBPC® per weight of oil, is accurately added to each unit serviced. If a different percentage of inhibitor is preferred, it is a very simple procedure to assure that this is accomplished. Keep in mind that today with the *proper* use of additives, reclaimed oil life may be extended to perhaps double that of a new oil. Care must be used not to inject large amounts of DBPC® in a localized area, as high concentrations of the inhibitor can lower the impulse strength of the oil.

The cycle of Hot Oil Cleaning is completed as the outgoing oil reenters the transformer at approximately 200°F.

7. Complete Desludging of Transformers (Sludgpurging)
When sludges begin to form in the solid cellulosic insulation (oil classes 4, 5, and 6), then the basic problem is with the transformer, not just the mineral oil (Figure 7.22). The problem is severe and extended Hot Oil Cleaning or complete desludging (SLUDGPURG®) procedures are required.

This procedure has two absolute requirements to accomplish the end result of complete desludging of the transformer. *First,* the aniline point of the transformer oil must be reached (160° - 180° F or 72° - 82° C) in the transformer and in the cellulose insulation in particular. *Second,* the transformer oil must be clean enough (40 dynes/cm and higher) to hold the redissolved sludges in solution to be stripped out in the fuller's earth filter beds.

It takes eight to ten recirculations (passes) before the temperature in the transformer windings and insulation even reaches the aniline point (160° - 180° F), where the oil begins to act as a solvent for its own decay products. At the aniline point, desludging actually begins. Badly sludged units require a sufficient contact time with the hot clean naphthenic oil to complete the process. The first stage of a complete desludging program requires 20 recirculations of hot oil through the transformer. Figure 7.35 illustrates the essentials for transformer desludging, pointing out the four criteria to be met.

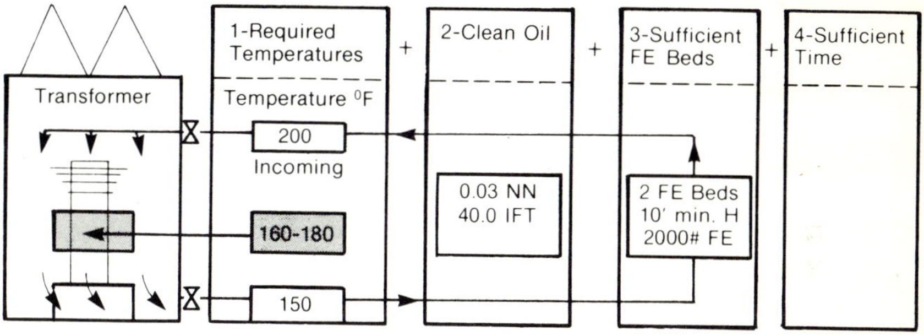

Figure 7.35 - No complete desludging of a transformer takes place unless all four criteria pictured are met.

Oil Reclamation Systems

Oil reclamation equipment incorporates one or more reconditioning/reclaiming methods combined into a portable, mobile, or stationary system, including pumps, heaters, and other necessary instruments.

A modern reclamation or purification system may consist of a settling tank, transfer pump, heater, prefilter or filter/water separator or both, clay towers, vacuum dehydrators, final filter, inhibitor injection system, and all necessary instrumentation and controls.

A complete permanent on-site oil reclamation system is illustrated by the flow diagram of Figure 7.36. These systems may run into high capacity installations—up to 300 gallons per minute for transformer manufacturing companies. Naturally, high capital costs are involved.

Such equipment could also be installed on a trailer unit so that it could be moved to the transformer site for oil processing—either reconditioning or reclaiming.

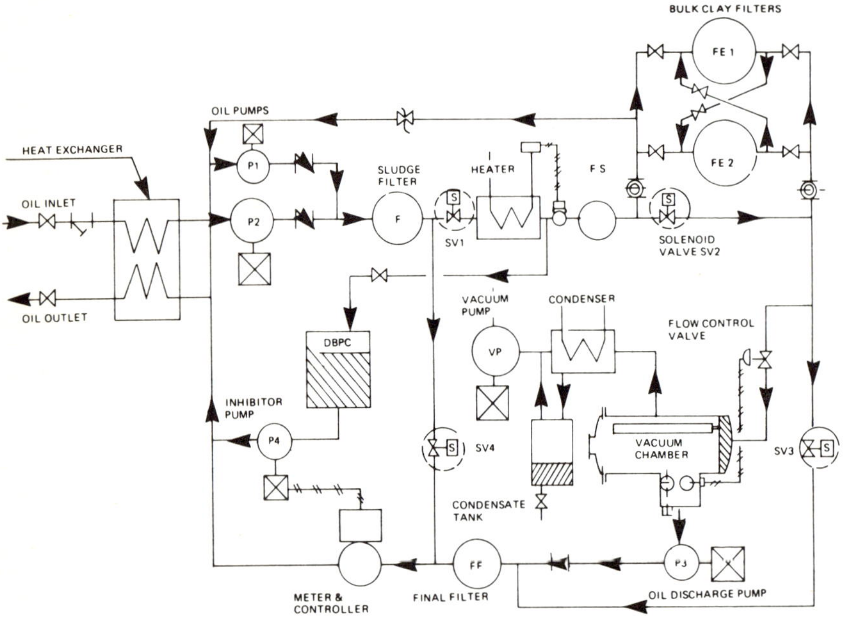

Figure 7.36 - Typical flow diagram for modern oil reclamation system.

Determining the Proper Location

1. Centralized Reclamation Site For Pooled Oil
   Utilities and large industries choose a location where large amounts of oil can be received and treated after it is collected. This method consists of a tower filled with an absorbent material such as fuller's earth, storage tanks for the contaminated pooled oil and for the clean reclaimed oil, and the necessary pumps and plumbing required to circulate the used oil through the system. The oil-clay *contact time* and the *depth*

of the clay bed need to be regulated to assure adequate upgrading of the oil. This type of reclamation can be done by either gravity or pressure percolation. Equipment is usually custom-built for the particular job.

2. Mobile Reclamation of Pooled Oil
   The method is similar to centralized reclamation, except that the oil-treating unit is mobile and can be transferred from one source of *pooled* oil to another. This system normally employs a pressure percolation method. Equipment for this method is commercially available.

3. On-Site Reclamation
   In on-site reclamation the oil is processed in the transformer. It is pumped from the bottom of a transformer, heated, and pumped through an adsorbent bed such as fuller's earth, filtered, degassed, and dehydrated before it is returned to the top of the unit. The process is continuous and is concluded when the oil meets prescribed test values. The processing equipment is commercially available.

Making the Final Choice of Reclamation Method

The choice of reclamation method that will prove the most practical and economical for a given system depends upon the geographical characteristics of the power system, the existing facilities available for application to such work, and the facts concerning the various types of reclaiming equipment and methods described in foregoing pages.

For example, on a system with large quantities of deteriorated oil or on a system where oil reconditioning has been done in the past at a central location, the gravity-percolation method of reclaiming has many advantages. It requires a minimum amount of new equipment, attention, and labor. On a system where the oils requiring attention are widely scattered in location, service outages are difficult to obtain and spare equipment is at a premium, some type of portable pressure percolator may be indicated for reclaiming in the field by recirculating the oil in the equipment.

Irrespective of the reclamation method, these four critical factors should be incorporated into the chosen oil reclamation equipment:

1. Dehydrate the oil before it contacts the adsorbent in order to prevent water from wetting the clay. Water will cause at least

partial and possibly complete blocking of the clay, thus making it necessary to discard that batch of clay.

2. Dehydrate the oil coming out of the clay treater (as clay can hold 10-13 percent moisture) to prevent water from being present in the finished product. This is particularly true when recirculating the oil in a transformer.

3. Filter oil through a ½ micron filter prior to its returning to the transformer, in order to remove any remaining clay, lint, fine ground iron, or other contaminant passed through the processing equipment (restore loss of oil dielectric strength and thus loss of dielectric strength of equipment).

4. Add an oxidation inhibitor to reclaimed oil in concentrations up to 0.3 percent by weight. Processing oil through fuller's earth at 70° C or higher will remove any oxidation inhibitors left in the oil prior to servicing. Extremely oxidized oil has also had its inhibitor consumed in the deteriorating process.

Conditions that have proven satisfactory for most inhibited mineral oil processing are found in Table 7.7. These are conditions which will not remove the inhibitor as the result of processing.

Needless to say, when energized reclamation is the choice, this procedure should be only done by experts in the field.

TABLE 7.7

ASTM RECOMMENDED PROCESSING CONDITIONS OF INHIBITED MINERAL OIL[1]

Temperature° C[2]	Minimum Pressure	
	Pa[3]	Torr Approximate
40	5	0.04
50	10	0.075
60	20	0.15
70	40	0.3
80	100	0.75
90	400	3.0
100	1000	7.5

[1]ASTM D-3487, Part 40.

[2]If higher temperatures are used, test the oil for inhibitor content and add as necessary.

[3]Pascals.

CAUTION: Attempts to dry apparatus containing appreciable amounts of free water may result in significant loss of inhibitor even at these recommended conditions.

On-Site Energized Complete Desludging of Transformers

Introduction

Several alternatives of solving the sludge problem have already been suggested. Some of these approaches were found to be useful in desludging transformer units. For example, service shop flushing can partially clean the insulation, while reconditioning the oil can help remove free water from transformer oil. Nonetheless, such alternatives fail to solve *completely* the problem of sludges. The answer therefore, lies in a complete desludging process. When mineral oil testing reveals a neutralization number greater than 0.30 mg KOH/gram and an interfacial tension of less than 17.9 dynes/cm, complete desluding procedures are required.

How the Complete Desludging Process Works

Figure 7.21 pictures the concept of layered sludge as seen in a cross section of a transformer coil. During the desludging process, layers 3, 4, and 5 can be effectively removed by hot naphthenic oil, redissolving the sludge and bringing it back into solution. This sludge brought back into solution is then returned to the reclaimer (Figure 7.31), where it contacts the fuller's earth beds and the sludges are removed.

The next layer of sludge (number 2), in some cases, is somewhat hardened by the continued application of heat over a long period of time. A resulting polymerization of the deposits renders greater resistance to solvents. If this is true, continued recirculation (over 20 passes) of hot clean transformer oil at the aniline temperature (160°-180° F) may not be an economical approach to the problem (more on this later).

The layer next to the coil (number 1) may have solidified and become chemically insoluble and is therefore a permanent part of the transformer. The old timers said, "That sludge is the only thing holding my transformer together." They may be right!

Consider Figure 7.37 as an illustration of what the desludging process is doing in a sludged transformer.

The sludge deposited on the interior of the transformer, in the presence of hot naphthenic oil, reverses chemically and goes from a solid into a liquid, reentering the oil. This oil is then processed through sufficient fuller's earth arranged in deep-bed configuration (up to 14 feet in depth), where the redissolved sludges are stripped out by adsorption and the clean hot oil is returned to the transformer. This hot naphthenic oil then penetrates into all parts of the transformer, into the sludge saturated insulating papers, the cooling radiators through the cooling ducts and in between the windings. This provides for a *continuous redissolving action* of the sludge deposits. The transformer is subjected to multiple passes of hot naphthenic oil throughout its entire structural system. This provides for sufficient time exposure of the hot oil to the sludges to effect removal.

Throughout the process almost perfect layering of the hot naphthenic oil occurs with the deteriorated oil because of differences of viscosity caused by temperature change.

In contrast, *on-site de-energized* Hot Oil Cleaning (150° - 160° F) fails to dissolve sludges in the innermost recesses of the insulating system. Moreover, the energized situation—the *mechanical vibration* of the entire transformer's structure under the influence of 60 Hertz power and *added heat* from copper losses and hysteresis—helps lower and keep the oil's viscosity (35 SSU) at the ideal point, where it will blend with the other oil into the innermost recesses of the transformer.

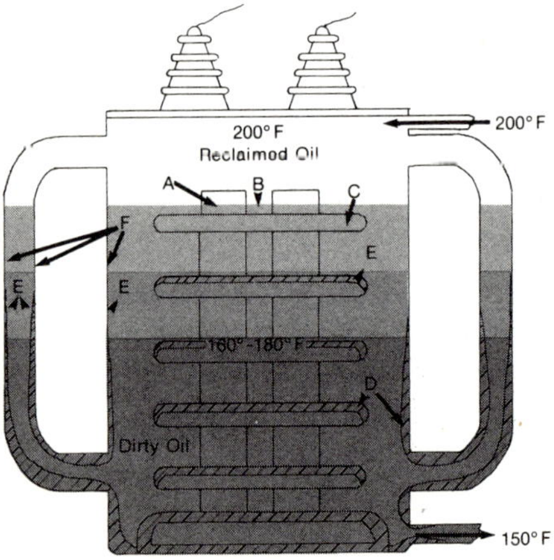

Figure 7.37 - Cross section of energized transformer undergoing sludgpurging. (A) Core, (B) Ventilating duct, (C) Coil, (D) Sludge to be removed, (E) Sludge being redissolved and brought into solution, (F) Surfaces already cleaned of sludge.

When Will a Desludged Transformer Require Additional Treatment?

Transformers are like people—they are all the same but they are all different. Therefore, precise scientific predictions of future behavior of a desludged transformer are not that simple.

Case histories obtained on eight typical units are shown in Figures 7.38 and 7.39. The bar graphs indicate that initial results were encouraging, while subsequent tests proved that the work was sufficient as the transformers were still operating at least in the "Prop A" category in 1979-1980. Even the worst cases (units number 4, number 6, and number 9) have responded to Hot Oil Cleaning after only one treatment and have for the past eight to ten years shown no need for further processing (Figures 7.40-number 4 and 7.41-number 9).

Experience has shown that any power transformer having a neutralization number greater than 0.30 mg KOH/gram and an interfacial tension of 18 dynes/cm or less (or oil quality index number below 60) is a prime candidate for the complete desludging program (20 recirculations of hot naphthenic oil) rather than a partial treatment of only ten passes. In such cases, oil decay products have permeated throughout the cellulosic fibers of the insulating materials of the transformer.

An ASTM report has stated:

> If one has allowed the transformer oil to go to the sludging point, it might be more economical to Rerefine only to 30 dynes and inhibit. Put this 30 dyne oil back in the transformer and allow it to dissolve the sludge left in the transformer. Then after a short period, as determined by tests, the oil is Rerefined again, assuming that all the sludge that is left in the transformer has been dissolved.

This then is the principle involved. While the initial desludging procedures (measured in hours) will remove large quantities of decay products, only a time exposure of months or years will effect the complete removal of residual sludges. Desludging procedures of 30 hours cannot correct fully a problem developed over a 30 year period. The longer reclamation has been put off, the more sludge accumulation dictates that more time is needed for the reclamation process. Therefore, as a guideline, monitor the transformer oil annually after completion of the first desludging stage, and re-service the unit when the acid number again reaches 0.15 mg KOH/gram and 24.0

dynes/cm. This may take from one to five years after the initial desludging, depending on the operating conditions (e.g., Unit number 7 in Figure 7.39).

Why are the numbers 0.15 mg and 24 dynes so important? For the simple reason that a NN of 0.15 mg KOH/gram and an IFT of 24 dynes/cm represent the maximum oil degradation beyond which the oil is unable to keep the sludges in solution. Furthermore, the oil itself at this point may actually begin to contribute more sludge to the unit!

Use Figure 7.42 to assist in answering the question, "When is it best for transformer servicing to be rescheduled?" As an example, complete desludging (20 passes) is selected for a transformer having oil specs of NN equal to 0.60 and IFT of 14 dynes; 3-5 years of "service free life" could then be guaranteed (Table 7.9).

In contrast, what could we expect from our example if the unit were serviced with insufficient treatment?

- If service used were only 15 passes (sludgpurging-SLP), the transformer may require complete desludging (20 passes) in nine months.
- If service used were only ten passes (Hot Oil Cleaning-HOC), the unit may require complete desludging in three to six months.
- If service used were only six passes (reclaiming-RE), complete desludging may be required in only one month!

Transformer oil test data on transformers that have been completely desludged using a SLUDGPURG® program shows that these procedures have effectively solved the problem of removing extensive sludges from the transformers. This data has been accumulated over the past 10 years on well over 3000 energized transformers.

The Limits of Energized Desludging

Some authorities still believe that oil with an acid number (NN) beyond the limit of 0.5 mg KOH/gram should be scrapped. But ongoing test records of a major servicing contractor confirm that transformer oils in much worse condition have been successfully desludged in transformers while energized, starting in 1964. The record shows that conserving such deteriorated oils is not just based on isolated cases. Consider a typical sampling of 1000 distribution and power transformers from ten different clients.

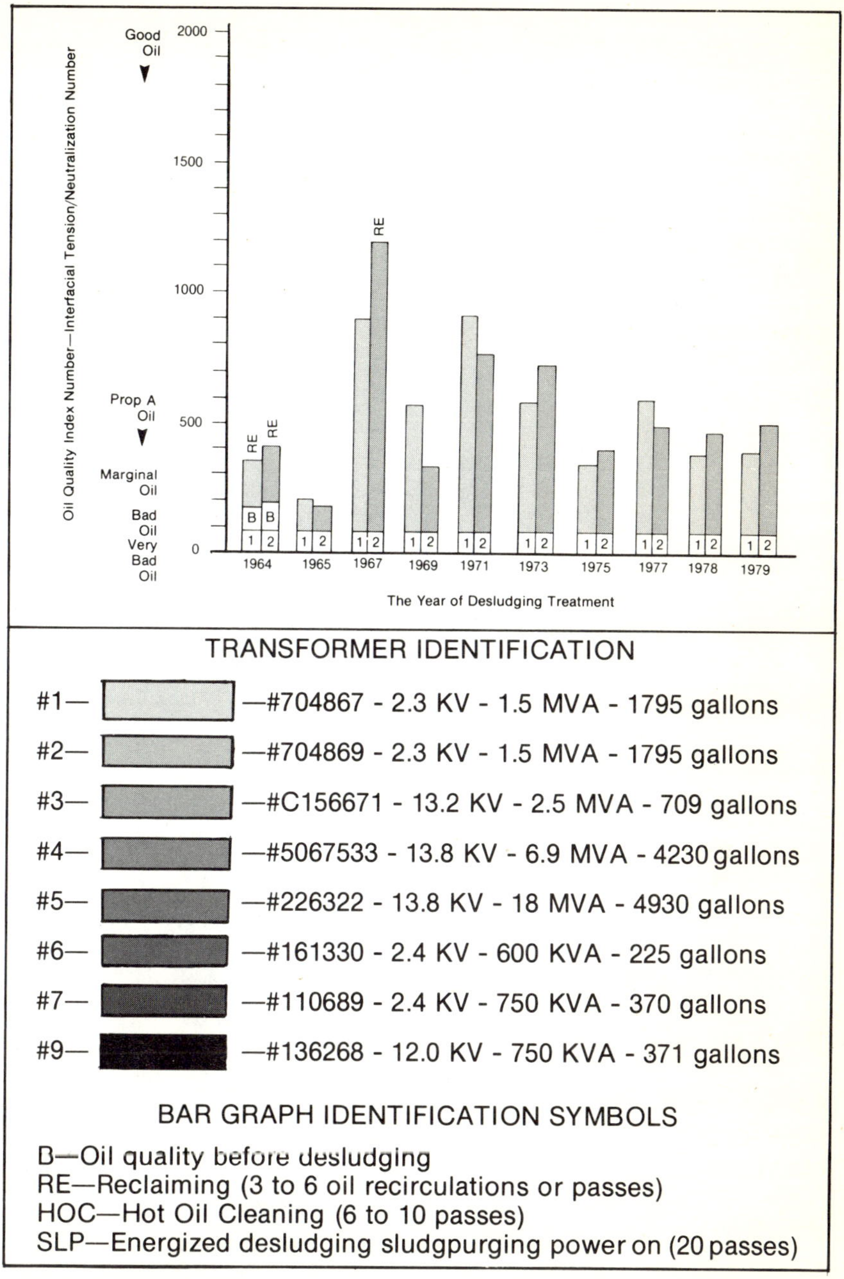

Figure 7.38 - The chronological history of typical desludged energized power transformers.

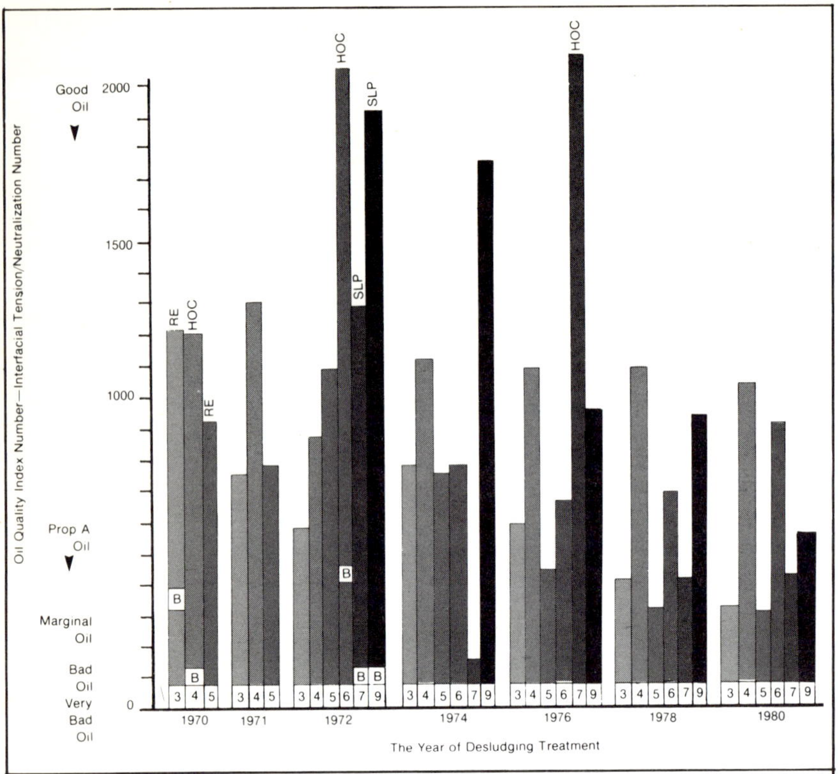

TRANSFORMER IDENTIFICATION

#1— —#704867 - 2.3 KV - 1.5 MVA - 1795 gallons
#2— —#704869 - 2.3 KV - 1.5 MVA - 1795 gallons
#3— —#C156671 - 13.2 KV - 2.5 MVA - 709 gallons
#4— —#5067533 - 13.8 KV - 6.9 MVA - 4230 gallons
#5— —#226322 - 13.8 KV - 18 MVA - 4930 gallons
#6— —#161330 - 2.4 KV - 600 KVA - 225 gallons
#7— —#110689 - 2.4 KV - 750 KVA - 370 gallons
#9— —#136268 - 12.0 KV - 750 KVA - 371 gallons

BAR GRAPH IDENTIFICATION SYMBOLS

B—Oil quality before desludging
RE—Reclaiming (3 to 6 oil recirculations or passes)
HOC—Hot Oil Cleaning (6 to 10 passes)
SLP—Energized desludging sludgpurging power on (20 passes)

Figure 7.39 - The chronological history of typical desludged energized power transformers.

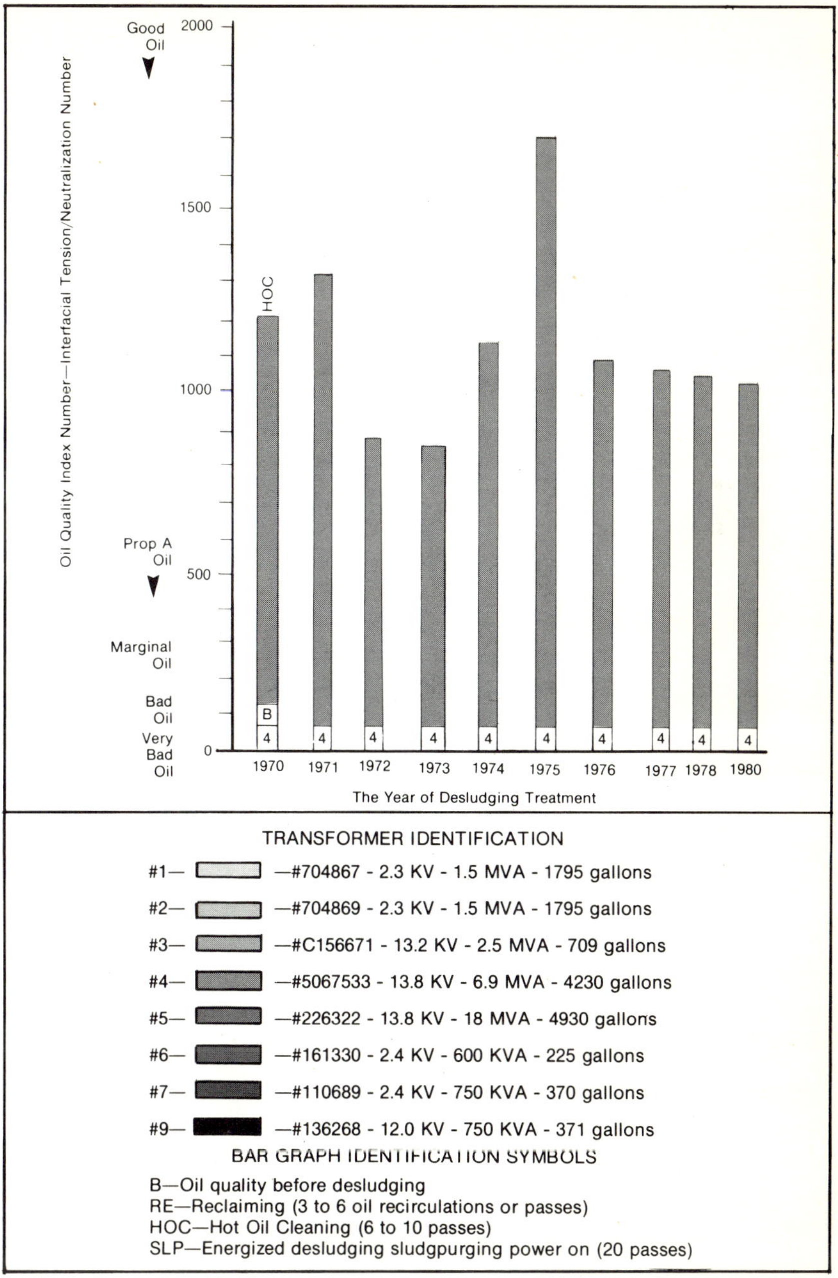

Figure 7.40 - The chronological history of a desludged energized power transformer (Unit number 4).

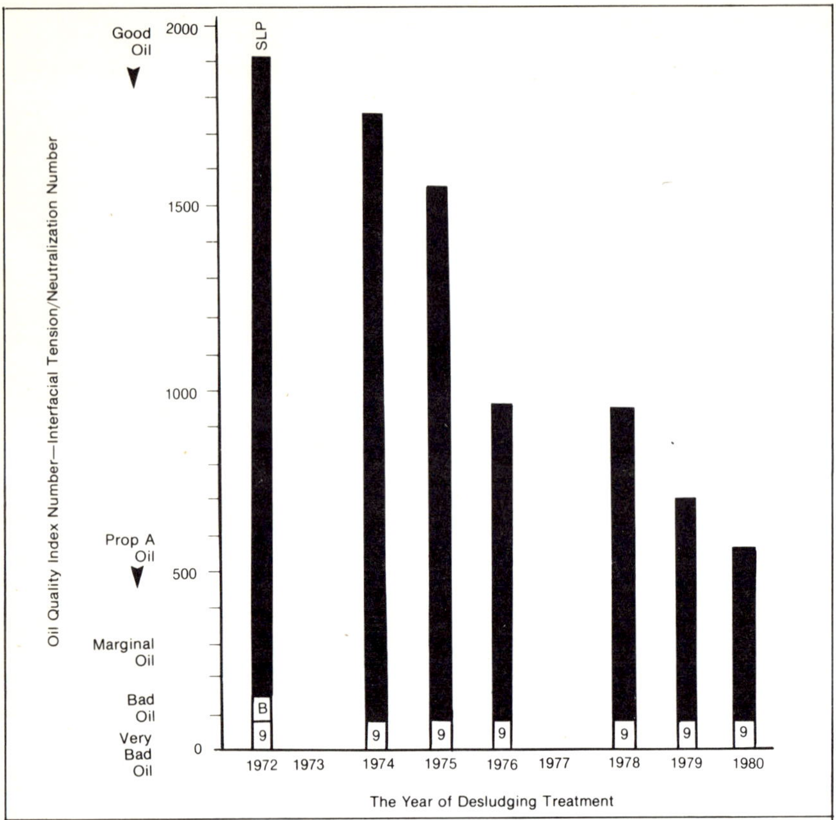

TRANSFORMER IDENTIFICATION

#1— —#704867 - 2.3 KV - 1.5 MVA - 1795 gallons

#2— —#704869 - 2.3 KV - 1.5 MVA - 1795 gallons

#3— —#C156671 - 13.2 KV - 2.5 MVA - 709 gallons

#4— —#5067533 - 13.8 KV - 6.9 MVA - 4230 gallons

#5— —#226322 - 13.8 KV - 18 MVA - 4930 gallons

#6— —#161330 - 2.4 KV - 600 KVA - 225 gallons

#7— —#110689 - 2.4 KV - 750 KVA - 370 gallons

#9— —#136268 - 12.0 KV - 750 KVA - 371 gallons

BAR GRAPH IDENTIFICATION SYMBOLS

B—Oil quality before desludging
RE—Reclaiming (3 to 6 oil recirculations or passes)
HOC—Hot Oil Cleaning (6 to 10 passes)
SLP—Energized desludging sludgpurging power on (20 passes)

Figure 7.41 - The chronological history of a desludged energized power transformer (Unit number 9).

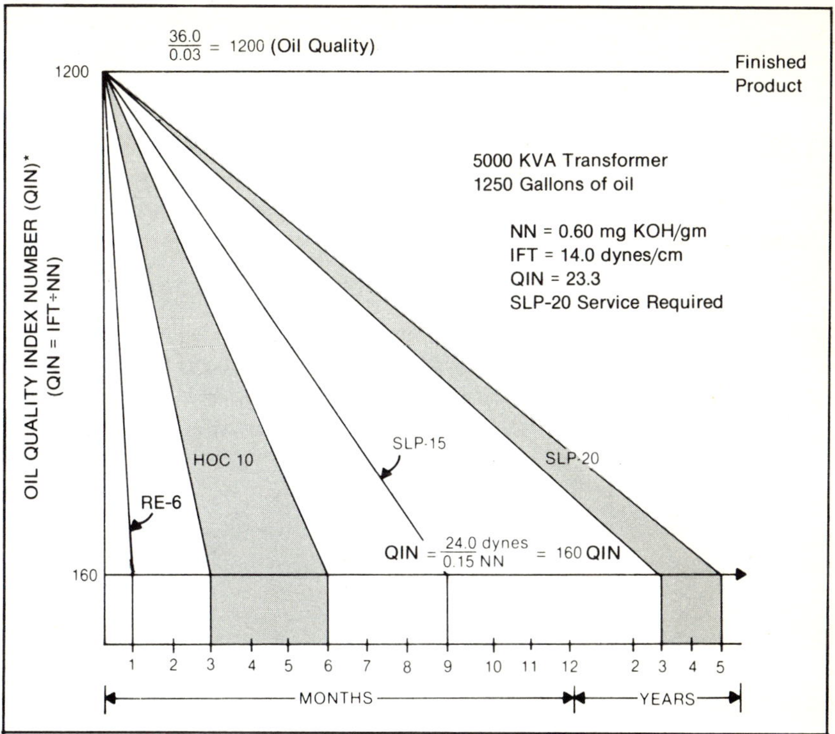

Reclamation Identification Symbols:

RE—Reclaiming (3 to 6 recirculations or passes)
HOC—Hot Oil Cleaning (6 to 10 passes)
SLP—Energized desludging or sludgpurging (15 to 20 passes)

*QIN is the same as Myers Index Number (MIN)

Figure 7.42 - Typical results obtained from various degrees of oil reclamation.

Data shows that 13 percent of the units (or 135), when initially tested (in 1971 and before), had oils with acid numbers greater than 0.5 mg KOH/gram. As the result of energized desludging, 95 percent of these severely sludged transformers (128) still remain in service and the oil is still in usuable condition. These units represent 858 transformer-years of operation. In addition, this desludging process has conserved over 25,000 gallons of valuable insulating oil (260 gallons/unit). This data certainly gives credibility to the higher standard of 0.85 mg KOH/gram published by the General Electric Company for the Electric Power Research Institute (EPRI). Furthermore, the experience of one major servicing contractor has shown that in the absence of heavy sediment, transformers having acid numbers as high as 1.50 mg KOH/gram can still be economically serviced.

Other Constraints of Energized Desludging

1. Evaluation of Oil Tests, Analysis, and Transformer Observations
   Every transformer is not a candidate for desluding while energized, and some transformers should not be desludged at all.

   When tests show that oil is in Class 7 ("Disaster City"), there is probably not much life left in the transformer! Energized desludging should not be attempted if the following conditions exist.

   a. Oil Screen Testing reveals:

      - Cloudy oil, indicating the possible presence of moisture.
      - Sediment containing excessive metallic content or bits of insulation.
      - ASTM D-877 dielectric test (flat disc cup) of 24 KV or below.

      Confirmation by a Karl Fischer moisture content of at least 50 ppm water indicates that the insulation is wet. In each case, first perform complete de-energized insulation tests, including power factor, turns ratio, and megohm resistance of all windings, prior to any servicing. These tests would determine the integrity of the insulation system.

   b. Oil Screen Testing reveals:
      ASTM D-877 dielectric test (flat disc cup) of less than 22 KV. Perform complete de-energized insulation tests and prepare to dehydrate the insulation system. The insulation is extremely wet.

   c. Gas-in-Oil Analysis by Gas Chromatography reveals:
      High concentration of combustible gases. Repair and de-energized degasification would first be required. Acetylene (C_2H_2) indicates arcing. Any transformer with arcing problems should not be treated at all using a mobile reclaimer. The unit should first be taken off the line, repaired, and degassed.

d. Observation:

- A terminal board (on the tap changer assembly) which has visible sludge should not be serviced while still on the line.
- A transformer over 7200 volts primary without filter press valves should be de-energized while its oil is being processed.

It is necessary under these conditions to introduce hoses through the manhole at the top of the transformer. This manhole normally is located between the high voltage bushings! Besides this obvious danger, other problems could be insufficient clearance (for the hose) between the windings and the case, and the possibility of free water in the unit with no way of detecting it.

2. Transformer Age
Fifty-year-old transformers have been successfully desludged while energized (Table 7.8, units number 1, number 2, and number 10). Therefore, it's difficult to make a generalization. An evaluation of any unit more than 35 years old having an acid number greater than 0.85 mg KOH/gram should be made. Consider the unit's critical value and its stress loading history to determine whether sufficient life is left to justify servicing the unit, even power off.

One contractor's experience has shown that when transformer oil has been ignored until NN exceeds 1.5 mg KOH/gram, the sludge will probably be so heavily accumulated in the unit that "Power On" desludging considered thus far would not be the answer. "Power Off" desludging procedures would be required. In one instance, a 333 KVA transformer with 185 gallons of oil yielded 45 gallons of sludge material. This transformer (unit number 8, Table 7.8) was a prime candidate for "Power Off" desluding procedures.

"Power Off" Desludging

In the "Power Off" treatment the problem is not the dirty oil but the vast accumulation of sludges. The transformer is drained of the oil (and the oil properly disposed of), its interior flushed with hot naphthenic oil to remove the accumulation of bulk sludges, and then refilled with new oil. An extended and controlled heat run is conducted to purge the transformer and its insulation of the decay products permeating the unit.

After this initial "Power Off" procedure, follow-up treatment will probably be required before all soluble sludge can be removed. The unit would then be a prime candidate for additional "Power On" desludging procedures. Annual or more frequent transformer oil testing is mandatory to monitor the transformer's response to treatment.

Recognizing That Sludge Problem Has Been Solved

The best evidence that energized desludging procedures are successful are simply the facts:

1. Early clients now only require annual oil testing and occasional polishing up of the oil with a light fuller's earth treatment.
2. Clients no longer have transformer failures at the rate they had before beginning a preventive maintenance program.

Table 7.8 shows with annual oil testing that even reclaimed oils have consistently remained in the "sludge-free range" year after year when they were initially serviced in either the "Prop A" or Marginal oil categories. One early client has reported that the desludging procedure on his transformers has resulted in a $200,000 savings over a 12-year period and that transformer failures during that time have been completely eliminated! More recently, a major manufacturing corporation saved 32 transformers valued at $1,000,000 using sludgpurging procedures; the company also received an additional benefit—continued coverage by its insurance carrier.

Many more companies report outstanding benefits from complete desludging. Typical comments have included:

1. "Transformers are running up to 20° F cooler after desludging."
2. "Before we used this type of service, we had three large transformer failures a year. Since instituting a regular maintenance program of oil testing and desludging of transformers using the combination of vacuum dehydrators and pressure percolators, we have not lost a transformer during the past seven years."
3. "Our testing on a large transformer, 5000 KVA, showed that the insulation power factor, megger readings, and other tests indicated the unit should be scrapped. Using a vacuum dehydra-

tion system and pressure percolator to desludge the transformer, we now expect to get 10 to 12 more years of operation from the unit. All of the insulation tests after the treatment were comparable to a new transformer."

TABLE 7.8

A TYPICAL RESULT OF TRANSFORMER OIL PROCESSING[1]

Unit	Rating / Date of Manufacture	Initial Oil Class[2]	Initial Treatment[3]	Year	Most Recent Treatment	Current Oil Class	Years Since Last Treatment
1	1.5 MVA/2.3 KV 1923	Bad	RE	1964	Not needed	Proposition A	15
2	1.5 MVA/2.3 KV 1923	Bad	RE	1964	1967-RE	Good	12
3	2.5 MVA/13.2 KVA 1954	Proposition A	RE	1970	Not needed	Proposition A	9
4	6.9 MVA/13.8 KV 1949	Very bad	HOC	1970	Not needed	Good	10
5	18 MVA/13.8 KV 1951	Bad	RE	1970	Not needed	Proposition A	10
6	600 KVA/2.4 KV 1961	Proposition A	HOC	1972	Not needed	Good	8
7	750 KVA/2.4 KV 1953	Very bad	SLP	1972	1976-HOC	Proposition A	4
8	333 KVA/12.0 KV 1942	Extremely bad	SLP-off	1971	1972-SLP 1976-SLP	Good (spare unit)	4
9	750 KVA/12.0 KV 1957	Bad	SLP	1972	Not needed	Good	8
10	10 MVA/65.0 KV 1925	Very bad	HOC	1975	1978-SLP	Proposition A	2

[1]Data taken from service records of ten units serviced by a major transformer servicing contractor for six clients.

[2]See oil classification chart, Figure 7.21.

[3]Servicing process identification.

RE—Simple oil reclaiming (3 to 6 recirculations of oil through reclaimer, Figure 7.31).
HOC—Hot Oil Cleaning (6 to 10 recirculations).
SLP—Energized desludging or sludgpurging (20 recirculations of hot oil).
SLP-off—De-energized desludging (oil with NN higher than 1.5 mg KOH/gram).

Figure 7.43 shows a case history of a 5000 KVA transformer, which was refilled with new oil in 1967, after the interfacial tension of the oil declined below 20 dynes/cm and the acidity increased above 0.4 mg KOH/gram of oil. New oil did *not* help, as after one year's operation the interfacial tension dropped below 25 dynes/cm and acidity increased slightly over 0.5 mg KOH/gram.* Desludging removed a bad actor from the oil by adsorption in 1969. The residual sludges were removed by multiple recirculations of hot naphthenic oil to dissolve the deterioration products and remove them. The transformer settled down to a nice operational level of interfacial tension equal to 30 dynes/cm and neutralization number equal to 0.05 mg KOH/gram for the next eight years, as confirmed by annual screening tests.

The Reliability of Energized Transformer Desludging

Fifteen years of field experience have proven the prediction by Frank Doble in 1952 that "Power On" servicing of transformers would be safely accomplished if the proper equipment is used. Figure 7.38 shows two typical examples of power transformers that have been desludged while energized successfully since 1964. These are not isolated examples, for the record shows that more than 35,000 energized units have been successfully serviced by one service company without incident when certain safety procedures are followed.

Safety Precautions

Safety precautions begin with oil testing and analysis to determine if there are any unusual conditions in the transformer that might make it inadvisable to desludge with the unit energized.

Basic safety precautions for personnel protection must include the following:

- Staying clear of energized conductors.
- Adequate grounding of the processing equipment to station ground.
- Using wire-braid reinforced high-pressure hose for all connections between the transformer and the mobile processing equipment (static discharge).

**Observers have witnessed this many times. We believe this to be true in this case also. If the client would have tested the oil 30 days after the new oil refill, he would have found that the IFT would have dropped to at least 25 dynes/cm and the acid number advanced to over 0.10 mg KOH/gram. Figure 7.43 should show this rapid degradation of the oil in a 30 day period, not 1 year. See dotted lines.*

- Providing the circuits serving booster pumps with ground-fault circuit interrupters (GFCIs) to protect personnel from shocks. (Booster pumps are required only when the reclaimer unit and the transformer are widely separated).

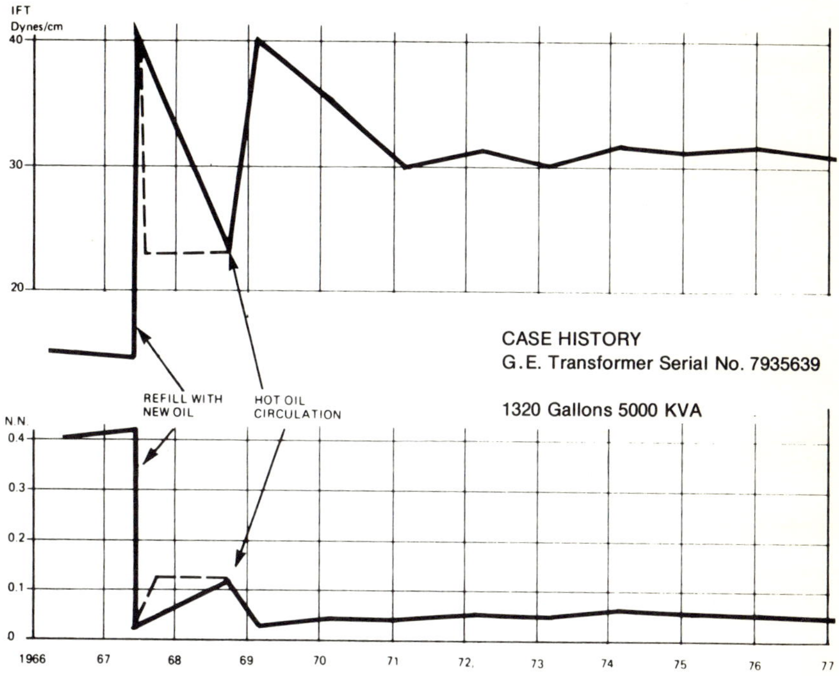

Figure 7.43 - Contrasting results of replacing sludged oil with new oil and use of hot oil processing.

Energized Processing Innovations

The following particular safety features should be built into the processing equipment system and their operation monitored by operating personnel:

- Liquid level detectors and controls to ensure that current-carrying internal transformer components will not be exposed to the atmosphere. These controls should be designed to maintain a constant oil level in both the transformer and the process equipment.

- Automatic heat regulation to maintain the oil at the aniline point and ensure that the temperature will not rise high enough to scorch the oil.
- Alarm devices to signal when pressure, temperature, and oil levels deviate from prescribed limits.
- Filters at the discharges of all pumps to catch metal cuttings that might be released from the pump impellers and gears.
- Design that ensures that vacuum is drawn only in the process equipment and not in the transformer.
- Instrumentation for monitoring flow rates to provide an evaluation of the fuller's earth filter efficiency. A log of flow rates should be maintained throughout the desludging process and oil must be sampled and tested to determine its condition.

These tests should be conducted using the latest ASTM laboratory test methods.

Energized Reclamation—Objections Answered

Some transformer authorities assert there is no advantage to energized desludging especially when you have zero risk with de-energized work. However, of the 92 companies responding to the 1977 EPRI Milestone Number 10, 73 recondition oil in transformers and 17 regularly service energized transformers.

Three risks are commonly cited as reasons for not servicing an energized transformer if it can be removed from service:

1. Negative transformer tank pressure (that is, less than atmospheric). If negative pressure could be drawn on the transformer, it could cause collapse of the tank or cooling fins. However, this is an invalid concern if the process equipment draws vacuum only in the process equipment itself. The transformer must be vented to the atmosphere throughout the process to permit it to function, thus eliminating this objection.

2. Development of air bubbles in the windings or in the reclaiming process, possibly causing air locks in the pumps and flashover in the transformer. A properly designed system, however, uses double protection against this possibility in the form of both an air trap and vacuum degasification coupled with a controlled flow rate.*

**The transformer oil returned to the transformer is "gas hungry" and will readily absorb gases, not produce gases or cause bubbling. An invalid complaint!*

3. Effect of stirred-up contamination in the transformer on the oil's dielectric breakdown strength. This concern is unjustified because of three fundamental features of the desludging process:

   a. There is absolutely *no forced* circulation in the process of servicing an energized transformer.

   b. Oil is removed from the bottom of the tank and returned at the top, and the desludging is done gradually over a period of time.

   c. There is a clear-cut interface (the process equipment) that eliminates the recirculation of contaminants in the transformer tank.

   Once contaminants in suspension have been drawn off at the bottom of the transformer tank they are gone forever. Only clean, dehydrated, degasified, particulate-free oil is returned to the tank. Desludging is a process of gradual dissolution of sludges over a period of time, with contaminants going into solution as they are liberated from the inner surfaces of the transformer. Therefore, there is absolutely no agitation of particulate matter during processing of energized transformers. There is only a continuous polishing of the oil in the reclaimer and the redissolving of soluble sludges—oil decay products.*

Concerning the Reclaiming Contract

Once oil screening tests have indicated a clear need for servicing and a price has been agreed upon, a formal *on-site* reclaiming contract is entered into with the servicing contractor (Figure 7.44). This contract defines the responsibility of both the contractor and the client, and spells out the cooperation expected of the client and the services (and results) that will be provided by the contractor. If there are any conflicts between policies of the contractor and the client, adjustments are

**In an actual experiment involving a nine gallon energized transformer (using an oil pumping rate of 600 gallons per hour), the only visible evidence on the oil surface was a very slight standing ripple effect that was set up by the pulsing of the gear pump, transmitted through the oil in the hoses. On the other hand, the interior of the transformer became visible as the oil brightened, and then the sludges were seen to redissolve and bright metal parts appeared as the process continued.*

made to resolve the conflicts before the contract is signed. A typical agreement will call for the client to extend the following cooperation to the contractor:

1. Information the Servicing Contractor Needs

   a. Current test results of transformer oil—a representative oil sample of current accurate ASTM test results of the equipment covered in the contract are needed to evaluate the oil and the subject transformer condition. The contractor reserves the right to confirm these test results with his own oil tests. Most contractors will accept their own test results for up to six months.

   b. Liaison man—the customer must assign a company contact man who assists in the location and setup of the equipment; orients their personnel in the plant concerning safety precautions, and points out other in-plant requirements.

   c. Electrical power—if the customer cannot provide the electrical power, the contractor needs to make provisions beforehand to provide for a mobile power generator set.

   d. Material storage location—a clean and dry location where reclaiming materials—fuller's earth, transformer oil, and oxidation inhibitor—may be stored prior to the arrival of personnel and equipment (particularly on large jobs).

   e. Access and availability, working hours—agreement provides prompt access to the transformer(s) to be serviced during the entire period of reclamation. If known beforehand, additional power cable (beyond 500 feet), additional hose and/or pipe, and extra oil pumps will be used if necessary to reach an "inaccessible" unit. Likewise, the contractor's right to plan a work schedule that may extend to 15-17 hours per day.

2. Responsibilities of the Contractor
   The contractor will provide all of the required equipment, supplies, and trained operating personnel, and will connect and operate the reclaiming system (Figure 7.45).

   a. Warm-up—at the beginning of each work day, the oil in the reclaimer is circulated throughout the system to assure proper dehydration of the reclaimer prior to connecting to the transformer. This can take up to two hours.

Figure 7.44 - ***On-site*** *energized desludging of substation transformers has been performed for more than 30 years. Only recently has the electrical industry formally recognized that successful operation is being done; and done safely, resulting in many benefits—less downtime, lower labor costs and no transportation or rigging costs.*

Figure 7.45 - The practice of ***on-site*** *desludging of energized transformers requries utilizing specially designed fail-safe equipment and trained personnel. Reclamation is therefore not only an economical and safe procedure, but it results in the conservation of two valuable resources—good (naphthenic) mineral oil and transformers.*

b. Operation—the desludging process will continue on each transformer until contract conditions have been met and the oxidation inhibitor (DBPC)*, added at the specific rate of 0.3 percent by weight, has been added to the newly-processed oil.

The finished results guaranteed by one major contractor are summarized in Table 7.9.

TABLE 7.9

GUARANTEED TRANSFORMER OIL SPECIFICATIONS[1]

ASTM Oil Screening Tests	Time Following Processing		
	A Immediately	B 30 Days	C 3 or 5 Years[3]
Acid Number (NN) D-974	0.03 mg KOH/gm maximum	0.05 mg KOH/gm maximum	0.15 mg KOH/gm maximum
Interfacial Tension (IFT) D-971	35.0 dynes / cm minimum	30.0 dynes / cm minimum	24.0 dynes / cm minimum
Dielectric Strength[2] D-877	35.0 KV minimun	—	—

[1]These test results will be obtained by conducting the current and updated ASTM laboratory test procedures. These guaranteed oil test results cannot exceed those oil specs furnished by the original oil refiner. Transformers having less than 100 gallons of oil—no guarantee normally offered.

[2]Under care, custody, and control of customer. Leaks in the tank which are not the result of natural aging or the reclamation process can change this.

[3]Depending upon original condition of oil.

The oil condition will be guaranteed for an extended period, ranging from 30 days to 5 years (Table 7.9). The contractor should re-service the transformers that do not meet the guarantee at no additional charge to the customer. For units containing more than 100 gallons of oil, this guarantee depends on the initial oil quality, the type of insulating material used, and the cleanliness of the manufacturing facility.

**DBPC is a registered trademark of the Koppers Company, Pittsburgh, Pa.*

An agreement contains these additional provisions, if the guarantee is to be valid:*

- The insulating oil must be tested on an annual basis by the same contractor.
- The transformer is not to exceed a normal operating top oil temperature of 60° C.

Special Liability Provisions

At times, the client's own oil tests may indicate a need for additional work under the warranty provisions of the contract, and these test results may conflict with those of the contractor. In such cases, the usual procedure is to agree to have a third party conduct tests. The cost will be borne by the contractor if the referee's test confirms for the client; otherwise, the oil testing by both parties would be charged to the customer.

As stated earlier, not every transformer should be treated energized. Therefore, the contract should also include a provision that the customer recognizes his responsibility in permitting the transformer to deteriorate to the "Corrective Maintenance" level; that the customer also acknowledges premature failure may occur as the result of an accumulation of contributing factors, such as age, overloading, extensive deterioration products and so on, which cannot be the fault of the contractors, or the energized desludging system.

One major transformer maintenance contractor makes available reclamation insurance in the client's name for *each* new purchase order, regardless of how much work is performed at a particular time. These matters are handled in a routine way, and in over 15 years of operation in all 48 contiguous States of America, there has been only a small amount of difficulty in satisfying the insurance department of the industrial complex to be served.

The Bonus of the Hot Oil Cleaning System

In speaking to the issue of the oil oxidation and its primary decay products, little has been said about moisture in the transformer or the

**Certain manufacturers during the past decade have used different types of insulating materials and these have played havoc with IFT. The cleanliness of some manufacturing facilities is also questionable, with the same effect on the IFT.*

combination of moisture with the other decay products. Yet complete transformer maintenance should deal with both chemical and electrical stability.

The following field case history illustrates this objective and the versatility of the Hot Oil Cleaning method to meet both criteria.

An industrial transformer rated 7.4 MVA experienced a water cooling coil rupture, releasing approximately 200 gallons of cooling water into the transformer's paper insulation system. Electrical test results as tabulated in Table 7.10 (Test Number 1) indicated high power factor and low insulation resistance. The logical choice was made to dehydrate the transformer oil and winding insulation with a Keene Bowser vacuum dehydration system. The Keene unit has a 150 cfm vacuum pump and 1500 cfm blower.

Dehydration began on December 11, 1976. Insulation resistance readings were taken periodically so that a determination could be made when the unit was dry. It was decided after three days to perform the same tests on a similar unit and use the results as a comparison (Test Number 2).

In analyzing the test results of both units, it could be seen that no unusual results were obtained in unit number 1. Therefore, further dehydration appeared to be the solution to the high power factor and low resistance readings.

After approximately 25 days of dehydrating the insulation, resistance still had not reached an acceptable level. During the same period, customer personnel had noticed that the oil had been gradually turning darker and that sludge was starting to appear in the samples being drawn. On January 4, 1977, electrical tests (Test Number 3) were made. The resistance measurement was still unacceptable, even though the power factor was approaching a good value.

Only heat and vacuum had been used thus far. It was then concluded that the reason for the bad readings was due to the high sludge content. Since the Keene unit is designed primarily to dehydrate windings, an additional process was necessary to rid the unit of sludges and decay products. On January 5, 1977, a mobile hot oil reclaimer was contracted to perform the complete desludging or SLUDGPURG® procedure.

On January 10, 1977, after 20 recirculations of hot naphthenic oil through the closed system, the final tests were performed (Test Number 4). The final results showed the oil was brightly polished, the

TABLE 7.10

TABULATED OIL AND ELECTRICAL TEST DATA (DEHYDRATION OF A MEDIUM POWER TRANSFORMER)						
UNIT IDENTIFICATION TYPE OF TRANSFORMER—RECTIFIER UNIT NUMBER 1 SERIAL NUMBER—9631 KVA RATING—7413 GALLONS OF OIL—1800						
Date	Power Factor at 20° C	Polarization Index (P.I.)		Insulation Resistance		Insulating Factor
		Pri	Sec	Pri	Sec	
Winding Test No. 1 12/11/76	45%	1.14	—	7.5 to 10 K ohms	—	Bad
Oil Tests Before Processing 12/11/76	Dielectric strength 20 KV Acid number 0.12 mg KOH/gm Interfacial tension 23.6 dynes/cm Oil quality index number 197 (marginal oil class)					
Winding[1] Test No. 2 12/14/76	—	—	—	—	—	—
Winding Test No. 3 1/4/77	4.3%	1.14	1.19	18.4 to 25 M ohms	16.5 to 23.6 M ohm	Deteriorating
Winding Test No. 4 1/10/77	0.37%	1.27	1.12	1659 to 2607 M ohms	2370 to 2800 M ohms	Good
Oil Tests After Processing (20 recirculations)	Dielectric strength 45 KV Acid number 0.04 mg KOH/gm Interfacial tension 36.4 dynes/cm Oil quality index number 910 (good oil class) Moisture content 29.9 ppm (parts per million)					

[1]Comparative test of rectifier unit number 8.

insulation was restored to "near-new" conditions, and the unit was clean and dry (a reading of 29.9 parts per million water was obtained). The transformer was then put back into service.

Follow-up tests were recommended in order to monitor the unit's operation including oil, combustible gas-in-oil analysis by gas chromatography, moisture ppm test, and dielectric strength.* The combination of dehydration and sludgpurging illustrates the point that transformer life can be extended when the services of oil reclamation are properly utilized.

Conclusions

ASTM and IEEE (circa 1980) are even now investigating how to apply ASTM tests, currently only applicable to new oils, to used and reclaimed oils, and to establish proper guidelines for the reclamation and use of reclaimed insulating oil.

Fifteen years of experience have established an in-depth data base on more than 3000 desludged transformers. This data base confirms that using the proper equipment and procedures, complete desludging of energized transformers is economical and a safe procedure that prolongs transformer life. It also results in preserving two diminishing assets—good transformer oils, as well as transformers.

To reiterate—transformer mineral oil itself can be reclaimed in *one* pass through a fuller's earth bed of proper size and utilization. Therefore, the additional passes of hot naphthenic oil required by Hot Oil Cleaning and SLUDGPURG® procedures are to redissolve decay products in and on transformer insulation, core and coils—to scrub out the vital parts of the unit.

The number of recirculations through the system required to accomplish the cleaning of a transformer depends on four factors:

- The equipment used
- The amount of fuller's earth
- The condition of the oil
- The age of the transformer

It is recommended that a minimum of 1500 pounds of earth be used and 200° F oil enter the transformer at the top filter press valve. To assure that sludge does not precipitate out of the oil onto the core and coil of a transformer and to effectively preserve the life of the unit, Hot Oil Cleaning should occur before a neutralization number of 0.15 mg KOH/gram and an interfacial tension of 24 dynes/cm are reached (an

**Chapter 4.*

oil quality index number above 160). However, when test results fall in ranges as in Table 7.11, the appropriate recommendation is made.

TABLE 7.11

SUGGESTED RECLAMATION GUIDELINES			
NN	IFT	Bad Range (45-159) Oil Quality Index Number*	Number of Passes Hot Oil Cleaning
0.19 mg KOH/gm	24.0 dynes/cm	127	6x
0.20-0.29	18.0 dynes/cm	62-90	10x
0.30 and above	17.9 dynes/cm below	Below 60	20x

**Oil quality index number is the same numeric as Myers Index Number.™*

Part 4—Use of a Transformer Cooling System

Introduction

An increasing practice of loading transformers beyond their nameplate rating has come about for several reasons. First, it is justified for economic reasons — it appears to be cheaper to overload than to buy a new higher capacity transformer. Many utilities, co-ops, and manufacturers have difficulty in raising money for new capital investment. Second, those who overload transformers do not understand the damage that results from overload and excessive heat. The result has been the abnormal sacrifice of transformer life through overheating due to overloading.* This is, without question, the most expensive solution available to solve the problem.

It is not surprising then, that IEEE Power Engineering Society (PES) reports that utility distribution and industrial power transformer overheating is a fact of life — and is increasing! Therefore, any worthwhile preventive maintenance program must solve this problem now — and solve it economically!

The Overheating Problem

Cause of the Problem

Transformer overheating is caused by both transformer manufacturers and transformer owner/operators.

1. Transformer manufacturers have for several decades designed transformers aimed at obtaining the maximum KVA per pound of material utilized. Transformers thus run hotter.†

   Since it is now recognized that transformers, particularly newer designs, are stressed more and therefore run hotter from overheating, it is imperative that preventive measures be taken to avoid their early retirement.

2. Many transformer owner/operators have often neglected their transformer units, or have seriously abused them. This

Chapter 1, Part 2, p. 26; Chapter 2, Part 4, pp. 204-210.

†Chapter 1, Part 3, p. 123.

has resulted in their running hot due to sludged oils, overloads, dark colored exterior paints, closed radiator valves, clogged water-cooling coils (either leaking or turned off), and malfunctioning temperature gauges.

Table 7.12 shows, for example, how transformers are operated by North American utilities; note especially the average top oil temperatures of transformers in-service.

TABLE 7.12

TOP OIL TEMPERATURE OF TRANSFORMERS IN-SERVICE[1]

Temperature °C	Transformers				
	Pole Type	Network	Substation	Transmission Line	Generator Station
High	150	140	150	150	150
Low	60	60	60	60	60
Average	105	98	95	95	94
Number of Responses	61	48	83	78	80

[1]EPRI Utility Survey in 1977.

Effect of Heat on Transformers

Overheating is extremely detrimental to the transformer as it affects both the oil and paper insulation system. The copper and iron are relatively unaffected.

1. Cellulosic Paper
   Insulating materials shrink and become more brittle. Certain secondary effects are also important, such as the production of "gases" and "free water" during the decomposition of the pressboard and paper. If the "free water" remains where it is generated, it further accelerates the degradation process. If the "gases" liberated during decomposition are unable to escape from the windings, bubbles may collect in regions of

high electrical stress, displace oil, and give rise to premature failure. Therefore, since a transformer is less able to withstand short circuits, impulse, and switching surges, transformer overloading should be limited to a hot spot temperature not exceeding 140°C.*

2. Mineral Oil

   Recall the accepted rule for cellulosic paper as the "8°C rule" while the transformer mineral oil operates on the "10°C rule.† Also, the transformer owner/operator must contend with two critical temperatures: 105° or 110°C for solid insulation; 60°C for transformer oil.

   Maximum useful transformer oil life can be expected if the top oil temperature does not exceed 60°C. The expected useful oil life in an energized transformer may be 20 years before the acid number reaches the critical point of 0.25 mg KOH/gram. Beyond this acid number, the aging rate changes from a linear to an exponential function.‡ The useful life of the oil is cut in half for each 10°C increase of temperature beyond 60°C, all other factors remaining constant. Table 7.13 shows the expected time period at various top oil operating temperatures to reach the critical acid number. While ANSI/IEEE has established criteria for the paper, there is no equivalent guideline for the insulating oil.§ Therefore, we recommend that the transformer owner/operator keep this important truth in mind: for maximum oil and paper life 60°C is the maximum top oil temperature permissible. When the temperature exceeds 60°C, immediate steps should be taken to correct or alleviate this problem. An ASTM publication reports that at 70°C even a nitrogen blanket transformer has sufficient oxygen available through breakdown of the cellulosic insulation to cause the **oil** to deteriorate as rapidly as the liquid in a free-breathing transformer!**

**Chapter 2, Part 4 (Loading Guides for Oil-Immersed Transformers), p. 207.*

†Chapter 2, Part 4 (Insulation Aging Laws), p. 200.

‡Chapter 6, Part 1, Figure 6.2, p. 438.

§Chapter 2, Part 4, (Loading Guides ...), p. 209.

***ASTM Symposium on Insulating Oils, Special Technical Publication #218 (1957).*

TABLE 7.13

ESTIMATED TIME FOR TRANSFORMER OIL TO REACH NEUTRALIZATION NUMBER OF 0.25 mg KOH/g[1]	
Operating Temperature	Transformer Oil Life
60° C	20 years
70° C	10 years
80° C	6 years
90° C	2 1/2 years
100° C	1 1/4 years
110° C	7 months

[1]Critical acid number 0.25 mg KOH/g of oil.

3. Electrical Losses
   Conductor I^2R losses and transformer core losses increase with increased temperature. This is wasted energy in the form of heat.

4. Polymeric Wire Coatings
   For some types of polymeric coatings conductor temperatures of 120°C and higher can produce significant dielectric losses.

Economic Considerations

Historically, overloading transformers in-service, thereby accepting the penalty of short insulation life, has been more economical then to relieve the transformers by purchasing larger or additional units. Today, as simply a matter of dollars and cents, the economics of overloading transformers is a subject of research and study. The higher cost of energy and the losses sustained by transformers in overload conditions might dictate "no overloads." This is especially true with utilities handling huge blocks of power. With the cost of electric power escalating by leaps and bounds, and looking at the cost of the overload losses for a 20-year period, the economics may mandate purchase of larger transformers and operating them "underloaded" rather than overloaded. By contrast, for industry, the choice is not as clear cut as for utilities and "overloading" is a way of life. Upon the purchase of any future transformers, it would be advantageous to question the manufacturer as to load losses under overloads in order to make a power calculation for a 25-year period, and then to convert the number to dollars. It will surprise you. It might even make the decision for you.

Studies now suggest avoiding overloading transformers more than the minimum peak load level. To achieve an ideal situation, all transformers in a system should be loaded up to 100 percent of nameplate on the basis of their self-cooled rating.

Examining the Problem

Transformer Insulation System

Transformer insulation systems normally age as a function of temperature. However, elevated temperature accelerates this aging process to the extent that we can conclude that "approximately 90 percent of cellulose deterioration is thermal in origin." Therefore, it is obvious that overheating results in deterioration, and if nothing is done to impede it, the ultimate destruction of the insulating paper will come much sooner than is expected or warranted.

1. Composition of a Typical System
   For the last five decades, transformer manufacturers have relied on oil and various types of paper such as cellulose, tapes, sheets, etc. to insulate transformers. Oil-immersed cellulose is still the principal insulating system utilized by the transformer manufacturer.*

2. Inherent Nature of the Cellulosic Molecule.
   It is the presence of secondary hydroxyls groups in the cellulosic molecule that constitutes the seeds of destruction that lead to limited life of the paper.†

Although oil-impregnated cellulosic paper provides the best available transformer insulation, its dielectric strength can be drastically

Chapter 1, Part 2, (The Insulation System).

†*Chapter 1, Part 2, Figure 1.45, p. 61.*

reduced when subjected to excessive heat. The glucose units deteriorate, resulting in the formation of water (H_2O) and carbon dioxide (CO_2), depending on the temperature, time exposed to temperature, and initial dryness of the transformer. If the transformer winding is not extremely dry, the decomposed cellulosic products act as a catalyst, promoting and accelerating further degredation of the cellulose, as evidenced by increased brittleness or carbonization. Such cellulosic decomposition may also affect other components of the insulation system, giving rise to premature failure of transformers.

Thus, normal aging of insulation takes place within an "in-service" transformer, resulting in an expected loss of transformer life calculated at one percent per year or four percent in any one emergency. Normal loss of life is estimated to two percent per year. Excessive heat accelerates the aging process by reducing the mechanical and dielectric strength of the oil-immersed cellulosic paper. Therefore, the life of the transformer **is** the life of the paper!

A Closer Look at NEMA TR-98

Transformer manufacturers rightfully boast of their products as being capable of operating at high temperatures. Therefore, much emphasis is put upon their technology in manufacturing units capable of operating with 65°C average winding rise above ambient temperature (resulting in a 95°C top oil temperature) without detriment to the insulating system. However, the underlying assumption is that the solid insulation will operate in a clean and dry liquid. This means the transformer oil must remain in the "very good" range throughout the unit's life in order to obtain maximum life for the cellulose. Thus, the life of the transformer consists of the life of both oil and paper!

The loss of one-half, or 50 percent of the initial tensile strength of the cellulose insulation is now considered to be "end-of-life." NEMA TR-98 (1971) Loading Guide provides the best approved and accepted transformer life assessment currently available.* Figure 2.28 shows this relationship between a transformer's hot spot operating temperature and the total in-service usage of the unit in hours or years.†

**The ANSI/IEEE proposed Loading Guide (Project No. 507, May, 1979) will likely follow NEMA TR-98. It should be published during 1981.*

†*Chapter 2, Part 4, p. 205.*

Transformer Loading Criteria

The Manufacturer's Design Criteria

Experience has shown that transformer manufacturers and users hold different criteria "parameters" concerning the same product. The manufacturer's design criteria are based on the assumption that their products are capable of operating at high temperatures continuously. Technically, this is not correct because there is no insulating material (neither oil nor cellulose paper) that will remain in a "very good" state when subjected to excessive heat.

Heat transfer (measured in kilowatts KW) is only one function of the oil and since 1965, by design, the oil has been worked to the extreme limits. Whether older or new designs, maintenance of the oil is essential to keep the cellulose in a good environment. Heat is bad enough. Since heat and oil decay products such as acid, sludge, alcohol, water and peroxide gases work together, the problems are compounded. The transformer boiled in oil is one thing. For it to be boiled in a high acid and sludged oil is something else. The newer units are stressed more and run hotter under the same load conditions. Notice in Table 7.14 the difference in transformer design criteria that illustrates the problem.

User's Operating Criteria

A proposed ANSI/IEEE Loading Guide suggests that a 65°C average winding rise power transformer operating continuously in a 30°C ambient at its 100 percent design rating (110°C hottest spot temperature) should have a normal life expectancy of 65,000 hours (or 7½ years). That's bad news. There is absolutely no reason for anyone to be satisfied with something as vital as a transformer with only a 7½ year life expectancy! Please note (Table 7.15) that by reducing the hot spot by 25°C or down to 85°C life expectancy skyrockets to 40 + years. Now, that's good news!

It is obvious to anyone who studies Figure 7.46 in the light of what has been already said, that the *"transformer design criteria" should never be related to "transformer operating criteria,"* that is, if the objective is to realize long trouble-free transformer operation.

TABLE 7.14

COMPARING OLD AND NEW TRANSFORMER DESIGN CRITERIA		
Temperature Criterion	Older Design Criteria (thru 1962) Temperature Degrees Celsius	Current Design Criteria (since 1963) Temperature Degrees Celsius
Ambient	40° C	30° C
Temperature Rise due to 100 percent load	55° C	65° C
Estimated allowance for hot spot[1]	10° C	15° C
Total hot spot	65° C	80° C
Hottest spot temperature	105° C	110° C
Gallons oil per KVA[2]	From 1 to 1/3	From 1/4 to 1/9

[1]The location of the hottest spot has proved most elusive.

[2]Tables 1.14 and 1.15, Chapter 1, Part 3.

CAUTION: The manufacturer's design criteria **should not** be the operating criteria.

Other Loading Considerations

A number of other loading factors must be considered, because the most recent Loading Guide does not take them into account or actual operating conditions may be different than are assumed for the standard basis of rating.

1. Maximum Winding Overload Capacities:
   IEC Loading Guide (354) states that overloading should not exceed 150 percent of the rated load (Hottest spot temperature less than or equal to 140°C; top oil less than 115°C).

TABLE 7.15

COMPARING CURRENT AND PROPOSED USER OPERATING CRITERIA		
Temperature Criterion	Current User Criterion (e.g. Utility) Degrees Celsius (Substation Transformer)	Proposed User Objective Degrees Celsius
Ambient	30° C	30° C
Temperature Rise Under Load	65° C	30° C
Top Oil Temperature	95° C (average)	60° C (ideal)

ANSI/IEEE NEMA Guides have said 200 percent, short-time loadings maximum top oil temperature, 110°C; maximum hottest-spot temperature of 150° or 180°C; maximum short time loading (½ hour) of 200 percent.*

2. Load carrying capability of external transformer parts or associated equipment:

   - Bushings
   - Tap changers
   - Current transformers
   - Electrical protective devices
   - Current carrying capacity of station conductors
   - Disconnect switches

3. Transformer Efficiency†
   Increasing emphasis on energy conservation as well as the rising cost of electrical energy has forced owner/operators to consider the cost of transformer losses. Even though power transformer efficiencies approach 99.5 percent, transformer losses still play a significant role in plant operating costs. (The higher the core and coil temperatures, the greater are these losses). The owner/operator should find out at

The forthcoming ANSI/IEEE Loading Guide will likely follow the IEC 140° C limitation.

†*Chapter 1, Part 2, p. 26.*

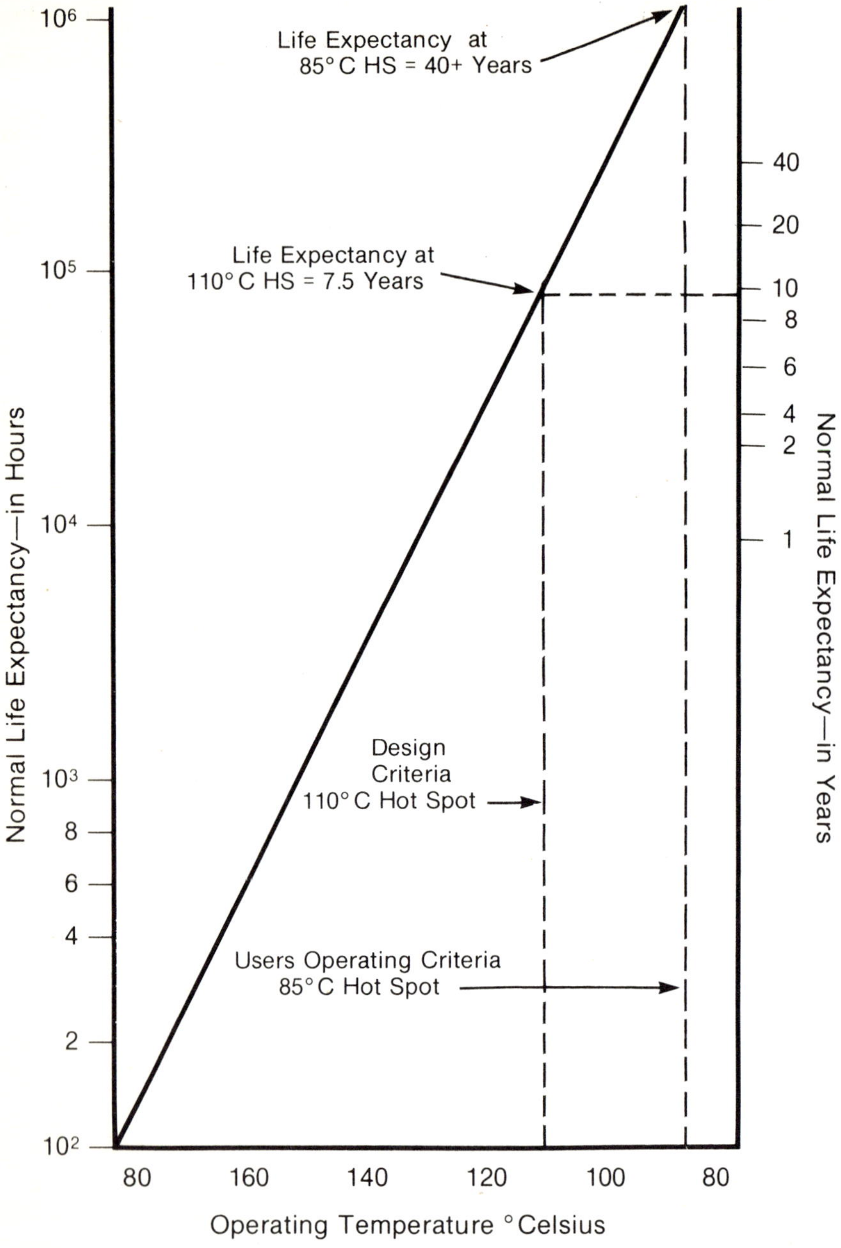

Figure 7.46 - Life of winding insulation versus temperature, 65° C rise designs. NOTE: Design Criteria of 110° C HS (hot spot) gives a life expectancy of only 7.5 years. Users Operating Criteria of 85° C gives a life expectancy of 40+ years.

what load the efficiency is maximum. In time past, we have assumed maximum efficiency at full load, but this may not always be true!

Energy losses in a transformer consist of no-load and load losses. Nothing can be done about the no-load losses since they are constant for a given primary voltage and frequency. However, load losses are essentially 1^2R losses due to flow of load current in both primary and secondary windings. As the load increases, so do the load losses — but note — as the current increases as a linear (straight line) function, the losses increase exponentially! For example:

> A transformer having only 50 amps of current flowing at one ohm of resistance has a total load loss of 2500 watts. When the current is doubled to 100 amps, the losses are squared (100^2 X 1 or 10,000 watts!)

Most authorities recognize that the maximum operating efficiency of any transformer exists when the load losses (a variable) equal the no-load losses (a constant).

4. Internal Condition of the Transformer

   a. Insulation Resistance

      - Insulation resistance is inversely proportional to temperature, and the resistance value of the insulation halves (divides by two) for every 10 percent increase in temperature.

   b. Power Factor

      - Indicative of moisture or particle contamination in transformer insulating system (increases with temperature).
      - This test can also be performed on bushings (hot collar test).*

   c. Oil Condition

      Sludges inhibit the heat transfer characteristics of the oil.

      Deteriorated oil will hold more moisture than a clean oil.

*Chapter 5.

While the dielectric strength of the oil using the flat disc D-877 test may not be affected, the integrity of the insulating system is adversely affected.

10°C Rule for transformer oil. For every 10°C increase in temperature above 60°C, the sludge free life of the oil is cut in half!

d. Sufficient gas/oil expansion space for load tap changer compartment and certain types of oil-filled bushings.

5. Temperatures Other Than 30°C Ambient
Table 7.16 covers the range of ambients between 0°C and 50°C for cooling air. Check with the manufacturer before loading on the basis of ambient air less than 0°C or greater than 50°C.

Ambient temperature for indoor installations should be controlled by adequate ventilation.

TABLE 7.16

LOADING ON BASIS OF TEMPERATURES[1]
(Ambient Other Than 30° C and Average Winding Rises Less Than Limiting Values, for Quick Approximation, Ambient Temperature Range 0° C to 50° C)

Type of Cooling[2]		Percent of Rating	
		Decrease Load for Each Degree C Higher Temperature	Increase Load for Each Degree C Lower Temperature
Self-cooled	OA	1.5	1.0
Water-cooled	OW	1.5	1.0
Forced-air cooled	OA/FA, OA/FA/FA	1.0	0.75
Forced-oil cooled	FOA, FOW or OA/FOA/FOA	1.0	0.75

[1]IEEE Project 507, May, 1979, p. 11. [2]Chapter 1, Part 2, Table 1.10.

6. Other Environmental Conditions
   Transformer operation is limited by altitudes greater than 1000 meters (3300 feet), abnormal wind velocity or by dark exterior coatings.*

7. Auxiliary Cooling
   The amount of safe additional loading varies widely, depending upon such factors as:

   - Design characteristics of the existing transformer
   - Type of cooling equipment
   - Permissible decrease in voltage regulation
   - Limitations on associated equipment

Coping With Transformer Overheating

General Remarks — Hot Spot Temperature

Temperature is not uniform throughout the transformer. As a rule of thumb, the "hot spot" temperature is located two-thirds of the way up the coil stack and one-third of the way into the high voltage winding. Approximately two-thirds of the heat is generated in the copper coils, and one-third in the steel core.

The major part of temperature rise is due to the 1^2R heat losses in the copper winding. Also contributing to this hot spot increase is reduced cooling because of different rates of oil flow within several key areas of the unit. If the hot test spot temperature exceeds 140°C, gases may be liberated in the oil-impregnated solid insulation. Gassing may produce a possible risk to the dielectric strength of the transformer since more stress is thrown on the oil than on the paper or pressboard.†

An important point needs to be emphasized regarding the "hot spot" temperature. The hot spot is now normally described at 15°C over and above top oil temperature. The transformer design engineer, however, combines the temperature rise at 100 percent load (65°C) with the hot spot allowance (15°C) and states:

> The hottest spot temperature should be estimated at 65° plus 15°C, or 80°C above ambient temperature. The hot

**Chapter 8, Part 2.*

†IEEE Project 507 (May, 1979), p. 2.

spot is "cooler" at less than full load, but "hotter" at overloads. It has been rightly determined that at overloads from 125 to 150 percent the hot spot can be up to 50°C higher than the top oil temperature. Figure 7.47 shows the relationship of hot spot to top oil temperature that can exist for medium power transformers.

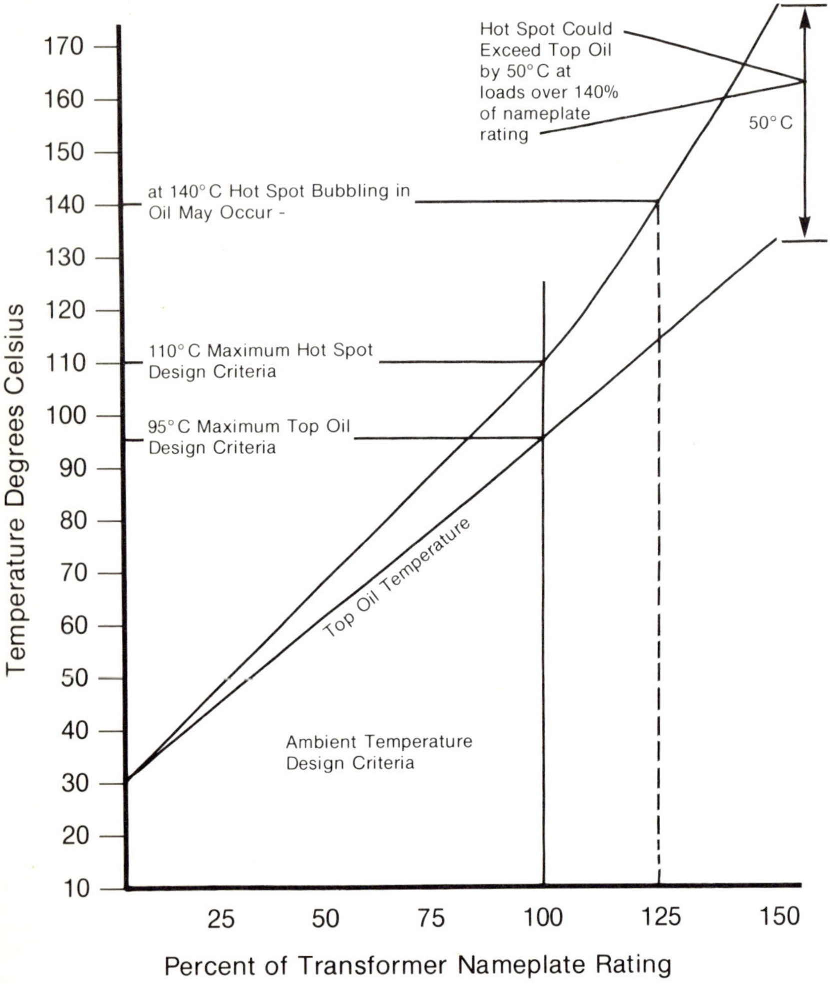

Figure 7.47 - Typical thermal relationships for medium power transformers. When loading goes much over 100% of nameplate, the hot spot and top oil curves can vary considerably for individual transformers, depending on the manufacturer's design. J.F. Brubaker, "Fault Protection and Indication on Substation Transformers" (1977).

Seven Alternative Approaches

As a transformer owner/operator, it is quite likely that you are either currently running units overloaded or contemplating an overload situation. Keep in mind that while you are operating overheated units, transformer life is being sacrificed, NEVER to be regained. You should, therefore, consider alternatives you can employ to confront transformer overheating. Several approaches have been suggested, among which are the following.

- Procrastination — ignore the situation
- Replacement of existing banks of tubes with more cooling capacity
- Purchase of several large forced-air cooling fans
- Placing of mobile transformers in parallel with existing units
- Purchase of new higher-rated transformers
- Spraying water on transformers
- Utilization of external forced-oil cooling systems

1. Procrastination — Ignore the Situation
   Procrastination is an answer, but not viable in today's economy. This decision is simply playing roulette where the only sure thing is that sooner or later, a premature transformer failure will occur.

2. Replacement of Existing Banks of Tubes with More Cooling Capacity
   Studies have shown that 40 percent of the nation's steel production is used to replace steel which has been destroyed by corrosive attack (at a total cost of $50 billion annually). Case histories have shown that in chlorine, caustic, and marine environments, transformer banks of tubes have to be repainted about every third year and/or replaced every seventh year. This is definitely very expensive and involves downtime.

3. Purchase of Several Large Forced-Air Cooling Fans
   Large forced-air cooling fans can be purchased and used to cool transformers during isolated emergencies where water is not available. Although this is a step in the right direction, it provides for only a limited amount of cooling.

4. Placing of Mobile Transformers in Parallel with Existing Units

Another alternative is adding mobile units in parallel with existing transformers so that they can share some of the increased loading. The problem, however, is that this approach is limited to situations where existing transformers are required to carry substantially large loads for only limited and infrequent periods of time.

5. Purchase of New Higher-Rated Transformers
Increasing energy costs indicate that for existing older equipment needing replacement, purchase of a new larger size transformer is certainly a valid alternative. Considerable savings can result through increased efficiency (where load losses can be adjusted to match no-load losses) and reduced energy consumption.

Additional benefits that can be realized by having larger size equipment include:

- Reduced voltage regulation
- Improved motor starting characteristics
- Increased system reliability
- Additional back-up capacity for emergency conditions

Nevertheless, for existing relatively new transformers (ten years old), this alternative would be an expensive one. It may also involve a long delivery schedule, as well as downtime during installment of the new equipment.

6. Spraying Water on Transformers
Spraying water on units, although it will probably never be accepted as a standard technique, has been utilized as another solution for cooling transformers on extremely hot days. The technique has obvious limitations. For instance, limited availability of water at the transformer site, as well as the water quality, can be a drawback. In addition, it is very likely that gradual corrosion or rusting of the unit may adversely affect the normal operation of the transformer.

7. Utilization of External Forced-Oil Cooling System
A final alternative involves the use of an external forced-oil cooling system designed to service an energized transformer so as to (1) dissipate undesired heat, and (2) bring the transformer down to a top oil temperature of 60°C under full and overload condition (Figure 7.48). An auxiliary cooling system can be a permanent or a temporary solution to the over-

heating problem, as it utilizes forced convection to optimize efficiency. It is easy to install, requires less maintenance and can be easily transported to the transformer site in case of emergency.

Figure 7.48 - A forced oil transformer cooling system in the field.

Best Possible Solution

External Forced-Oil Cooling System

Of the seven oil cooling alternatives, option 7 appears to have all of the "plus" features. The system, to be effective, must be designed to supply sufficient cooling capacity to cool down an existing overload situation so that the transformer can operate in the 60°C top oil temperature range.

1. Definition of External Forced-Oil Cooling System
   The external forced-oil cooling system is a heat exchanger uniquely designed to remove heat from insulating liquids used in energized transformers. As a heat exchanger, it has the capacity to transfer thermal or heat energy. Examples of heat exchangers include automobile radiators and heaters, air conditioners, boilers, etc. In short, the system is a heat exchanger modified into a transformer cooling system for the purpose of cooling the oil safely.

2. What the System Does
As a heat exchanger, the external forced-oil cooling system's primary function is to remove heat from the cooling liquids in energized transformers. It has been designed and engineered to keep transformer units operating at 60°C top oil temperature range at an ambient temperature of 30°C under full load condition. At this operating temperature the life of the transformer is maximized. This does not mean that the transformer is "fail-safe," but since HEAT (with almost infinitesimal amounts of oxygen and water) is the great destroyer of transformer insulation (cellulose paper impregnated with transformer cooling liquid), it only stands to reason that keeping transformers running at a maximum of 60°C top oil temperature will yield great dividends.

3. How a Typical System Works (Figure 7.49)

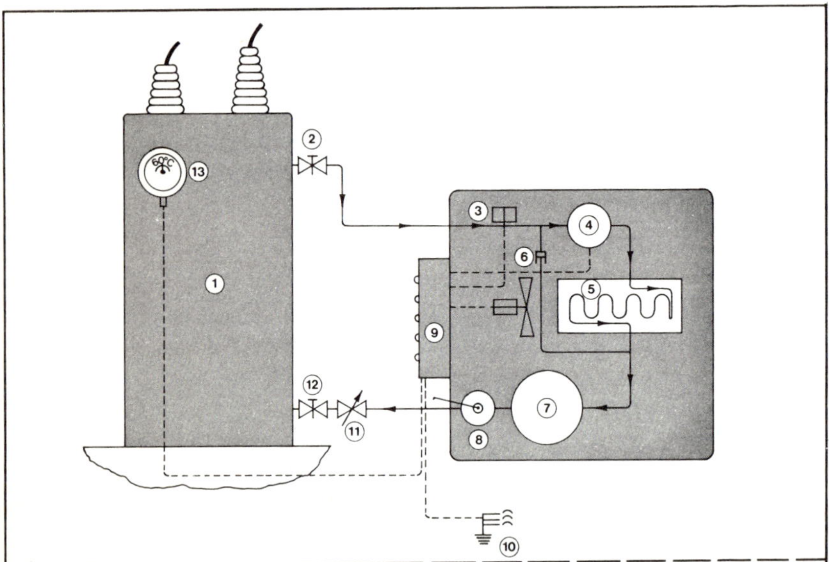

1 Transformer
2 Top filter press valve
3 Flow switch
4 Positive displacement pump
5 Heat exchanger
6 Pressure relief
7 Automatic air bleed
8 Edge wound filter
9 Control box with indicating lights
10 Customer power
11 Check valve
12 Bottom drain valve
13 Thermostatic controller

Figure 7.49 - How a forced-oil cooling system works.

a. General operation

Oil is pumped through the system's heat exchanger (5), and cooled by blowing ambient air through the finned coils. The hot transformer oil is pulled from the top filter press valve (2), and circulated through the heat exchanger (5), where oil is cooled by forcing ambient air over the finned tubes. The oil is then sent through the air trap (7) to remove any incidental air. Before the oil is sent back through the bottom valve of the transformer (12), it is safely screened through an edgewound filter (8).

b. Experience—A Case History

Typical operation on a furnace or rectifier type transformer can be seen in Figures 7.50 and 7.51. Here, the transformer loading is not continuous, but serves a batch process. The top oil and hot spot temperatures are kept low by permitting the oil transformer load to rise over its duty cycle, without heat build-up in the oil.

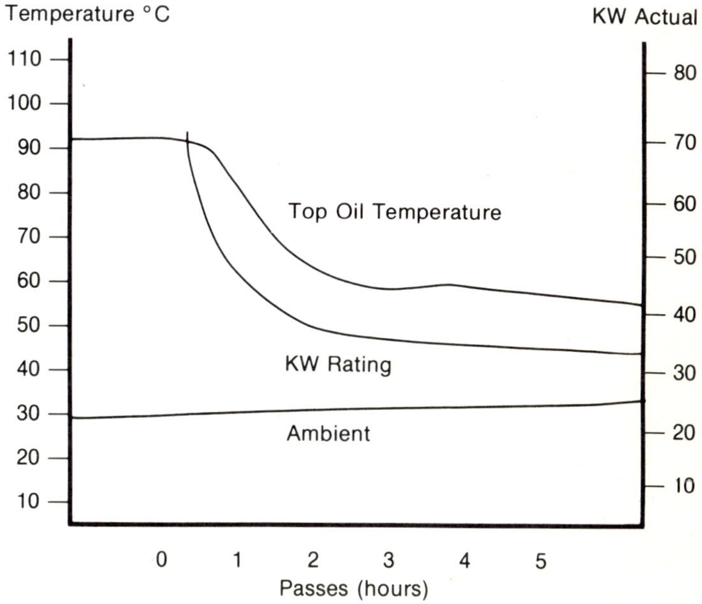

Figure 7.50 - An actual 30 KW forced oil cooling system in operation with a 4650 KVA furnace transformer located indoors containing 920 gallons.

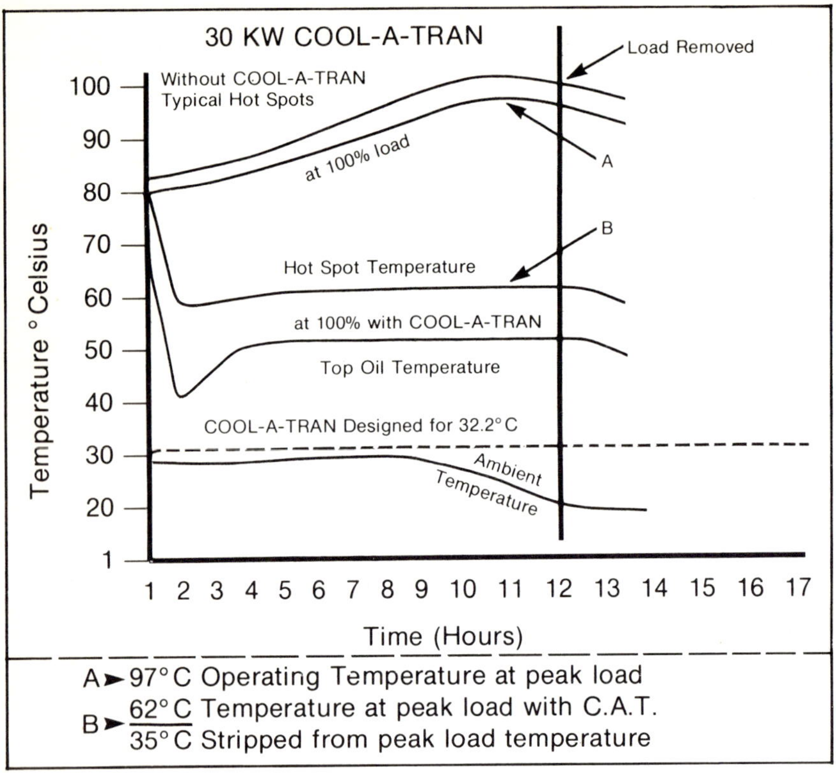

Figure 7.51 - A 30 KW forced-oil cooling system (COOL-A-TRAN™) reduced hot spot operating temperature 35°C (from 97°C to 62°C).

Oil Cooling System Sizing Charts

Loading to Nameplate Rating

For loading transformers up to the nameplate rating (100 percent load application), Figure 7.52 may be applied. If a given transformer of X-MVA and Y-gallons of oil capacity is overheating due to load, then Figure 7.52 suggests the size cooling system to reduce the temperature from the manufacturer's recommended operating temperature to the 60°C range.

As an example, consider a 1.0 MVA unit with 333 gallons of oil at an ambient temperature of 24°C. A 15 KW forced-oil cooler was then selected.

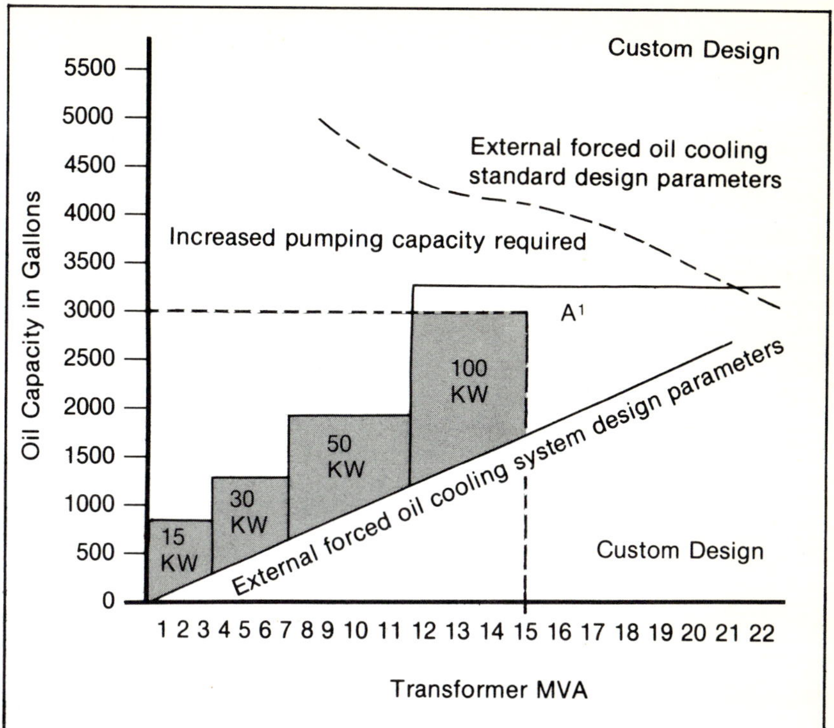

[1]Example Point A - 15,000 KVA transformer with 3000 gallons of oil requires a 100 KW unit to maintain top oil at 60° C at 30° C ambient at 100 percent load.

- Standard External Forced Oil Cooling System

Figure 7.52 - Full load sizing chart.

Field data have shown that within 1½ hours the steady state transformer operating temperature had dropped from 70°C to 50°C and held for two hours at which time the test was terminated.

In addition, the relationship between the amount of oil and the rate it is circulated is critical for proper heat transfer in KW (Figure 7.52). *Using the MVA nameplate rating alone is insufficient.* For if the oil circulation rate is slow and the volume of oil is high, there will not be maximum heat transfer. Using our example, if the 1.0 MVA unit had 1000 gallons of oil (an older design), then Figure 7.52 shows that a higher rated gear pump, in gallons per minute (gpm), would be needed.

Loading Beyond Nameplate Rating

Overload sizing, as expected, is much different than full load operation (Figure 7.53).

Overload has been defined as transformer loading beyond the maximum nameplate rating. One suggested guideline for medium power overloaded transformer conditions was given in Figure 7.47. These typical thermal relationships take into account the accelerated hot spot temperatures beyond full load capacity.

Finally, verify the system selected by the following engineering calculations.

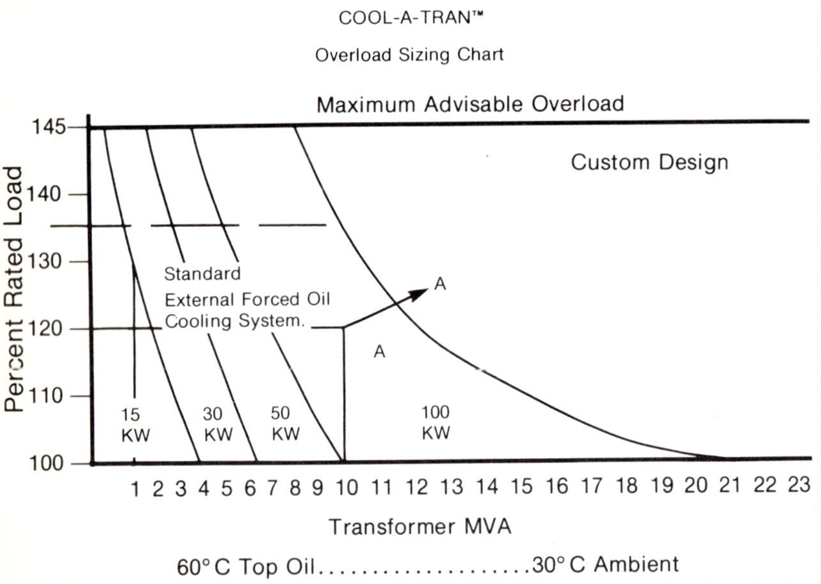

A 10,000 KVA transformer can be operated at 60° C top oil temperature, 30° C ambient with a 100 KW External Forced Oil Cooling System.

Figure 7.53 - Overload sizing chart. ***Caution:*** *check with Figure 7.52 to make sure cooling system has sufficient gear pump capacity for maximum dissipation of heat.*

Engineering Formulas

Quantitative Equations for Up to Full Load Sizing

Equations used in forming the application charts are given in Figures 7.54 and 7.55. The equations for fin replacement will be needed on the job to "ball park" the cooling system size needed for that application.

Up to Full Load Application

$$KW = (0.5)\ (.01)\ (KVA) \left(\frac{\text{Ambient } ^\circ C}{32}\right)$$

Full Load Fin Replacement

$$KW = .01\ (KVA) \left(\frac{\text{Ambient } ^\circ C}{32}\right)$$

Figure 7.54 - Full load sizing equations.

Full load losses are calculated from the following relationship:

$$KW_{FL} = \frac{KVA\ (1 - \text{efficiency})}{(.01)}$$

To arrive at the proper cooler size for a hot running transformer whose manufactured top oil temperature is 95°C, we assume a linear relationship between the Δ T's of the 95°C oil temperature and the desired 60°C top oil.*

Quantitative Equations for Overload Sizing

In sizing the cooling system for units that are loaded beyond nameplate, two things are important to remember.

- All heat due to overload should be removed.
- Be sure the transformer is operating in the 60°C top oil range or in the case of high (145 percent) overloads even lower to allow for hot spots.

*ΔT = *change in temperatures.*

Overload Application

$$KW = .01\ (KVA)\ (OL^2 - .5)\ \left(\frac{\text{Ambient}\ {}^\circ C}{32}\right)$$

Overload Fin Replacement

$$KW = .01\ (KVA)\ (OL^2 + .5)\ \left(\frac{\text{Ambient}\ {}^\circ C}{32}\right)$$

Figure 7.55 - Overload sizing equations.

Removing the heat due to overload (OL) is removing the losses due to the overload or:

$$KW_{OL} = \begin{matrix}\text{Full Load}\\ \text{Losses}\\ \text{in KW}\end{matrix} \left(\frac{\text{New KVA}}{\text{Old KVA}}\right)^2 - \begin{matrix}\text{Full Load}\\ \text{losses}\\ \text{in KW}\end{matrix}$$

$$= \begin{matrix}\text{Full Load}\\ \text{losses}\\ \text{in KW}\end{matrix} (O\ L\ )^2 - \begin{matrix}\text{Full Load}\\ \text{losses}\\ \text{in KW}\end{matrix}$$

$$= \begin{matrix}\text{Full Load}\\ \text{Losses}\\ \text{in KW}\end{matrix} (OL^2-1)$$

$$KW_{OL} = KVA(.01)\quad (OL^2-1)$$

Now suppose we removed only heat due to overload. The transformer would still be operating at 95°C level and this is not acceptable. Thus, the full load heat should also be removed yielding a total of:

$$\text{Overload} = KVA\ (.01)\ (OL^2-1)$$

$$\text{Full Load to } 60^\circ C = KVA\ (.01)\ (.5)$$

$$KW_{OL} = KVA\ (.01)\ (OL^2-1 + .5)$$

$$= KVA\ (.01)\ (OL^2 - .5)$$

and finally, adjusting for the actual ambient temperature yields this expression:

$$KW_{OL} = .01\ (KVA)\ (OL^2 - .5)\ \frac{\text{Ambient}^\circ C}{32}$$

$$KW_{OL} = .01\,(KVA)\,(OL^2 + .5)\,\frac{Ambient^0C}{32}$$

FIN REPLACEMENT

Advantages of an External Forced-Oil Cooling System

1. The external forced-oil cooling system has been designed to limit transformer top oil operating temperature to 60°C — a temperature that will maximize transformer life. However, for a given application, consider a cost-benefit analysis to determine whether increased unit life would be worth the additional cost in energy consumption; and whether a transformer owner/operator would set a 60° or 70°C goal as top oil temperature. Thus, a properly designed forced-oil cooling system can become a permanent solution to overheating problems.

2. The second advantage of the external forced-oil cooling system is that its unique high efficiency design calls for less power usage. For example, a 15 KW unit pulls only 2.2KW —a great saving both in energy and money. In addition, compared to the use of additional fans, this system would be used for much shorter periods of time, thus minimizing power cost.

3. By acting as a heat transfer device, the external forced-oil cooling system can help the transformer to achieve an even distribution of heat, thus avoiding the condition whereby the unit always has very hot oil at the top and very cool oil at the bottom. In addition, the system helps to reduce to a minimum the temperature differences that might exist between the coil surface and transformer oil.

4. This external forced-oil cooling system is equipped with a flow monitoring device that can detect any flow malfunction of the whole cooling system. Each unit is also equipped with a "red" warning light on the control panel which can be used to locate the problem with ease. Every external forced-oil cooling system has as standard equipment an Auto Klean Filter which protects the transformer from any material passing into the cooler. Should this device be in need of service, it can be easily cleaned by moving the ratchet handle back

and forth 10 to 12 times. Other safety features include the pressure relief system for the oil circulation pump. This device will de-energize the unit if the pressure exceeds the normal operating range, while the latter is equipped with mechanical seals that resist oil leakage.

5. Another advantage of the external forced-oil cooling system is the ease with which it can be installed. Emergency installation can be done within 2 to 4 hours. Permanent installations would take longer depending on each location. The site can be prepared while the unit is in production or in the process of being shipped. The manufacturer of this cooling system can provide the services, of (a) complete installation, (b) supervision of installation by plant operating personnel and (c) annual maintenance, including steam cleaning, painting, leak repairs, new pumps, etc., where necessary.

6. Experience has shown that in chlorine, caustic and marine environments, transformer fins and tubes may have to be repainted every third year, and new banks of fins purchased every seventh year. Obviously, this is expensive and involves downtime. In such cases, the external forced-oil cooling system would minimize, if not virtually eliminate, this expense.

The Bottom Line — Warranty

Finally, if you decide that a forced-oil cooling system is the way to go, check whether the unit carries a warranty that expressly guarantees it will keep the operating temperature at the desired level as well as for defects in materials and workmanship for a specific period of time. In addition, it is the customer's responsibility to make sure that the transformer to be serviced by the cooling system is in fact a dependable transformer, and that the unit is not at the point of failure prior to applying a cooling system.

Transformer Cooling System — A Solution Not a Cure-All

An external forced-oil cooling system cannot be used to cure all types of transformer problems. The following criteria must **all** be met before a cooling system should be applied to a transformer. The goal is to do the job right, the first time.

1. Is the transformer oil "good" — clean, dry and nearly oxygen free? These eight ASTM oil screening tests should be performed: dielectric strength; acid number; interfacial tension; color; specific gravity; sediment; visual (cloudy or clear?) and water content in parts per million (bottom and top oil samples).*

2. If the oil is heavily sludged with an acid number greater than 0.4 mg KOH/gram of oil, the unit should be completely desludged prior to cooling system installation.†

3. Is the transformer oil free of significant amounts of combustible gases? First, a dissolved gas-in-oil analysis by gas chromatography should be performed to establish the base condition of the oil to determine the combustible gas content. The condition causing the generation of combustible gases should be corrected. The oil should then be degassed prior to cooling system installation.‡

4. Is the transformer painted a light color? A unit painted with a dark color in areas of high sun exposure can operate up to 15°C hotter than a light color under identical conditions (of primary concern in the Sun Belt area of the USA).§

5. Is the tap changer capable of handling an overload condition? It is generally acknowledged that transformers will operate at 125 percent of nameplate rating without special precautions. Contact the manufacturer concerning its overload capacity. Perform turns ratio and DC resistance tests on each step of the tap changer.

6. Are the bushings and leads capable of handling an overload condition? Again, contact the manufacturer. They may have to be upgraded. Run an infrared "hot spot" survey on the bushings before overload and during operation at overload.**

Caution

The transformer must provide sufficient gas/oil expansion space for the load tap changer compartment and certain types of oil-filled bushings.

Chapter 3, Part 1.

†*Chapter 7, Part 3, Table 7.11.*

‡*Chapter 4.*

§*Chapter 8, Part 2.*

***Chapter 8, Part 4.*

7. Is the transformer insulation satisfactory for overload conditions? A battery of insulation tests should be run, including power factor series, insulation resistance series, and turns ratio. Higher operating temperatures may cause a very hazardous situation because of the rapid drop in the insulation resistance with a rise in temperature.*

8. Will there be any external provisions required for an "alarm" and/or buzzer as a hot spot indicator? Is a higher temperature scale provided for both top oil and existing hot spot temperature gauges? Keep in mind the hot spot temperature will rise from top oil temperature by the square of the overload over and above the normal 15°C rise for the 100 percent load level. Recent IEEE reports recommend that, for safety's sake, the hottest spot temperature should not go higher than 140°C.

A forced-oil cooler is just that — a cooler! It will not correct the conditions listed above. These become preliminary steps prior to purchasing the cooling system. Once a unit is in operation, dissolved gas-in-oil analysis by gas chromatography should be made periodically to determine if any combustible gases are being generated; the rate at which these combustibles are being generated; the nature of the combustible gases; and to determine the effects of the overloads on the internal components of the transformer. An old adage states — solving one problem creates other problems. In this case the temperature problem has been solved for overloaded transformers. The only way to monitor the transformer to see if all is going well is to use the dissolved combustible gas-in-oil analysis.† We also recommend use of the Karl Fischer water content analysis in "ppm" (parts per million), as well as an annual infrared bushing inspection, since overloads can liberate water from cellulose and produce hot spots in undersized bushings or connections.‡

**Chapter 5.*

†Chapter 4.

‡Chapter 3, Part 3; Chapter 8, Part 4.

Chapter 8

Preventing Unscheduled Substation Outages

Introduction

On November 9, 1965, between 5:15 and 5:30 PM, a vast area of Northeastern United States and parts of Eastern Canada experienced a devastating power failure. Over 30 million people, living in an area of approximately 80 thousand square miles were without electricity for 8-13 hours, which produced an economic loss estimated at $100 million. The reported cause of this "Great Northeastern Blackout" was a protective relay that was wrongly set. Because the relay had not been coordinated with the system as conditions had changed, succeeding "dominos" fell. In another unrelated but devastating example, about 40,000 customers including eight hospitals were left without power for about five hours. It was their third power failure within one year (1979) attributed to underground cable problems.

These and most other unscheduled power failures could have been prevented — by a **complete** power system maintenance program.

Most unscheduled power failures are either initiated by natural phenomenon such as wind, storms, ice, flooding, etc., or contact accidents affecting electrical wires and power poles. The cutting of underground cables by hand-operated and mechanical digging equipment, overheating of transformers and the improper operation of switchgear are also likely to disrupt power flow. Since plant operators cannot eliminate all unpredictable incidents, engineers have designed electrical protective devices (switchgear) such as fuses, circuit breakers, and relays to reduce the damage sustained when an abnormal condition occurs. These protective devices act as "safety valves" to reduce electrical "pressure" on faulty electrical conductors.*

**H.S. Orth, "Your Plant Can Survive an Abnormal Electrical Condition if..." (1969).*

Electrical conductors should possess a safety current capacity which helps them to dissipate electrical current or heat. Then every electrical apparatus possesses "a damage curve." This simply means that a given insulation can only resist a certain temperature for a given period of time; and the higher the current, the shorter the time required to damage insulation (Figure 8.1).

In order to prevent this damage during overload or short circuit conditions, various protective devices are installed, which possess the same operating conditions as those described in the damage curve. It is important to select protective devices which can sense the current and operate effectively before an abnormal condition occurs. The devices should be set in such a way that they not only protect the device they are designed to protect, but can also be coordinated to remove the faulty device from the electrical power system. Figure 8.2 illustrates this concept.

For electrical protective devices to work efficiently and maintain the safety of electrical equipment for which they were designed to protect, they need to be checked periodically, to make sure that they are not impaired by exposure to dusty, smoky, or corrosive atmospheres, and other mechanical accidents. Thus, a regular inspection and maintenance of switchgear is of paramount importance in protecting electrical systems from unscheduled outages. The maintenance of switchgear should include thorough cleaning of fuses, circuit breakers, and relays to eliminate all the contaminants that are likely to come in contact with electrical power. In addition, all electrical connections on protective devices should be tightened, lubricated, tested, and calibrated in accordance with manufacturers' test manuals. Failure to perform such a maintenance program may lead to unplanned electrical faults.

It should be noted that there is more than ample evidence to illustrate that most protective devices are not being maintained on a regular basis. There is one case on record where an inspection of protective relays was made after nine years of operation, and it was discovered that the original shipping blocks were still intact.* Such shipping blocks are used to prevent the movement and possible damage of mechanical parts from the factory to the installation site. If protective relays cannot move mechanically, then obviously they will not provide protection.

**W.H. Mapes, "Electrical Maintenance or Repair?" (1976).*

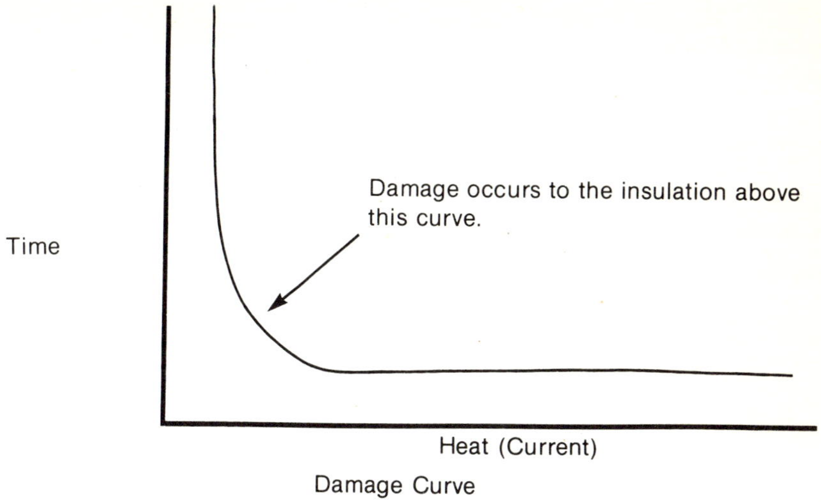

Figure 8.1 - A typical damage curve for electrical apparatus. BES, Inc., "Protective System Coordination," ***The Energizer*** *(September 1980).*

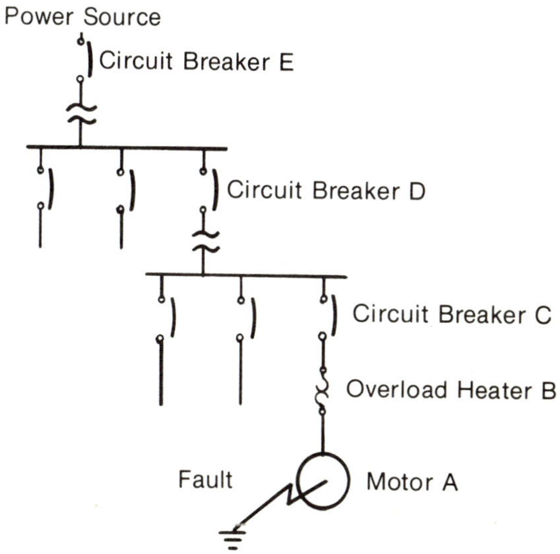

Figure 8.2 - Coordination of protective devices is a fundamental necessity. If all the relays which trip circuit breakers C, D, and E are not coordinated, then a fault which should be cleared by B and C could actually be cleared by D or E, causing loss of more of the power system than is necessary. In extreme cases, it is possible to lose an entire plant in a case that should only cause a motor to be taken out of service.

It is, therefore, imperative to regularly inspect and maintain electrical protective devices because even though all maintenance can be planned, Murphy's Law is still operating — "When something can go wrong, it will go wrong at the worst possible time". Putting it differently, plant operators cannot afford to wait until there is an "electrical accident" without knowing whether or not the electrical "safety valve" will function properly when the need arises.

All unscheduled electrical faults are not caused by defective switchgear. Some may result from poor protection by the paint system, or from contamination of transformer bushings and insulation structures, or buried cable failures.

The application of protective coatings or paints to steel surfaces can help to protect electrical apparatus from severe deterioration caused by destructive atmospheric elements. Because transformers generate internal heat, their coating system must permit heat escape and prolong their lives from the effects of corrosion.

Since corrosion is impractical to eliminate, the secret to sound engineering is to detect and control rust through barrier coatings, or preventive painting. This type of painting should, in fact, be executed even before rust begins to appear. Procrastinating such an action will result in remedial painting, which is both time consuming and costly. Neglecting to paint steel surfaces of electrical equipment may also lead to unplanned electrical shut-downs, or unnecessary replacement of the equipment.

Apart from the waste that may arise from poor protective coatings, contamination flashovers can be a major problem of high voltage lines, due to natural deposits and industrial pollutants on transformer bushings and standoff insulators. The flashover problem is so serious that an IEEE-EEI joint committee concluded that it is the second major cause of line outages in the United States and Canada.* Transformer bushings and insulator structures should, therefore, be cleaned even before there is evidence of contamination, simply because there is no scientific way of testing such devices to determine the "time to clean."

Footnote: **IEEE Committee Report, "A Survey of the Problems of Insulator Contamination in the United States and Canada. Part 1," (1971). Interruptions in utility service due to over voltages was the number one cause of line outages. EEI - Edison Electric Institute, New York, NY.*

Electrical faults that plague contaminated bushings and insulator structures, as well as steel coatings can be detected either visually, manually, or thermographically using infrared (IR) technology. The latter technique is a relatively recent addition to many plant maintenance programs. It is becoming an accepted plant strategy for preventing unscheduled plant shutdowns and energy losses. Visual and manual inspection of electrical equipment, such as control panels, connections, overhead feeders and substations, is not possible at all times and can be expensive and time consuming. By using infrared scanning while the electrical equipment is operating to detect potential failures, net maintenance dollars can be saved.

Part 1—Switchgear and Associated Equipment

Introduction

An electric power system is so constructed to ensure the availability of electrical energy without interruption to every load connected to the system. In all practicality, this can not be done at 100 percent reliability because of several limiting factors. Power lines are generally installed overhead and therefore are subject to breakdown due to storms, falling objects, vandalism, overload, neglect, etc. Even lines that are installed underground are not immune to failure. Problems occur due to environmental hazards, accidental cutting, overloads, neglect, and so forth. These problems not only cause mechanical failures but electrical failures as well.

These electrical failures sometimes cause violent changes in the system operation and place a severe strain on the switching and protective devices used to interrupt fault current and thereby remove the fault from the line.

The switching and protective devices are generally classified as switchgear which can further be defined as switching and interrupting devices and their combination with associated control, metering, protective and regulating devices. They are used in all phases of generation, transmission, distribution, and conversion of electric power.*

Types of Circuit Breakers and Other Protective Devices

Switching and interrupting devices come in many types, sizes and shapes, the most common being circuit breakers and fuses. General classification of circuit breakers in common use today are (1) oil (2) air blast (3) air magnetic and (4) molded case. Newer types being used at the higher transmission voltages include oil cutout, vacuum switches and SF_6 gas switches.

Fuses can be classified as (1) one-time (2) lag (3) dual element (4) expulsion and (5) current limiting. Automatic oil reclosing devices

Chapter 1, Part 2 (Transformer Auxiliary Equipment), p.90.

restore power circuits to operation, as soon as temporary electrical disturbances (such as fallen tree limbs, kites or other foreign objects) are removed.

The process of oil disintegration is a very effective way to absorb energy and oil is an excellent dielectric. One chief disadvantage of oil breakers is that oil has to be maintained — tested and reclaimed.* Oil is also a fire hazard causing some people to be hesitant about using oil.†

Oil circuit breakers for below 13.8 KV and interrupting capacities are usually frame mounted with a single tank for all three phases.‡ At the higher voltages and interrupting capacities, (such as 3000 or 6000 amperes) each phase is in a separate tank. Some breakers are supported from the top so the tank can be lowered for internal inspection and adjustment.

Auxiliary equipment for OCBs consist mainly of the power supply for the operating circuit. Batteries have been favored for solenoid operation (tripping a circuit) and have a high degree of reliability if properly maintained. Pneumatic operations require a source of air, usually provided as part of the circuit breaker unit. Air reservoir is sufficient for numerous operations and can be replenished by a compressor driven by an ac motor.

Relays are designed to monitor a host of complex circuit conditions and if an abnormal condition is detected, trip or signal the associated circuit breaker in order to remove the fault from the line. More about relays later.

1. Oil Circuit Breakers
   (OCBs) are the oldest type of breaker and today are still the leading type by a wide margin, in the United States, for voltages above 13.8 KV (Figure 8.3).

   OCBs interrupt circuits while current is flowing through them. The make-and-break contacts are the bayonet type, fitting into a tulip-type stationary contact (Figure 8.4). Arcing

**Chapter 3, Part 1 (Testing) and Chapter 7, Part 3 (Oil Reclamation).*

†Being open to the atmosphere, the water and carbon can accumulate in the bottom of the tank. Contact surfaces must be kept free from polymer coatings from the oil or from "coking."

‡Metal clad switchgear are normally used below 13.8 KV.

takes place between the tip of the bayonet and beveled flange on the tulip contact. Then spring action moves the contacts.*

Figure 8.3 - A typical oil circuit breaker (OCB).

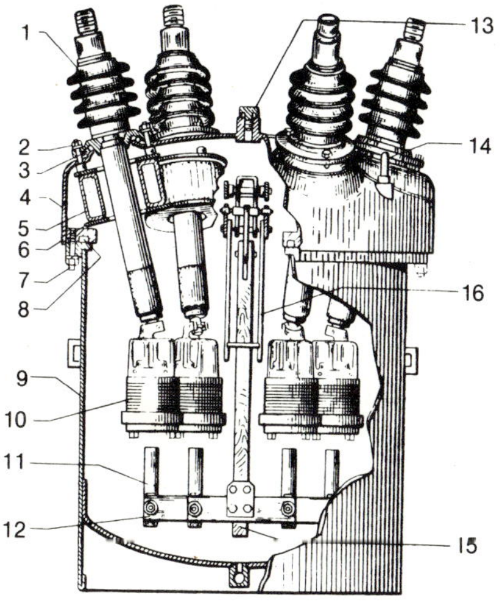

1. Bushing
2. Assembly Nut
3. "O" Ring Gasket
4. Dome
5. Bushing Current Transformer
6. BCT Support Plate
7. Tank Nut
8. Tank Gasket
9. Tank
10. Interrupter
11. Contact Rod
12. Crossarm
13. Analyzer Rod Access Hole Cover
14. Oil Gauge
15. Lift Rod
16. Lift Rod Guide

Figure 8.4 - Pictorial diagram of a typical oil circuit breaker (OCB).

*Switchgear is a very specialized area. For further details, consult Lythall, ***J & P Switchgear Book*** (1972), or Fink and Beaty, ***Standard Handbook for Electrical Engineers*** (1978).*

Once the bayonet has completed its upward travel (when the contact closes), low resistance contact is obtained between sides of the bayonet and the inner surface of the tulip. This provides an excellent current path.

Higher power applications often call for the use of OCBs. Circuit breaker protection normally always involves these four basic components: (1) A sensing device, such as a current transformer, which measures the quantity to be controlled; (2) a relay which interprets the signal from the sensing device and initiates the proper circuit breaker response; (3) an actuating mechanism, which opens and closes the contacts, such as a solenoid and linkages; and (4) an interrupter consisting of the contact and arc-controlling parts.

2. Air Blast Breakers
Their high speed operation takes over high voltage and current interrupting applications where an oil circuit breaker is limited. The burning of the contacts of compressed air breakers is usually less than that of OCBs because of the short arcing time and the consequent limited arc energy available for burning.

   Air type circuit breakers can be divided into two broad groups: air magnetic breakers and molded case breakers.

3. Molded Case Breakers
These breakers are characterized by a molded plastic case (Figure 8.5). Extreme compactness is one of their most important features. The smaller sizes are generally riveted together and cannot be repaired, but the larger sizes may be opened. Their chief application is for protection of small motors, lighting and appliance circuits. Molded case breakers are built in frame sizes from 30 up to 3000 amperes.

   The most popular kind of molded case breaker is the thermal-magnetic type. Long time delay is provided by a **thermal** element (bimetallic strip) and instantaneous trip is provided by a **magnetic** element. It is conceivable that a breaker may trip from 90 to 110 percent of rated current.

   Current limiting fuses can be used to back up molded case breakers where the available short-circuit current would otherwise be beyond the capacity of the breaker. The practice has become important in recent years because of substantial increase in available short circuit current in low voltage.

If a fault occurs, the breaker will trip via its own internal thermal and instantaneous elements, and the fuses will blow only if the magnitude of fault current approaches a value which would damage the breaker.

Large air circuit breakers are used at voltages as high as 13.8 KV. They are usually applied as metal-enclosed unit type switchgear, with drawout arrangement and medium voltage ratings at 2.4, 4.16, 7.2 and 13.8 KV. Corresponding continuous current ratings vary from 1200 amps to 3000 amps. Interrupting MVA vary widely from 75 MVA to 1000 MVA.

The drawout feature is an important advantage of metal clad switchgear (for 4.16 to 13.8 KV). Large circuit breakers are "racked out" from the cubicle on wheels after they are disconnected. "Stab" connections are plug type so when the breaker is installed and locked in place, the power and control circuits are connected properly. Removing the unit simultaneously disconnects the circuits. Most switchgear provide a test position in which control circuitry operates, but main contacts are not energized.

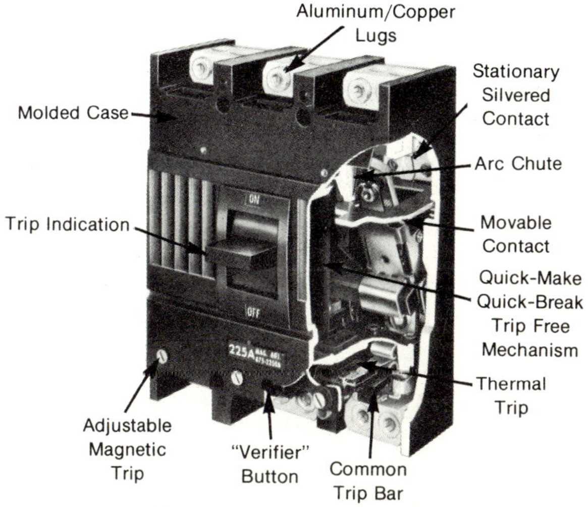

Figure 8.5 - A typical molded case circuit breaker. (Courtesy of General Electric Co.).

Low voltage air circuit breakers, in the range below 600 volts, are quite popular as individually mounted units. Nominal voltage ratings are 240, 480 and 600 volts. Continuous current ratings run from 15 to 4000 amperes.

In small circuit breakers such as low voltage power circuit breakers all these components may be included in one unit and be an integral part of the breaker. Larger breakers such as oil circuit breakers have separate relays and sensing devices which add greatly to their flexibility.

4. Protective relays

These devices are designed to sense different types of circuit conditions, such as current, voltage, frequency, impedance, direction of current flow, and so forth. When a fault condition is detected, the relay contacts close completing the trip coil circuit of the breaker.

The breaker will then trip removing the fault from the system (Figure 8.2). The trip circuit voltage is usually dc supplied by batteries; however, occasionally ac voltage is used for the trip circuit. It may be more convenient for that particular installation.

A person must study the kind of protection that is available, and then select the type of relay best suited to his needs.

Fuses are the simplest type of circuit protector but unlike circuit breakers, combine the functions of fault detection and interruption in one element. Fuses, like other protective devices, must carry rated load current continuously, and must safely interrupt the fault current for which they are intended. For voltages under 600 volts, one-time fuses are the simplest and least expensive but have the shortest time delay.

Renewable lag fuses introduce more time delay by special construction of the fuseable link. A typical lag link has narrow and wide element sections in series. Wide portions act as heat sinks. Under short-time overloads these sections absorb heat from the surrounding thin elements, preventing the fuse from blowing. If overload continues, wide elements cannot absorb all the heat and the narrow elements melt.

On short circuits the thin sections melt immediately because the wide elements cannot draw off the heat rapidly enough.

Dual-element fuses make use of fusible metals with different melting points. Fault current is interrupted rapidly by a fuse link similar to conventional links. Overloads are handled by a different element such as a solder pot under spring tension. Fusible material (eutectic materials) of this element has a lower melting point than that of the fuse link, but a large mass takes time to heat up. Overloads of short duration do not melt the eutectic material, but if sustained, the material softens and spring pulls the contact loose, opening the circuit. Thus, time delays are possible.

The leading types of fuses are expulsion and current limiting. The expulsion fuse (used over 600 volts) has a fusible element held under spring tension and contact is maintained through a plunger. When a fuse element melts, the plunger draws the arc through the tube at high speed. The arc causes compressed boric acid or similar compounds to generate high pressure gas which in turn extinguishes the arc.

Current limiting fuses are used both below and above 600 volts. At the lower voltage, this type of fuse is being used increasingly as back up protection for molded case circuit breakers. In the higher voltage range, they are being used with load interrupter switches.

The fuse element, for a current limiting fuse, is made of silver and surrounded by a fine quartz sand heat dissipation sink. This construction produces a fast-acting fuse with the element melting in ¼ cycle or less, before current has time to rise to maximum value.

Maintenance of Protective Devices

Relays and circuit breakers do not give any obvious indication they will not work properly when called upon. Since circuit breakers function only when power is supplied to their trip coils, system protection can be no better than the day-to-day condition of the breakers and relays. Only careful inspection and testing will assure you that your protection system is alive and ready to operate when the need arises.

Relay and circuit breaker maintenance is not complicated, but must be done carefully and with some understanding of construction and operating principles * (Figures 8.6 and 8.7).

Since relays consist of contacts, springs, solenoids, and insulated wire, age alone can have its effect. Magnets can lose strength. Insulation can deteriorate and cause a short. Rust can prevent movement. Corrosion can insulate contacts. Relays should be cleaned, lubricated and calibrated periodically in accordance with manufacturer's instructions. Larger plants usually find that it is a good investment to purchase their own test equipment and train their own personnel for equipment maintenance. Smaller industrial plants cannot justify such a program, so they contract with service organizations to do the testing and maintenance at a reasonable fee.

Frequency of testing depends on the type of equipment, how critical it is to the overall system performance and environmental conditions. These include dust, moisture, vibration, heat, and corrosive atmosphere. These conditions can cause the relays to trip early, late, or not at all.

A time interval of one year is the most common in today's well maintained industrial plant maintenance program. The value of many tests lies in comparing them with results of preceding tests on the same piece of equipment. One major public utility requires that once a year all protective relays be taken out of service, given a close visual inspection to look for signs of damage, and then be operated to see that they trip or perform their function (tripping, reclosing, alarm, etc.). On drawout type relays, they should be removed from their case for a better look. Even on simple auxiliary relays, the cover should be removed for a visual exam. In these inspections we not only look for the obvious burned, broken, or corroded elements: we also look for sluggish or stuck moving elements, loose connections, and similar problems. The tap setting should be verified against the setting information. Make sure that all parts of the circuit are working. Examples of the details one large public utility requires include proof that both coils of a dual trip coil circuit operate, that carrier relays send a blocking signal, that a blocking signal from the other end actually will block tripping, that the circuit breaker antipump relays work, and similar details to prove that all functions of the circuit are working.

Again, for details see such publications as Lythall,* ***J & P Switchgear Book**, (1972).*

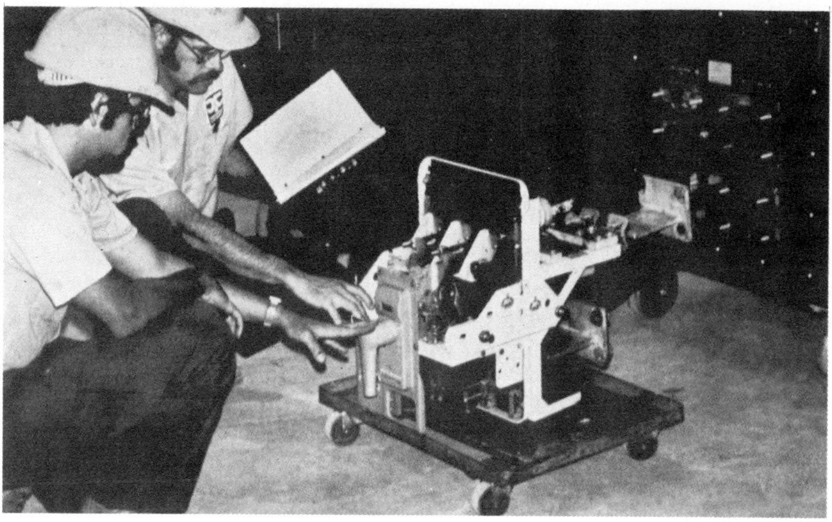

Figure 8.6 - 40,000 ampere circuit breaker test set.

Figure 8.7 - Protective relay test set.

Supplement regularly scheduled tests with special ones made whenever you suspect that the equipment is not functioning properly. Observant operating personnel can also prevent further trouble by reporting abnormal performance. Again, this same major utility gives a computerized status report every quarter.

Surge Arrester Testing

The surge arrester is one of the most important protective devices in use on electric systems ensuring continuity of operation despite repeated overvoltage resulting from lightning and switching.* Its function may be likened to a reverse circuit breaker, normally open, but closing to discharge transient currents accompanying a disturbance. After discharging transient currents, it must reopen to prevent the flow of system power which would be destructive to itself and result in system disturbances. It must function as an insulator under normal conditions, but at the instant a disturbance occurs it must become a conductor of low enough resistance to prevent the development of dangerous voltages which would destroy the apparatus it is designed to protect. With the passing of the disturbance, the arrester must revert to its role as an insulator. All this must be accomplished within microseconds without the aid of operators or relays.

Despite its importance, the surge arrester is probably the least attended of protective devices. This, in part, can be attributed to developments that have made these devices very reliable and almost trouble-free.

Failures of modern arresters are relatively few. Experience has shown that failures caused by damaged, defective or contaminated units are the most prevalent. Acceptance and routine maintenance testing in the field can minimize this type of failure.

The latest standard for arresters is ANSI/IEEE Standard For Surge Arresters for Alternating-Current Power Circuits C62.1-1975. The ANSI Standard deals primarily with design and factory tests, while we are presenting material that deals with field testing of surge arresters.

These field testing procedures are not tests of the complete protective characteristics of a surge arrester but are tests of the mechanical condition and insulating qualities of an arrester.

**Chapter 1, Part 2, Figure 1.58, p. 82 and p. 89.*

Experience has shown that certain routine field tests can detect defects that would affect the ability of an arrester to function as a protective device. Three of the routine tests are visual inspection, power frequency dielectric loss, and dc insulation resistance. Other tests can be performed in the field but they are more difficult and require more expensive equipment.

- Visual inspection gives an indication of the external physical condition of the arrester. No test equipment is necessary. Maintenance personnel should check for loose hardware, cracked porcelain, cracked cement, surface dirt, and signs of flashover.
- Power frequency dielectric loss measurements are effective for detecting moisture, foreign deposits, corrosion, broken resistors, punctured disks, etc.

  A dielectric loss or power factor test set is needed. Compare watts-loss at test voltage with previous test data and/or table of average normal values for similar units.
- DC insulation resistance measurements also give an indication of the condition of the arrester housing and elements with regard to moisture, corrosion, broken resistors, punctured disks, etc.*

High or low voltage dc test equipment is necessary for this test. Record the insulation resistance and/or leakage current and applied voltage (below sparkover). Compare test results with previous data or data recorded for similar units.

Tabulated dielectric loss (watts loss) data is available to serve as a guide to the test operator in establishing a range of losses for the various makes and models of arresters. Once a range of losses has been established, any deviation either higher or lower should be investigated. Because surge arresters are basically resistive in nature and rated primarily on the basis of the losses obtained, power factors need not be calculated. Correction factors are not necessary throughout the range of temperatures in which surge arresters are normally tested.

Test results on arresters are affected to varying degrees by surface leakage. These losses can usually be minimized by wiping the porcelain with a clean, dry cloth; however, it may be necessary to use cleaning agents and waxes.

Chapter 5.

Where the effects of surface leakage can be discounted, abnormal losses can then be attributed to one or more of the following.

1. Higher-Than-Normal Losses

   - Contamination by moisture and/or dirt or dust deposits on the inside surfaces of the porcelain housings or on the outside surfaces of sealed-gap housings
   - Corroded gaps
   - Salt deposits
   - Cracked porcelain

2. Lower-Than-Normal Losses

   - Broken shunting resistors
   - Broken pre-ionizing elements
   - Mis-assembly

A single arrester can be tested only by the routine ungrounded specimen test where the high voltage lead connects to the top of the arrester and the low voltage lead connects to the bottom of the arrester which is ground. Before testing, make sure the associated bus is de-energized and removed.

Power Cable Fault Locating

During the 1930's the larger metropolitan areas in the U.S. became networked with underground distribution and subtransmission power systems.* Naturally, this involved the use of electrical "pipes" or cable for moving electricity from the source to the load. A cable in its simplest form is merely a conductor surrounded by an insulating material. Copper and aluminum have been the most common conductor materials, while typical insulating materials include oil-impregnated paper, varnished cloth, and thermoplastic. Specifications are given in the *National Electrical Code.*

While power cable is the most rugged of all electrical apparatus, it tends to be greatly abused. The majority of cable failures occur at terminations or splices.

**A sizable increase in the case of underground cable took place in larger industrial plants during 1945-1955. Then in the 1960's a great expansion of residential and commercial surrounding metro areas also utilized underground residential distribution (URD).*

A fault which develops in a power cable will generally fall into two classifications: an open/high resistance fault or a dead short/low resistance fault. The high resistance fault is the most common and usually consists of a break or gap in the insulation. This, in effect, causes a high resistance to develop between conductors in the cable or between a conductor and the outer shielding or casing of the cable. The low resistance fault is essentially the same except that the conductors in question may actually be in physical contact with the outer shielding or casing or with each other; or, the high resistance fault may come in contact with water. Either type of fault is normally located in a shielded cable with the use of an impulse generator, Thumper, or radar to time the signal.*

The principle of fault location is simple, in which a high dc voltage is applied to a capacitor, which will charge to a predetermined value, as established by the setting of an adjustable sphere gap. The cable under test is in series with the sphere gap and capacitor. When the voltage in the gap reached the present value, the capacitor instantaneously discharges across the gap and a pulse of energy with a very steep wave front is sent down the cable to bombard the fault. If the potential of the pulse is of sufficient magnitude, it will break across the fault and return to the low side of the capacitor. If the pulse is not of sufficient magnitude to break down the fault, the arc current will merely serve to charge the cable capacitance.

The voltage of the energy pulse may be increased by increasing the sphere gap setting until the voltage becomes of sufficient magnitude to break down the fault. The maximum voltage applied to the cable is limited by the adjustment of the gap, and the minimum voltage is determined by the resistance of the fault.

In most cases, when the energy pulse flashes across the fault, there is a loud audible report, which will reveal its exact location. It is only necessary, therefore, to traverse the route of the cable and listen for the fault to reveal itself.

In the case of a dead short or where other conditions limit noise, a pickup coil with suitable indicating device or earphones may be used. The magnetic field created in the earth at the fault can also be used to detect the fault.

Digital radar fault location is similar to cable thumping. In this method a pulse of energy is sent out over a line or cable and a

**Megohm meter measurement may suffice for non-metallic shielded cables.*

portion of the energy is reflected by an impedance discontinuity back to the receiver. The transmitted and received pulses are displayed on a cathode-ray tube and a ranging device is provided to measure the time interval. Time is in direct proportion to distance of the fault.

Part 2—Protective Apparatus Coatings

Introduction

Protective coatings or paints are used for various purposes. In most homes and office interiors, for example, they are primarily used for aesthetics, whereas, in steel structures such as chemical plants, coastal bridges, electric transmission towers, and apparatus surfaces, they are used for corrosion resistance to the atmospheric environment.* Without coatings, such steel surfaces would corrode rapidly and revert to iron oxide.

Coatings provide protection against deterioration by forming a very thin protective layer (or film) that must resist all destructive atmospheric elements.† Proper maintenance and repair of this thin film can extend the life of protected equipment. Such maintenance and repair of coated surfaces can also result in enormous financial savings, if performed at the appropriate time. Thus, a minor touch up of the protective coating may save apparatus from disintegrating beyond repair to a state where the whole structure will require a complete new coating system — a task involving thorough surface preparation, priming, and finish coating. Therefore, in selecting and applying protective coatings, it is important to consider the quality of coatings to be employed.

Historically, any quality coating available was applied to electrical apparatus substrates. This practice is fallible for today's computerized transformers. High performance coatings must serve a dual purpose. First, the paint system must allow for rapid radiation of heat, and be capable of reflecting solar light rays. Second, the system must protect against localized atmospheric conditions.

Since transformers generate internal heat, they must be properly coated to help prolong their lives. Even color affects a transformer's ability to do this. Therefore, transformers require a different evaluation of paint systems because of the multiplicity of problems to be solved.‡ At least 20 different systems of painting transformers are

**Apparatus includes OCBs, current transformers, switchgear housings, as well as transformers.*

†Thin, protective dry layer of 6-11 mills.

‡ANSI Std. C57.12.22 (1969). Among the requirements for pad-mounted transformers an undercoating over the regular finish shall be applied to all surfaces that are in contact with the pad to minimize corrosion.

now on the market. Considering the variables of each of these systems, an infinite number of combinations are available from which a choice can be made. As a matter of interest, 30 percent of all current research (Circa 1981) is devoted to developing new paint systems. A layman can scarcely take time to research all of these new products as they come on the market. It is, therefore, important to seek professional help from coating control engineers who can help to control corrosion in transformers.

The Corrosion Problem

Of all the metals available, the principal metallic substrate of electrical apparatus is steel, because of economy, flexibility, and ease of fabrication; nevertheless, it is the most likely metal to corrode.

The most minute form of coating deterioration often leads to pin hole rust. This is followed by underfilm cutting, which destroys the coatings surrounding the rust blooms. These rust blooms can quickly develop into isolated pitted rust if there is a direct chemical attack on steel surfaces. With time, rust penetration is accelerated, leading to oil-leaking problems, particularly in the area of the cooling fins, tubes, and weldment areas.

When such a condition occurs, the overall coating may not be salvageable. The delay which occurred in preventive maintenance will be very costly if repair to cooling radiators and downtime is needed. Therefore, the basis for corrosion control is attacking the problem when the initial signals of deterioration occur (Figure 8.8)*.

Since the cost of corrosion and corrosion control to the United States has been estimated at nearly 70 billion annually, corrosion control of apparatus should be an indispensable goal of engineering. In the electrical power industry alone, corrosion is said to be responsible for 50 percent of all electrical outages. The economic impact of this problem is so acute that the utility industry is spending about ten million dollars annually on corrosion research.†

Corrosion is normally caused by chemical and electrochemical oxidation, and occurs in the presence of both oxygen and water.

**Figure 8.8 is located in the Color Section following Chapter 10.*

†NSB Publication 511-2, 511-1 (Circa 1978).

Dampness — The Overriding Factor in Corrosion

Most contaminants will have little or no effect in the absence of moisture. A film of dew, saturated with corrodents, provides a very aggressive electrolyte for the promotion of corrosion.

Humidity above 80 percent causes rapid rise of corrosion. This is due to the function of the hygroscopicity of contaminants. A constant source of wetting is provided to cracks, crevices, cooling fin bracing, and under top-lids of transformers. It has been said that two-thirds of all continental corrosion occurs in humid climates near oceanic bodies.

Thus, a damp surface attracts corrosion products from the air, and provides a basis for accelerated corrosion. Among the major types of corrosion are the following: bimetallic, crevice, pitting, thermogalvanic, spalling, and poultice (internal, organic acidic sludges) corrosion. Such forms of corrosion are often associated with rural, marine, caustic, and chlorinated environments. In some cases, bad transformer oil acts as an internal corroding agent.

Since corrosion is impractical to eliminate, the secret to sound engineering is its early detection and control through barrier coatings, rather than preventing it.

Preventive Versus Remedial Painting

Traditionally, transformers have been the most neglected piece of equipment in the plant. Today, they are the easiest to maintain and to protect, and the hardest to replace. Keep in mind that "the condition of the coating and surface is undoubtedly the most important in any repair or maintenance program."

Preventive painting begins at the first phase of deterioration, that is, before rust begins to appear. The longer the decision to repaint is put off, the higher the costs. If the decision is put off too long, you are playing roulette with your electrical equipment — and finally oil leaks in the cooling radiators. In fact, you are into *remedial* painting, which costs a minimum of 25 percent more than preventive painting.

Three basic questions should be answered concerning an on-going maintenance program.

1. What is the condition of the original coating, and how much disintegration or corrosion has already occurred?

2. What is the condition of the basic surface? Both the type of repair and the type of coating vary with the underlying surface condition.

3. What is the existing type of coating on the surface?
If the ASTM Standard (ASTM D-610), which was developed to rate the condition of a surface, is recognized and followed, the condition of the surface can be rated by several people. The evaluations of paint deterioration should be reasonably consistent.

   The first phase of such deterioration normally shows up as changes in the top coat appearance with no significant effect on its protective qualities. These changes include a combination of soiling, color change, loss of gloss, rust, bleeding, or chalking. The vital signposts of paint deterioration are briefly described below.

Signposts of Paint Deterioration

Continuous surface inspection should focus on the following trouble indicators.

1. Chalking
The result of weathering of epoxy paint at the coating surface. Cause: sunlight. Contrary to "house paint" claims, controlled chalking can NOT be an asset as a self-cleaner on a transformer. It may even attract additional air-borne particulate matter.

2. Checking and Cracking
Describes the breaks in the paint film which are formed as the paint becomes hard and brittle. Major cause: temperature variation (expansion and contraction of the steel surface).

3. Flaking and Peeling
As penetrating moisture loosens relatively large coating areas, the paint curls slightly, exposing more of the basic surface, and finally flakes off. Cause: Wrong choice of paint or improper cleaning of surface.

4. Mildew
A fungus grown in the presence of moisture which feeds on the coatings. Factor generally only associated with house paints.

5. Rust (Corrosion)
   Deterioration of a substance (usually a metal) because of reaction with its environment; especially with oxygen, moisture and the oxides of carbon, nitrogen, and sulfur (ASTM D-610 Rust Guide).

6. Permeability
   The process of soaking up water by the paint film and the transmission of the water through to the metal surface. Then, hydrogen (which is at the surface) develops embrittlement.

7. Osmosis
   The *accelerated* blistering of the paint film due to the formation of soluble salt solutions beneath the coating caused by moisture (water vapor transmission — W.V.T.).

8. Spalling
   Separation or coat loosening adhesion and peeling away, that is, paints won't tolerate excessive surface temperature.

The above coating failures can be easily located in areas such as the bases of transformers, particularly where units sit on concrete without air vents; sharp edges such as cooling fins; angle iron bracings welded onto fins for rigidity; the lower third of cooling fins, close to the manifold; underneath of top-lid, close to explosion tubes, and bottoms of radiators due to corrodents so migrating. Coating failures of this nature need immediate attention, if deterioration is to be checked.

Solution to Paint Deterioration

The application of protective coatings to steel surfaces affords a solution to the corrosion problem. Such a solution involves a thorough preparation of steel surfaces, followed by the application of primer and finish coats.

Surface Preparation

Correct surface preparation is the key to the lifelong performance of any coating system. Studies have shown that about 80 percent of all coating failures are a result of inadequate or incomplete surface preparation.

Quality surface preparation can be costly, but one should never reject it because the longer repainting interval will offset the additional cost. Since labor is the major cost factor, comprising approximately 70 percent of the job price, it is much cheaper to do it right than to do it all over.

There are three principal methods of preparing steel surfaces for painting.

Scraping and Wire Brushing

Scraping and chipping of very thick layers of old paint and other contaminants are often used as a preliminary to other preparation techniques such as wire brushing or power tool cleaning. Wire brushing or hand tool cleaning is one of the most widely used techniques for removing loosely adhering mill scale, deteriorated coatings, and rust. Unfortunately, wire brushing does not usually reach lower contours of the steel, except the upper contours of surfaces. Thus, sound coatings which are glossy or very hard to clean should be "sweet-blasted" to slightly roughen the surfaces.* This will help maximize the adhesion capacity of the new coating. Wire brushing equipment is varied, and includes a wide selection of brush sizes and shapes made from different types of bristles.

Removal of heavy rust scale or previous coatings over extensive areas can be done by electrically powered equipment. Power tool cleaning usually results in a better job compared to hand tool cleaning. This technique is much faster, easier, and more economical.

Powder Blasting

Blasting is perhaps the most effective method of removing mill scale, old paint, rust, hardened askarel, and other surface layers of the steel substrate. This can be achieved by blasting with pulverized limestone or corn cob.†

For best results, the following sequence of operations should be followed:

**Sweet blasting is very light sand blasting just enough to "pickle" the surface.*

†Sand is not recommended since it will damage porcelains and is too harsh for cooling radiators or glass insulators.

1. Before blasting, remove spilled oil or askarel, salt, chemicals, dust and other contaminants by solvent wash.
2. Before blasting, remove hardened oil and grease by spraying with a chemical degreaser to soften.*
3. Blast to the desired grade such as: white metal blasting in which all mill scale, rust, rust scale, previous coating, etc., are completely removed, leaving the surface a uniform gray-white color; and near-white blasting in which the metal is cleaned until at least 95 percent of each square inch of surface area is free of all visible residues.
4. After blasting, all grit and dust should be removed with a clean brush, vacuum cleaner or dry and clean compressed air. Before using a primer coat, all areas close to the surfaces to be painted should be wetted down to prevent wind-blown debris from contaminating the paint.
5. In areas of severe corrosion, solvent wash the substrate with mineral spirits or naptha reducers.
6. Painting of the blasted surfaces should be done as soon as possible, that is, before contaminants settle on the cleaned surface.

Even with proper surface preparation, it is safe to predict that rust will recur **again** in the same areas, as shown in Figure 8.9. What then can be done? Years of experience have led one major contractor to apply a special pre-primer to help neutralize rust in such uncontrollable areas.†

More severe rust-pitting — may result in oil leaks if mechanical blasting is used (Figure 8.10). Cooling radiators are particularly thin and subject to greater corrosion from within (by oil decay products) and from without (by atmospheric pollution and weather). Also, radiators do not have as much internal heat to dry the surface as does the main tank. As a matter of interest to show how thin some of the radiators are, many reported cases show that they will actually collapse under only four psi vacuum.

Caution: Rust forms **under the oil and leaks may have already developed prior to surface preparation.*

†Pre-primers are helpers only—not absolute neutralizers.

Other important considerations in abrasive blasting include the type of equipment used, such as nozzles and pressure hoses. The size of nozzles and distance from substrate are also important as they can affect the quality of the finished blast. During the blasting process, operators should wear a protective face mask to protect themselves from the hazards of fine dust. To avoid such hazards, wet blasting is recommended.

This involves the use of water soluble inhibitors, such as phosphates and chromates — which can effectively retard superficial corrosion after sandblasting.

Figure 8.9 - A special pre-primer will prevent recurring rust in the same areas.

Chemical Cleaning (De-energized)

Chemical cleaning or acid pickling is probably the most satisfactory technique of preparing surfaces for a protective coating. The method involves the use of dilute acids (up to ten percent) such as sulfuric, phosphoric and hydrochloric solutions at temperatures ranging from 70°-200°F. These acids dissolve rust, mill scale, and other contaminants from steel surfaces and convert them into soluble salts.

One practical method of chemically stripping paint from transformer surfaces is the flow procedure. In carrying out this procedure, the

Figure 8.10 - Pitted rust (lower left-hand corner) needs immediate attention.

first step is to remove fans, gauges, and other accessories from the transformer. Those that cannot be removed, such as tap changers, bushings, and any hardware should be masked with rubber sheeting. The second step is to arrange drip pans around the base of the transformer, and mask its bottom in a way that will direct the flow of solution and stripped coating into them as illustrated in Figure 8.11.

The most widely used chemical stripper is an alkali solution to which a wetting agent is added. Since this type of solution works more effectively when hot, it should be kept in an external tank at a temperature of about 180°F (82°C). The solution is pumped through a hose and then through a nozzle to the transformer surface being stripped. The alkali solution tends to blister the paint, which peels off and drips into the drip pans. The solution can be reused, provided it has been screened and poured back into the heating tank. After cleaning with the alkali solution, the stripped surface should then be scrubbed with a strong commercial detergent solution. After scrubbing, the surface should be flushed thoroughly with fresh water, and then dried before it is coated. Any paint, grease, oil or salt that does not come off can be removed by steam cleaning.

The final step in surface preparation is to thoroughly dry the surface in readiness for coating. This step also involves the selection of the right type of coating system to be used.

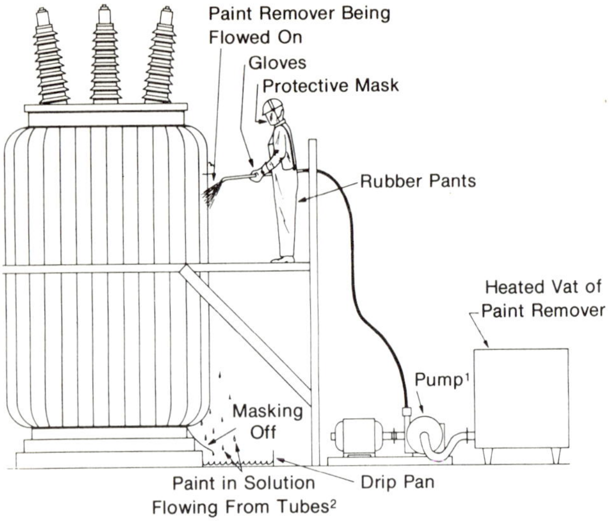

[1]Pump requires special seals.
[2]Overflow paint is returned to the pump.

Figure 8.11 - Flow method: chemical removal of old paint. General Electric Co., "Repainting Transformers" (1971).

Selecting the Right Coating System

Two major factors come into play in deciding what kind of coating system should be chosen — environment and color.

Choosing the right coating system for electrical equipment is not an easy task. It may be better described as a process of **elimination** rather than selection. Today, we are swamped with paint choices — epoxies, vinyls, oil coatings, alkyd resin paints, phenolic resin paints, urethanes, etc. These are only a few of the many different types of coatings available for maintenance painting. Remember, there are over 20 different systems with variables in each system. Therefore, in choosing a coating system, particularly for transformer units, one should be guided by certain criteria. Table 8.1 illustrates criteria that can be employed in selecting coatings for transformers.

Environmental Criteria

Two additional factors should also be taken into consideration —the amount of surface preparation required, and the environment in

TABLE 8.1

CRITERIA FOR CHOOSING TRANSFORMER COATINGS[1]

Barrier (or finish coats) Coating Type	Gloss retention	Color retention	Fading	Water resistant	Acid resistant	Alkali resistant	Solvent resistant	Marine resistant	Adhesion	Hardness	Temperature 190° F
Alkyd	A	A	A	A	NR	NR	NR	NR	V	A	V
Alkyd silicone	V	V	V	V	NR	NR	A	V	V	A	V
Vinyl	A	A	A	V	V	V	A	V	A	A	NR
Polyester epoxy	A	V	A	V	A	V	V	V	V	V	V
Polyamide epoxy	NR	A	A	V	A	V	V	V	V	A	V
Amine epoxy	NR	NR	NR	V	V	V	V	V	V	V	NR
Oil-modified epoxy	NR	NR	NR	V	A	A	A	A	V	A	V
Chlorinated rubber	A	A	A	A	V	V	NR	A	A	A	NR
Urethane (ASTM Class V)	V	V	V	V	V	V	V	V	A	V	A
Oil-modified urethanes	A	A	A	A	A	NR	A	A	A	A	NR

[1]Relative *resistance* properties of commonly used transformer coatings.

Primers Coating Type	Adhesion	Chemical resistant	Hardness
Long oil	V	NR	NR
Vinyl alkyd	V	A	A
Epoxy	V	V	V
Water	A	A	A

Legend

V - very good

A - average

NR - not recommended

which the equipment is located. Selecting protective coatings for different environments may be determined by the geographical location (Figure 8.12). This is important because there is a direct relationship between color and operating temperature. In other words, color can have some effect on transformers in service.

Effect of Color on Paint System

In choosing a surface color, one should consider the effect of internal heat loads on the paint system. This is important since certain colors will increase temperature rise of transformers in sunlight.

Traditionally, electrical equipment has been painted dark colors such as black, blue, green, and gray. The 50 year old theory held that while white bodies absorbed less color heat than black bodies, black bodies tended to radiate internal heat better than white bodies. The practical assumption was that the two effects neutralize each other. An operating temperature rise of a white-painted transformer was only about 10-20 percent lower than a black-painted unit. So, the actual difference of no more than 2°C advantage could not justify the use of special paint, meaning other than the standard black. However, today's operating temperatures have increased by a factor of two or more, making that 10-20 percent reduction more attractive.

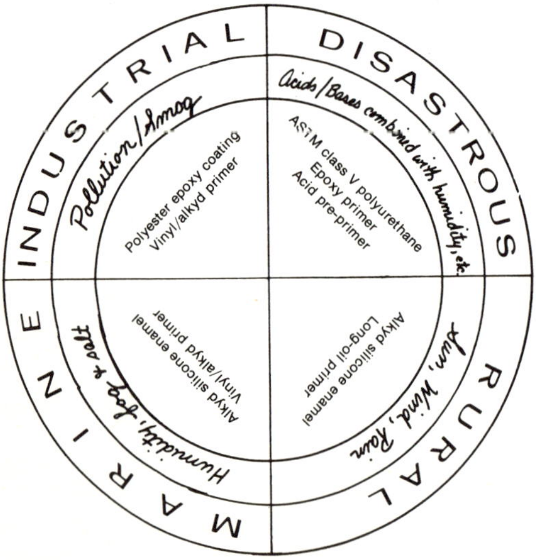

Figure 8.12 - Geographical product guidelines.

In addition, white bodies have proven to radiate low temperature (below 200°C) heat equally as well as black bodies.*

Figure 8.13 shows that white bodies do not absorb heat well at high temperatures. Thus, contrary to past theory (black body emission), a white body will operate several degrees cooler than a dark-colored surface. Figure 8.14† illustrates this factor of reflectance, or the ability of the top coat to reflect sunlight, and consequently heat. Gloss reflectivity can only achieve its maximum potential by using hard and smooth-surfaced finish coating.

Silicone alkyds, polyester epoxy, and ASTM class V polyurethane exhibit the highest order of gloss and color retention.‡ Table 8.2 shows how the various colors compare in terms of light reflection.

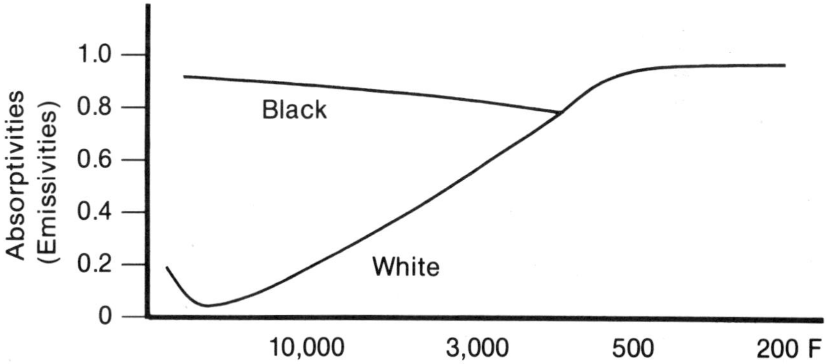

Figure 8.13 - Comparing the emissivity of black and white paints.

To illustrate the effect of color on paint systems, an example will suffice. In climates with direct sun radiation, a black-painted transformer under extremes of solar radiation would increase the temperature 15°C above the ambient temperature generated by the equipment. On the other hand, a white-painted unit will absorb a little radiation (about 15 percent of 15°C or 2°C). Therefore, a white transformer operates 13°C cooler. Table 8.3 compares various colors and their radiation absorption coefficients.

**Up to 200° C both dark and white colors absorb 95 percent heat. Hence, in the range transformers normally operate white will conduct with the same rapidity as a dark color.*

†*Figure 8.14 is located in Color Section following Chapter 10.*

‡ *Some modified formulas do* ***not*** *exhibit* ***high gloss.***

TABLE 8.2

Color	% Color Reflection
White	80-97%
Very Light Tints	70-80%
Light Tints	60-70%
Medium Tints	25-60%
Aluminum	41%
Dark Blue	3-15%
Black	1-2%

TABLE 8.3

Paint Color	Solar Radiation Absorption Coefficient
White	0.14
Cream	0.25[1]
Yellow	0.30[1]
Light gray	0.35[3]
Light gray, green, blue	0.50[1]
Medium gray, green, blue	0.75[1]
Dark gray, green, blue	0.95[1]
Black	0.97[2]

[1]Approximate—because of wide variations of the color spectrum.

[2]Variance in black colors.

[3]Estimate.

In choosing a "light-colored" paint, one must not make the mistake of picking aluminum paint. Studies have shown that although aluminum has good sun reflectance, (up to 100°F) the aluminum flakes from such paint overlap like leaves, and forms an insulating barrier which slows dissipation of internal heat (aluminum emissivity is 0.55 at normal temperatures). A plain aluminum-painted tank will run approximately 30 percent higher temperature rise than if painted with a non-metallic paint.*

The Primer Coat

Primers are important insofar as exterior finishes are concerned because of the exposure of the top coat to the outside elements, to variations in temperature and humidity, and to internally generated heat.

Primers play a major role in protective coatings. First, a selective primer acts as a glue, providing an adhesive bond between the prepared surface and the finish coat. Universal primers will not lift existing aged paint, but will accept active oxygenated solvent high performance coatings such as epoxies and polyurethanes. Lack of primer adhesion will result in the coating system peeling off in "sheets." Second, a primer acts as a rust inhibitor. A passive rust-inhibiting primer helps to neutralize hidden rust, control under film cutting, and adds strength to the coating system. Typical rust inhibitors include zinc chromate, iron, red lead, aluminum dust, barium metaborate, and zinc dust.

Oil-based primers such as tung oil and linseed oil contain one or a combination of metallic rust inhibitors. These metallic inhibitors like zinc and flaking aluminum also form heat barriers. Since transformer oil may ruin any product containing zinc, zinc-rich paint is not advised for painting pad-mounted electrical apparatus. Note that oil-based primers are slow-drying, and therefore require a longer waiting period before the finish coat is applied. Of course, the warmer the apparatus, the faster the drying period.

Third, the primer must be compatible with the prepared surface, as well as the top coat, to assure top-coat adhesion. As examples, in

**ANSI/IEEE Standard C57.12.00-1973. "Metallic flake paints such as aluminum, zinc, etc., have properties which increase the temperature rise of transformers except in direct sunlight."*

non-corrosive or mild environments, oil-lead, phenolic, or epoxy ester primer may be used in conjunction with a silicone alkyd finish coating. In corrosive areas, vinyl alkyd (heavily rust-inhibitive pigmented), or epoxy primer may be employed along with an active solvent finish coat (that is, epoxies). Finally, in aggressively corrosive areas, a high build epoxy primer coupled with a Class V urethane finish may be chosen.

On the other hand, use of vinyl, zinc-pigmented, or chlorinated rubber primers should be avoided. A vinyl primer requires a bright metal surface, and has poor bonding (adhesion) qualities. Zinc-pigmented primers, as mentioned earlier, may be attacked by transformer oils. In addition, they are high build, very heavy (about 26 pounds per gallon), cannot be flow-coated and may add a lot of weight on the cooling radiators resulting in the impairment of heat transfer (internal to external). Chlorinated rubber primers are limited by temperature exposure (no more than 135^0F), as well as poor adhesion and cohesion (the attraction between atoms and molecules within a substance). Even the best of today's paints — urethanes — are difficult to apply in damp, humid areas.

Other conditions that dictate the final composite choice include:

1. Recoat time — that time after an application when a coating achieves maximum hardness.
2. Basic surface condition.
3. Chemical and abrasion resistance.
4. Type of application, such as:
   a. Immersion in liquids — continuous direct exposure to various liquids.
   b. Marine and chemical environment — underfilm corrosion because of affinity of coating to moisture, oxygen, or chemicals.
   c. Supplementary pigments — such as for the reduction of water vapor transmission that leads to blistering.

Thus, it can be realized that painting transformers is a unique service, requiring coating specialists who are familiar with the types of primer and finish coats. The next section deals with the third and final step in a coating system.

The Finish Coat

The finish coat performs several functions.

1. Aesthetics
   This is a very important area, because the first reaction to a paint job is "How does it look?" The use of high gloss coatings is functional as well as decorative.*

2. Safety
   Color coding, indicating various standard locations and hazards, as well as increased visibility. For example, some painting manufacturers use yellow to designate caution against physical hazards, and purple to designate radiation hazards (OSHA color coding).

3. Protection

   a. Weather: Sunlight, wind, abrasion pollution, heat, and humidity. For example, in winter an energized unit could be hot at the top, very cold at the bottom, with quite a differential between internally generated heat and external temperature (thermogalvanic corrosion). For comparative purposes, consider how much heat difference various coating systems will accept. High continuous temperature tolerances are found in alkyd silicones (300°F) and catalyzed epoxies (225°F), polyurethane (200°F), while low temperature tolerances are found in chlorinated rubber (140°F), and vinyl (140°/150°F).

   b. Water: Dampness may stain, harm or destroy a coating. Keep in mind that water from industrial cooling towers contains acids with a low pH factor. Therefore, water will capture corrosive droplets and contain them on the apparatus surface long enough to promote electrochemical reactions.

   c. Chemicals: Acids and alkalies like sulfuric, hydrochloric, nitric, phosphoric, formic, caustic soda, lye, lime, as well as fertilizers, tree sprays and road salt will attack any paint system. Therefore, by choosing the right paint system, longer life for the transformer protective coating can

High gloss finishes show scuffs and effects of air-borne particulate matter more easily than low gloss finishes.

be assured, even under the most adverse conditions. Nevertheless, under severe conditions no paint is assured of longevity.

In applying the finish coat, the following points must be taken into consideration. First, the finish coat must be compatible with the primer coat. Adhesion and cohesion are qualities of top priority. Several quality universal primers are available, and can minimize this problem. Second, tailor the coating system for the best chemical resistance by using catalyzed epoxy or ASTM Class V polyurethane. Figure 8.14 illustrates the final coat on apparatus.* Third, if corrosive fallout or splashing exists, then any system will fail prematurely.

For instance, the most critical painting problems exist in chemical plants in and around the cooling towers. Two alternatives are possible. Either paint when the cooling tower is not operating, or build a covering for the transformer while various phases of painting are going on. Only a few minutes exposure of the prepared surface to over-spray before the primer is applied programs the paint system to failure.

As you think about color, think ASTM Class V urethane, for it exhibits the following qualities:

- Resistance to moisture permeation
- Temperature flexibility
- Corrosion resistance
- Ninety-five percent light reflection (white only) and excellent color retention (fade resistance).
- Hard finish. Fungus cannot start a growth in humid areas.
- Curing time (36 hours) when corrosive resistance takes effect. Note that the solids content may vary among different paint formulators. The foregoing criteria apply to ASTM Class V, catalyzed TDI isocyanate with ultraviolet filter (or HDI modified with acrylic resins).

Fourth, finish coatings need not be thick to be effective. Too much millage is detrimental leading to early flaking, adds excessive weight to the cooling fins, and impairs radiation, convection, and ventilation. Six to ten mills of paint is the accepted average.

Finally, the methods of applying the finish coating are a significant factor to the longevity of the paint system.

**Figure 8.14 is located in the Color Section following Chapter 10.*

Methods of Applying Paint

There are three methods used for applying a finish coating to steel surfaces — brush or roller (hot stick), flow method, and spraying. These methods may be used separately or in combination during the energized operation. This choice is not a simple decision. The economics of the three methods is weighed against the suitability of the system to each type of application.

1. Brushing
   Brushing is perhaps the oldest known method of applying paint to power transformers and other accessories. Although the method is no longer used to any great extent (it is a very slow process), it is still employed extensively for spot touching. Roller application, which was introduced in the early 1950's, is used in conjuction with brushing. Its advantage over brush application is that it is much faster and can cover large areas within minutes.

2. Flow Method
   New application has been used successfully for painting or repainting electrical apparatus in the field. The method can be used on self-cooled or fan-cooled power transformers, requires minimum equipment, and is rather inexpensive. The equipment, as shown in Figure 8.11, might include improvised sheet metal pans with flared edges to hold access paint from the tank; a gear pump with a pressure adjustment system; and a neoprene rubber hose with a flattened aluminum tubing for applying the paint.

   The coating process itself is fairly simple. Before painting, accessories such as bushings, nameplates, gauges, and breathers should be covered with masking tape to protect them against paint splashes.* Where the technique is used by some utilities, painters have found it practical to use brushes for painting the cover and base of transformer units.

3. Spraying
   High pressure (not airless) spray application is considered the most effective atomization process for coating electrical apparatus with radiators, such as forced-oil air-cooled units.

Fan blades are usually stainless steel or tempered. Thus painting is not recommended.

Spraying equipment is both complex and costly, and ranges from simple aerosol spray bombs that can be purchased from hardware stores to high pressure gun assemblies. The latter technique offers many advantages such as speed and efficiency, and allows the painter to get close to bushings, gauges, and valves.

Where spraying is chosen for coating steel surfaces, external-mix spray guns are often recommended. Before any spraying is done, all accessories should be masked to protect them against unnecessary coating.

Precautionary Measures

In any high quality protective coating, it is important that precautions be taken to assure professional results. First, do not insist on blast cleaning to white metal, especially in fin areas. Keep mill thickness under ten, or expect heat buildup. Second, it is always advisable not to purchase the cheapest paint, and to avoid vinyl, zinc-pigmented, or chlorinated coatings. Third, avoid purchase of paint without guarantee of mutual compatibility between primer and finish coatings. Fourth, do not paint apparatus using dark colors or aluminum paint. Finally, painting should be done only when ambient temperatures are suitable. Where painting is being performed on or near energized equipment, safety personnel should always be provided.

Exploring High-Performance Coatings

Since no paint is suitable for all applications, it is necessary to tailor the product to specific degrees of corrosion.

1. Mild, Marine or Rural
The use of alkyd-silicone enamels is very evident where lasting appearance, resistance to mild chemical exposure and marine atmosphere prevail. The initial gloss approaching 90 percent provides for the reflectance of solar light rays. The presence of 30 percent silicone in the preferred formula resists normal weathering, and elevates the constant temperature limit to 300°F (150°C) without harming the finish.

2. Heavy or Chemical
Catalyzed polyester epoxies are formulated for optimum

gloss and color retention qualities in their generic family. The hard and glossy finish results in a coating which is resistant to chemical attack, abrasion and high heat. While some epoxies will quickly fade and discolor, the polyester formula will retain its appearance longer in constant temperature ranges exceeding 200°F (~93°C).

3. Corrosive Chemical Atmosphere
   ASTM Class V polyurethanes, when coupled with a catalyzed epoxy primer, are gaining wide usage wherever corrosion engineers specify the most physically durable, flexible coating.

   They are used by airlines to protect their jets, by railroads for their diesel-electrics, and the transportation industry selects this coating in corrosive areas where maximum resistance to chemical attack is mandatory. No coating is without several negatives. ASTM Class V polyurethane is no exception.

   - Oxygenated solvents (including alcohols).
   - TDI or HDI (Solvents in formula possibly may cause bronchial discomfort and if inhaled for extended periods may cause bronchialnoma or cancer of the lungs).
   - Difficulty in flow-coating.
   - When being sprayed eye irritation and tears will result.

Ongoing Research

Environmental demands and governmental concern for human health have intensified research and development efforts of the coating industry.

One focus is on the refinement of water-soluble coatings. At this point, water-borne coatings do not have the proven durability of solvent-based coatings. We are confident that this issue will be resolved because 30 percent of all chemists today are involved in research solutions to such coating problems. During the interim, however, proper corrosion control starts with preventive maintenance — the recoating of units before rust becomes too apparent.*

Even* ***new* *transformers should be carefully examined after installation. Scratches and scuffs frequently received should be touch-up painted.*

Part 3—Energized Cleaning of Transformer Bushings and Substation Insulators

Introduction

At least since 1902, the problem of transformer bushings and standoff insulator contamination due to natural deposits (such as early morning dew, salt fog in sea coast areas, and more recently, smog) and subsequent industrial pollution has plagued the electrical industry. Such contamination has often resulted in noisy substations, damage to insulating surfaces, partial discharge, tracking flashover, and loss of power.

It is possible to prevent insulators from flashing over by periodic hand wiping (de-energized), periodic washing, or periodic dry cleaning using such techniques as dry powder blasting with an insulated "hot stick" or periodic coating with grease compounds (energized or de-energized).

The Contamination Problem

Bushing, insulator and lightning (surge) arrester structures in the substation are subject to the same contamination problems as those that plague transformers and steel coatings; some authorities say they are even more so because of electrostatic attraction for airborne contaminations (Figure 8.15).

Build-up of contamination can, however, cause more problems to porcelain insulator structures than to metal parts of a transformer system. This is echoed by A.D. Lantz of Ohio Brass, who in 1956 stated that:

> Surface contamination and leakage currents caused operating difficulties eighty years ago . . . and exist today as one of the most serious insulator problems. Bushings and insulators are still not maintenance-free, and still need regularly scheduled cleaning to rid them of different types of contaminants.

Types of Contaminants

Contamination material comes in different sizes and varieties. Table 8.4 below shows a range of typical pollution-particle sizes.

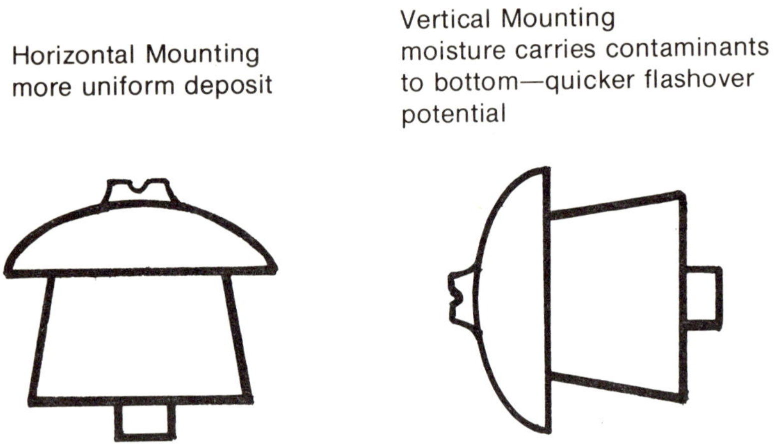

Figure 8.15 - The type of insulator mounting determines the degree of contamination. Two positions of pin or cap insulators. Compare with Figure 1.64B.

TABLE 8.4

RANGE OF TYPICAL POLLUTION-PARTICLE SIZES		
Nature of suspended matter		Diameter
Inorganic	smokes	0.001-0.3
	fumes	0.01-1.0
	dust	1.0-100.0
Organic	bacteria	1.0-10.0
	plant spores	10.0-20.0
	pollens	15.0-50.0
Water	fog	1.0-50.0
	mist	100.0-100.0
	drizzle	50.0-400.0
	rain	400.0-4000.0

Airborne contaminants can be divided into two general classifications: Conductive and non-conductive or inert. Common conductive pollutants include carbon dust, fly ash, metallic particles, some ores, volcanic ash, and some chemical salts such as sodium chloride, sodium sulphate, magnesium chloride, and some acids. When in solution, these ionic salts can cause partial discharges and flashovers by providing a conductive coating on the surface of an insulator structure. The severity of pollution caused by these salts is generally measured in terms of the equivalent salt deposit density.

Non-conducting contaminants are composed of solid material that does not go into solution as ions. They include clays such as kaolin and bentonite, inorganic cements, silicone dioxide as well as fertilizer and lime dust in farming areas. These inert contaminants may be hydrophilic, and thus increase the wetting rate of the insulator. On the other hand, other inert materials such as grease and oil may be hydrophobic, and often decrease the wetting rate of the insulator. Once grease and oils have reached their absorbent limits the oils/grease harden and then support surface wetting. Non-conducting pollutants such as cement dust do not cause trouble when dry, but when the humidity is high, they absorb water and change from inert to conducting. When some critical value is reached, arcing over occurs.

Figure 8.16 shows the relative distribution of the most common contaminants. This data is based on seven years experience of a

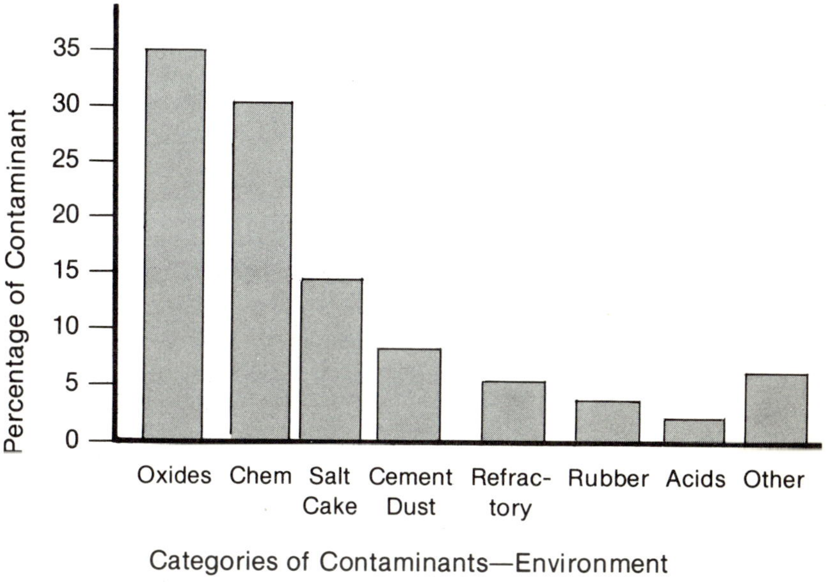

Figure 8.16 - Most common airborne contaminants affecting useful service life of bushing insulators.

major transformer maintenance contractor and indicates that the highest percentage of service work has occurred in the areas of metallic and chemical contamination.*

Field experience has further shown that the worst contaminants are metallic oxides. Such oxides may even have a more destructive effect on transformer bushings because of the magnetic fields present (Figure 8.17). In extreme cases, internal heat coming up through the bushing from the transformer may increase the formation rate of glazed deposits. These two factors may help explain why the failure rate of bushings is much higher than that of standoff insulators.

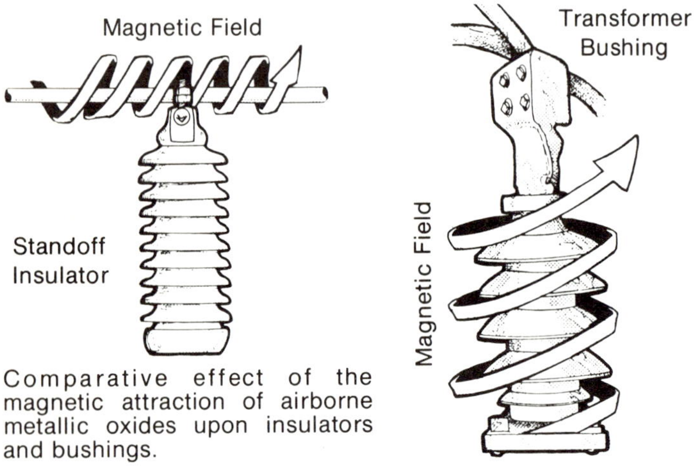

Figure 8.17 - Effect of metallic oxides on bushings/insulators. Bushings are even more critical because a bushing is a conductor, as well as an insulator (Figure 1.64). Thus, a magnetic field exists along the entire length of the bushing. Heat from the tank also compounds the rate of conducting potential.

Dynamics of Contamination Collection

1. Air Movement
   Many researchers have investigated the dynamics of contamination collection on bushings and insulators. Results of such studies have not been conclusive on the factors that can be employed to accurately predict the collection rate.

**In addition, such sea-coast areas as the California and Gulf coasts, New Jersey and Maryland which frequently (daily in California) have salt fog but no rain for many months.*

It is, however, generally accepted that air movement is responsible for the contamination of insulator structures. These contaminants can be easily observed in areas where wind direction is fairly constant, such as 138 KV substations and 34.5 KV lines in Colorado Springs, Colorado where air movement is a significant factor six months out of the year.

Most airborne particles are collected in areas of high stress such as the pin region on suspension insulators, where the electrostatic or electromagnetic field has some significant influence. This field also holds particles through a process of dielectric polarization once the particles are settled on the surface.

In industrial regions, especially near oil refineries and chemical plants, insulator surfaces may be subject to etching or erosion if they are in the presence of a diluted acidic solution. The effect of such a condition is a reduction in natural washing, such as heavy rain and an exposure to greater accumulation of wind-borne pollutants.

2. Moisture Deposits

Moisture deposits on insulator surfaces are another way through which contamination is collected. Insulator surfaces can accumulate moisture through condensation, fog and impingement of water particles. Condensation takes place when the surface temperature of the insulator is reduced below the dew point temperature. Studies have revealed that flashovers on in-service insulators often occur in the early morning hours — the time when dew condensation is at its peak.

Impingement of water mist (droplets), which usually occur in fog and rain, is another way in which contamination is collected on the under surfaces of insulators. Changes occur as droplets absorb chemical and metallic components. It should be noted that the direction of wind has a great influence on moisture deposition on insulator surfaces.

In cold areas, snow and ice have been observed to cause flashovers on insulators, particularly when they help to superimpose contamination. Studies have demonstrated that flashover voltage increases considerably before snow melts down from the insulator surface, decreases as snow melts, and increases again upon complete disappearance of snow. The degree of contamination depends on distance from the sea. In other words, the closer to the seashore insulators are, the more serious the contamination, particularly for those areas that have long periods with no rainfall, such as southern California, Hawaii, and so forth.

Because of new construction, many heavily traveled highways are located near powerlines and substations. During the winter, a combination of snow, ice, rock salt, rust inhibitors and calcium chloride can create situations leading to bushing problems.* Other substations adjacent to railroad tracks used by oil burning locomotives are also subject to similar problems. Another environmentally critical area is near cooling towers, which produce a fine water mist that could blow into nearby lines and substations. The mist absorbs conducting dioxides en route to the porcelains. In each case, the contaminants are different, but the results — tracking and failure — are the same.

Theory of Contamination Flashover

Over the past fifty years, researchers have conducted studies to investigate physical processes involved in the movement of a discharge over a moist surface. Since a completely satisfactory theory of contamination flashover has not yet been developed, various theories have far been advanced to explain flashover mechanism in bushing and insulator structures;† as summarized by an IEEE Working Group as follows:

> As current flows through the conducting surface film, heat is dissipated evaporating some of the moisture, causing dry bands to form . . . Because these dry bands have much greater resistance than the wet surface, the voltage stress concentrates across them, occasionally leading to small, intermittent, scintillating electric discharges (from a few milliamperes to one ampere peak). The surface is wetted further, the discharge

**Snow will absorb excessive amounts of pollutants, such as carbon dioxide and carbon monoxide. Then after melting a heavy film is left on porcelains, which can lead to run away voltage leakage. Murphy's Law activates and porcelains explode under severe electrical stress.*

†*Circa 1981.*

current increases until at some point the discharges elongate, join together to bridge the entire insulator and trigger a power arc. Because the flashover discharges grow along the insulator surface, the hot power arc which follows their path may damage the insulator (Figure 8.18). Figures 8.19 and 8.20 show actual results of ongoing contamination.*

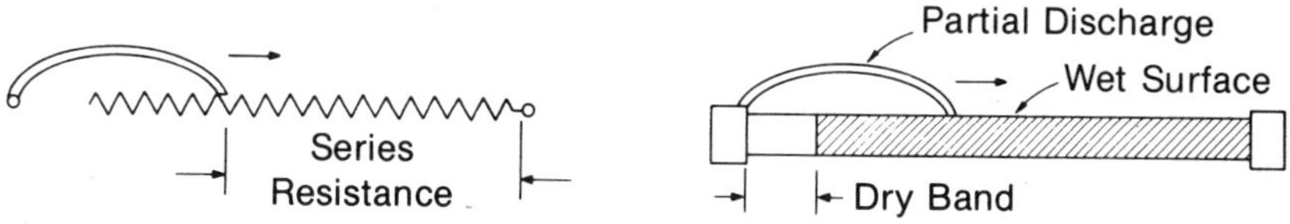

Figure 8.18 - Schematic representation of a discharge on a rod-type insulator.

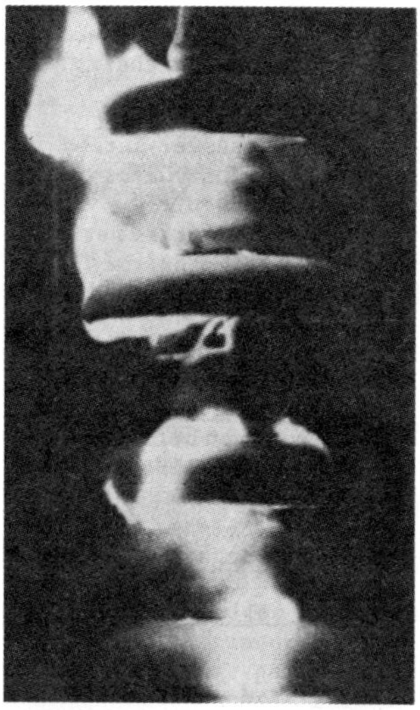

Figure 8.19 - Typical flashover of a contaminated suspension string showing the power arc which occurs after each individual unit is bridged by low current discharges.

Alternative Theoretical Approaches

Other proponents have also shown that their theories agree with the data from flashover tests. Hampton, for example, based his flashover

**Figure 8.20 is located in the Color Section, following Chapter 10.*

theory on an experiment in which he used a water jet to simulate a contaminated long rod insulator. Results of his study led him to conclude that flashover is a stability problem, which occurs when one of the electrical discharges crosses the dry bands and extends across the wet part of an insulator surface. Nacke and Wilkins also believe that contamination flashover is an electrical stability problem.

Although most of these flashover theories seem to suggest that flashover mechanism is a stability problem, Jolly believed that flashover is primarily an electrical breakdown process caused by the field concentration at the discharge tip. Such a process occurs in the following series of steps.

The Flashover Process

P.J. Lambeth has said that flashover of an insulator occurs when most of the surface is covered with a wet contaminant layer of low resistivity; and often takes place under any of the following conditions:

- When an insulator is being wetted after it has been energized at normal working voltage for a prolonged period of time.
- When a wet contaminated insulator has just been energized after its normal working voltage.
- When a wet contaminated insulator is subjected to a transient voltage.

When any of the above conditions are present, then flashover takes place. In brief, the flashover process follows a sequence (Figure 8.21) of four stages.

1. Contamination — Low Level Ionization
   As impurities are deposited on the insulator surface, a low shortened resistance path is created through which an electrical current can flow. A high voltage stress occurs at these points, leading to the breakdown of the air dielectric.

2. Water Filming State
   During hot, dry weather, impurities bake on the porcelain surface due to magnetic field and heat generated by electricity and solar affluency. Rain merely moistens the hardened impurities and tracking begins. The insulating bushing eventually becomes a conductor.

3. Severe Ionization
As the voltage stress at points of contamination increases, the phenomenon of severe ionization occurs. This condition makes its presence known in the form of rays, streamers or fingers.

4. Flashover
The ionization phenomenon merges into the electrical overstress type of glow discharge called corona.* Conductivity increases and at some critical voltage stress, arcing-over or flashover occurs. Flashover could also occur between phases (Figure 8.22).

It should be noted that flashover is only but one aspect of bushing insulator problems. Other undesirable effects are known to exist.

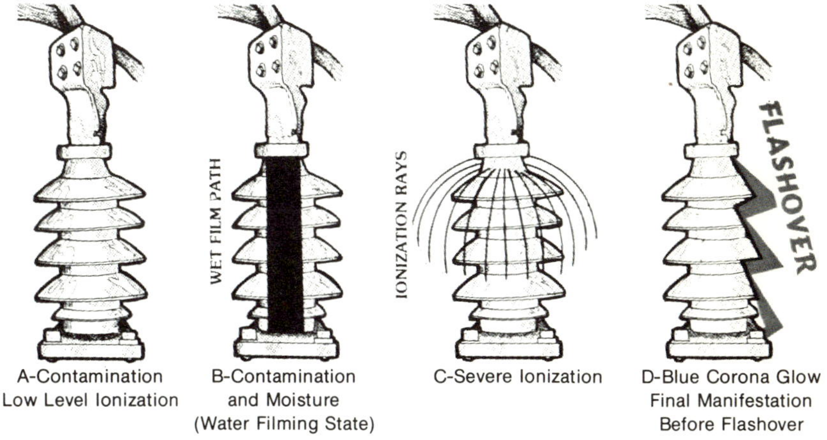

Figure 8.21 - The Flashover Process: A four-stage sequence.

Other Undesirable Effects of Insulator Contamination

Insulator contamination can be detected by our external senses. Such effects include corona manifestations, and corrosion of metal

**Corona for transmission lines and power stations is defined as a "luminous discharge due to ionization at the air surrounding a conductor around which exists a voltage gradient exceeding a certain critical value."*

work. Thus, it is clear that few bushing failures occur without some warning.

Corona Manifestations

Four types of corona manifestations are generally associated with electrical overstress.* "Visual-corona" is perhaps the most common one, and can be observed in the predawn darkness (or at dawn) as blue-violet light dancing (so-called scintillation activity) over insulator surfaces (corona glow). The second type is "audible corona" or "audible frying" which sounds like running water or frying grease. A smell of ozone indicates discharges are present. Finally, and perhaps the most serious type of corona discharges is the electrical effect which causes "radio" (RI) and "television interference." It often results from field distortion caused, for instance, by an incomplete wet-contamination layer. When such signs occur, bushings should be checked more frequently because flashover is imminent, and power shut-down not far behind.

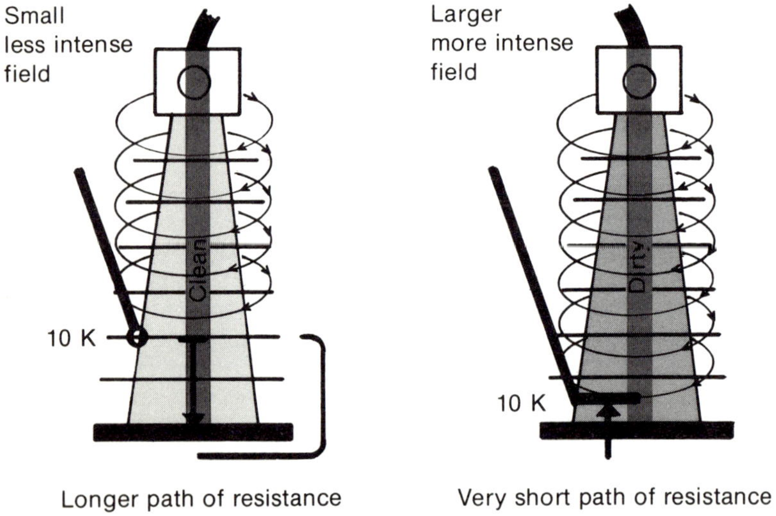

Figure 8.22 - Potential flashover between transformer bushings in different phases. If two phases were heavily contaminated (conductants) the potential for phase-to-phase flashover exists.

"Pre-corona" under a ultra-steep wave immediately preceeds steep wave flashover.

Corrosion of Metalwork

Contamination causing flashover is also likely to influence the metalwork. High leakage currents may cause corrosion through the process of electrolysis. This phenomenon seems to be significant only for dc insulators.

The above undesirable effects of corrosion can be checked by on-site periodic maintenance, which can consist of infrared inspection or through traditional visual inspection while insulator structures are de-energized, and power factor testing. Areas to look for during visual inspection include: cracked, broken or chipped porcelain, deterioration of cemented joints, gasket on solder seal leaks, moisture, evidence of arcing, broken mechanical connections, discolored oil or oil leaks (in oil-filled units), and contamination build-up on the porcelain.

Preventive Maintenance

Insulator problems can be detected in two ways — by visual inspection and by infrared inspection (IR).* Items requiring attention on a visual inspection include the following:†

1. Porcelain
   Porcelain can be chipped, cracked, or broken. Glass finish can be roughened or eroded. Chips, if small, are of minor importance and can be painted with Glyptal paint or the like to obtain a glossy finish. Minor cracks can be sealed with paint or epoxy provided the mechanical strength of the unit has not been impaired. If this is the case, the insulator must be replaced.

2. Metal Parts
   These can be checked for corrosion, and painted as necessary.
3. Gaskets
   Gasket materials will deteriorate with age, depending on the temperature surrounding the gasket surface and the time at that temperature. In most cases, the physical properties of

**Chapter 8, Part 4.*

†ANSI/IEEE Guide for Loading Power Apparatus Bushings, C76.1-1976 (Proposed revision, p. 757-9/15/80).

gaskets will change progressively if high temperatures are maintained for long periods. Such changes could result in the reduction of dielectric strength through loss of gasket seal. Most gasket materials are badly affected by ozone which is present with corona. Therefore, it is important to inspect gaskets because they can become major sources of leaks. Care must be taken to obtain the right material and the proper thickness (for compression).

4. Capacitance Taps
   These should be examined for proper gasketing to prevent the entrance of moisture. The tap compartment should be filled with an insulating compound (oil, petroleum). The ground should be inspected on grounded taps.

5. Oil Level
   The oil level should be checked by inspecting the level indicators. If there is a leak, it should be repaired, and oil added. If there is no sign of a leak but the oil level is low, the problem may be internal, and it may be more prudent to replace the bushing.

6. Cement
   Cement is used to attach metal parts to porcelain. It may also be used to join insulators together. Crumbling or chipping cement can be repaired in the field.

7. Lower End
   The oil level in the apparatus should be checked to ensure that it covers the bushing end to the proper operating point.

8. Contamination
   The bushings and standoff insulators are subject to the same contamination as the transformer and steel paints. Build-up of contamination can cause more problems with porcelain than with metal parts. These contaminants should, therefore, be removed.

Alternative Methods of Protective Maintenance of Transformer Bushings, Surge Arresters and Standoff Insulators

Several practical methods of taking the offensive in bushing/insulator maintenance are available, and have been used with varying success. These methods help keep conventional porcelain and glass insulator surfaces clean, and thus prevent contamination flashover. Periodic bushing and insulator maintenance can be achieved through one or more of the following methods:

- Hand wiping when the units are de-energized;
- Washing when units are either energized or de-energized;
- Dry cleaning with air-blast materials, when units are either energized or de-energized, and
- Coating with silicone compounds, either energized or de-energized.

Overinsulation can also be considered as another technique for preventing bushing and insulator contamination.*

Hand Wiping

Hand cleaning involves cleaning insulator surfaces covered with dust, and other surface pollutants. The practice is normally used in areas where insulators cannot be cleaned by high pressure washing or are inaccessible, or are too close to energized equipment. Although hand wiping is an efficient method of cleaning insulators, it is tedious, time consuming, and expensive in that equipment has to be de-energized.

Materials used for hand wiping may include dry wiping rags for soft, loose deposits, and wet or kerosene-soaked cloth, solvents, chemicals, and brushes or steel wool for insulator surfaces covered with hand caked deposits.

Because hand wiping has to be done on de-energized bushings and insulators, there is always the problem of downtime. In addition, two factors beyond the price of premium time labor are: workers are exposed to the possibility of falling, and cleanliness of the equipment is in direct proportion to the fatigue of the workers. Cement deposits also require the use of tools to chip and scrape the coating loose which can naturally cause damage to the surface being cleaned.

Periodic Insulator Washing

Cleaning bushings and insulators with a high pressure water stream has been found to be a safe and effective technique of removing

**Chapter 8, Part 4 (Infrared Scanning), and Chapter 5 (Bushing Power Factor Testing).*

only water reducible non-adherent films from insulator surfaces. The method is often used in areas where natural washing by rain is inadequate to clean insulator contamination, such as California and other coastal areas. The frequency with which insulators should be washed depends on the environment in which the structures are located. Nonetheless, it is important that insulators be washed before they reach a critical level of contamination. As a general rule, such a level can be estimated from past experience on periods between flashovers, from results of equivalent salt-deposit density (ESDD) measurement, and the presence of corrosion and corona manifestations.

1. Washing Equipment
   Most washing equipment in current use has been designed by equipment manufacturers, utility companies, or by both groups. Such equipment include different types of nozzles, hoses, and trucks.

   Three major types of nozzles used for insulator washing are:

   - Portable, hand-type sprayer
   - Fixed-spray nozzle
   - Remote control jet nozzle

   The first can be easily carried around, and are used for either regular maintenance, or for emergency washing. Portable nozzles, which consist of a stainless tip and a brass section, are usually designed for individual use. For example, one utility uses a hand-hold nozzle sprayer on its energized equipment which resembles the type used to apply herbicides to weeds and brush. In most transmission line washing jobs, nozzles with stream pressures of over 500 psi are preferred. Fixed nozzles are usually designed for specific spray patterns and can be obtained commercially.

   The type of hose used for insulator washing is usually ¾ of an inch or one inch in diameter, with a minimum bursting pressure of 3,200 psi. When using the hose, it is advisable to pull it off completely from the reel so as to maintain the expected pressure.

   Tank trucks are also a major aspect of insulator washing equipment. Two types are often employed — the portable nozzle washer truck, (Figure 8.23) which has no boom and the remote control nozzle washer truck (Figure 8.24). The latter vehicle is usually equipped with a boom that is hydraulically operated, and makes it possible to position the nozzle within desired washing distance from the insulator structures. Both types of trucks are often equipped with 300 to 2900 gallon water tanks. The tanks are usually made of stainless steel, aluminum, or fiberglass material.

2. Portable Hot-line Washing Equipment
Hot-line washing equipment has proven to be a safe, effective and economical technique for removing pollutants from energized insulators. Hot-line washing of bushings/insulators should not simply be considered as a mere washing practice. Washing cannot be done in freezing weather, and the water must be continually tested to make sure that it is free of minerals, and has a resistivity of 150,000 ohms or better. Live-line washing should, therefore, be performed by qualified personnel, so as to avoid serious flashover faults.

When conducting hot-line washing, it is important that the established procedures and parameters be strictly adhered to. These include:

- Verification of safety criteria by each company.
- Provision of a safe level of leakage current in the wash water stream. Leakage current is often influenced by water parameters such as nozzle-conductor distance, water resistivity, water pressure and nozzle orifice diameter.
- Provision of maximum safety for personnel involved in the washing process, in case a flashover takes place.

It should, however, be noted that the method is fairly safe, because in 1937, an investigation was made into fire fighting on energized systems which revealed that by using a diffused stream (hollow core) of water, the leakage current was quite small (0.12 milliamps). One company successfully tested their spray nozzles by spraying water on hardware cloth energized to 70,000 volts to ground.

3. Washing the Insulators
One method used for hot-line washing is fixed spray washing, in which low pressures (50 to 100 psi) are employed in areas where regular washing is required. The method is effective in preventing sea salt contamination flashover faults. In employing this method, local parameters that

influence washing, such as precipitation, wind, water resistivity, and contamination severity should be taken into consideration (Figure 8.25).

Another method involves the washing of insulator strings in which a stream of water is directed first at the insulator nearest the energized conductor, so as to take advantage of both the impact and the swirling action of the water to remove contamination deposits. After washing the line insulators in the string, the wash stream is then moved directly back on the clean units below to re-rinse them. The procedure is repeated until all the units in the string are thoroughly cleaned.

Live-line high pressure washing at 60-day intervals is effective only in removing water soluble and non-adhering contaminants, but not insoluble chemicals such as hardened salt or cement dust. Additional cleaning of protective devices will probably be required. In fact, hard crusts under live water washing may cause flashover.

Dry Air-Blasting With Non-Abrasive Material

Powder blasting of energized or de-energized transformers is another very effective and economical technique of cleaning insulators. It does not damage the equipment; yet it will remove all insoluble particulate contamination — including metallic oxides, cement dust, chemicals, salt-cake, acid, smog, and other pollutants. One of the most common abrasive materials used for removing old crusted silicone grease and other soft pollutants from insulators is ground-up corn-cob. The other common medium is limestone powder with a Rockwell hardness of number 4 (Note that diamond has a Rockwell hardness of number 10; porcelain glaze, number 6). Since limestone powder is a non-conductive material, the possibility of flashover during energized work is practically impossible, except during rainfall (Figure 8.26). A light dust film left after cleaning will wash away at the first rain shower, leaving absolutely no residue. Limestone powder is not harmful as it cleans and polishes the bushing insulator surfaces to a factory-fresh finish. Other materials such as potter's clay, crushed coconut shell or ground walnut or pecan shells are sometimes used for removing hard deposits such as compacted cement dust.

Figure 8.23 - Portable washing truck equipment. (125 psi, 50 GPM using demineralized water).

Figure 8.24 - Remote-control nozzle ("peashooter") washing truck.

Figure 8.25 - Close up view of fixed spray nozzle washing as widely practiced in Japan.

Powder blast cleaning of bushings and insulators can be safely performed up to 161 KV, using a bucket truck or a cherry picker with an insulated boom (Figure 8.27). Such a truck is usually equipped with air pumps capable of carrying air pressure of about 150 psi, a storage tank in which the abrasive materials are held, and a reel for holding the hose. Powder blasting can also be performed from other aerial locations such as substation structures or off-towers (Figure 8.28). Working from the ground level, insulators can be cleaned up to 230 KV using a "hot stick."

Consumers of electrical power have traditionally used two methods to protect bushings and insulators from contamination flashovers, coating the insulator structures with silicone or other compounds, and overinsulation.

Use of Silicone Dielectric Compounds

Grease compounds applied on insulator surfaces serve two major functions. First, they tend to "encapsulate" conducting contaminants and render them harmless. When contaminating particles land on a grease-coated surface, the grease compound fluid, by osmosis "reaches out" and engulfs the particle. This prevents a path of contamination. Second, the silicone compound prevents water-filming and leakage currents over extended periods under wet service conditions. The silicone causes water to bead-up on its surface, thus preventing a conductive film from foaming (its hydrophobic characteristic). Figure 8.29 illustrates the effect of silicone on contaminating particles.

It is for these two reasons that silicone compounds have been extensively used (since 1953) for coating insulator surfaces. In the United States, silicone compounds are generally used, whereas petrolatums (petroleum jellies) are extensively used in Europe, and on a limited scale in the United States.

Conventional silicone compounds are usually made from a mixture of silica filler and a silicone fluid — an active component which provides surface mobility and water repellency. Silicones maintain virtually the same viscosity, and therefore, a temperature range of -50°C to 200°C. Petroleum jellies are made up of hydrocarbon oils and wax. They usually soften with an increase in temperature, and begin to lose their property of cohesion and water repellency. Furthermore, when petrolatums become loaded with contamination, they begin to lose their effectiveness. Silicone has been found to be more effective than petroleum jelly for coating bushings and insulators (Table 8.5).

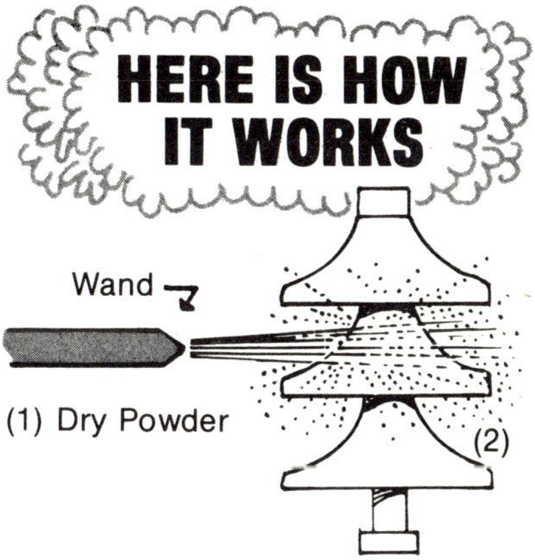

Figure 8.26 - How dry powder blasting works. Dry powder (1) under air pressure blasts the ceramic insulator, and (2) this combination penetrates the contaminant, rebounding from the insulator surface carrying the contaminant with it.

Figure 8.27 - Powder blast cleaning of transformer bushings from a bucket truck.

Figure 8.28 - Powder blast cleaning of standoff insulators from an extended bucket truck.

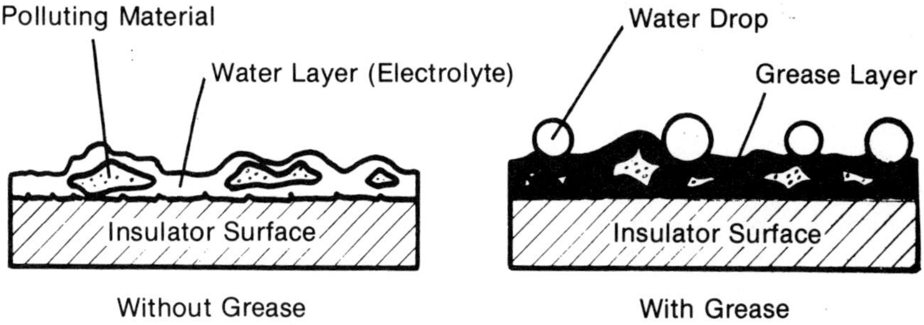

Figure 8.29 - Effect of silicone compound on insulator surface. The only function that silicone basically performs is preserving insulator dielectric surface integrity by preventing moisture from reaching engulfed pollutants. Thus, silicone is a water-repellent compound.

TABLE 8.5

COMPARISON OF PETROLEUM JELLY AND SILICONE COMPOUND		
Basic Constituents	Petroleum Jelly	Silicone Compound
Useful Temperature Range	0° to 60°C	-50° to 200°C
Recommended Spraying Temperature	90° to 115°C	Ambient Temperature
Encapsulation Rate, Ambient	Slow	Rapid
Ease of Application	Difficult, especially in cold weather	Good
Material Costs	Low	High
Application Costs	High	Moderate (compared to petroleum jelly)
Cleaning Costs	High	Moderate (compared to petroleum jelly)

High silicone effectiveness requires that insulator surfaces be thoroughly cleaned and dried. Insufficient cleaning will result in tracking underneath the grease compound. This, in turn, can lead to permanent damage to the surface glaze and under severe stress, to cracking, breaking, chipping, or insulator disintegration.

Often an insulator surface has been thoroughly cleaned. *New* silicone compound should be applied evenly over the entire insulator surface to a thickness of $^{1}/_{16}$ to $^{3}/_{16}$ of an inch. The thickness of the silicone coating will generally depend on the nature of the contaminants and the geographical location of the substation. Where petroleum jellies are applied on insulator surfaces, the coating should be much thicker than that of silicone compound. A major disadvantage of using a thinner layer of compound is that the surface of the insulator might be damaged when sparking takes place. This often results in the cracking of the bushings, and insulators; and the eroding of glazed porcelain surfaces, toughened-glass insulators are especially vulnerable. It is for this reason that silicone compound is not recommended for use on glass insulators.

There are several problems associated with using silicone — on bushings and standoff insulators. The compound is very expensive, and costs about $8.50 per pound (See Table 8.6 below).* Once the silicone has been applied, it is difficult to remove, and can only be removed by a rag soaked in a solvent, or by corn-cob blasting (energized) and in some cases, even this is totally ineffective. Water washing is even less effective. In fact, if the silicone is not removed when "used up," the problem can be worse than if no compound has been used at all.

Another problem is how to tell when the silicone has reached the end of its usefulness. Unfortunately, the "best" method to date is the "finger test." This is done by wiping one's finger across the silicone surface (de-energized) to see two things — how clean the subsurface is, and how hard the coating has become. Silicone must be applied to a clean surface. When it is reapplied, it must also be put on a clean surface. Care must be taken in applying the compound as it is very slippery, and should only be put on insulators and insulator caps and slightly on metal heads of oil filled bushings — not steel.

**Circa 1980.*

TABLE 8.6

TYPICAL QUANTITIES OF SILICONE GREASE REQUIRED TO COST A SPECIFIC INSULATOR[1]			
Insulator Type and Description[2]	Area (Square Feet)	Amount of Compound Required	
		Pounds	Ounces
13 KV one-piece pintype	0.70	0	3.6
34.5 KV line-post insulator	2.32	0	12.1
Standard 10-inch suspension unit	1.82	0	9.5
Smogtype 10-inch suspension unit	2.47	0	12.5
34.5 KV pin-cap apparatus insulator	4.24	1	6
69 KV pin-cap unit used in two unit stacks	4.96	1	10
High voltage pin-cap apparatus unit	7.73	2	9
69 KV bushing	9.1	3	—
138 KV bushing	21.4	7	—
230 KV bushing	42.4	16	—

[1]If a given insulator were coated with a 1/16 inch coat of compound, each insulator would require the amount of compound listed.

[2]Courtesy of Ohio Brass Co., Mansfield, Ohio.

The major advantage of silicone compound in protecting against contamination buildup is this: if an insulator protected with the compound flashes over, the porcelain glaze is damaged less than if it were unprotected. The arc over product of silicone is silica, which is non-conductive, whereas, the arc over product of petroleum jelly is carbon, which can be conductive.*

In addition, the bushing surfaces should be clean before a valid insulator power factor test is made. While a slow increase of power factor may be normal, degradation of that portion of the insulation which operates at greatly elevated temperature will result in a substantial increase in power factor. An unusual increase in power factor may become an indicator of the detrimental effects of loading beyond nameplate rating. Bushings which have been loaded beyond nameplate rating should be measured more frequently. Since power factor varies with temperature all measurements should be made at 20° to 25°C or corrected to the reference temperature.†

Grease compounds can be applied to de-energized insulators by hand, using rubber gloves, pads or brushes, or by air-spray equipment. In order to apply the grease compound more uniformly, the air spray method is generally recommended.

Where insulators are energized, "hot stick" air spray equipment is generally used, and should meet all the state and OSHA requirements for hot-line maintenance. Figure 8.30 shows air spray silicone compound being applied to transformer bushings.

The effective service life of silicone grease can vary from one month to five years, depending on the thickness and evenness of silicone layer applied, and the geographical location in which the insulators are situated.‡ For severely polluted environments, one major utility has reported an annual or semi-annual scheduling for use of silicone compound. Another facility, located in a nonpolluted area, has gone to no more than two years between energized cleaning. Our twelve years' experience shows that the majority of clients choose to do so on a yearly basis, regardless of contamination, and the process used.

**Pure carbon is a good resistor; contaminated carbon is not.*

†*Chapter 5, Part 3, p. 406.*

‡*The evenness of the film is as critical as its thickness.*

Figure 8.30 - Air spray silicone grease applications to 138 KV bushings.

It should thus be appreciated that even with absolutely effective powder blast cleaning, grease compounds will gradually lose their water repellency and arc resistance due to extended exposure to high voltage corona, ultra-violet light, water erosion, and/or particulate contamination. When this condition occurs, grease compounds should be removed by energized blasting, using corn-cob.*

Overinsulation

Overinsulation as a method of checking contamination flashover is advantageous in that only infrequent maintenance is required. In addition, the bushings and insulators are standard items which can be readily obtained from manufacturers or from suppliers. Thus, overinsulation, consisting of additional stacks of insulators, at best only moderately decreases maintenance cost. Since airborne particles will cover a large insulator at the same rate as stacked small ones, this method is not as effective as reasoning would indicate.

Increasing insulator size or creepage length has proved not to be the ultimate answer to flashover problems in severely contaminated areas. Nonetheless, vertical mounting of insulators — in contrast to horizontal — has proven beneficial in minimizing contamination

**A bushing will dry the silicone compound at a faster rate than an insulator due to rising heat and elongated magnetic field (Figure 8.17).*

problems (Figure 8.15). Again, common sense and experience reminds us that preventive maintenance is the ultimate solution to clean bushings and insulators.

The Final Decision

Each substation presents a unique problem dependent upon many factors — the length of dry season, fog, salt spray, industrial contaminants, design and prevailing winds.

Clients have recognized that regularly scheduled powder blast cleaning is the best approach in eliminating unscheduled power outages. They also recognize that a high voltage silicone insulating compound coating is an integral part of their bushing-insulator maintenance system in 40 percent of all cleaning situations, regardless of the environment. Silicone has been most often chosen to combat environments of salt-cake and cement dust.

Nevertheless, the use of silicone is not beneficial in every situation. For example, one substation is located near a cooling tower which should be silicone treated, while another substation located only 1,000 feet further away would not require treatment. The primary parameter for silicone usage is actually simple — if moisture is your problem, (historically) water-repellent silicone is your likely answer.

The decision on the type of maintenance program and its frequency depends mainly on the effluency of airborne contamination. Criteria that will also assist in making a decision include construction type, condition and age of bushings/insulators in-service.

Keep in mind — "Surface contamination and leakage currents caused operating difficulties eighty years ago . . . and exist today as one of the most serious insulator problems." Bushings and insulators still are not maintenance-free. In fact, maintenance-free insulators have been the subject of research for years. But as an IEEE conference paper has reported, "Contamination-proof insulator designs do not exist and are unlikely." At this point, the classification and tabulation of more than 40 design styles of insulators support this view. Even in light of new insulator designs — light weight rubber, epoxies, synthetic resins, and glass-flake — bushing and insulator contamination, will likely remain a problem plant maintenance personnel cannot avoid.

Part 4—Infrared (IR) Scanning

Introduction

Infrared is basically light which is invisible to the human eye. It acts like visible light and its radiation is detected in much the same manner, but not with the eye. So instruments have been developed.

One of the laws of physics states that all bodies with temperatures above absolute zero (-273°C) emit infrared radiation. The quality of this radiant energy is often described in terms of its wave length, and can be measured in microns or thousandths of a millimeter. The wave length spectrum of electromagnetic radiation ranges from 10^{-13} millimeters to 10^{5} millimeters, and is broken in bands as illustrated in Figure 8.31*. Visible light falls in the very narrow band of wave lengths from 0.38 microns to 0.76 microns (micron = 0.001 milimeters).†

In the electrical industry, infrared scanning or thermography has been employed as a tool to measure the temperature of various types of surfaces, thereby detecting potential defects in transformers, high amperage connections, and other current-carrying devices.

Development of Infrared Technology

The development of infrared instrumentation gained impetus in the early days of World War II, where it was employed for military night vision purposes. From this humble beginning, the technology has continued to grow, making great inroads throughout industry.

In 1964, the AGA Corporation, an international Swedish concern that manufactures infrared imaging systems, began to investigate the use of thermography as a tool for detecting potential trouble spots in overheating electrical apparatus. In 1965, the firm introduced to the United States its Thermovision® System on a commercial basis. The technology received a tremendous boost following the 1965 Northeastern blackout.

**Figure 8.31 is found in the Color Section, following Chapter 10.*

†An IR camera operates from 1.2 to 22 microns, and IR photographic film is from 0.7 to 1.2 microns.

Today, several infrared equipment manufacturers are producing such instruments as infrared research spectrophotometers, infrared horizon detectors, infrared communication equipment, night time surveillance equipment, space navigation systems, remote sensing equipment, anti-camouflage thermography, and other diagnostic thermographic systems. All these instruments operate on the same general principle, utilizing the infrared rays to detect temperature gradients.

Types of Commercially Available Systems

A variety of infrared thermographic systems are available for detecting heat related problems in energized electrical equipment and other industrial environments. The instruments range from radiometers such as hand-held viewers and remote thermometers, to television-type imaging systems or scanners which offer better resolution and more flexibility.

Radiometers

These are portable non contact infrared thermometers (IRT) with meter readouts. Several manufacturers such as Mikron, Wahl, AGA, and Raytek produce a variety of these long-range thermometers, some of which are mounted on gun stocks, and even include telescopic sights (Figure 8.32). The instruments are ideal for spot checking overhead transmission and distribution lines — but your aim must be good.

Similar devices without optical magnification are also used for spot checking exposed wiring and other electrical apparatus in industrial plants. Infrared thermometers usually require no cooling since they use thermistor detectors. They are quite compact, weigh less than five pounds, and operate on battery packs.

Scanners

Scanners are radiometers that use optical components to deflect the radiometer's small field of view through a televisionlike scan pattern (Figure 8.33). The imaging systems surveyed below are only a representative of scanners currently in use and available commercially.

Figure 8.32 - Mikron® Infrared thermometer with telescopic sight. (Courtesy of Mikron Instrument Co.).

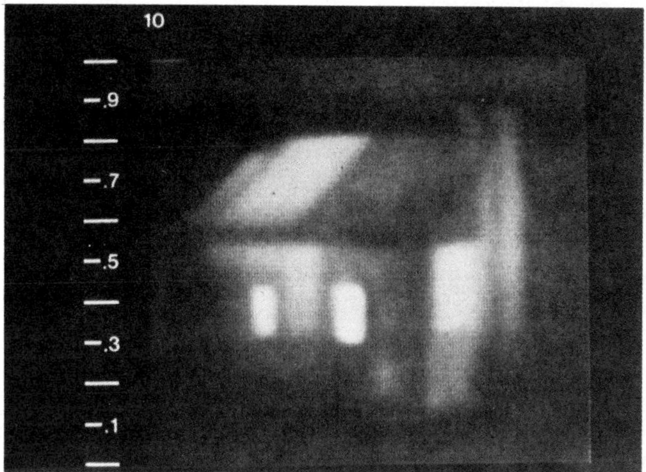

Figure 8.33 - In this infrared picture of a home, the areas of higher temperature appear lighter while areas of lower temperature appear darker. (Courtesy of ThermoTest Infra-Red Surveys).

1. AGA Thermovision®
   This imaging system is designed to detect energy radiation from an object, and displays the image on a cathode-ray tube (CRT) similar to a black and white or color television picture tube (Figure 8.34). Knowing the amount of energy permits knowing the temperature. Model 720 utilizes an

isotherm function that measures temperature differences existing in the object viewed. This model also uses liquid nitrogen to cool the detector in order to allow for rapid scanning. Table 8.7 compares three models, such as Figure 8.35, as well as other selected infrared scanning systems.*

Additional accessories include lenses, extension rings, spectral range filters, photo-recording attachments, video-recording equipment, and camera extension cables.

2. Hughes Probeye™
   This is a lightweight, hand-held, highly accurate infrared viewer which employs a light emitting diode (LED) display. Figure 8.36 shows the external components of the device, while Figure 8.37 illustrates the light paths through the instrument, as well as its principle of operation.

   The Probeye uses a self-contained rechargeable nickle-cadmium battery. Carefully focused, the device can detect temperature differences as little as 0.1°C between adjacent objects. Moreover, it can accurately locate sources of heat in total darkness, and through smoke and dust. The Probeye requires only minor maintenance, such as keeping the lens clean and the power pack and argon cylinder recharged.

3. XS-410
   As a product of Xedar Corporation, this portable infrared TV viewer operates on a rechargeable, replaceable battery pack with a running time of 2½ hours. It can be easily connected to a TV transmitter or videotape recorder for ultimate versatility. Because the instrument does not require gas for cooling, there is no need to disrupt monitoring in order to refill cylinders with liquid nitrogen, or high pressure argon gas. Figure 8.38 illustrates the camera being used to monitor electric faults in a substation.

4. Barnes ThermAtrace™
   This is a single-line thermal scanning device developed by Barnes Engineering Company (Figure 8.39). Its battery pack will operate the device for about four hours continuously. It has a pyrolectric detector that requires no cooling.

**Since IR has become an intensely competitive area, consult the various manufacturers for the latest information.*

During operation, the instrument can be mounted on a tripod to make it steady. Since the baseline of this device automatically clamps on to the coolest object in the scan, the unit can be easily used for measuring temperature differences.

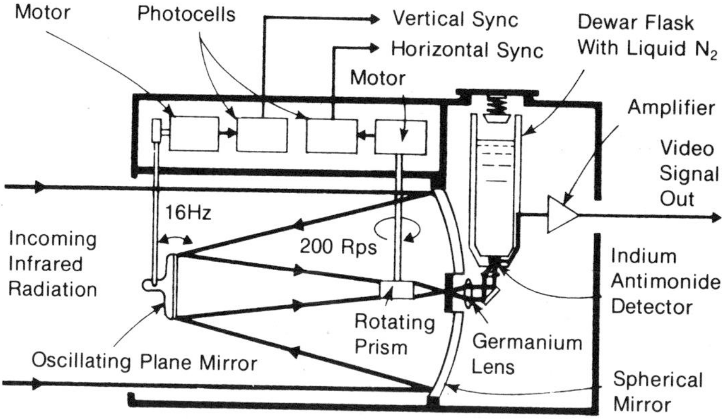

Figure 8.34 - The sensing head of AGA Corp.'s unit uses a liquid-nitrogen flask to keep the detector at maximum sensitivity and resolving power. Video amplifier, controls, and viewing screen are in separate chest- or cart-carried module. (Models 720, 780, and 782).

Figure 8.35 - AGA Thermovision® 110 uses no liquid cooling gas, allowing the viewer to be used anytime, nearly anywhere for six hours.

TABLE 8.7

COMPARISON OF SELECTED INFRARED (IR) SCANNING SYSTEMS[1]					
Type	Model Number	Description	Power Requirement	Weight	Estimated Cost
AGA Thermovision®	110	Handheld, non-temperature measurement thermal viewer. All electric operation. Optional Polaroid® photo recording. Object temperature range: -30° C to 800° C. Minimum sensitivity: 0.1° C at 30° C. Operating environment: -20° C to 55° C.	3-NiCd batteries (6 hr.); Charger 110/220 Vac. 50/60 Hz. 10 W	6.6 lbs. (3 kg)	$13,000
	720	Handheld portable Polaroid® camera unit with IR refractive prism scanning, selectable temperature range apertures. Display unit is portable with B/W TV monitor with automatic preset photographic recording mode. Object temperature range: -20° C to 200° C in 9 sensitivity steps. Sensitivity: 0.1° C at 30° C. System operating temperature: -15° C to 55° C	Power supply battery charger unit for 100, 220, 240 Vac, 50/60 Hz. 35 VA or separate battery for 8-15 Vdc, 20 W.	14 lbs. (6.3 kg)	$26,000
	780, 782	Similar to 720 system plus object temperature range: -20° C to 1600° C: Temperature measurement, color monitor, analog and digital thermal analysis, etc.	Variety	Varies	$35,000 to $60,000
Hughes Probeye™	649 and 650	Conveniently portable and completely self-contained handheld imager. Can reveal temperature differences as little as 0.1° C at 22° C between closely adjacent objects and between objects and their background. Contains an argon gas-filled cylinder used for the detector coolant. Meets particular demands of mining, industrial and utility facilities. Object temperature range: -20° C to 900° C, and extending to 2000° C with additional filters.	1.5 W from rechargeable battery	7.5 lbs.	$8,000 to $10,000
Xedar	XS-410	Versatile portable infrared TV camera. Has a 3 inch display aperture which allows viewing with both eyes to avoid eye fatigue. Has a built-in filter for single step attenuation. Temperature sensitivity is less than 0.2° C for 60 lines of resolution. Object temperature range: Extension up to 1000° C. One inch pyroelectric vidicon tube requires no gas coolant. Device can be used for research, industrial and utility monitoring, as well as for fire fighting and prevention.	Battery pack (2.5 hr.) ac-dc charger (3 hr.), 110 Vac	6.9 lbs. (3.11 kg)	$12,000
Barnes ThermAtrace II™	12-750	Portable infrared scanner than can plot a video graph of the temperature distribution along a line. The scene is then viewed in visible light with the temperature trace superimposed. Device is designed for laboratories, utility plants and other field work. Temperature span adjustable in five ranges from 10° C to 1000° C. Minimum detectable temperature difference is 0.5° C at 25° C.	2 W from rechargeable battery, (4 hrs.) or 120 Vac	9 (4.1 kg)	$8,000 to $10,000

[1]Since IR is a highly competitive area, current information may vary from this data. Check with the manufacturer.

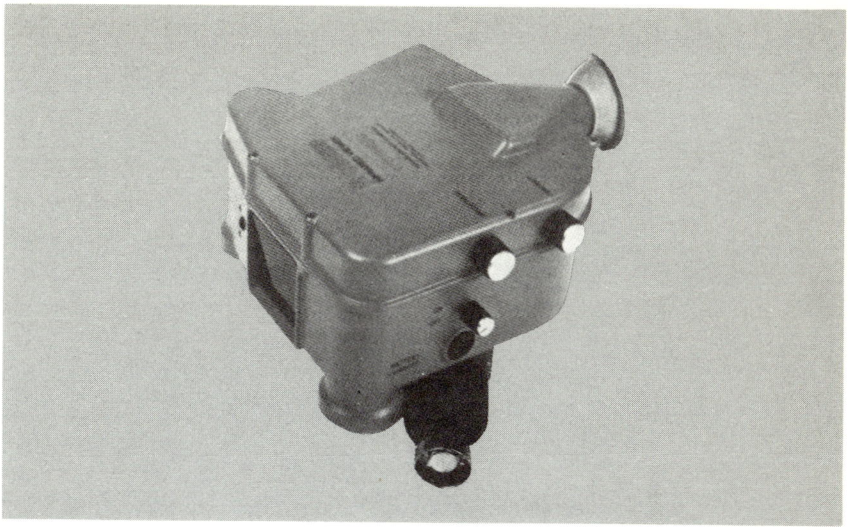

Figure 8.36 - Low-cost Probeye™ infrared viewer is hand-held. Note tank of argon used as a coolant at 5000 psi at bottom, with shutoff valve protruding. (Courtesy of Hughes Aircraft Corp.).

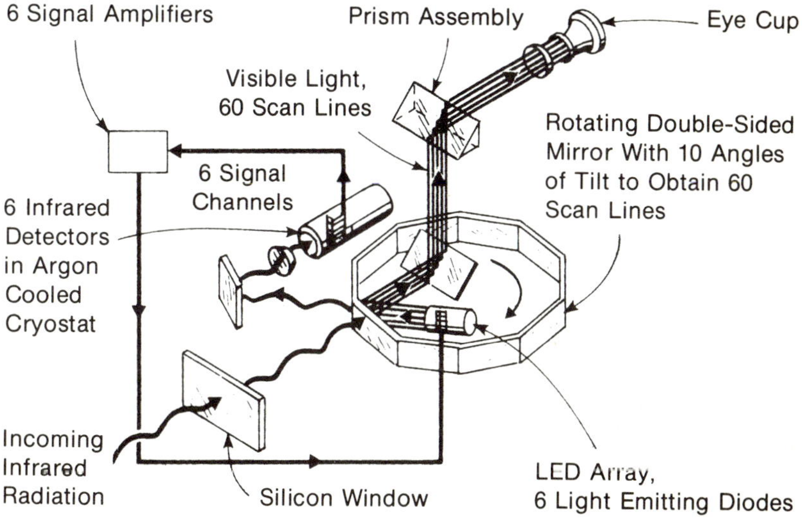

Figure 8.37 - Hughes Aircraft unit weighs 7.2 pounds with argon cylinder (for cooling the detectors) and uses a 10-segment, double-sided, rotating mirror with 6 LEDs to give 60-line scan image. The hottest objects appear bright red, cooler areas black. Field of view is 18° horizontal x 7.5° vertical. Temperature resolution is 0.5.

Figure 8.38 - Portable infrared scanner monitoring utility substations for electrical faults. (Courtesy of Xedar Corp.).

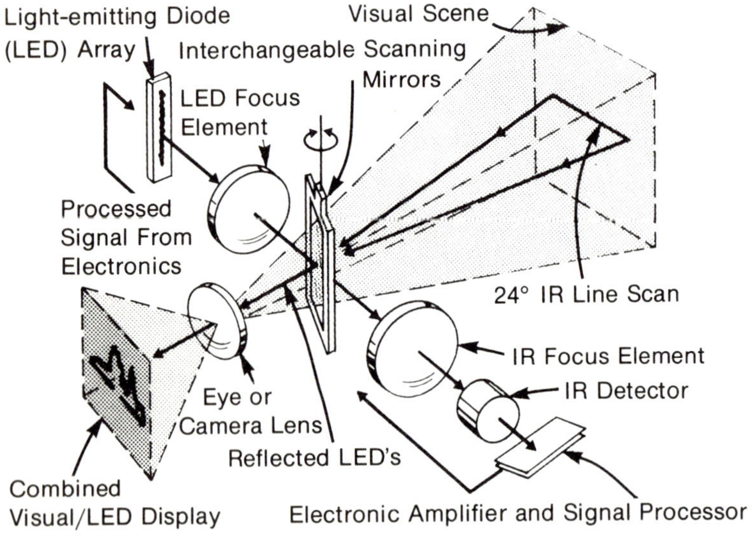

Figure 8.39 - ThermAtrace™ one piece 9 pound unit scans with an oscillating reflecting transmitting mirror which is transparent to visible light, but reflects infrared energy. By collecting IR energy, the unit is capable of focusing that energy on a sensitive IR detector where it is converted into an electronic signal. (Courtesy of Barnes Engineering Co.).

The Role of Infrared Scanning in Today's Economy

Advantages of Infrared Scanning

Commercially, infrared scanning has several economic advantages, one of which is the ability to detect and diagnose electrical problems in their early stages, enabling the plant maintenance personnel to plan and take corrective measures before a major problem occurs. Other additional benefits are:

- Measurements are obtained quickly and with immediate results. Such data would be far more expensive and time consuming if obtained visually or manually.
- Measurements are made remotely, that is, without contact between the instrument and areas being inspected. Consequently no service interruption is required.
- The equipment is easy to use in confined areas and other inaccessible places.

Applications in Electrical Industry

Because of the speed, accuracy, ease and safety with which energized electrical apparatus can be checked with infrared imaging systems, it is possible to inspect almost all electrical equipment in the plant. Today, the types of electrical equipment that are normally scanned for possible hot spots include: bushings, insulators, switches, bus bars, motor controllers, motor bearings, compression connectors, splices, bolted clamps, braided wire, capacitors, power lines, oil pumps, circuit breakers, as well as transformers. Electrical systems run hot in such situations as loose compression, fatigue, overload, or dirt accumulation.

Today, more and more electrical and utility plants are employing infrared scanning inspection as part of their ongoing preventive maintenance program, without any production shutdown. This has been made possible since IR scanning is best done under **heavy** electrical loads, since certain hot spots cannot be detected while the equipment is at partial load.

Tests by the Electrical Power Research Institute (EPRI) and a survey published by the 1976 Annual Conference of Doble Engineering

Clients have clearly established the reliability and economic justification of using infrared survey for locating hot spots in electrical equipment. One utility company has used the following temperature variations above ambient as a guideline for scheduling repair:

Up to 20°C	-	indicates trouble, but servicing is not urgent; schedule at normal downtime.
20°C to 40°C	-	indicates servicing is urgent; schedule servicing within 30 days.
40°C and Up	-	indicates an emergency condition; schedule servicing immediately.

Other companies believe that temperature readings can be misleading. One example is a damaged component creating a temporary weld that will yield a low heat reading itself. Therefore, it is suggested that any connection showing abnormal heating might constitute a potential danger which requires some investigation.

Use in Energy Conservation

Infrared scanning is being used extensively in energy conservation heat studies. The National Bureau of Standards (NBS) researchers have made quantitative heat measurements on buildings using IR thermography in conjunction with a heat flow reference manual (Figure 8.33), as well as to evaluate industrial plants.

Other companies have also have been involved in energy conservation techniques using infrared thermography. The Texas Division Dow Chemical, U.S.A., has, for example, used the Model 750 Thermovision® system to find defective spots of insulation and other general insulation qualities of pipes, vessels, etc. In January of 1975, Daedalus Enterprises, Inc. of Michigan performed an infrared survey over Cornell University campus at Ithaca, New York. The survey not only revealed major faults in steam lines, but also detected leaks as small as ¼ inch in diameter, in areas where pipes were buried six feet or more beneath soil, concrete and blacktop. Thus, infrared thermography has been used as a great tool for the detection of insulation faults and other industrial related problems.

Other Applications

Infrared thermography has also been employed in quality control for tracking down elusive "flares" in equipment that was either manufactured or installed defectively. For instance, some utilities now require what amounts to "acceptance test" scanning of major distribution equipment installations as soon as the substations are energized and loaded (See Figure 8.44).

In the mining industry, infrared thermography is now used to detect concealed fires and other potential spontaneous combustion sources such as hot spots in coal beds and shotholes containing undetonated charges in underground blasting sites. When accidents occur, infrared equipment is used by rescue workers to quickly locate fire sources and lost or injured miners in smoke-filled mining tunnels. Other uses include scanning of thermal pollution and geological formations as well as remote sensing of agricultural land.

The practical applications of infrared inspection are many, and will continue to increase with the demands placed upon the technology by informed and responsible people.

A Typical IR Survey

Although many infrared instruments are used for IR surveys, most scanning so far reported has been performed using AGA Thermovision®, Barnes Model T-101, or the Hughes Probeye. With proper planning, a surprisingly large number of points can be checked in a very short time using IR cameras (Figure 8.40).

When a customer requests an IR survey, the contractor should plan an itinerary before the actual scanning. The plant maintenance supervisor should also provide a liaison person to assist the contractor's equipment operator through the plant. During the actual survey, the test-set operator will scan a large number of surfaces. When a hot spot is detected, the unit will provide an "instant" response. The nature of this response will be determined by the magnitude of the heat source and the degree of sophistication employed by the particular type of test equipment used.

As the survey progresses, records are kept and a report of all objects exhibiting a temperature of 5°C higher than ambient is made. In a three phase system for example, when one phase has 5°C variation from ambient, it would be reported; but if all three phases have 5°C rise, it would **not** be considered abnormal.

Whenever the surface temperature of a load tap-changer (LTC) exceeds the surface temperature of the main tank (>1°C), the LTC

Figure 8.40 - Test technician conducting a substation infrared survey.

has a problem. In cases of severe problems, the ITC may be 3^0 or 4^0C above the main tank.

Because of the interference of sunlight, it may be necessary to make certain critical readings on cloudy days or at night.

Most infrared survey contractors will provide the customer with a positive identification of areas indicating faults, and a daily log of activities during periods of work in the plant. These field reports are usually supplemented with a formal engineering report and thermographic pictures showing the hot spots. If available, graduations on the vertical and horizontal axis, and special tables will help both the customer and equipment operators to interpret the thermal image to determine the temperature rise. Thermographs illustrate trouble spots that might have gone unnoticed until failure occurred, had IR equipment not been used (Figures 8.41 through 8.44).

Making The Right Choice

A choice as to which IR device is best for a given application is not always readily apparent. Certainly, the benefits derived from even the most expensive equipment could justify its purchase many times over. Furthermore, the more sophisticated data gained through their use has real advantages. The added facility of both quantitative temperature differences and a permanent pictorial record makes an attractive package.

To determine the level of performance required to accomplish the desired task, the following basic information will be helpful.

- Copper melts at under 1085°C; aluminum melts at 650°C.
- Most wire and cable insulations are rated for maximum operating temperatures of 60°-90°C.
- Copper will anneal, over long periods of time at temperatures in excess of 100°C.
- Tensile strength of copper starts to decline at 150°C.

Thus, one may conclude that there is really little need for high temperature monitoring capabilities when the melted copper will tell it all. The ability to read lower temperature ranges, below which electrical equipment normally experiences damage may be unnecessary. The less sophisticated, and hence less expensive, IR device may be quite adequate.

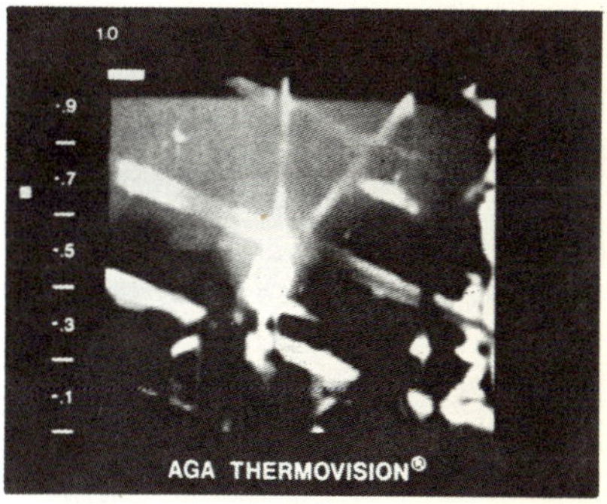

Figure 8.41 - Contrasting visual (L) and IR inspection (R).

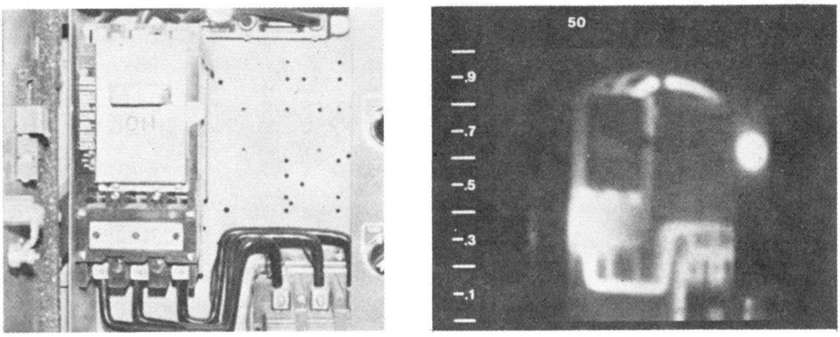

Figure 8.42 - Thermogram clearly pinpoints wiring with bad breaker/wire connections. (Courtesy of Thermotest Infra-Red Surveys).

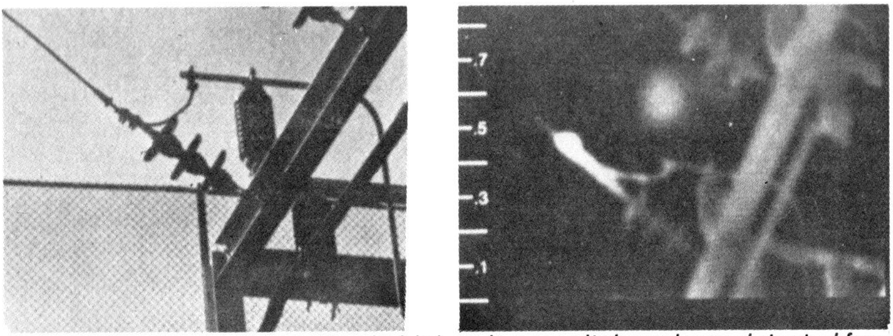

*Figure 8.43 - Loose connection in high-voltage switchyard was detected from ground level. (Courtesy of **Plant Engineering®** and AGA Corporation).*

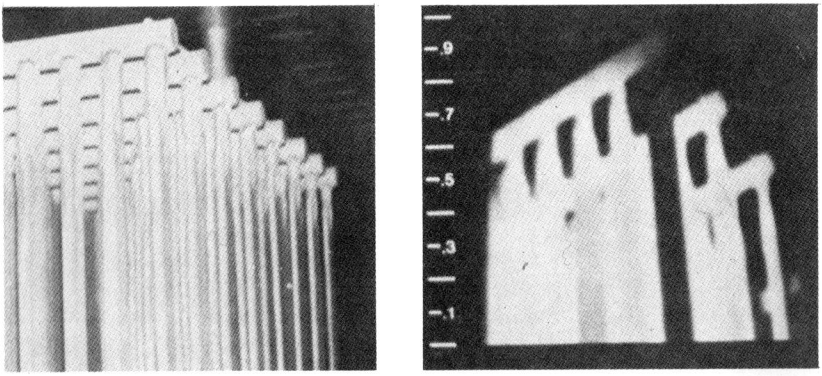

*Figure 8.44 - Blocked transformer cooling radiator was detected because of lack of heating in the radiator. In another similar case history IR thermograms enabled a utility to prove certain radiators on a new transformer were either blocked or improperly designed. (Courtesy of **Plant Engineering®**).*

Chapter 9

What To Do About Askarel (PCB) Transformers (Past, Present, and Future)

Part 1—Straight Talk On Askarels

Introduction

Fluids containing polychlorinated biphenyl (PCB), known generically as "askarel," have been used for nearly 50 years for heat transfer and insulating electrical equipment in applications requiring a highly fire resistant liquid. Thus, transformers could be used indoors close to the load center.

PCB is exceptionally stable and resistant to oxidation from acids, bases, and other chemical agents. This same stability has caused the liquid to persist and so in the late 1970's the U.S. Environmental Protection Agency (EPA) established regulations governing PCB use, maintenance, handling, storage, and disposal (Figure 9.1). These regulations are not voluntary, but mandatory, and therefore an understanding of these rules must precede our primary focus—maintenance of service-aged transformers, in this case, askarel-filled equipment.

In addition, the EPA banned the manufacture of the fluid, and thus the production of *new* askarel-insulated transformers has ceased.

However, current U.S. government regulations do not require earlier than normal removal of PCBs from service in totally enclosed industrial and utility locations ("railroad transformers"—only 1000 units—are required to be retrofilled to alternative fluids.)* It is anticipated that many of the estimated 140,000 in-service askarel-filled transformers and 35 million oil-filled service-aged units (assumed by the EPA to be PCB-contaminated) may be around for many years to come.

**Pending legislation may ban continued use of PCBs in the food and feed industries.*

Figure 9.1 - PCBs: On their way out. (Courtesy of Plant Engineering®).

PCB History

Invention and Development (Primary Uses)

Askarel was first used in power capacitors (Figures 9.2 and 9.3) and then in power transformers (Figure 9.4). PCB was invented in 1929 as a fire-resistant dielectric fluid for capacitors by the Swann Chemical Company. The company was later bought out by Monsanto Chemical, St. Louis, Missouri. The late Dr. Frank M. Clark patented its application for use as a transformer insulating fluid for the General Electric Company in 1931. The first transformers were put in-service in 1933.

As summarized by Underwriters Laboratories (UL) Incorporated (Table 9.1):

> Transformer askarels are considered non-flammable at ordinary temperatures. Under practical conditions, formation of combustible or explosive mixtures is regarded extremely unlikely. The fire hazard is very small.

As noted in Table 9.2, certain applications such as the pulp and paper industry have particularly relied upon askarel-filled equipment.

TABLE 9.1

FLAMMABILITY RATING UNDERWRITER'S LABORATORY (UL)*	
Water	0
Transformer Askarel	2-3
Silicone	4-5
Transformer Oil	10-20
Ether	100

TABLE 9.2

ESTIMATED ASKAREL TRANSFORMER POPULATION

Classification	PCB Units Per Total Units
All Units	1 of 250 (140,000 of 35,000,000)
Power Units	1 of 36 (140,000 of 5,000,000)
Pulp and Paper Industry	2 of 5[1]

[1]Non-random sampling of ten clients

See Chapter 2, Part 1, Table 2.1.

Figure 9.2 - A typical power capacitor bank. (Courtesy of McGraw-Edison Company).

Figure 9.3 - A typical capacitor as used in electronic circuits.

Figure 9.4 - A typical askarel transformer bank located inside a factory.

Secondary Uses

Unique fire resistant capabilities for PCB use extended far beyond its initial purpose, encompassing the following diverse applications (listed in order of predominant uses):

- **Capacitors,** including fluorescent light ballasts
- Plasticizers in synthethic resins and in rubbers
- Carbonless carbon paper (NCR)
- **Transformer liquids**
- Hydraulic fluids
- Lubricants, cutting oils
- Heat-transfer liquids
- Paint pigments
- Sealants
- Adhesives
- Printing inks
- Floor waxes
- Caulking compounds
- Dedusting agents
- Frying pans (heat transfer)
- Mounting medium for microscopic slides

Table 9.3 shows that over a billion pounds of PCB were produced between 1931 and 1977. Virtually all PCBs in existence today were produced synthetically. About 40 percent of this synthetic fluid has entered the environment (Table 9.3).

TABLE 9.3

PCB HISTORY IN THE UNITED STATES

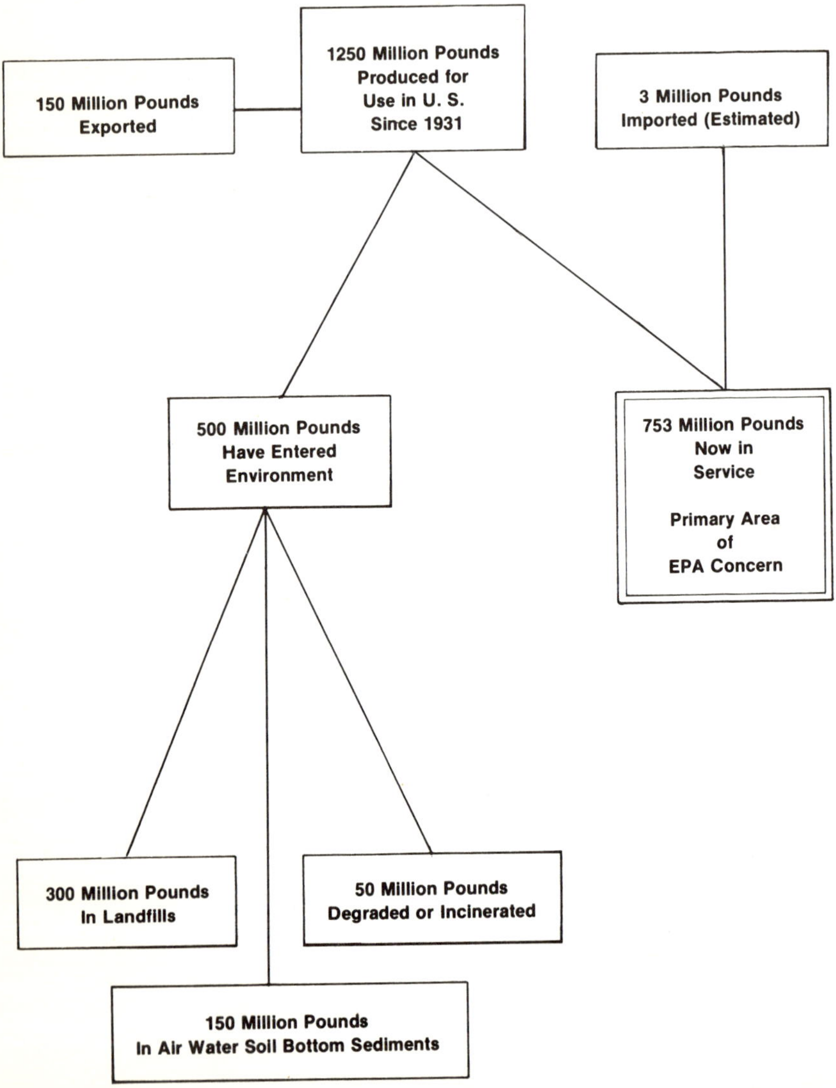

How to Identify PCBs

Throughout its history, PCB has been manufactured in the United States by Monsanto, under the trade name Aroclor®. It was then commonly wholesaled to second parties who would designate the generic "askarel" by a private label trademark for resale to end-users. Table 9.4 gives the more common registered trademarks used in the U.S. and overseas.

Most authorities state that some European PCBs are more toxic than U.S. Aroclors. In any event, designations are most likely to be used to identify the coolant in a high fire point transformer. The designations "askarel" and "PCB" are rarely found on a unit's nameplate.

PCB Characteristics

Molecular Structure

The PCB molecule consists of two phenyl molecules joined together, with two or more hydrogen atoms replaced by chlorine atoms (Figure 9.5). Two hundred and nine chlorine-substituted biphenyls can be created by the replacement of chlorine atoms at the various corners of the carbon rings. However, only three types of PCB are normally used in electrical transformers—1242, 1254, and 1260.* The first two digits, (12), designate the number of carbon atoms in the molecule and 42, 54, and 60 stand for the weight percent of chlorine respectively in each type. Other grades are used in capacitors.

Chemical Composition

Commonly used transformer askarels are essentially a blend—mixtures of chlorinated biphenyl and chlorinated benzene (Table 9.5). The PCBs are mixed to give particular viscosity characteristics. Nevertheless, transformer askarels are generally interchangeable.

In Table 9.5 reference is made to additive(s). The scavenger acts as a "getter" for free hydrogen chloride gas should any be present as a result of spurious arcs.

An earlier scavenger, tin tetraphenyl (1944), had a tendency to precipitate out on top of askarel as a milky-white crystalline material.

**ASTM D-2283 lists seven types of askarel that have been used in transformers over the years by various manufacturers.*

TABLE 9.4

ASKARELS CONTAINING PCBs IN U.S.A.	
Trademark	Manufacturer
Aroclor	Monsanto
Asbestol	American Corporation
Askarel	See Note 1
Chlorextol	Allis Chalmers
Diaclor	Sangamo Electric
Dykanol	Cornell Dubilier
Elemex	McGraw Edison
Hyvol	Aerovox
Inerteen	Westinghouse Electric
No-Flamol	Wagner Electric
Pyranol	General Electric
Saf-T-Kuhl	Kuhlman Electric

[1]Generic name used for insulating liquids in capacitors and transformers; may contain PCBs.*

**According to ANSI/IEEE/ASTM, "askarel" is "a generic term for a group of synthetic, fire resistant, chlorinated aromatic hydrocarbons used as electrical insulating liquids. They have a property under arcing conditions such that any gases produced will consist predominantly of noncombustible hydrogen chloride with lesser amounts of combustible gases." (IEEE Standard 591-1977; ANSI C57.12.80; ASTM D-2864).*

OVERSEAS	
Trademark	Manufacturer
Pyroclor	Monsanto (England)
Clophen	Bayer (Germany)
DK	Caffaro (Italy)
Fenclor	Caffaro (Italy)
Kennechlor	Mitsubishi (Japan)
Phenoclor	Prodelec (France)
Pyralene	Prodelec (France)
Santotherm	Mitsubishi (Japan)
Kanechlor	Kangegafuchi (Japan)
?	Sovol (U.S.S.R.)
?	Chemko (Czechoslovakia)

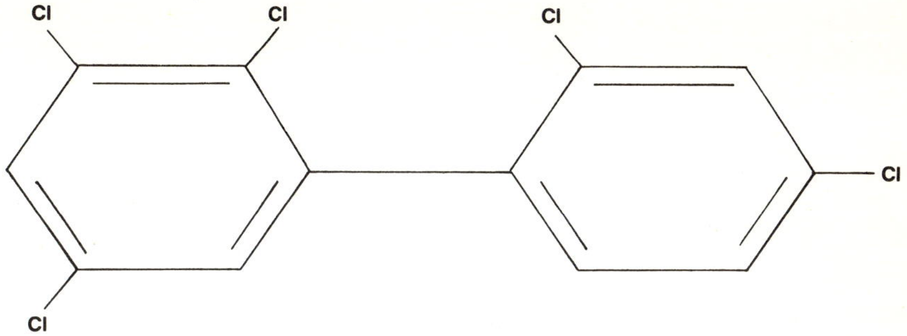

Figure 9.5 - PCB chemical structure.

TABLE 9.5

THE COMPOSITION OF TYPICAL TRANSFORMER ASKARELS[1]			
	Method ASTM D-2283		
Trade Names	Type D Inerteen® 70-30	Type E Inerteen® 100-42	Type G Pyranol® A13B3B3
Ingredients (% by Weight)	-	-	-
Aroclor® 1254, Chlorinated Biphenyl (54% Chlorine by Weight)	70	-	60
Aroclor® 1242, Chlorinated Biphenyl (42% Chlorine by Weight)	-	100	-
Trichlorobenzene	30	-	40
Phenoxypropene Oxide Scavenger	0.18 to 0.22	0.18 to 0.22	-
Diepoxide Scavenger	-	-	0.115 to 0.135
[1]Monsanto Chemical Co.			

This cloudy condition should be not interpreted as the presence of moisture in the early askarels.

Monsanto has not found evidence of reaction between the different scavengers when various askarels are mixed in any proportion.

Electrical Capability

Askarel, though one of the best insulating fluids developed by science, has been limited in usage. It has never been used on large transformers (beyond 69 KV, 15 MVA). In fact, during the early 1950's, transformer authorities recognized the practical askarel design limit as 34.5 KV (7.5 MVA). Most PCB units, therefore, fall in the 750-7500 KVA range.

Whereas mineral oil has an excellent impulse strength characteristic, askarel is limited in application because of its lightning strike capability (Figure 9.6).

On the other hand, oil must *never* be mixed with askarel. "Over two percent of oil by volume in askarel begins to lower its fire resistance."

Relationship to DDT

In some of the early analytical instrument studies PCB so resembled DDT (Dichlorodiphenyltrichloroethane) that part of the past impact of DDT in some situations may have been ascribed to PCBs. While similarities exist, the biggest difference is crucial; whereas DDT is a known acute toxic substance (physical effect apparent in a matter of hours), PCB by contrast is thought to be toxic because it is *bioaccumulative* (that is, its effect is *slow* accumulation in tissues of living matter over a period of years).

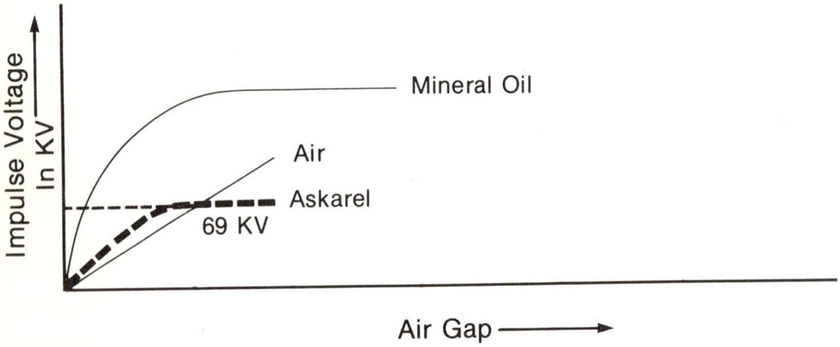

Figure 9.6 - Comparing lightning strike capability of insulating media.

The PCB Dilemma

U.S. Hazardous Classification

Only after many years of use did the PCB ecological problem become apparent. Table 9.6 contrasts insulating fluid criteria between 1968 and 1976. Obviously, sacrificial trade-offs have taken place.

PCB has been the most publicized item in U.S. environmental regulations; it has received the most press because of its assumed toxicity and as one of the most abundant pollutants in the world. Unfortunately, however, because of its abundance, the mass media has wrongly called PCB a "deadly poison"; they have even gone as far as to classify PCB (along with dioxin) as one of the two most dangerous contaminants on the planet earth!

What is not sufficiently known is that the EPA alone has at least six different classes of hazardous substances ("X", "A", "B", "C", "D", and "Priority Pollutant"); the Department of Transportation (DOT) has at least four ("Class A" and "Class B Poisons", and "ORM-A and -E Substances"); and the Occupational Safety and Health Administration (OSHA) has at least one more grouping. One substance may also receive multiple classifications.

EPA lists PCB both as a "Category A" hazardous substance (a legal spill, 10 pounds or more), and as "a priority pollutant" but not as a "poison". For comparison's sake, the "worst possible case" would be a hazardous substance, listed as "Category X" (legal spill, one pound or more) and "Class A poison" under the DOT criteria. Furthermore, PCB is nowhere listed in these various classifications as an "extremely hazardous" substance and has yet to be proven as "toxic" to public health.

DOT lists PCB as an ORM-E Substance (other regulated material), a material that is not included in any other hazard class. DOT only considers PCB as a hazardous substance when its reportable quantity (RQ) is equal to or greater than 10 pounds. When such a quantity is shipped, the container must be marked with a label designating "polychlorinated biphenyls," its DOT classification (ORM-E), its DOT identification number (UN 2315) and the letters RQ (reportable quantity) 49 CFR 172.316 (a) (7), 172.324 (b). The 10 pound reportable quantity follows the EPA's definition of a reportable spill near a waterway (*Federal Register*, Vol. 44, No. 169, August 29, 1979, p. 50777).

TABLE 9.6

CHANGING TRANSFORMER FLUID PRIORITIES[1]

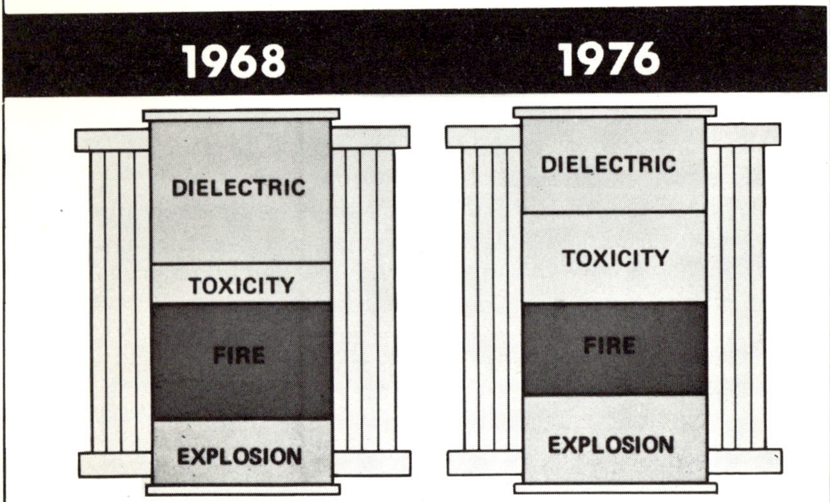

[1]*Courtesy of Dow Corning Corporation.*

Recent studies by the National Research Council conducted for the Environmental Protection Agency (Office of Research and Development) included these remarks:

> Based on anticipated (PCB) use patterns, the expected human and environmental exposure to the chemical would be primarily from protective coatings (lacquers and varnishes). Some risk would be anticipated in occupational situations, but these exposures could be carefully controlled. If the anticipated (PCB) use had been limited to heat exchangers and closed electrical equipment (capacitors and sealed transformers), the testing data might not have suggested potential hazard to human health or the environment. In such a case, PCBs might have "passed" the toxic substance evaluation.

Cost-Benefit Evaluation

A dilemma becomes especially apparent when benefits are compared with social costs (Tables 9.7 and 9.8). As one example of the environmental dilemma, food packaging materials made from *recycled* paper containing carbonless NCR copy paper and printing inks, are even sources of some contamination.

Effect on the Ecological Food Chain

The same characteristic of non-biodegradeability that has made PCBs useful for many industrial purposes has made them harmful in the food chain. This ecological process proceeds from the plankton (the simplest plant or animal), to fish, birds, and so on through to man himself (Figure 9.7). In fact, PCBs may biomagnify in the food chain—that is, as they move up the food chain towards man, their concentration may increase.

PCBs were first discovered in the flesh of Baltic fish in 1966. In humans the appearance of symptons of alleged PCB poisoning was observed in Japan in 1968 after inadvertent consumption of approximately 0.5 grams (2000 ppm) PCB in rice oil (from leaking heat transfer piping).

Several authorities now believe that the real culprit in this "Yusho" incident was not "normal" PCB at all, but one by-product of heated PCB (600° F)—polychlorinated *dibenzofurans*, which may be 500 times more toxic than PCBs!* This by-product in effect caused certain acute or immediate physical effects.

TABLE 9.7

COST-BENEFIT EVALUATION BENEFITS OF PCBs
Transformer Safety Transformer Reliability Reduction In: Property loss Loss of life Medical costs Employment loss Reduction in air pollution (PCB used in electrostatic precipitators used to control particulate matter) Reduction in cost of controlling air pollution

**This by-product may also result from insufficient incineration and may substantiate the closing of all PCB incineration sites as of April 18, 1978.*

TABLE 9.8

COST-BENEFIT EVALUATION SOCIAL COSTS OF PCBs (ENVIRONMENTAL DAMAGE COSTS)	
Direct:	
	Commercial fisheries
	Recreational industry
	Contaminated food
	Non-recycling of paper
	Worker lost time & medical (?)
Indirect:	
	Recreational fishing & hunting
	Loss of birds
	Human loss of physical & mental well-being (?)

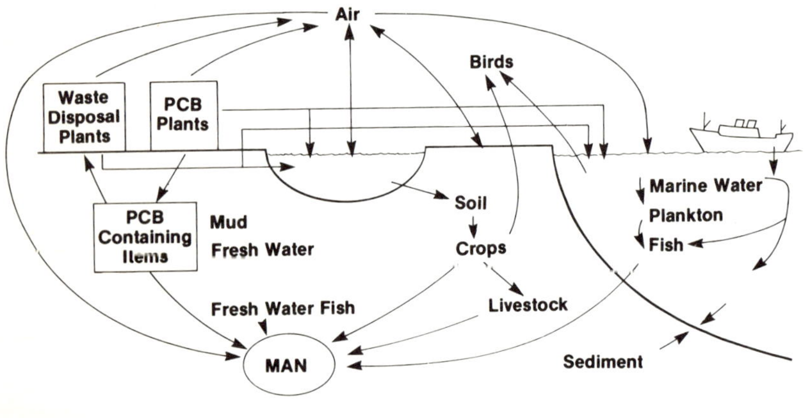

Figure 9.7 - Fate of PCBs in the environment. (K. Higuchi, ***PCB Poisoning and Pollution****, Academic Press, 1976).*

In the past, the EPA has stated that "...it is probably inappropriate to use the data from the Yusho episode to make quantitive estimates of the toxic hazard posed by PCBs to humans."* It would be like comparing apples with oranges! Keep in mind also the real possibility that non-U.S. PCB may be more "toxic" than American types.

*****Federal Register,*** *Vol. 42, No. 22 (Feb. 2, 1977), p. 6537.*

The all too obvious conclusion is that more extensive testing is required before any definite conclusions can be made. It is fair to say that the long term effect of PCB on the food chain may not be known for 20 to 30 years. Therefore, during this time of uncertainty, the best practice would be to minimize PCB exposure to the environment.

ASSESSMENT OF PCB HAZARD

An analysis of PCB data compiled following the criteria of FIFRA and TSCA* proposed guidelines, leads to the conclusion that PCBs are persistent, and are likely to accumulate. PCBs do not appear particularly toxic for short-term exposure, but results are subject to interpretation...

FIFRA Testing Results
The conclusion reached regarding the potential hazard posed to the environment by PCBs based on FIFRA requirements is **questionable.**† Acute toxicity is observed *only* for some aquatic invertebrates, but these species are not required test animals in the FIFRA guidelines.

**FIFRA - Federal Insecticide, Fungicide and Rodenticide Act. (1972)*
TSCA - Toxic Substances Control Act. (1976)

† - TMI emphasis

Precaution in Handling PCB in the Field

Because PCB remains an excellent dielectric fluid, such equipment should be around for many years to come (or in retrofilled PCB units). Personnel working with PCB should take precautions to protect the environment.

Limit exposure to PCB! No chemical is safe all the time; for example, the sulfuric acid in a car battery, if not handled properly, will burn the body and if dumped into a waterway, it will kill fish.

Human PCB exposure normally consists of sampling fluids and testing and repairing transformers. Thus, when working with PCB or solvent rinses used in PCB retrofill (many of these are also on EPA or DOT hazardous material lists) please note:

- Avoid direct skin contact
- Avoid direct eye contact by use of goggles
- Avoid breathing of vapors for extended periods of time
- Avoid vapors from a severely arced askarel transformer

Seeking Our Way Out of the PCB Woods

Confronted with this dilemma, America began its efforts "to seek a way out of the PCB woods" (Figure 9.8).

When the harmful effects of PCBs became known, five events took place. In 1972:

1. Standards were set by U.S. Food and Drug Administration for food contamination.
2. Standards were proposed by the U.S. EPA for water contamination.
3. Monsanto voluntarily withdrew PCB from all markets except totally enclosed systems—that is, sealed transformers and capacitors.

In 1977:

4. Westinghouse and General Electric ceased production of PCB transformers.
5. Monsanto voluntarily discontinued manufacture of askarel, after acceptable substitutes had been developed.

The transition to using non-PCB high fire point transformer and capacitor dielectric liquids, took place in the mid-1970's. All transformers that were manufactured before the mid-1970's—and that meet the requirements of a high fire point transformer as defined in Section 450-23 of the 1981 *National Electrical Code (NEC)*—invariably contain PCBs. With some rare, special-application exceptions, capacitors that contain more than three gallons of coolant, and that qualify for indoor use without vaults under Section 460-2 (a) of the *NEC* will contain PCBs if they were manufactured before the mid-1970's. Such equipment manufactured in the latter half of the 1970's may or may not contain PCBs; however, no PCB-insulated electrical equipment is manufactured today.

Even with these several steps, voluntary regulation proved ineffective. Therefore, U.S. governmental regulations have been published.

Figure 9.8 - A PCB transformer in a residential neighborhood. (Courtesy of ***Chemical Week****).*

Part 2—A Primer of U.S. PCB Regulations

Introduction

The first step toward PCB control in the U.S. was the adoption of the Toxic Substances Control Act (Public Law number 94-469, known as TSCA and pronounced TOSCA) which took effect January 1, 1977 (Figure 9.9). The intent of TSCA has been to protect in a reasonable and prudent manner the public health and the environment from *further* contamination from toxic chemical substances and mixtures. Though PCB is only one of the chemicals to be regulated, because of its abundance, it is the only one specifically mentioned (Figure 9.10). Another reason may be political reaction to the EPA's four-year delay in publishing the first PCB regulation (1972-1975).

PUBLIC LAW 94-469—Oct. 11, 1976 **90 STAT. 2003**

Public Law 94-469
94th Congress

An Act

Oct. 11, 1976
(S. 3149)

To regulate commerce and protect human health and the environment by requiring testing and necessary use restrictions on certain chemical substances, and for other purposes.

Toxic Substances Control Act.
15 USC 2601 note.

Be it enacted by the Senate and House of Representatives of the United States of America in Congress assembled,

SECTION 1. SHORT TITLE AND TABLE OF CONTENTS.

This Act may be cited as the "Toxic Substances Control Act".

TABLE OF CONTENTS

Figure 9.9 - The Toxic Substance Control Act as the authority for current PCB regulation.

PUBLIC LAW 94-469—Oct. 11, 1976 | 90 STAT. 2025

(e) POLYCHLORINATED BIPHENYLS.—(1) Within six months after the effective date of this Act the Administrator shall promulgate rules to— | Rules

(A) prescribe methods for the disposal of polychlorinated biphenyls, and

(B) require polychlorinated biphenyls to be marked with clear and adequate warnings, and instructions with respect to their processing, distribution in commerce, use, or disposal or with respect to any combination of such activities.

Requirements prescribed by rules under this paragraph shall be consistent with the requirements of paragraphs (2) and (3).

(2) (A) Except as provided under subparagraph (B), effective one year after the effective date of this Act no person may manufacture, process, or distribute in commerce or use any polychlorinated biphenyl in any manner other than in a totally enclosed manner.

(B) The Administrator may by rule authorize the manufacture, processing, distribution in commerce or use (or any combination of such activities) of any polychlorinated biphenyl in a manner other than in a totally enclosed manner if the Administrator finds that such manufacture, processing, distribution in commerce, or use (or combination of such activities) will not present an unreasonable risk of injury to health or the environment.

Figure 9.10 - Polychlorinated biphenyls as singled out in TSCA.

Congress also mandated that detailed regulations be published to implement certain sections of TSCA. The "Disposal and Marking Rule" took effect on April 18, 1978, while the 1979 "PCB Ban" law took effect on July 2, 1979.* The 1979 regulation completely superseded the 1978 version.

The published title "PCB Ban" is really a misnomer. A more appropriate title would be the "PCB Control Law." Source:* ***Federal Register**, Vol. 44, No. 106 Part VI (May 31, 1979), pp. 31514-31568; March 10, 1981, pp. 16090-16098.*

Unraveling the EPA "PCB Ban" Regulation

The law is clear. Choosing to keep PCB transformers in-service requires proper identification; choosing to dispose of units requires the same identification plus proper storage and disposal procedures. Either option requires accurate recordkeeping.

Definitions

Notable among the PCB associated terminology defined in Table 9.9 is a PCB item. A PCB item is any PCB associated thing that deliberately or unintentionally contains or has as a part of it any PCB or PCBs at a concentration of 50 parts per million or more (0.005 percent on a dry weight basis) and is therefore subject to EPA regulations.

Someone has illustrated the meaning of quantity in parts per million (ppm) using this analogy: one part in a million is comparable to one minute in 1.9 years! The definition of PCB liquid includes transformer oil and other liquid coolants that have been inadvertently contaminated.

Major transformer manufacturers have acknowledged that "some oil-filled transformers may contain varying concentrations of PCB". In this regard, it is illegal to *deliberately* dilute PCBs with mineral oil.

At the same time, if the oil level has dropped over the passage of time below the transformer radiators, the addition of mineral oil as a maintenance step only, will not violate the intent of the PCB law.

Current PCB regulations apply to all persons who manufacture, process, distribute in commerce, use, or dispose of PCBs. Our focus will be upon those who *use* and *dispose* of PCBs.

In-Service PCB Associated Items

Most existing totally enclosed transformers may continue indefinitely in-service.* Proper labeling, recordkeeping, *voluntary* inspection, and maintenance are the only necessary guidelines.

**An exception may be the food and feed industry (Federal Register, May 9, 1980, p. 30989); March 6, 1981, p. 15512; March 10, 1981, p. 16091 (required inspection).*

Certain transformers may be phased out by 1985. These units consist of railroad transformers (Figure 9.11), with few industrial units involved.*

Figure 9.11 - A Japanese railroad transformer. (Courtesy of Dow Corning Corporation).

1. Identification
   Under the current PCB regulation, all items listed in Table 9.10 should now display the appropriate PCB label (Figure 9.12).

2. Recordkeeping

   a. Who Must Comply
      Since July 2, 1978, an inventory must be kept concerning the status of your PCB items. The requirements apply to each facility owner or operator who has on his premises:

      - One or more PCB transformers and/or
      - 50 or more PCB large high or low voltage capacitors (see Table 9.9) and/or
      - At least 45 kilograms (99.4 pounds) of PCBs.

**Transportation locomotive transformers must be disposed of, or must meet the three-step time table for conversion to meet the definition of a PCB-contaminated transformer by 1985. Conversion calls for a series of draining of the liquid and refilling the unit with non-PCB fluid; the drained fluid must be disposed of in the manner prescribed for PCB liquids.*

TABLE 9.9

ENVIRONMENTAL PROTECTION AGENCY PCB REGULATION TERMINOLOGY[1]		
Term	Definition	Examples
PCB, PCBs	Any chemical substance containing only a biphenyl molecule that has been chlorinated to varying degrees, or any combination of substances that contain such substance	PCB liquids and nonliquids
PCB Transformer	Any transformer that contains PCB in a concentration of 500 ppm or greater	Askarel-insulated unit; some retrofilled units
PCB Contaminated Transformer[2]	Any transformer that contains 50 ppm or more, but less than 500 ppm, of PCB	Some oil-filled units; typical retrofilled unit
PCB Large High-voltage Capacitor	A capacitor containing more than 3 pounds (approximately 1 quart) of PCB dielectric fluid and operating at 2000 volts ac or above	Main substation power factor improvement capacitor
PCB Large Low-voltage Capacitor	A capacitor containing more than 3 pounds (approximately 1 quart) of PCB dielectric fluid and operating at less than 2000 volts ac	Power factor improvement capacitor installed at motor control center
PCB Small Capacitor	A capacitor containing less than 3 pounds (approximately 1 quart) of PCB dielectric fluid	Waveshaping capacitors in static inverter
PCB Article	Any manufactured item, other than a PCB container, that contains PCBs and whose surface(s) has been in direct contact with PCBs	PCB large high and low-voltage capacitors; PCB transformer; PCB-cooled motor
PCB Container	Any device used to contain PCBs or PCB articles, and whose surface(s) has been in contact with PCBs	Bottle, barrel, drum, or box
PCB Article Container	Any device used to contain PCB articles or PCB equipment and whose surface(s) has not been in contact with PCBs	Shipping or storage carton for capacitors
PCB Equipment	Any manufactured item other than a PCB container or PCB article container that contains a PCB article or other PCB equipment	Microwave oven; power factor-corrected lighting ballast
PCB Item	Any PCB article, PCB article container, PCB container or PCB equipment that, because of deliberate or unintentional action, contains or has as any part of it PCB or PCBs at a concentration of 50 ppm or more	Aroclor, Inerteen, Pyranol, contaminated transformer (mineral) oil, or coolants retrofilled to transformers formerly cooled with askarel

[1]Source: *Federal Register*, Vol. 44, No. 106, Part VI, May 31, 1979.

[2]Non-PCB transformer—any transformer that contains less than 50 ppm of PCB; the term is not defined in TSCA regulations.

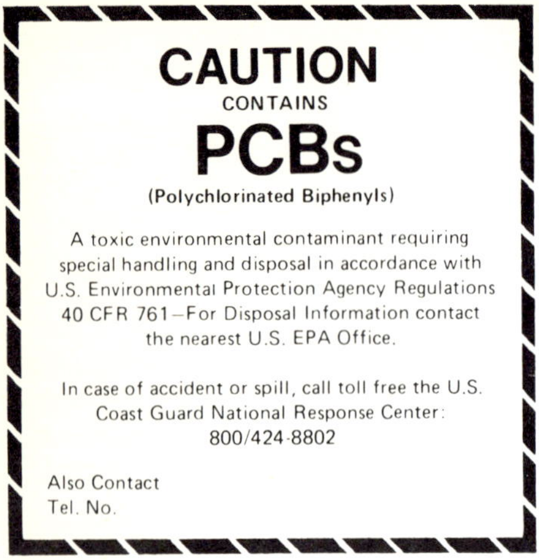

Label Identified as M_L

Label Identified as M_S

Figure 9.12 - Required Labeling Under TSCA. PCB items requiring labeling must be identified by the official PCB large form Mark M_L (above) or the official small mark M_S (above); permissible alternatives for labeling are given on pages 31543 and 31556 of the Federal Register, May 31, 1979. The Federal Register also gives detailed requirements for size, color, and durability. The M_L must measure at least 6 by 6" if the labeled item will accommodate a label of that size. The size of M_L can be reduced for items too small to accommodate that size; the largest possible size must be used. M_L may not be smaller than 2 by 2". The M_S label is to be used only for items too small to accommodate the smallest permissible size of M_L. M_S must measure 2.5 by 5 cm (1 by 2") if possible; it can be as small as 1 by 2 cm (0.4 by 0.8 inches) if necessary.

These records will form the basis for a required annual document prepared by each facility for the previous calendar year. It should have been available on July 1, 1979, and each July 1 thereafter. For multiplant companies in scattered locations, *one* record for the entire corporation could be maintained at a central location. The law requires that records are to be maintained in the individual company's own file, subject to request of copies or on-site EPA inspection. At the time of inspection, EPA representatives will show appropriate credentials, along with written notice of the facility to be inspected.

b. What Data is Required:

- Total number of PCB transformers and total weight in

kilograms of any PCB liquids contained in transformer.

- Total number of PCB large high or low voltage capacitors.
- Location of PCB items.
- Total weight in kilograms of PCBs and PCB items in containers including identification of container contents, such as liquids of capacitors (total weight means *excluding* the weight of the containers).

Use nameplate information to determine PCB weight requirements. The weight of PCB can also be determined by calculation, provided the calculations are based on valid assumption. For this purpose, the weight of pure PCB can be assumed to be approximately 12.5 pounds/gallon. Known or presumed concentrations can be used to calculate the weight of fluid in oil-filled units contaminated with 500 ppm PCB or more, using approximately 7.33 pounds/gallon as the weight of pure uncontaminated mineral insulating oil. Your annual reports must be kept for at least five years after a facility ceases to own or operate PCB related items in the prescribed quantities.

Failure to keep proper records is a violation of the law. *Civil* penalty of up to $25,000 could be imposed for violation of TSCA; each day the violation continues is considered a separate violation.

3. Generating Acceptable Records
With the exception of railroad transformers, TSCA does not require testing transformers to determine either the presence or concentration of PCBs. Oil-filled transformers containing less than 50 ppm of PCB contaminants do not fall under TSCA but disposal of the oil does (Table 9.10) and oil-filled transformers containing less than 500 ppm do not require labeling or records of their status.

Information gathered by the EPA indicates that very few oil-filled transformers will contain concentrations of PCB beyond 500 ppm, and the EPA expects that most owners of oil-filled transformers will assume that their units are contaminated to a concentration of less than 500 ppm. Such an assumption does not remove the onus from the owner in the event of an accidental spill of contaminated mineral oil. Ultimate disposal of the unit must, in any event, presume that the unit is contaminated

TABLE 9.10

REQUIREMENTS FOR LABELING PCB ITEMS[1]
PCB transformer (all askarel-cooled transformers, askarel-cooled transformers that have been retrofilled with non-PCB coolants and whose PCB content is 500 ppm or greater, and mineral oil-cooled transformers contaminated with PCB in concentrations of 500 ppm or more).
PCB large high-voltage capacitor.[2] For capacitors mounted on a pole or structure, or behind a fence, only one label need be posted on the pole, structure, or fence.
PCB equipment (any equipment containing as an integral component a PCB transformer or PCB high or low-voltage capacitor—unitized automatic power factor controller, etc.).
Other PCB articles—PCB-cooled electric motors, PCB hydraulic equipment and systems, PCB heat-transfer equipment and systems.
PCB storage areas, PCB containers, PCB article containers.
PCB transport vehicles (motor vehicles and railcars) carrying more than 45 kg (99.4 pounds) of PCBs.

[1]Items containing 50 ppm or more of PCB were required to have been labeled, unless otherwise indicated. Transformers containing less than 500 ppm of PCB need not be labeled; transformers containing 500 ppm or more were required to have been labeled by January 1, 1979. See Figure 9.12 for required label format.

[2]PCB large low-voltage capacitors are required to be labeled only at time of removal from service, unless they were installed after April 17, 1978. PCB large low-voltage capacitors installed after April 17, 1978 were required to be labeled at the time of installation.

beyond 50 ppm unless proved otherwise, calling for disposal in accordance with TSCA regulations.

A prudent course of action is to conduct a one-time test for all oil-filled transformers. Tests can be conducted over a period of time as the budget permits, with priority given to those units that expose the owner to maximum liability (such as those located along waterways) and to those nearing the end of life (thereby minimizing disposal costs).

Concentrations of PCBs in mineral insulating oil in excess of 100 ppm may reduce the oil's impulse strength. Interestingly, this fact has spawned research by several organizations to find ways of removing PCBs from mineral oil. This research was not spawned by TSCA, but was undertaken to find a way to improve the lightning-strike capacity of PCB-contaminated oil-cooled transformers. Hence, testing would indicate suspect units.

In addition, most major repair facilities will not accept a transformer for repair until it has been proven what the PCB content is.

Although TSCA requires neither testing nor labeling of oil-filled transformers having PCB concentrations below 500 ppm, testing is advisable. Units found to contain 50 ppm or more of PCBs should be identified to ensure that they receive special treatment. If tests confirm that the unit contains less than 50 ppm of PCBs, the owner is assured to have no potential liability under TSCA (Table 9.10). Tests must be performed in accordance with EPA-approved procedures.* Instruments used must be extremely sensitive and capable of measuring very small concentrations of PCBs.

Although testing for PCB content in oil by electron capture gas chromatography (GC) is the most prevalent laboratory method for low concentrations, the Beilstein field test can be a preliminary test to detect PCBs at a level of five percent (50,000 ppm) or greater. Very few units are this grossly contaminated.

Testing services are available from major transformer maintenance contractors.

4. Useful Comments Concerning PCB Test Results
Recall that the *Federal Register* (Vol. 44, No. 106, May 31, 1979, p. 31517) lists these three classes of transformers:

   a. PCB transformers (those units containing 500 or more ppm PCBs)

**An approved testing procedure has not yet been developed nearly two years after the effective date of the PCB law.*

b. PCB-contaminated transformers (those units containing 50-500 ppm PCB)

c. Non-PCB transformers (those units containing less than 50 ppm PCB)

The *Federal Register* also states that PCB testing shall be done by an "approved method". No formal approval of any method has thus far been made. The American Society for Testing and Materials (ASTM) has published a *proposed* standard test procedure and hopefully it will soon receive approval.

The proposed method uses the gas chromatograph equipped with an electron capture detector (reference *Federal Register* Section XII, "Test Procedures for PCB", May 31, 1979, pp. 31537-31538).

Both EPA and ASTM recognize this procedure is subject to "experimental deviation". Through experimental data and by interlaboratory cross-checking, a deviation of ± 10 percent may be an acceptable confidence level.

In order to be most confident in assigning a transformer classification, the test result must fall into the desired class, including the deviation difference. For example, a unit that tests at 46 ppm PCBs with a 10 percent increase of 4.6 would total 50.6 ppm. This exceeds the non-PCB classification and would be included in the "PCB-contaminated" class. Thus, to maintain the reliability of a classification system, determine your confidence limits and apply them to the test results.

The *Federal Register* states:

"EPA will not consider it to be good judgment to assume that the sample has less than 50 ppm PCB because the experimental error of the procedure overlaps the cut-off point" (May 31, 1979, p. 31358).

Out-of-Service PCB Associated Items

All out-of-service PCB items (Table 9.10) must also be properly labeled, whether spare units or those awaiting disposal.

Since April 18, 1978, PCB liquids (≥ 500 ppm) must be disposed of in only an EPA-approved incinerator. All nonliquid PCBs may be disposed of in an EPA-approved chemical waste landfill.

The various PCB items must be labeled and stored as summarized in Table 9.11 for PCB items containing more than 50 ppm PCB, unless otherwise stated.

A transformer declared as a spare may remain in a substation. If the unit is declared for disposal, then it must be moved to an EPA-approved storage facility pending appropriate disposal (Table 9.12, Note 2).

Permanent Disposal and Temporary Storage

1. Permanent Disposal
   One of the most practical ways to permanently remove PCBs from the environment is high temperature (at least 1200° C) incineration with proper dwell time. EPA regulations no longer

TABLE 9.11

PERMANENT DISPOSAL OR LONG TERM STORAGE OF PCBs PENDING DISPOSAL

(FOR PCB ITEMS OF 50 PPM OR MORE, UNLESS OTHERWISE STATED)

DISPOSAL MEANS THE INTENTIONAL OR ACCIDENTAL DISCHARGING OR OTHER COMPLETION OR TERMINATION OF THE USEFUL LIFE OF AN OBJECT OR SUBSTANCE.

Term	Disposal Provisions
PCB liquids (≥ 500 ppm)	Disposal in EPA-approved hi-temperature incinerator. No storage permitted.
PCB contaminated mineral oil dielectric fluid (50-500 ppm)	Disposal in EPA-approved hi-temperature incinerator; EPA-approved chemical waste landfill or a high-efficiency boiler.[1]
Waste oil (≤ 50 ppm PCB)	Such liquids may be used as a fuel, as a feedstock in the production of rerefined oils and lubricants or any other purpose, except as a sealant, coating or dust control agent (including dumping in a field).[2]
PCB-contaminated waste liquid (50-500 ppm) such as hydraulic or heat transfer fluids.	Disposal in EPA-approved hi-temperature incinerator, EPA-approved chemical waste landfill or an EPA-approved high-efficiency boiler.[1]

Table 9.11 continued on next page

Table 9.11 continued

Nonliquid PCBs; contaminated soil, rags, or other debris from uncontrolled spill (trash, trees, lumber).	Disposal in EPA-approved hi-temperature incinerator, or EPA-approved chemical waste landfill.[1] However, items may be considered to have been disposed of if they were deposited in a chemical waste landfill before April 18, 1978.
Nonliquid PCB waste; dredge spoils, municipal sewage-treatment sludges.	Disposal in EPA-approved hi-temperature incinerator, EPA-approved chemical waste landfill, or by other EPA-approved method.[1] The digging up of debris and other items disposed of in a landfill prior to April 18, 1978 is not required but the intent of the rule is that you can't add to an existing site and let it sit.
PCB transformers (≥ 500 ppm)	Disposal in EPA-approved hi-temperature incinerator or EPA-approved chemical waste landfill, following procedure given in text.[1]
PCB-contaminated transformer (50-500 ppm)	Disposal in municipal landfill or sold for salvage following procedure given in text.
PCB large high or low voltage capacitors (more than 3 pounds or 1 quart of PCB liquid)	Disposal in EPA-approved hi-temperature incinerator; disposal permitted until March 1, 1981 in EPA-approved chemical waste landfill.[1-3]
PCB small capacitors (less than 3 pounds or 1 quart of PCB liquid)	Capacitors owned by capacitor manufacturers or PCB article manufacturers (production quality-control rejects, etc.) to be disposed of in hi-temperature incinerator; disposal permitted in EPA-approved chemical waste landfill until March 1, 1981.[1-3] All other small capacitors mav be disposed of as municipal solid waste.
PCB hydraulic machine (die casting)	Disposal as municipal solid waste or sold for salvage following procedures given in text.
Other PCB articles	Disposal in EPA-approved hi-temperature incinerator, EPA-approved chemical waste landfill; following thorough draining and proper disposal of PCB liquid.[1]
PCB container (> 500 ppm)	Disposal in EPA-approved hi-temperature incinerator or EPA-approved chemical waste landfill (following temporary storage at an EPA-approved facility until properly drained and PCB liquid has been properly disposed of).[1]
PCB container (50-500 ppm)	Disposal as municipal solid waste, following proper draining and disposal of PCB-contaminated liquid.

[1]Long term storage of this item is permitted in EPA-DOT-approved containers and/or located in an EPA-approved facility until January 1, 1984. (Table 9.12, Note 2).

[2]*Federal Register*, May 31, 1979, pp. 31525, 31549.

[3]The capacitor disposal provision may not be extended by EPA since one EPA-approved incinerator for solids may be operating.

permit storage of unusable PCB liquids > 500 ppm. The first EPA-approved incinerator is not yet operational.

Several companies have applied for approval to operate incinerators; however, two of the sites are being held up because of public resistance out of fear of the unknown! As a point of fact, national environmental groups are not directly involved in these disputes; more than most, they realize the need to properly dispose of PCBs some place! A properly designed high temperature incinerator is the most logical choice. An obvious alternative is that liquid PCB "disappears" through the efforts of the illegal "midnight dumper".

Because of the delay in obtaining various permits and delivery of equipment, it will probably be several years before sufficient incineration capacity will be available.

Until such time as EPA-approved PCB liquid incineration becomes feasible, hang loose! Even when sites are open, practical logistics—distance and backlog—will make immediate incineration difficult. Keep the liquid in a good transformer, hold on to the liquid in EPA-DOT-approved containers. Store and inspect it weekly until disposal.* A spill prevention control program and counter measure plan should be in effect.†

Fortunately, the EPA recognizes the limitations of technical and economic feasibility to incinerate all PCB items. Therefore, the law permits disposal of PCB solids and contaminated liquids (50-500 ppm) at EPA-approved chemical waste landfills or high efficiency chemical boilers. At least eight approved chemical waste landfills are now "open". Since disposal in landfills will not result in PCB destruction, other new approaches now in the prototype stage may become available; for example, the use of microwave radiation or chemical transformation. Experimentation with various reagents to dechlorinate the PCB in which the askarel will yield a soluble by-product would certainly be an ideal solution. Because of the practical incentive, some form of detoxification may soon prove feasible.

The term "high-efficiency boiler" as used in TSCA regulations means a boiler capable of efficiently destroying low concen-

It has been reported that on 3/12/81, incineration began at two locations.

†*The "PCB Ban" law makes no provision for the temporary storage of PCB liquids.*

trations of PCBs; the term does not describe boiler efficiency in the sense of steam output per Btu of energy input. The *Federal Register* (May 31, 1979, pp. 31545 and 31546) contains highly detailed specifications of what constitutes a high-efficiency boiler and describes rigid procedures for boiler operation while PCBs are being incinerated.

Significant among the many criteria are a boiler rating of a minimum of 50 million Btu/hour, rigid limits on permissible stack emissions, requirements for continuous monitoring of stack emissions, prohibitions against introducing PCBs during boiler startup or shutdown, limits on PCB feed rates, and requirements for detailed documentation of the disposal operation.

PCB transformers, notably, can be permanently disposed of in a chemical waste landfill if all these conditions are met:

- The unit must be drained of all free-flowing liquid and the liquid must be treated in the manner prescribed for PCB liquids in Table 9.11.
- It must then be filled with a suitable solvent (such as kerosene, xylene, toluene) in which PCBs are readily soluble (some of these liquids are also "hazardous").
- The solvent must be left in the unit for at least 18 hours, and the unit must then be drained thoroughly.
- The drained solvent—which is now a PCB item—must be treated in the same manner prescribed for PCB liquids in Table 9.11.

Disposal of any *PCB-contaminated transformer* (50-500 ppm PCB) requires these steps:

- Transformer must be drained of all free flowing liquid.
- The dielectric fluid must be disposed of in any EPA-approved high-temperature incinerator, chemical waste landfill, or high-efficiency boiler.
- Drained transformer carcass may then be disposed of in a municipal landfill, or sold for salvage.

2. Interim Storage Pending Disposal
Temporary storage of high concentrations of unusable PCB is not permitted (temporary storage is 30 days or less). Table 9.12 lists the provisions for temporary storage of PCB articles, equipment or containers.

TABLE 9.12

CRITERIA FOR TEMPORARY STORAGE OF PCB ITEMS PENDING DISPOSAL	
Nonleaking PCB Article or Equipment[1]	*Temporary* storage permitted in any convenient location where the article or equipment is safe from damage. *Long-term* storage (required after 30 days) must be in an EPA-approved storage area.[2] Item must be identified as to date it was removed from service and checks for leaks must be made at least once every 30 days.
Nonleaking, Structurally Undamaged PCB High-Voltage Capacitors and PCB-Contaminated Transformers	Long-term storage permitted on skids adjacent to an EPA-approved storage area until 1/1/83. The intent is to have EPA-approved storage area immediately available in event of leakage. Inspection for leakage must be made weekly.
Leaking PCB Articles and Equipment	Storage permitted for up to 30 days in EPA-DOT-approved containers in a location at which the containers will not be subjected to damage. Enough absorbents must be added to convert the liquid to a nonflowing mixture. Long-term storage is not permitted; item must be disposed of as soon as it is practicable.
PCB Containers Containing Nonliquid PCBs (Soil, Rags, Debris) or PCB-Contaminated Liquids in PCB Concentrations of 50 Through 500 ppm[3]	Storage permitted in temporary storage area for up to 30 days, provided that a spill-prevention-control and counter measure plan has been prepared.[4] Long-term storage is not permitted.

[1]The terms leaks, leakage, or leaking (seeping or weeping) apply to any PCB article, container, or equipment that has PCB on any portion of its exterior surface.

[2]An EPA-approved storage facility is defined as a roofed, walled, and diked enclosure with a continuous 6-inch high curb having an impermeable floor such as concrete or steel. No such storage facility may be located where the elevation is below the 100 year flood water elevation.

[3]It is not permissible to store unusable liquids (liquids with a PCB concentration of 500 ppm or more) pending disposal. It is permissible to store PCB for topping off in-service PCB transformers in an EPA-approved storage facility.

[4]Suggested guidelines are given under "Spill Prevention Plan" in the September 1, 1978 *Federal Register* (p. 39276). The guidelines are proposed for incorporation into federal regulations as mandatory requirements.

When long-term storage is desired prior to disposal, contaminated PCB liquids (50-500 ppm) and nonliquid PCBs must be stored in an EPA-approved storage facility.

No storage facility may be located at a site which is below the 100-year flood water elevation (that is, a flood that has an average frequency of occurrence in the order of once in 100 years). Contact your regional office of the U.S. Geological Survey for a hydrological map that documents whether or not your present or proposed site is adequate (Figure 9.13).

3. Recordkeeping
   In conjunction with disposal requirements, the transformer owner must keep the following data for the annual composite record, documenting the status of PCB items slated for disposal:

   - Date removed from service.
   - Date placed into storage pending disposal.
   - Date placed in transport for disposal.
   - Total weight (in kilograms) and identification of PCBs in PCB containers.
   - Total number of PCB transformers and total weight (in kilograms) of PCBs contained in the transformers.
   - Total number of large high or large low voltage PCB capacitors.
   - Location and name of owner of initial storage or disposal facility.

   It is again permissible to determine the weight of PCB by calculation (using valid assumptions of the density of the PCB chemical substance or mixture) if the volume of the container or transformer is known.

   Failure to keep these proper records is a *civil* violation of the TSCA law.

4. Legal Responsibilities Concerning PCB Discharges
   Spills, accidental or otherwise, and other uncontrolled discharges of liquid PCBs are defined as disposal; therefore, proper procedures must be undertaken.

   Ignoring or attempting to conceal a leak, a spill, or other uncontrolled discharge of PCB is engaging in unlawful disposal and subject to criminal prosecution.

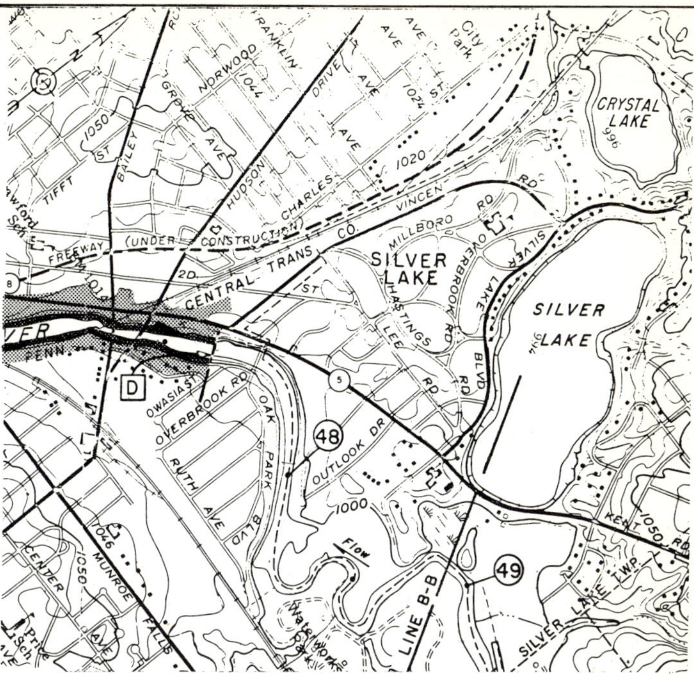

LEGEND:

STANDARD PROJECT FLOOD (500')

INTERMEDIATE REGIONAL FLOOD (100')

(49) DISTANCE FROM MOUTH IN MILES

[D] LOCATION OF VALLEY CROSS SECTION

Limits of overflow indicated may vary some from actual locations on ground, as explained in this report.

Figure 9.13 - A typical hydrological map.

a. Preventive Action

Weekly inspection for all items temporarily stored in non-EPA-approved facilities or on skids (temporary storage—30 days or less); 30-day minimum inspection for all items stored in an EPA-approved facility. Experience has already shown that inspection for leakage should be made more often than every 30 days—even with *new* PCB containers. Therefore, spill prevention is the only sure way to treat a potential spill situation.

b. Corrective Action

1) In the event of minor PCB leakage, required legal actions consist of:

   - Transferal of leaking containers and articles, and their contents, to properly marked, nonleaking EPA-DOT-approved containers.
   - Use of absorbents such as imbiber beads to clean up any spilled or leaked material immediately.
   - Disposal of (or store for future disposal) residue as indicated in Tables 9.11 or 9.12.

2) In the event of major spill:
   The "PCB Ban" law defines a PCB spill as any discharge of sufficient magnitude to give reason to suspect that it has produced, at any point in the surrounding soil, gravel, sludge, fill, rubble, or other land-based substances, a PCB contamination level exceeding 50 ppm. Section 311 of the Federal Clean Water Act has defined a spill as 10 pounds (4.54 kilograms) of PCB.

The following legal measures must be taken in the event of a PCB spill:

- When near a waterway, report spill to U.S. Coast Guard National Response Center, 800-424-8802.*
- Transfer leaking containers and articles and their contents to properly marked, nonleaking EPA-DOT-approved containers.
- Use absorbents to clean up any spilled or leaked material immediately.
- Dispose of (or store) residue as indicated in Tables 9.11 or 9.12.
- Conduct tests, using EPA-approved sampling techniques and laboratory methods, to determine if PCB concentration of soil, etc. exceeds 50 ppm.
- Decontaminate areas containing more than 50 ppm of PCB in accordance with your regional EPA. Improper cleanup may bring liability under both TSCA and the Clean Water Act.

**Reporting of a spill near a waterway is required under the Clean Water Act, Section 311. Failure to report a spill involves up to a $10,000 fine or imprisonment for not more than one year, or both. A discharge of a "reportable quantity may give rise to a penalty of up to $5000 per violation."*

Several companies have already established their own special teams that are sent out to handle PCB spills and clean up. Contracted cleanup service is certainly a viable option.

Other PCB-Associated Laws

Compliance with TSCA does not ensure that all laws governing PCBs will be met. In fact, the intent of the various U.S. Federal statutes (Table 9.13) is the integration of all facets of the PCB problem without doing so at cross-purposes.

Owners of PCB-related equipment should also continue to familiarize themselves with local and state laws controlling the use, handling, storage, transportation, and disposal of hazardous substances.

Local ordinances governing PCB disposal are few in number and, as mentioned earlier, typically come about due to public opposition to a particular existing or proposed PCB disposal facility.

On the other hand, some state laws are more stringent than the U.S. Law (Michigan and Oregon). Other states, such as Wisconsin, Minnesota and Connecticut, have reinforced or expanded the current federal regulation.

Notable among other federal laws that should be checked include the Resource Conservation and Recovery Act (RCRA)—Public Law 94-580, 1976 and section 311 of the Federal Water Pollution Control Act (FWPCA), Public Law 92-500 (or Clean Water Act, 1972) and U.S. Department of Transportation (DOT) regulations governing transportation of hazardous materials. (See DOT regulation 49 CFR Parts 171-173, May 22, 1980 for required PCB shipping instructions.)

RCRA is a "cradle-to-grave" manifest control, requiring that the generator of hazardous wastes maintain records and be responsible for all such wastes from the point of generation, through the transportation process, to the point of final disposal. It is no longer legal to merely turn hazardous wastes over to a contract hauler, saying that "I don't want to see them again."*

**Even though complete final rules of RCRA may not yet be published, several existing state laws indicate that the prudent way to go would be to prepare now for inevitable full U.S. implementation. Final disposal is the only certain time when the generator of hazardous waste no longer has any liability concerning the status of PCB items. The final decision will ultimately be decided in the courts (Federal Register, May 31, 1979, p. 31539).*

TABLE 9.13

PCB ASSOCIATED FEDERAL LAWS

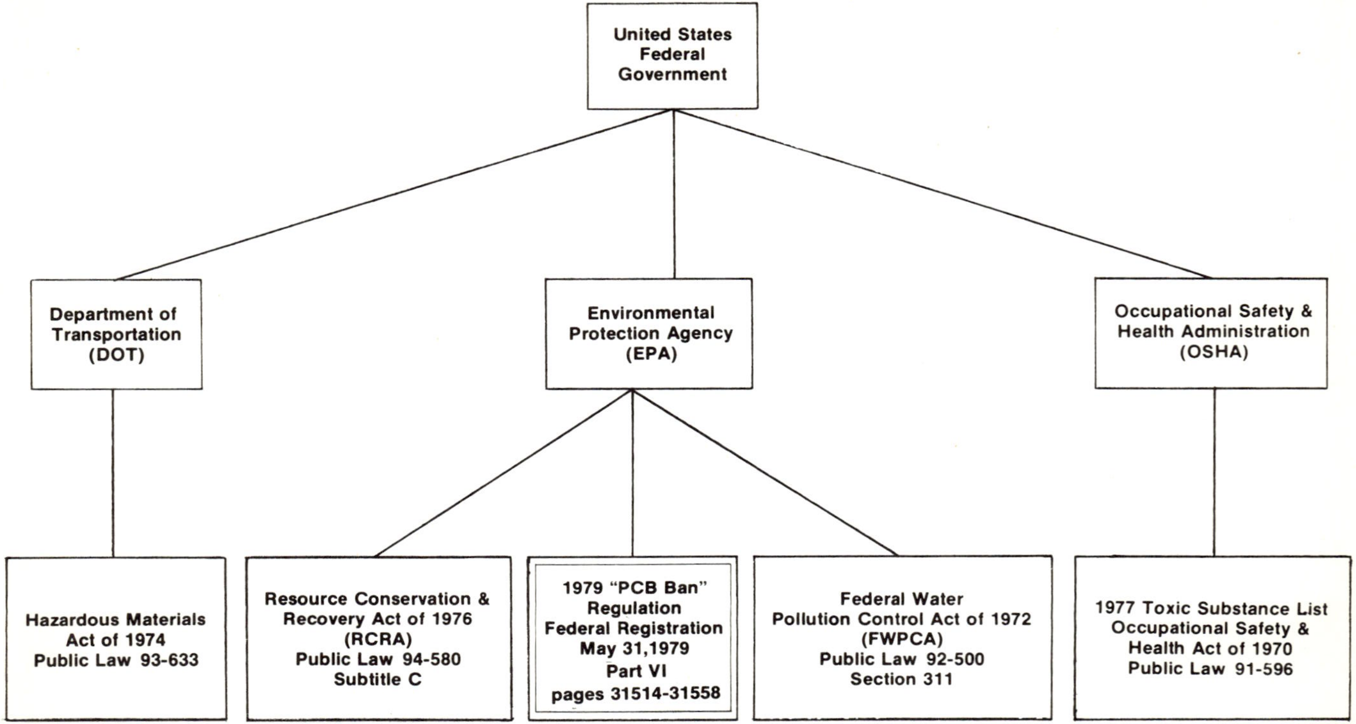

Enforcement—Federal Clout Behind PCB Regulations

Provisions for enforcement of the "PCB Ban" rule were included as part of the Toxic Substances Control Act. The law states that it is illegal for any company to fail or refuse to comply with any published provision pursuant to Section 6 (e) (1) of TSCA (as found in the *Federal Register* of May 31, 1979) and any subsequent revision.

Civil penalties of up to $25,000 can be imposed for violation of TSCA; each day the violation continues is considered a separate offense. In addition, violators may be subject to $25,000 as a *criminal* penalty and/or imprisonment of up to one year. It shall be unlawful for any company to:

- Fail or refuse to establish and maintain records concerning status of PCBs that are either in storage, disposed of, and/or still in-service.
- Fail or refuse to permit access to or copying of records.
- Fail or refuse to permit entry or inspection upon the presentation of appropriate credentials and written notice of particular area of inspection.

Penalties are prescribed under TSCA, Section 6 (e) 1, 11, 15, 16, and 17, P.L. 94-469, October 11, 1976.

Again, enforcement extends beyond TSCA, to include the Clean Water Act revision of August 29, 1979, concerning failure to report a spill (10 pounds or 4.54 kilograms) near a waterway.

An "Enforcement Proceedings Manual" drafted by the Pesticide and Toxic Substances Enforcement Division of EPA describes the following courses of action that may be taken against an alleged violator.

- Notice of non-compliance
- Administrative civil penalty
- Civil court action
- Criminal court action

Most violations will likely be handled through administrative civil action (that is, fines).

In its guidelines on determining the size of a civil penalty, the manual also sets down *four* levels of violations. They are as follows:

- Level I—This violation is of the type, such as failure to promptly dispose of PCB-contaminated rags, minor leaks in stored transformers, etc. The initial assessment for a Level I violation will be $1000 per day per violation. Where such a violation meets the criteria for issuance of a Notice of Non-compliance, the penalty may be waived.
- Level II—These violations are generally serious in nature, and may include general record keeping and marking violations, and storage violations where exposure is more hazardous than at Level I. The initial assessment for a Level II violation is $5000 per day per violation.
- Level III—This level applies to violations of a very serious nature, such as failure of incinerators to maintain proper temperatures for sufficient periods of time, and leakage from a chemical waste landfill not resulting in permanent environmental damage. The initial assessment for a Level III violation is $15,000 per day per violation.
- Level IV—These violations are the most egregious (flagrant), both in terms of blatant disregard for the requirements of the standard and the damage caused. Such violations would include significant spilling and dumping of PCBs resulting in large-scale contamination, and operating an unapproved disposal facility. The initial assessment for Level IV is $25,000 per day per violation.

The manual concludes that, upon determining the level of violation, the indicated penalty should be used as a base from which adjustments should be made. These additional statutory criteria consist of: a violator's ability to pay; inability of a violator to continue in business; history of prior violation; and the degree of capability of the violator.

Finally, Table 9.14 lists typical examples of violations of U.S. Federal PCB regulations.

A Final Word

Because of all the red tape and hassle, much criticism has been directed toward the expense of EPA regulation. But keep in mind that the long-term effect of PCB on the food chain may not be known for many years. Therefore, public support for costly environmental protection remains strong. That's because of the high cost of the alternative—not regulating!

In any event, while the issue of PCB control continues in a state of unending frustration, the best route to go is to keep informed of the latest state and local regulations and to make a "reasonable effort" to comply with current federal regulations (required marking, record keeping, voluntary inspection and maintenance, and suggested one-time PCB analysis of transformer mineral oils).

Interim Measures Program (Federal Register**, March 10, 1981, p. 16090); quarterly inspection and recordkeeping of all PCB transformers; and for all PCB and PCB-contaminated transformers used in the feed and food industry, weekly inspection and recordkeeping, "moderate leak" reporting to EPA Regional office, and immediate maintenance are required (Effective date: 5/11/81).*

TABLE 9.14

ENFORCEMENT THE CLOUT BEHIND THE U.S. PCB LAW				
Date of Occurrence	Regulatory Authority	Quantity and/or Violation	Penalty $	Total Cost $
March 5, 1973	Clean Water Act, 1972; also state of Tennessee	1500 gallon intentional spill	—	Over $2 million clean up and disposal, exclusive of pending law suits
Late April, 1978	TSCA 1976, 1978	Improper incineration of PCB liquid	$25,000	$25,000
July 28 thru August 2 1978	TSCA 1976, 1978	Deliberate dumping of 30K gallons along 210 miles of roadway	Up to $125,000 in fines and 5 years in prison	At least $2.5 million for cleanup and disposal (at taxpayer's expense)
October 24, 1978	TSCA 1976, 1978	Small accidental leak from syringe; improper packaging - labeling	$11,000[1]	$11,000[1]
April 25, 1979	TSCA 1976, 1978 Clean Water Act Amendment 1977 (PCB)	Improper disposal of PCB (some into a river)	$10,000[1]	$10,000[1]
August 7, 1979	OSHA (1970) 1977 Toxic Substances List (PCB)	Mishandling of recognized hazardous material that causes or is likely to cause physical "harm"	$46,800[1]	$46,800[1]
February 4, 1980	TSCA 1976, 1979	Incomplete recordkeeping etc.	$64,800[1]	$64,800[1]

[1]Pending appeal.

Part 3—Available Options to Owners of PCB Transformers

Introduction

Three basic options are available to the owners of PCB transformers:

1. The units can be retained as PCB transformers.
2. The units can be drained, flushed and filled with a non-PCB coolant.
3. They can be disposed of and replaced with new non-PCB transformers.

If a transformer is relatively old, the best choice may be (especially near waterways) to replace it now, but if new, then to retain or retrofill seems more likely.*

Following the EPA-required inventory of in-service askarel transformers, "prioritize" equipment for retaining, retrofilling, or replacing (say, within five and ten years). Transformer liquid screening tests would be a basis for this equipment evaluation. Then this evaluation should be included as part of a plant PCB manual. A loose-leaf binder should be indexed for quick reference. Federal, state, and local regulations should be included in sections concerning PCB use, marking, recordkeeping, storage and/or disposal, handling, maintenance, and spill prevention and control procedures.

Keeping "On-Line" Askarel Transformers

Requirements For Equipment Remaining In-Service

So long as no transformer or capacitor leakage occurs, the only requirements for in-service PCB units are that they be properly labeled and that records are maintained.

Spill Prevention Control and Countermeasure Plan (SPCC)

Begin to establish a spill prevention and spill control plan. Under Section 311 of the Clean Water Act certain steps are required (40CFR, Part 112, pp. 22-34).

**Pending legislation may ban continued use of PCBs in the food and feed industries.*

1. General Considerations

- Designation of one person at each plant location to see that every step of the plan is effectively implemented.
- Training factory and field personnel concerning the importance and use of SPCC plan (such training shall be held not less than once per year).
- Written description of any spill incident occurring during the prior 12 months. A spill is defined as 10 or more pounds of PCB (4.54 kilograms).
- Estimate of the effects of a potential spill from the facility, based on local drainage patterns and storage capacity.
- Keep records to document and identify all actions taken, persons involved, and dates. Among these records, keep an ongoing inventory of outgoing PCB items, including final disposal site. Satisfy state manifest requirements.
- Establish written procedures for action to be taken when a PCB incident occurs. This plan should be readily available for inspection.
- SPCC plan reviewed and certified as good engineering practice by a Registered Professional Engineer.

2. Spill Prevention Control
Spill prevention involves a series of operation procedures to be followed in the performance of a PCB activity (operating or testing apparatus):

- Appropriate containment and drainage controls, including dikes, berms, or retaining walls; curbing, culverting, or other drainage control (for units near waterways); absorbent materials (such as sawdust, imbiber material or "floor dry"); disposable clothing, polyethylene gloves, boots and goggles for personnel; shovels, extra empty EPA-DOT-approved drums; solvent rinses such as kerosene, or xylene; fire extinguishers: and self-contained breathing apparatus.
- Preventive maintenance and housecleaning—establish a checklist for at least monthly inspections. Insure that units are structurally sound. Check for gasket leaks (covers, handhole, or bushings). Check sampling valves and other fittings. Check pumps and cooling radiators for leaks. Take immediate corrective action as needed.
- Proper storage and/or disposal of contaminated materials per *Federal Register,* May 31, 1979, p. 31555.

- Locate so that areas are isolated—containers are not vulnerable to damage or puncture from vehicular, forklift, or other material handling equipment.
- Locate areas away from floor drains or crack-deteriorated floor areas.
- Mark EPA-DOT-approved drums as containing PCB-contaminated oil, nonliquid PCBs (rags, etc.) or PCB fluids per *Federal Register*, May 31, 1979, p. 31555, Section 761.42 (c) (1) (IV).

3. Countermeasure Plan
Spill control involves these measures to be followed when a release of PCBs occurs during any PCB activity.

a. PCB Leakage or Minor Spill
(Anything less than 10 pounds, where 12.5 pounds equal approximately one gallon). When even a small leak is detected through regular inspection, take these steps:

- Notify immediate supervisor at once.
- Find the source.
- Rectify the cause of leak or minor spill.
- Clean up area with sawdust, imbiber material, and solvent; rinse as appropriate.
- Transfer leaking PCB items to properly marked, non-leaking EPA-DOT-approved containers.
- Dispose of (or store for future disposal) contaminated materials per *Federal Register*, May 31, 1979, pp. 31545, 31555.

b. Major Spill
(Anything of 10 pounds PCB or more, per *Federal Register*, August 29, 1979, p. 50777). In the case of a major spill, take the following steps:

- Contain spill.
- Notify immediate supervisor at once.
- Rectify (if possible) the cause of the spill.
- Report a spill as required by the Clean Water Act.
- Begin clean up process. Using disposable clothing, boots, polyethylene gloves and goggles, take these precautions: avoid extended exposure (vapors); avoid direct skin contact; avoid eye contact.
- Use sawdust, imbiber material, or "floor dry" to clean up immediately the spilled material.

- Clean up a PCB spill in a manner similar to any oil spill, using an approved cleaning solvent such as kerosene, or xylene. A "floor dry" product may be used.
- Place debris in approved EPA-DOT containers.
- Transfer leaking containers and their contents into properly marked nonleaking EPA-DOT-approved containers.
- Dispose of (or store) contaminated materials per *Federal Register*, May 31, 1979, pp. 31545, 31555 or any more stringent state or local requirements.
- Conduct tests, using EPA-approved sampling and laboratory methods to determine if PCB concentration of soil exceeds 50 ppm PCB.
- Decontaminate areas containing more than 50 ppm PCB in accordance with regional EPA.
- Annotate records as to any change in the category of the substance and its container.
- Attach to every EPA-DOT-approved drum filled with PCB items, a document showing description of the item, destination (all bill of lading info) and directions for actions to be taken by the carrier, in the event of a spill, to confine it and to clean it up.

In-Service Maintenance

Routine maintenance (fluid testing, filtering and so forth) by the owner or outside contractor is permitted. However, the contractor performing transformer servicing is required to obtain a special annual EPA exemption to the ban on processing and distribution in commerce of PCBs.

PCB-contaminated transformers (50-500 ppm PCB) may be topped off, refilled or rebuilt. Existing askarel-filled units may be topped off with PCB dielectric fluid and minor repairs, such as gaskets, tap changers or bushings, may be performed. However, major transformer repair, including rewinding of coils, is not permitted on units of 500 or more ppm PCB. PCB transformers retrofilled and reclassified to PCB-contaminated units (prior to failure) may be rebuilt.

Because of askarel's high solvency characteristic, a common concern involves leak assessment, though no hard and fast rule exists whether a unit should be repaired or discarded. Other than to try to determine where the leak(s) is located, each situation should be evaluated on its own individual merit.* One clear evidence of leakage appears as sugar-like crystals on a unit's explosion vent or pressure relief device. These crystals are deposited at the time of evaporation of the trichlorobenzene solvent. Evidence of a leak may also be in the form of damp or wet spots on the concrete mounting pads. If the leak is from a gasket area (cover, handhole or bushing), then replacement of the leaking gasket with a synthetic polymer gasket, properly compressed, usually corrects the situation. Gaskets can be sealed temporarily by painting over the leaking area with epoxy cement.

A number of users keep their early-built askarel units in continuous service, in critical installations, by equipping them with compound pressure gauges for reading above and below atmospheric pressure. Positive pressure is maintained in the shell by introducing nitrogen at two to three pounds above the atmospheric pressure to prevent ingress of moisture (not to prevent oxidation). Transformer operating temperature and pressure are recorded daily. If a *sudden pressure drop* is noted or excessive nitrogen is consumed, the gaskets are checked for leaks with a soap solution and the leaks are repaired. If a leak is from a weldment seam or connection (more recent askarel units), try to stop "weeping" by using an appropriate epoxy repair kit. Whenever leaks occur, check also the liquid level; a low liquid level can impede cooling and result in a hotter running unit.

Evaporation of the trichlorobenzene solvent causes a thickening of the liquid and a resultant loss of heat transfer capability. When askarel becomes thickened, the unit should be considered for retirement. Thus, on the basis of such external evidence, older askarel units should be tested to see whether they should be replaced rather than to have a spill prevention plan.

Askarel Testing

Askarel transformers have given most reliable service down through the years. But this reliability is not an excuse to eliminate preventive maintenance.

The well-known screen tests used to determine transformer mineral oil quality may mean quite different things when applied to askarels. Askarel consisting of relatively simple and constant chemical composition may be tested readily.

Because of askarel's high specific gravity (1.5 average), water and some other impurities are most likely to be found at or near the surface.

**Chapter 9, Part 2, Table 9.12, Note 1.*

Therefore, the top sample is considered the best representative specimen (ASTM sampling procedure D-923). If the askarel is in good condition, a top sample should appear clean, free from carbon, clear or faint yellow, and contain no obvious water.

The most valuable tests for in-service askarel filled transformers at the same location include dielectric strength, water content, color, neutralization number and visual test (Table 9.15).

Any small amount of acidity is traceable to the dissolving of certain materials with which it comes into contact. Such small values (0.09 mg KOH/gram of askarel) do not affect the quality of PCB. Likewise, since askarel is not affected by oxidation, neither the fluid power factor nor the interfacial tension are used as guides in determining the condition of the askarel.

The key to insuring long-time use for in-service PCB transformers is keeping the askarel water-free. As the late Dr. F.M. Clark remarked on his many years of experience with askarel filled transformers, "The important thing is to keep them dry—otherwise leave them alone." This observation proved especially applicable with newer equipment when welding shut (new units) rather than gasketing (older units) became the primary method of sealing askarel transformers.

Contrary to mineral oil, the dielectric breakdown voltage is the most important maintenance test for askarel. Although some authorities say that the dielectric test of askarel is a good indicator of water and other impurities, this is not totally true. A low dielectric strength value can indicate the presence of free water or conducting particles, and possibly a high water content of the liquid of the order of 75 percent saturation or more. Therefore, an owner/operator should specify both a dielectric strength test and the ASTM D-1533 Karl Fischer water content test to determine the exact percentage of saturation. This is only an indication of the relative wetness or dryness of the liquid insulation, and not the solid insulation.

The dielectric test is not sensitive to ordinary dissolved polar materials. Table 9.16 shows this relationship of dielectric breakdown voltage versus moisture.

When the dielectric has dropped below 26 KV, moisture content should be determined, using the Karl Fischer method. If the water content value is 70 ppm or below, the liquid is satisfactory for continued service; if higher, the liquid should be dried and the transformer insulation tested to see if it needs dehydration.

TABLE 9.15

BASIC TESTS FOR ASKAREL FILLED TRANSFORMERS				
ASTM Test Method	Status of Units	Frequency	Criteria for Continued Service	Information Provided by Test
Dielectric Breakdown Strength D-877	In-service Unit relocated (before) Retrofill-before & after (12 hour soak)	6, 9, or 12 month intervals	> 26 KV (minimum)	Moisture pickup or arcing
Moisture (PPM) D-1533	In-service Unit relocated (before) Retrofill-before & after (12 hour soak)	When dielectric < 26 KV or at 6, 9, or 12 month intervals	70 ppm (maximum)	Water content in fluid > 70 ppm; dehydration required
Color D-2129	In-service	6, 9, or 12 month intervals	From faint yellow to light brown[1]	Grey-Black indicates arcing
Visual D-1702	In-service Unit relocated (before) Retrofill-before & after (12 hour soak)	6, 9, or 12 month intervals	No carbon sediment or cloudiness	Degree of arcing
PCB Content	After retrofill	Analysis as the fluid establishes equilibrium (as paper/ fluid PCB content changes)	—	Determine EPA labeling and disposal regulations (40 CFR 761)

[1]Light pink, green, or blue are sometimes noted but are not cause for concern.

TABLE 9.16

RELATION OF DIELECTRIC BREAKDOWN VOLTAGE TO AMOUNT OF DISSOLVED WATER IN ASKAREL AND MINERAL OIL

	Breakdown Voltage (ASTM D-877)	
Water Content (PPM)	Askarel	Mineral Oil
0	70 KV	50 KV
20	55	39
40	47	30
60	40	26
80	38	22
110	10	5

If water is found below 70 ppm by the Karl Fischer method, the transformer can be considered to be dry enough for continued operation. If water is found in excess of 70 ppm, then the unit should definitely be dried, assuming all other tests are satisfactory.

If the moisture is near the saturation level (about 125 ppm at room temperature) a thorough inspection should be made for water droplets in the transformer tank, and even for "globules" of water floating on the askarel surface. If found, the unit should be repaired by a maintenance contractor who has the required EPA exemption to the 1979 "PCB Ban" rule.

Sometimes, the dielectric breakdown voltage can also be lowered by severe arcing. This turns the askarel noticeably black or grey. Such a rapid change in color is usually caused by carbon particles in suspension which are an arc-over product.

If the acidity of the askarel were to increase *suddenly* to at least 0.09 mg KOH/gram (from a typical 0.02), arcing in the unit could be the cause. This arcing condition would release hydrogen chloride (HCL) gas which combines with water, producing hydrochloric acid. This is titrated as an acid component, and HCL deteriorates the insulation system and other internal parts of the transfomer. Therefore, HCL should be removed from the unit by bubbling dry nitrogen through the askarel after the transformer is de-energized.

At least one authority has said if a serious arcing condition exists, the best route may be to scrap the unit.

Askarel is a clear water-white to faint yellow liquid. After long-term use this color may gradually intensify to bright yellow or light brown. As long as the fluid remains clear, is free from contamination or cloudiness, has a low acid number, and has at least a 26 KV dielectric, it is satisfactory for continued use.

Any color change—such as to a green, red, or blue cast—indicates extraction of impurities from colored construction materials, such as from certain types of gaskets, pressboard, core bindings, coatings, or adhesives, not fully compatible with the askarel. If a distinct foreign color pick-up is noted, check the complete range of electrical characteristics. Such a condition, however, is not known to have had an adverse effect on askarel's dielectric strength.

In the transformer problems that arise, servicing of the askarel is usually not the solution, since it does not oxidize, sludge, or deteriorate. When any of the foregoing tests indicate poor or questionable conditions, a thorough check of the transformer itself is in order. The unit then should be repaired (refer to "In-service Maintenance" in foregoing pages to determine if such repair is permissible under the law).

Power factor and resistivity tests are of little value in determining the condition of an askarel. Whereas high power factor and low volume resistivity in mineral oils are commonly regarded as "danger signals" that the oil has deteriorated, this is not true of askarel liquid insulation, unless the dielectric is low or the moisture content is high (Table 9.17).

Since askarel has greater solvent action than mineral oil, trace contaminants from commonly used construction materials can vary the power factor and volume resistivity of askarel, without influencing its dielectric breakdown voltage (Table 9.18).

Retrofill of Askarel Transformers

Definitions

The transformer industry has gone full circle. Askarel as an insulating fluid and coolant was developed in the early 1930's to replace mineral oil in situations where fire resistance was important. The high stability of askarel in transformers made it incomparable in many locations, such as schools, hospitals, and high rise buildings. Yet it is this very stability and, therefore, its persistance in the environment which has begun the phase out of PCBs.

TABLE 9.17[1]

APPROXIMATE RELATIONSHIP SHOWING THE INSIGNIFICANT EFFECT OF POWER FACTOR AND VOLUME-RESISTIVITY ON DIELECTRIC BREAKDOWN VOLTAGE OF TRANSFORMER ASKAREL

Power Factor (60 cyc.)		Volume Resistivity x 10^9 ohm-cm (at 100° C, 500 Volts DC, 0.1″ gap)	Breakdown Voltage 25° C, 0.1″ gap
100° C	25° C		
2%	0.05%	1500	35KV
55%	0.1%	500	35
15%	0.7%	100	35
20-25%	2.0%	60-70	35
40-50%	—	25	35

[1]Monsanto Chemical Co.

TABLE 9.18[1]

EFFECT OF COMMON INSULATION MATERIALS ON POWER FACTOR AND DIELECTRIC STRENGTH (HEAT AGED 96 HOURS IN ASKAREL AT 100° C)

	Askarel After Exposure	
Material Immersed	Power Factor, Percent at 60 cyc., 100° C	Dielectric Strength 25° C
None (control)	1.0	35KV
Black varnished cloth	85.0	42
Copper	1.5	40
Pressboard	2.0	37
Manila paper	1.5	39
Phenol formaldehyde resins	1.6	41
Shellac	6.0	36
Iron	5.0	39
Synthetic rubber	70.0	39

[1]Monsanto Chemical Co.

A viable method for phasing out askarel-filled equipment is retrofill. Transformer retrofilling is the procedure used to remove the askarel, flush the unit, and refill with a replacement fluid as the coolant and dielectric. Table 9.19 shows various comparisons between commercially available PCB replacement fluids and askarel.

Two basic methods of retrofill have been used in practice. One method is the service shop retrofill, wherein the transformer is removed from service, transported to a service facility, retrofilled, and then returned to its installation site. This method requires extensive downtime, and exposes the unit to the additional hazard of damage while in transit and the possibility of a PCB spill during shipment.

Field retrofill, done at the transformer location, is the second method. This procedure is essentially as effective as a shop retrofill, if properly planned and executed.

Our focus here will be on the field method and will include pros and cons of retrofill, determining retrofill candidates, the actual planning, procedure, and necessary hardware to successfully perform a retrofill.

The obvious advantage of a field retrofill is that it can be accomplished with a minimum of service interruption. It is also much less costly than the shop retrofill, as costs for rigging and transporting the transformer are not involved. The efficiency of the retrofill procedure is also approximately the same. Using the field system, two 400 gallon units can be retrofilled per each eight to ten hour day. At the same time it is evident from experience that it is economically impractical to completely remove all PCBs from askarel transformers via the drain and flush method.

The Retrofill Option

While the EPA encourages retrofilling, it is not true that this decision would always be the most prudent. Table 9.20 lists the pros and cons concerning converting PCB to non-PCB units.

In its simplest concept, the retrofilling of askarel transformers with a non-askarel fluid involves the following steps:

1. Drain the askarel.
2. Flush the transformer with an appropriate solvent to remove as much askarel as possible from the unit's windings, insulation, and elsewhere (97 percent removed during first drain/flush cycle).
3. Fill the unit with the new fluid.

TABLE 9.19

CANDIDATE FLUIDS FOR PCB RETROFILL						
Typical Fluid Properties (ASTM Test Method)	Candidate Fluids				Reference	
	Silicone Fluid	Modified Hydro-carbon	Syn-thetic Hydro-carbon	Chlori-nated Benzene[1] (TCB)	Mineral Oil	Askarel PCB & TCB
COMPARISON OF ELECTRICAL TEST PROPERTIES						
Dielectric Strength (D-877), KV	43	43	43	30	43	43
Dielectric Constant (Permittivity)	2.7	2.2	2.2	—	2.5	4.3
Power Factor (D-924) % @ 25°C	0.02	0.002	< 0.001	—	0.002	0.1
% @ 100°C	0.10	0.515	< 0.001	—	0.10	0.5
Dissipation Factor (D-924), % @ 25°C	0.004	0.01	—	—	0.01	0.03
Volume Resistivity (D-1169), ohm-cm	$1x10^{15}$	$8x10^{12}$	$6x10^{16}$	$8x10^{11}$	$1x10^{15}$	$2x10^{12}$
COMPARISON OF THERMAL TEST PROPERTIES						
Flash Point (D-92), in °C	283	276	291	none	150	none
Fire Point (D-92), in °C	342	310	307	none	160	none
Pour Point (D-97), in °C	-70	-18 to -30	-50	-30	-40	-37
Thermal Conductivity 25°C cal/(sec cm^2C)/cm	$3.6x10^{-4}$	$3.2x10^{-4}$	—	—	$2.85x10^{-4}$	$2.8x10^{-4}$
Specific Heat 25°C (cal/gm/°C)	0.34	0.46	0.50	0.28	0.503	0.264
Coefficient of Volume Expansion (cc/cc °C 25°C)	$10.4x10^{-4}$	$8.0x10^{-4}$	$7.5x10^{-4}$	$8.0x10^{-4}$	$8.6x10^{-4}$	$7.0x10^{-4}$
COMPARISON OF PHYSICAL TEST PROPERTIES						
Specific Gravity (D-1298)	0.960	0.887	0.837	1.37	0.886	1.50
Neutralization Number (D-974), mg KOH/gm	0.01	< 0.01	0.01	—	< 0.01	< 0.02
Viscosity (D-445), Centistokes 25°C	50	480	200	2.5	16	15
100°C	16	15	14	0.7	2.3	2.2

[1]Chlorinated benzene may not obtain EPA or OSHA approval because of possible toxicity.

TABLE 9.20

PROS AND CONS OF RETROFILL PRACTICE		
Pro	Problem	Con
EPA encourages practice	Current EPA stance	Practice not mandatory but it seems to be at odds with energy (incineration) and resource conservation (disposal of PCB)
Continuing control of PCB units foreseen	Future EPA regulation	Counter productive to planning; anticipate future PCB use ban in all equipment, especially feed & food industry
Choice preferable to replacement. On-site service, one day downtime vs. months or more (delivery of new unit) plus time for installation	Cost	Replace vs. series of retrofills (drain & flush) required to bring fluid level down to non-PCB classification: cost of PCB analysis (for up to 3 year period)
Any degree of retrofill reduces potential liability in case of spill	Hazard	With retrofill 50% more fluid to handle, store and/or eventually incinerate; possibility of spill during retrofill
EPA permits rebuilding of unit converted to PCB-contaminated one (before failure); later sell failed carcass for salvage; reduced ultimate disposal cost of PCB-contaminated liquid and income from salvage.	Disposal	Very limited incineration facilities - cost of storage until incineration feasible (a failed unit must be scrapped: required disposal, cost of complete transformer plus cost new unit and time delay). Disposal of additional flushing fluid volume
Several satisfactory substitutes per Factory Mutual tests; continuing R&D	Liquid dielectric & coolant	No present liquid a one-to-one substitute. Effect on certain physical properties; leave units alone
To date no such problem has occurred	Partitioning effect at interface of 2 different fluids, if drain and flush is not complete	Possible lower dielectric strength and premature failure

Transformer Retrofill Candidates

Even after a tentative decision is made to retrofill a given PCB transformer, *not* every unit should be retrofilled. A selective evaluation would involve these considerations:

1. Retrofill ruled out for these units:

   - Small (500 KVA or less) transformers (relatively low replacement cost).
   - Units at least 25 years old.
   - Units that test faulty or marginal per Table 9.15 (replacement preferred).
   - Units that can be readily replaced (no physical size limitations).
   - Units located near waterways (regular inspection, leak repair, and replacement or move as soon as possible).
   - Units where EPA-approved diking would be environmentally sufficient (retaining the unit).
   - Units located in environmentally non-critical location (retain units with proper inspection and maintenance).

2. Retrofill accepted choice for these units:

   - Large transformers.
   - Relatively new transformers (less than 10 years old).
   - Special transformers (unusual voltage or frequency).
   - Units requiring extensive rigging and building modification to remove and replace.
   - Units located in environmentally critical areas, *if* above criteria also apply; otherwise, replacement may still be the better way to go.

3. Final unit evaluation before a decision is made to retrofill a PCB transformer:

   - Electrically test unit for winding power factor, bushing power factor (15 KV and above), insulation resistance, dielectric absorption, polarization index, and turns ratio at all tap positions.
   - Contact manufacturer of each unit which gave "good" test results, requesting information on the type of material used for gasket seals and internal wiring insulation. Determine the vacuum and pressure capability of the transformer tank and accessories, and obtain any loading and temperature data that may be available.

Choosing to Retrofill

1. Spill and Safety Considerations
During retrofill the potential for an accidental PCB spill (10 pounds or more) into the environment is great. But this risk and the various cost considerations are judged to be worth the long range benefits to be gained.

A contractual decision must be made as to who is to assume the responsibility and ownership of all askarel and PCB-contaminated waste, that would be accumulated during the retrofill (the service company performing the retrofill or the transformer owner/operator). Until sufficient incineration is available, this decision may vary. Special precautions prior to starting the retrofill include:

a. Special drip pans containing imbiber material should be placed under all askarel drums, pumps, filters, and hose and tool storage areas.

b. The job site area should be roped off and warning signs should be posted to minimize pedestrian traffic adjacent to the site.

c. Any floor drains in the area should be temporarily plugged.

d. A temporary dike may have to be placed around the transformer.

e. The transportation of bulk askarel should be in a tank trailer made of heavy gauge steel with double welded seams. The tank should have no bottom openings. All loading and unloading operations should be done from the top of the tank. The tank should be equipped with splash proof vents. Each hose connection should be made with a connector that can be safely wired to prevent accidental breaking of the coupling. All fittings must be absolutely leakproof.

f. All contaminated rags and gasket material, as well as spent imbiber material, must be drummed for proper disposal.

g. Make available to all personnel the necessary disposable equipment to minimize contamination of the environment.

h. Disposable clothing, goggles, boots, and gloves should be worn while handling the askarel to prevent secondary laundry contamination.

i. The process of spray cleaning of the core and coils from the top of the unit can be more comfortably accomplished if the operator is using a full face mask type of air breathing apparatus.

j. A portable eyewash fountain and emergency shower facility should be available.

Although all precautions will be taken to prevent spills, a backup or contingency plan should be developed to cope with unforeseen possibilities, such as transformer tube or drum damage.

2. Supplies and Equipment Required
All equipment used to transfer, pump, or hold askarel or PCB-contaminated liquid, shall be of material that meets with all local, state, and federal regulations.

Any pump and hoses used should be compatible in construction and materials with both askarel and the flushing solvent. Any waste storage containers shall be EPA-DOT-approved. In the following check-lists, silicone is the PCB replacement fluid; however, the same basic procedures are required for retrofilling with mineral oil, high fire point hydrocarbon fluids, or some forthcoming satisfactory PCB substitute.

a. Safety Equipment

- Hard hats
- Safety goggles
- Organic vapor breathing mask
- Plastic film (Polyethylene)
- Sorbent material
- Floor and crack sealer
- Drip pans
- Fiber board disposal drums with plastic liners
- Metal drums (DOT-17E-Gauge number 18; 55 gallons)
- Drum handling equipment
- Portable fan
- Diking material
- Safety barricades
- Portable eye wash fountain
- Portable shower

b. Retrofill Fluids

- Silicone
- Flush solvent

c. Liquid Handling Equipment, Askarel

- Pump
- Hose
- Dip tube
- Spray nozzles with wands
- Tank trailer (approved)
- Rope (to tie off hoses)

d. Liquid Handling Equipment, Silicone

- Filters, 10 micron
- Degassing equipment capable of reducing dissolved gas content of silicone from 16 to 5-8 percent at a flow rate of 600 gallons per hour (gph)
- Hose
- Pumps (special to handle silicone)

e. Tools

- Miscellaneous hand tools
- Nylon cord to tie off tools
- Safety drop cord light

f. Testing Equipment

- Insulation resistance test set
- 50 KV liquid dielectric strength test set
- Transformer turn ratio test set
- Laboratory support to run dissolved water content
- Laboratory support to run dissolved gas content

g. Miscellaneous Equipment

- Gasket cutting set
- Gasket material (silicone gaskets not compatible with silicone fluid)

3. The Retrofill Procedure (Capsule Summary)
The retrofill of a transformer in the field is complex and must be carried out in a precise, orderly manner to insure a dielectrically acceptable unit. Listed below are the necessary major steps.

a. Advance planning

b. Removal of unit from service

c. Pretesting of unit

- Electrical
- Liquid

d. Checking for leaks and repairing

e. Preparing site for askarel transfer

f. Draining of askarel from unit

g. Regasketing

h. Solvent flushing of unit

i. Final silicone flush of unit

j. Refill of unit with silicone

k. Post test of unit

- Electrical
- Liquid

l. Return unit to service

m. Make a record of the retrofill and allow for a follow-up retrofill after one or two years, using activated charcoal or some other type of filtration; meanwhile, monitor the PCB content.

A more detailed discussion now follows.

4. Advance Planning
This is actually a two-part preparation, because the transformer owner and the retrofill service company (if one is used) have definite areas that must be explored before a retrofill can take place.

a. Transformer Owner/Operator

1) Check with fire insurance carrier, legal department, local electrical inspector, for compliance with code, state regulations, and OSHA.

2) Once approval is obtained, make arrangements for rerouting electricity to de-energize the unit and provide an uninterrupted supply of power elsewhere.

3) Provide an auxiliary source of electric power required for this retrofill.

4) Make site arrangements such that minimum hose length is needed for the fluid handling equipment.

5) Check with transformer manufacturer as necessary, and supply service company with complete unit data, such as:

- Nameplate data
- Location
- KVA load specifications
- Pressure/vacuum capability of tank
- Types of seals
- Service history
- Other information the manufacturer feels may be important for retrofilling.

6) Coordinate activities with all pertinent company departments.

b. Service Contractor

1) Make arrangements for a representative to visit job site to inspect the area and identify potential risk areas, such as existing drains, waterways, and other places PCB could enter the environment.

2) Have necessary specialized test equipment and knowledge to perform and properly evaluate all liquid screening and electrical testing required for the retrofill process.

3) Evaluate transformer on a unit-by-unit basis per manufacturer's data and application. Select gasket material to be used (some materials must be changed as they are not compatible with silicone fluid). Loading and thermal characteristics (derating if gas space problem exists).

4) Procure a new nameplate identifying dielectric fluid and date of retrofill with allowance for derating of the unit if it runs too hot after conversion.

5) Check unit for other maintenance that should be done simultaneously with PCB conversion (such items as

gauges, sampling ports, pipe fittings, manhole covers, valve and gasket packing, welds, and sampling devices). Order all necessary materials.

6) Determine if existing float level gauge will work in silicone fluid. This will have to be replaced if the float sinks in the (different specific gravity) silicone fluid.

7) Make arrangements for the silicone and solvent to be shipped to the job site and properly stored and protected until needed.

8) Test all fluids to insure that they meet purchase specifications.

9) Arrange details per contractual agreement—responsibility for handling, storage and disposal of askarel, and possible spill cleanup.

5. The Day of the Retrofill

Upon arrival at the job site all personnel involved, both service company and owners, should be briefed on their activities. The spill and control measure plans should be reviewed.

At this time it will be necessary to determine the exact time schedule of the retrofill so there will be no unnecessary downtime. All safety rules should be reviewed and an inspection tour of the work area made so that all personnel know the location of safety equipment.

All supplies—liquid handling equipment, safety devices, and spill prevention materials—should be in place before starting the retrofill.

The drains in the area of the retrofill must be plugged or diked prior to draining of askarel from the transformer. Drip pans or trays must be placed under all connections involved in liquid transfer and these connections must be safety wired to prevent accidental opening. In the event that the transformer is in an area near a watershed, the immediate area shall be covered with polyethylene film and diked with sandbags. The containment area shall be of a size such that it will hold the entire liquid content of the transformer in the event of a spill. The work area shall be barricaded to prevent unauthorized persons and vehicles from coming on the job site. A supply of imbiber absorbing material and waste containers shall be placed within the barricaded areas (Figure 9.14).

The transformer should now be de-energized and safety grounds applied. Samples of the askarel should be drawn from the top sample valve and tested to assure that it is not contaminated or wet. The transformer should now be electrically tested, using the insulation resistance test and turn-to-turn ratio, to assure that the unit is worthy of retrofill.

At this time the unit should be drained into the tank trailer. All hose runs must be as short as possible and no sag should be allowed in the hose. At no time should any hose be left unattended during liquid transfer. The receiving tank must be of adequate size to contain all the askarel without danger of overflow. This tank shall also meet all EPA-DOT requirements for a PCB container and shall be both top loaded and unloaded.

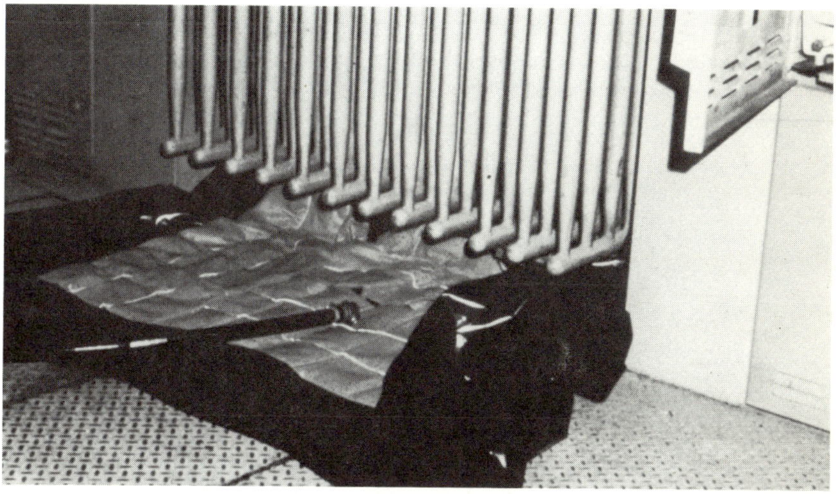

Figure 9.14 - Complete physical preparation of site for PCB retrofill procedure includes placing of adequate absorbent materials within a barricaded area.

While the transformer is being drained, the liquid should be replaced with dry air from an auxiliary supply or a drying type breather.

The unit must be opened to complete the draining of the transformer. A dip tube may be required to remove any liquid remaining below the bottom drain valve. Caution must be exercised to not damage the paper insulation during this operation. Prior to opening the transformer, any possibilities of foreign material entering the unit from overhead must be eliminated. Any tools to be used on top of the transformer must be tied off prior to removing inspection cover(s). All nuts, bolts, and washers must be counted and removed from the top of the unit before opening the cover. Anyone working on top of the unit shall have removed all loose articles from his person. At this time an inspection of the transformer should be made and any loose connections should be tightened. Any regasketing and/or leak repairs should now be completed (Figure 9.15).

The unit is now flushed with an electrical grade solvent to lower the PCB level. The flushing is accomplished by installing and recirculating a quantity of solvent using a small pump and spray nozzle. The spray nozzle must have a wide pattern in order to reach into all exposed areas of the transformer interior. The flow should be low pressure to avoid subjecting the windings to any harsh mechanical pressure. The nozzle and wand should be of such a size as to allow the operator to reach into the transformer and spray solvent into such areas as radiator tubes, tap changers, etc. (Figure 9.16).

At the completion of this flush, the unit is again drained completely and then flushed a second time with fresh solvent. After draining the second flush, a third flush using silicone fluid is used. All flush fluids are to be disposed of as PCB liquids.

The silicone handling equipment is now hooked up in a closed loop, and silicone is recirculated and tested to be sure it meets specifications. These tests are to include dielectric strength, water content, and dissolved gas content. The processing equipment is now hooked up to the bottom valve of the transformer and silicone liquid installed into the transformer. By filling in this manner, the degassed liquid will absorb air that tends to remain in the windings during the fill.

Upon completion of the fill, the unit should be sealed and let sit for at least 12 hours before energizing. This will allow entrapped air to rise to the surface and provide additional insulation impregnation.

Before energizing the unit, samples of the liquid should be drawn and tested for dielectric strength and water content. The unit should be electrically tested—insulation resistance and turn-to-turn ratio—to ascertain that no damage was done to the unit during retrofill. If all test results are good, then the unit should be energized, but not loaded. Approximately four hours after energizing, the unit may be loaded.

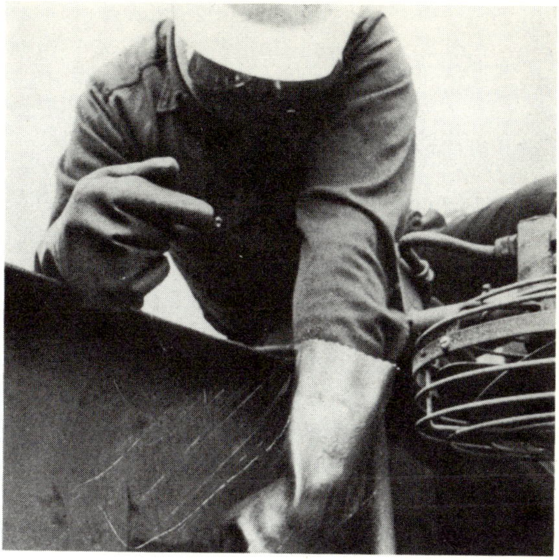

Figure 9.15 - Preparations prior to the flushing of a drained askarel transformer include regasketing and/or leak repairs.

Figure 9.16 - Each complete flushing of a drained askarel transformer requires spraying solvent into all exposed areas, including radiator tubes and tap changers.

In addition to the transformer test data, record all other pertinent information such as weather conditions (temperature and humidity), local environment, and other factors.

6. Getting the PCBs Out!
Following the first drain/flush/fill retrofill cycle, an askarel-filled transformer will still be classified as a PCB unit (probably in the range of 7000 ppm or 0.7 percent PCB). Conversion to a PCB-contaminated item (50-500 ppm) will require at least one additional step. Before a unit can be officially reclassified as a PCB-contaminated unit, the transformer must be tested and found to contain less than 500 ppm PCB after at least three months of in-service use (per "PCB Ban" regulation). A transformer can be reclassified to a non-PCB transformer when tested and found to contain less than 50 ppm. These levels can be achieved only by further processing of the dielectric fluid after a period of time has elapsed to where the residual PCBs in the windings have leached out into the fluid.

A recent method of PCB reduction, when the retrofill involves changing to a mineral oil based fluid (HTH-high temperature hydrocarbon), uses a low volume recirculating system attached to the retrofilled unit. This system continuously percolates the insulating fluid through an absorption medium formed by polychloroprene. Using changes as needed, the system may permanently reduce the *PCB concentration* below 50 ppm permanently in about 20 months. This need to reduce PCB concentration should be an economic incentive for development of other methods.

7. Ongoing Retrofill Evaluation and Maintenance
Retrofilled transformers should be monitored closely immediately following retrofill and thereafter on a scheduled basis to plot the concentration changes (Tables 9.21 or 9.22).

The frequency of inspection and testing depends upon the service to which the transformer is subjected. "Silicone liquid-filled (or retrofilled) transformers which are operated under extremely heavy loads may require more frequent inspection than those in normal or light service."

Color change alone (except for gray or black which indicates arcing) is not normally a danger signal, since the contamination is not likely to impair the dielectric strength.

ASTM D-2225 "Standard Method of Testing Silicone Fluids Used for Electrical Insulation" specifies the test that can be used to determine the condition of silicone fluids.

A modified Karl Fischer Method will thus be necessary to analyze water in ppm in the fluid. Silicone liquid may contain compounds which will interfere with the standard ASTM method D-1533.

Insufficient in-service data prevents any specific recommendations for basic maintenance tests for high fire point hydrocarbon retrofilled askarel units.

The basis of Table 9.22 (recommended tests for high fire point liquids) is the dielectric quality and water content of conventional mineral oil.

TABLE 9.21

RECOMMENDED MAINTENANCE TESTS FOR SILICONE RETROFILLED ASKAREL TRANSFORMERS

ASTM Test Method	Criteria for Continued Service	Unacceptable Values Indicate	Comments
Dielectric Breakdown D-877	25 KV (minimum)	Particulates, water	—
Visual	Sparkling clear-free of particles	Particulates, free water	Reconditioning by vac-degasifier or cartridge filter
Color D-2129	Water white		—
Power Factor D-924	0.5% (maximum)	Polar/ionic contamination	—
Water Content ppm D-1533 (Modified)	120 ppm (maximum)	Total water	Modified Karl Fischer Method
Viscosity D-445	50CS (± 5 CS) at 25° C	Fluid degradation, contamination	—
Fire Point D-92	300° C (minimum)	Contamination by volatile material	—
PCB Content	Analysis as the fluid establishes equilibrium (as paper/fluid PCB content changes); determine EPA labeling and disposal requirements.		

TABLE 9.22

RECOMMENDED MAINTENANCE TEST FOR ASKAREL TRANSFORMERS RETROFILLED WITH HIGH FIRE POINT HYDROCARBONS[1] (HIGH TEMPERATURE HYDROCARBON-HTH)		
ASTM Test Method	Criteria for Continued Service	Information Provided by Test
Dielectric breakdown D-877	25 KV (minimum)	Conductive contaminants present in the fluid, such as metallic cuttings or free water
Neutralization number D-974	0.10 mg KOH/g of oil (maximum)	Acid present in fluid
Interfacial tension D-971[2]	28 dynes/cm (minimum)	—
Water content ppm D-1533	50 ppm (maximum)	Total water content; leak or cellulosic deterioration
Viscosity D-445	No change by more than ± 10 percent	Fluid degradation, contamination
Fire point D-92	300° C (minimum)	Contamination by volatile liquid
PCB content	Analysis as fluid establishes equilibrium; determine EPA labeling and disposal requirements.	

[1]Until such time as specific criteria are determined, insufficient current information available.

[2]This test may not apply, since ASTM specifies D-971 for mineral oil only.

Decision-Making Guidelines on Phasing Out Askarel Transformers

Certain expenses are incurred, regardless of whether a decision is made to retain, retrofill, or replace PCB transformers. Basic cost factors are listed in Table 9.23 to assist you—a PCB transformer owner/operator—in evaluating your equipment and on making the final decision. In any event, that choice should be made as soon as possible for each PCB unit in your inventory. The total cost may at first appear prohibitive, but recall Table 9.24 and the possible financial spill liability (even though it may be accidental) under the "PCB Ban" regulation.

TABLE 9.23

REQUIRED COST CONSIDERATION IN UNDERTAKING RETROFILL OF AN ASKAREL TRANSFORMER
First Retrofill Cycle
Cost of downtime (normally, one day per unit) Cost of testing askarel liquid before removed Cost of electrical testing askarel transformer
Cost of refill dielectric fluid ($7 to $17 per nameplate gallon). Cost of testing the refill dielectric as received Cost of flushing solvent ($1 to $17 per gallon which would be approximately 25% of the nameplate volume) The higher cost fluid used as a solvent could be reclaimed, and used again, as dielectric fluid.
Cost of labor (3 men-8 hours each or 24 man-hours)
Cost of post retrofill testing—refill fluid and after 12 hours soak. Cost of testing retrofilled transformer prior to energizing. Cost of monitoring PCB content in refill fluid
Disposal of spent askarel (at least $6 per gallon for incineration plus transportation cost of $2 per mile) Disposal of flush solvent (incineration and transportation) Disposal in chemical waste landfill of PCB-contaminated support materials ($5 per cubic foot and transportation cost of $2 per mile) Cost of liability insurance (if available)
Subsequent PCB Removal Thru Filtration
Second decision whether to achieve <500 ppm or <50 ppm PCB content Cost of filtration system (activated charcoal, or polychloroprene medium, etc.) Cost of labor Cost of disposal, including transportation

Many of these costs are the same if the choice is to remove and replace an askarel unit with a PCB substitute. On the other hand, while present costs will be minimal for units retained, all those costs are only deferred, and will obviously be much higher in the future.

With these basic cost considerations as a guide, we suggest that you perform your own economic analysis, using current costs and numbers that are in line with your individual budgeting situation.

PCB Transformer Replacements

The Issues Involved

The third option available to owner/operators of askarel transformers involves replacement of the PCB unit with a non-PCB transformer. Where a severe hazard exists—such as lightning-prone areas or where PCB-filled units are located near streams, sewers, ditches, or other waterways—strong consideration should be given to replacing these PCB units. This choice is especially advisable if the unit is nearing the end of its useful life, or is relatively small with comparatively low replacement cost.

Traditionally, the purpose of the PCB transformer has been to provide safety in any indoor location such as apartment buildings, hospitals, factories, especially where concentrations of people are found. However, since PCB has been such "an outstanding performer" for so long, it's been difficult to find a satisfactory safe replacement. In fact, until manufacturing of new askarel-filled transformers was formally banned in 1979, nothing else was needed, since PCB-filled units satisfied the desired safety requirements.

Advances in transformer technology have thus begun again out of necessity. The need is especially critical for medium power units ranging in power rating to 15 MVA and primary voltage class below 34.5 KV. A number of new designs and materials have been introduced and promoted as reliable fire-safe alternatives to askarel transformers.* Table 9.24 lists commercially available replacement fluids for PCBs.

Discussions continue among the IEEE, users, manufacturers, and insurers on topics ranging from the basic cause of fires to the evaluation of high fire point liquids.

Assessment Criteria For High Fire Point Liquids

As expected, the basic causes of fire on or from transformers are internal or external. Internal origins are those due to winding or conductor failure, resulting from insulation breakdown. Such a failure may result

**No transformer alternative at this point in time matches the performance of askarel (Circa 1980).*

TABLE 9.24

COMMERCIALLY AVAILABLE REPLACEMENT FLUIDS FOR PCB

FLUIDS	Electric Strength 60HZ KV	Electric Strength Impulse Fullwave KV	Viscosity 25°C cSt	Viscosity 100°C cSt	Coefficiency Thermal Expansion cc/cc/°C X10-3	Specific Heat	Specific Gravity	Pour Point °C	Boiling Point °C	Fire Point °C	Typical Arc Formed Gases[3]	Cost	Availability
Mineral Oil[1]	43	300	16	2.3	0.86	0.503	0.886	-40	125	160	H_2, CH_4, C_2H_4, C_2H_2	Low	Unlimited(?)
Silicone Fluid	43	300	50	16	1.04	0.340	0.960	-70	250	340	H_2, CH_4, C_2H_4, C_2H_2	High	Unlimited
Synthetic Hydrocarbon (High Temperature Hydrocarbon)	43	300	33	5.7	0.63	0.500	0.837	-50	240	307	H_2, CH_4, C_2H_4	Moderate	Unlimited
Highly Refined Hydrocarbon (High Temperature Hydrocarbon)	43	300	294	13	0.64	0.460	0.887	-18 to -30	240	310	H_2, CH_4, C_2H_4	Moderate	Unlimited
Trichloro-Trifluoroethane	37	——	41	——	1.64	0.213	1.565	-35	48	NTB[2]	F_2, CL_2, CF_4	High	Unlimited
Chlorinated Benzene[4]	30	——	1.4	——	——	0.280	1.365	-30	NTB[2]	NTB[2]	HCL	Moderate	Unlimited
Askarel[1]	43	100	15	3.0	0.70	0.264	1.500	-37	NTB[2]	NTB[2]	HCL	Moderate	Unavailable

Table 9.24 continued on next page

Table 9.24 continued

[1]Included for comparison

[2]None to boiling

[3]Respective chemical symbols:

H_2 — Hydrogen	C_2H_6 — Ethane
CH_4 — Methane	C_2H_4 — Ethylene
CO — Carbon Monoxide	C_2H_2 — Acetylene
F_2 — Fluorine	CL_2 — Chlorine
CF_4 — Carbon Tetrafluoride	HCL — Hydrogen Chloride

[4]Chlorinated benzene may not obtain EPA or OSHA approval because of possible toxicity. (*Federal Register*, July 18, 1980, p. 48524 ff.)

from poor workmanship, surge voltage, excessive temperature, lack of proper maintenance, or misoperation, such as overload or misapplication. However, "more likely" are fires from external exposures due to known or unknown hazards. While known hazards such as combustible buildings (wood frame, etc.) can be protected against by proper installation, the unknown hazards "defy prediction by even the most imaginative installation engineers."

The 1981 *National Electrical Code® (NEC)* Article 450-23 has defined a high fire point insulating liquid to be non-propagating and to have a fire point of greater than 300°C.* However, the flammability characteristic alone is not sufficient. "Fire hazard" should require a much broader investigation of basic chemical and physical properties consisting of these various studies:

1. Ease of ignition—Cleveland open cup flash and fire point (ASTM D-92) and auto-ignition (ASTM D-2155) are well-defined test methods.
2. Fire growth—large scale pan burn test gives realistic fire growth information.
3. Flame spread—the propagation of a flame across the surface of a liquid is only valid for large scale tests. Large pan burns are best for measuring this property.
4. Heat release—heat of combustion (ASTM D-240) with burning rate.

**A registered trademark of National Fire Protection Association. Section 450-24, concerning nonflammable fluid-insulated transformers, has been added in the 1981 NEC (Table 9.1).*

5. Extinguishment—basic testing of extinguishment methods.
6. Smoke—modified NBS (National Bureau of Standards) smoke chamber test.
7. Fire gases—the toxicity of pyrolysis and combustion products is the key to human safety.
8. Oxygen depletion—limiting oxygen index and large scale fire testing.

The *initial* fluid properties, relative to fire hazard, would be well-defined by this series of tests.

The *rate* of heat release is considered to be the major fire hazard characteristic of large burning liquid pools. The *rate* of heat release rather than total heat release is the more significant standard. As an illustration, a gallon of gasoline or 10.4 pounds of lump coal will release 34,000 kcalories on total combustion. Gasoline, however, is more hazardous because it is likely to release its energy at a *faster* rate.

Fires release their combustion energy in two major ways: convective—the heat release rate of the fire that affects the exposure to the roof's structural members; and radiative—the heat release rate that affects the exposure to nearby exterior walls.

Table 9.25 lists the heat release rates for "less flammable" transformer liquids. These fluids have higher fire points and lower heat release rates than ordinary transformer oil. Transformers containing fluids having a fire point of at least 300° C (572° F) may be installed as long as certain physical and electrical protection is provided. Naturally, this requirement would include retrofilled transformers as well as new equipment.

Physical protection involves separation in terms of distance between a transformer and nearby high value equipment or combustible structures. In addition, separation consists of sufficient curbs, drains, and one-hour fire rating vaults and barriers. Physical protection may also include automatic sprinklers or water spray.

Electrical protection may require over-current devices (fuses or circuit breakers), a differential or gas detector relay, and/or transformer temperature detector.*

**Because the rules of the 1981 NEC (Section 450-23) are general in nature and lend themselves to a variety of interpretations, rule application should depend heavily on consultation with inspection authorities.*

TABLE 9.25

HEAT RELEASE RATE FOR LESS FLAMMABLE TRANSFORMER LIQUIDS (FACTORY MUTUAL APPROVAL GUIDE, MARCH, 1980)[1] (HEAT RELEASE RATE IN BTU/ft^2-min. KW/m^2)		
Type Fluid	Radiative	Convective
Dow Corning Midland, MI 48640 Dow Corning 561 *Silicone* transformer liquid	132 (25)	280 (53)
General Electric Company Silicone Products Department Waterford, NY 12188 *Silicone* dielectric fluid SF97	137 (26)	349 (66)
General Electric Company Large Transformer Division 100 Woodlawn Avenue Pittsfield, MA 01201 Iralec T-1[2] *(chlorinated aromatic hydrocarbon)*	1259 (238)	1957 (370)
General Electric Company Medium Transformer Department Redmond Circle Rome, GA 30161 Transformer Fluid R-113[3] *(trichlorotrifluoroethane)*	< 132 (< 25)	< 264 (< 50)
Gulf Oil Chemical Company P.O. Box 3766 Houston, TX 77001 Gulf FR Electrical Insulating Fluid *(high temperature hydrocarbon)*	2190 (414)	2983 (564)
RTE Corporation 1900 East North Street Waukesha, WI 53186 RTEmp Fluid *(high temperature hydrocarbon)*	1909 (361)	2888 (546)
Union Carbide Corporation P.O. Box 65 Tarrytown, NY 10591 Transformer Fluid L-305 *(silicone)*	137 (26)	349 (66)

Table 9.25 continued on next page

Table 9.25 continued

Westinghouse Electric Corporation Transformer Division 469 Sharpsville Avenue Sharon, PA 16146 Insulating Oil Wemco FR (*high temperature hydrocarbon)*	2333 (441)	2845 (538)
[1]None of these fluids to date matches the fire resistance performance of askarel. However, these liquids are considered "less flammable" than typical transformer oil. [2]Chlorinated benzene may not obtain EPA or OSHA approval because of possible toxicity. [3]Freon R-113 is now considered as a **non-flammable** liquid (December 1980).		

Needless to say, the other four liquid fire hazard studies are needed to relate the *total* fire hazard to people, as well as structures. A transformer under even normal in-service conditions produces a degree of aging in the insulation system. Therefore, a testing method for *long-term* thermal endurance should also be required.

Making a Tentative Selection

A review of all possible transformer requirements, including ANSI (American National Standards Institute), and/or IEC standards (International Electrotechnical Commission)* should be made before a decision concerning the preferred type of unit for a given installation is reached. Table 9.26 gives a check-list of transformer and dielectric fluid requirements.

The elimination of askarel as a cooling and insulating medium in new transformers, however, has had little effect on the basic factors affecting the two primary transformer options—liquid filled and dry type.

No one transformer type meets all the desired safety and environmental demands at a most reasonable cost. Table 9.27 gives a comparative summary of typical characteristics of secondary unit substation transformers available.

A Closer Look at PCB Replacements

Utility, industrial, and commercial users require safe, fire and explosion resistant transformers, preferably at a cost comparable with that of mineral oil-filled units.

**ANSI/IEEE Standards C57.12.00-1980, C57.12.00 a and b (1980), and C57.12.01-1979: IEC Standard-Selection and Application Guide (1978).*

TABLE 9.26

CHECK-LIST OF SIGNIFICANT POWER TRANSFORMER REQUIREMENTS	
Transformer	Dielectric (Liquids Only)
Power rating (KVA range)	Flammability
Primary voltage class (KV range)	Fire hazard tests (5)
Weight and physical dimensions	Electrical stability
Dielectric (BIL) rating (reliability)	Thermal stability
Withstand lightning/short-time overloads	Toxicity
Short circuit strength (through faults)	Biodegradability
Thermal rating (temperature rise capability)	Pour point
Sound level (decibels)	Viscosity
Cooling method: dry or liquid	Supplier sources available
Energy efficiency (unit losses)	OSHA acceptance
Transformer coolant resistant to —	NFPA acceptance
Fire	NEC acceptance
Explosion	UL acceptance
Environmental contamination	FDA acceptance
Total installed cost	Fire insurer acceptance
(Indoor/outdoor application)	
Annual maintenance cost	
Ease of maintenance	

Consider a brief discussion in terms of traditional versus recent commercial options, focusing on recent "less flammable" transformer dielectric fluids.

1. Traditional Options

   a. Mineral Oil Insulation Liquids
   If conventional oil-filled transformers were fire resistant, there would be no need to continue in this discussion. The long life, reliability, and low cost per KVA of oil-filled equipment testify to their outstanding service since the turn of the century. However, life as we come to know it is one of a series of choices. In selecting a transformer, the decision can be a difficult one.

TABLE 9.27

COMPARISON OF TYPICAL CHARACTERISTICS OF SECONDARY SUBSTATION TRANSFORMERS
(EVALUATION BASIS: STANDARD 1000 KVA 15 KV UNIT)

	Traditional						New in service				
Characteristics	Mineral Oil	Askarel (PCB + TCB)	Venti-lated Dry	Totally Enclosed Non-Vent	Gas-Filled Sealed (Sulfur-Hexa-fluoride)	Cast Coil (1976 Design)	Silicone Fluid	Hi-Temperature Hydro-Carbon Modified/ Synthetic	Vaporization Cooled (1978 Design)	Gas Insulated Vapor & Mineral Oil[2] (1979 Design)	Synthetic Ester (Retro-filled Unit)
First Year In Service	1887	1932	1884	1930 (est.)	1956	1960 (older design)	1972	1972	1963 (older design)	1960 (older design)	mid 1978 (England)
Primary Voltage Class (maximum)	Unlimited	34.5 KV	34.5 KV	34.5 KV	34.5 KV	34.5 KV	34.5 KV	34.5 KV	15 KV	34.5 KV	
Impulse Strength BIL (e.g.)[3]	95 KV	95 KV	95 KV	95 KV	95 KV	95 KV	95 KV	95 KV	95 KV	95 KV	95 KV
Thermal Rating (average rise)	65°C	65°C	150°C	150°C	150°C	80°C	65°C	65°C	45°C[4,5,6]	65°C	65°C
Weights	100%	112%	118%	140%	156%	124%	102%	103%	125%	100%	112%
Dimensions: Floor Space	100%	100%	122%	122%	156%	117%	100%	103%	110%	100%	100%
Height	100%	102%	107%	107%	114%	107%	102%	102%	109%	100%	102%
Application Indoor	Indoor (vault only)	Indoor	Indoor	Indoor	Indoor	Indoor	Indoor[5,6]	Limited[5,6]	Indoor[5,6]	Indoor[5,6]	Indoor[5,6]
Outdoor	Outdoor	Outdoor	Limited	Limited	Outdoor	Outdoor	Outdoor	Outdoor[6] No low ambient temperature	Outdoor	Outdoor	Outdoor
Sound Level - Decibels	58	58	64	64	63	63	58	58	58	Unknown	Unknown
Normal Resistance to: Fire	No	Yes	Yes	Yes	Yes	Yes	Yes[5,6]	Yes[5,6]	Yes	Yes[5,6]	Yes[5,6]
Explosion	No	Yes	Yes	Yes	Yes	Yes	Yes[5,6]	Yes[5,6]	Yes	Yes[5,6]	Yes[5,6]
Environmental Contamination	Yes	No	Yes	Yes	Yes	Yes	Yes[5]	Yes	Yes	Yes	Yes
Maintenance (See Table 9.30)	Medium	Low	High	Medium	Low	Low	Medium	Medium	Low[5,6]	Unknown	Unknown
Sources of Supply	Several	New Units	Several	Several	Several	Several	Several	Three	Two	One	One (?)
Unit Cost	100%	Out of	133%	175%	200%	170%	138%	120%	140%	110%	Unknown
Total Installed Cost	Up to[7] 233%	Produc-tion	133%	175%	200%	170%	140%[7]	120%	140%	110%	Unknown

Table 9.27 continued on next page

Table 9.27 continued

[1] Polychlorinated biphenyl plus trichlorobenzene.

[2] Use of perchloroethylene in combination with mineral oil.

[3] Reduced BIL available - responsibility of the purchaser.

[4] Lowest operating temperature because of greater cooling effect.

[5] Conclusion subject to interpretation of published data.

[6] Insufficient in-service use data.

[7] Extended cost includes provision for vault fire protection or outside vent.

If an oil-filled transformer below 35,000 volts is chosen for applications where fire resistance is required, Article 450-26 of the *National Electrical Code* demands that the unit be enclosed in a vault constructed for a one hour fire rating. This is a cost item, but costs escalate when the vault has to be torn down to remove the unit for repair or replacement.

One option that has often been neglected is that of removing any transformer from the proximity of people and buildings. Physical separation (distance) must be provided between the unit and the combustible or high valued occupancy, such as automatic production machinery, rolling mills, and the like. Of course, land must be available, longer cable runs would be required, and a larger rated transformer might also be required to compensate for the voltage line drop from the transformer to the use point. Nonetheless, this choice would permit use of conventional mineral oil with safety, and with the bonus of providing maximum cooling efficiency due to exposure to outside air.

b. Dry Type Equipment

The second traditional option involves use of the dry type transformer, where there is less danger from fire and explosion or to the environment.* However, its usage has fluctuated over the years. In addition, each dry type unit is generally designed larger and heavier, is more costly than liquid-filled units, has a limitation with regard to voltage rating and size, and finally life expectancy may be less. But with the demise of PCB transformers, the use of newly designed dry type equipment may prove suitable in many applications.

**Dry type units cannot be classified as nonflammable, Article 450-24, NEC.*

The first type—the open ventilated—should not be used in outdoor locations because of its sensitivity to moisture and air-borne pollutants, or indoors, in a humid atmosphere (paper mills) or a dusty environment (cement mills). But these objections are partially overcome in the totally enclosed non-ventilated (T.E.N.V.) design. As the name implies, the unit is enclosed in a steel case and therefore can be more reliably applied in contaminated environments.

At the same time, this design requires a large enclosure surface area to obtain sufficient heat transfer, so the actual KVA value may be less than the equivalent open-ventilated rating of the core and coil assembly.

The gas-filled sealed transformer enables users in the automotive and mining industry to have superior heat transfer and dielectric properties in the midst of a harsh environment. As expected, while it is the safest transformer manufactured, it is also the most expensive.

Finally, vying for a place in the transformer replacement market is the updated design of the cast coil dry type design (Figure 9.17). One kind of cast coil has its core and coil assembly totally encapsulated in epoxy resin to seal out all dirt and moisture. In a still newer design, the vacuum cast unit, the coil is placed into a stable environment all its own. The completely embedded winding conductors are locked into a solid mass.

Among notable benefits of this latest design are its high short circuit strength (exceeding both liquid-filled and conventional dry type transformers) and its higher short-time overload capability (Table 9.28). The vacuum cast coil type may also be the most energy efficient unit (minimim KV load losses) available for heavily loaded units, that is 75-100 percent load (Figure 9.18). However, data also shows that liquid-filled units may be the most efficient when lightly loaded.

2. Recent Commercial Options

a. Silicone Insulating Fluids
The transformer industry turned first to silicone liquid as a PCB substitute. Since electrical characteristics of silicone are comparable to mineral oil (see Table 9.1), the liquid has

been specified in many new transformers since 1974 and used successfully in retrofilled PCB units in Japan and the U.S. since 1972.

Today, silicone liquid is being used in at least 7000 transformers worldwide. One manufacturer has enlisted the cooperation of the owners of 81 retrofilled units. These transformers form the active base of a service life test program (since 1974). This statistical data of over one million hours of operation of silicone-filled units is establishing the reliability of such equipment in order to meet *National Electrical Code* (Article 450-23) requirements.

Figure 9.17 - Most recent POWER-CAST™ coil dry-type transformer is well-suited for use in harshly polluted environments. (Courtesy of Square D Company).

1) Development and Manufacture

Silicone is the generic term for a family of linear polymer liquids between the inorganic silicates and organic polymers. Some 27 candidate silicones have acceptable electrical breakdown characteristics.

The basic structure of the silicone chosen for use in transformers is depicted in Figure 9.19 where R groups are methyl groups or CH_3. Its chemical designation is polydimethylsiloxane (PDMS). PDMS remains fluid

TABLE 9.28

COMPARATIVE SHORT-TIME OVERLOAD CAPABILITY FOR VARIOUS TRANSFORMER TYPES		
Transformer Type		Time[1] Constant Hours
Liquid	55° C or 65° C	1.5
Dry	80° C	1.5
Dry	150° C	1.0
Cast	80° C	2.1

[1]Time constant is defined as the number of hours required to raise the hot spot temperature to 63% of its maximum rated hot spot, with rated load applied, starting from no load condition at 30° C ambient.

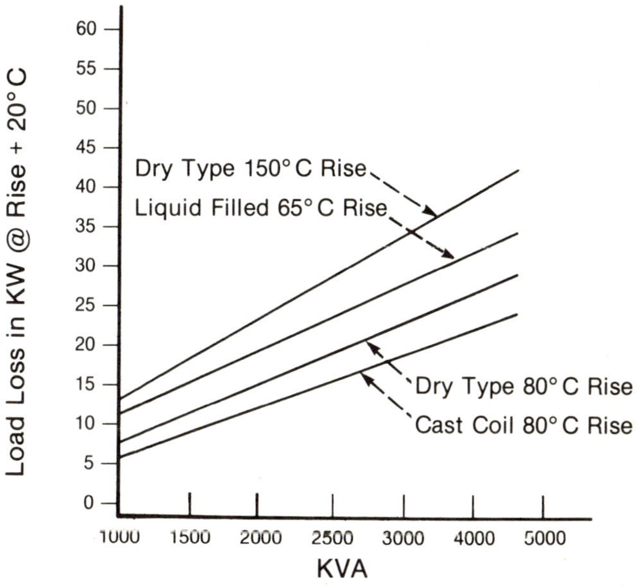

Figure 9.18 - Comparative energy conservation-load losses. This data is not intended to show any specific manufacturer's performance characteristics, since they will vary from one another, but a typical cross-section of various manufacturers. (Courtesy of GTE Sylvania).

over a wide range of temperatures, even though "X" may vary from 1 to 95 units.

One current method used to produce PDMS fluid involves a six-step process.

a) The reaction of a starting material (normally sand, SiO_2) with carbon at a temperature of 3000° C in an arc furnace, yielding metallurgical grade silicone and carbon dioxide gas.

b) The reaction of silicone with excess methyl chloride (CH_3Cl) produces various methyl silanes.

c) The fractional distillation of the foregoing end products in order to obtain dimethyldichlorosilane.

d) The reaction of water with dimethyl dichlorosilane to yield a mixture of 50-80 percent polydimethylsiloxane and 20 percent polydimethylcyclosilane.*

e) The refluxing of the mixture (d) in an acidic medium yields polydimethylsiloxanes.

f) The purification of final products (e).

Other methods are also used to manufacture various silicone fluids. Although the cost of silicone fluid is expensive, keep in mind the ever-increasing cost of mineral oil versus silicone's versatile physical properties.

2) Where Else Are Silicones Used?
Dimethylsiloxane fluids have been used in products that involve human and animal exposure for 30 years. They have been examined by the Federal Food and Drug Administration and are currently being used in a variety of medical and chemical applications; for example, the

The volume of water used determines the length of the polymeric chain. It is the fluid's viscosity that depends upon the polymer length. The more water used, the more the yield of the longer chain polymers, and in turn the higher the viscosity (measured in centistokes) of the rated oil. The higher the viscosity, the more limited the use of the fluid becomes. A maximum limit for transformer application would be 1000 centistokes (CS). Compounds approaching 1000 CS could not be employed in cold weather situations: 11 CS fluid (similar to mineral oil) could also be used, but by choosing 50 CS the desired fire point was achieved.

"Simethicone" used in the product "Digel". Other uses include:

- Skin creams and other cosmetics
- Pharmaceuticals
- Food packaging and processing
- Car polishes
- Hydraulic fluids
- Photography
- Water repellents
- Mold release agents
- Painting and coating additives
- Gas chromatography
- Rust prevention
- Power transmission
- Damping fluids
- Rubber and plastic additives
- Textile finishing

Extensive toxicological studies have shown silicone materials are among the least hazardous of all industrial chemicals.

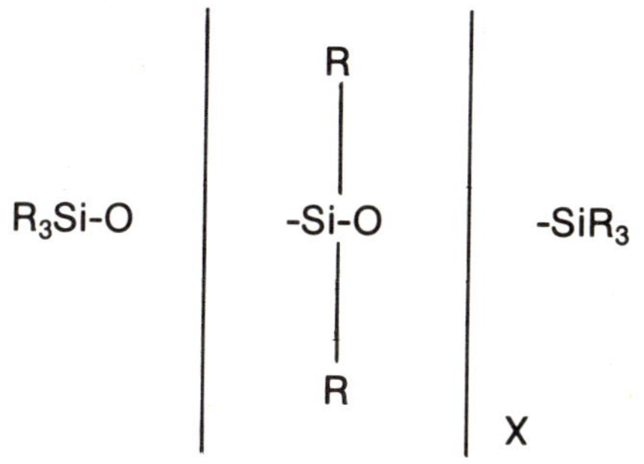

$R = CH_3$ $\quad$ X = 1 through 95 units

Figure 9.19 - Basic structure of the silicone fluid used in transformers.

Although silicones are only slightly biodegradable, they pass unchanged and thus do not collect in the food chain. They break down slowly in soil, eventually reverting to pure SiO_2 (sand).

3) Silicone Performance Characteristics

a) Flammability and Explodability
Recall that Underwriter's Lab (UL) provides a

classification service for flammabililty ratings (Table 9.1). UL does not classify silicone as an "inflammable" liquid; Factory Mutual (Table 9.25) lists the various liquids only in terms of "less flammable" fluids.

> Comparison of ignition and fire development in silicone liquid and mineral oil under extended hot arc conditions shows that silicone liquid ceases to burn when arc current is stopped. Mineral oil on the other hand continues to burn and thereby constitutes a fire hazard.

Since silicone ignites at 310°-340°C (cf. with mineral oil at 160°C), it is not classified as fire resistant; however, silicone fluid is self-extinguishing (*NEC*, Article 450-23).

Silicone fluid has the lowest heat release rate of any material used as a coolant for a **standard** liquid-filled transformer. The initial higher cost of silicone may be offset when no additional electrical or physical protection is required for silicone fluids in contrast to the other substitutes (Table 9.25).

b) Electrical Stability
Fine electrical properties stem from silicone fluid's unique chemical structure. Comprehensive published reports on silicone fluids show that they exhibit the same high dielectric properties as mineral oil-paper systems over a very wide range of temperatures, frequencies, and voltage stress levels. As with all likely fluids, the electrical characteristics of silicone will be reduced if grossly contaminated or wet.

c) Pressure Due to Expansion Coefficient
Silicone transformer liquid expands more with temperature increases than do askarels or mineral oil. This increases the probability of transformer "breathing" problems through aging gaskets. However, in field experience the problem has been far less serious than theory had predicted. Silicone liquid will dissolve more gas than will oil or askarel,

so the pressure buildup is thus compensated. The simple addition of a pressure/vacuum regulator will prevent gasket damage.

d) Temperature Rise/Heat Transfer
Data would tend to indicate that silicone liquid in a PCB transformer would cause the unit to run warmer than askarel-filled equipment. The higher viscosity of the silicone fluid does tend to reduce the liquid flow, but at the same time, the density difference between cool and warm fluid is greater, thus improving the thermal syphon convection flow of the liquid in the transformer. Thus, silicone fluid has heat transfer advantages as well as disadvantages—a liquid that cools well, but not quite as well as askarel (or mineral oil) when run in heavily loaded transformers. The cooling of silicone under low ambient operating conditions is superior to that of askarels and transformer oils.

Subsequent data has shown that temperature increase is not a constant due to thermal characteristics of the fluid, but influenced more by the type of transformer and its cooling tube characteristics. The transformer's head space, for example, can be increased to provide better cooling and more space for thermal expansion. This increased tank size may be offset by reduction in the amount of liquid needed for the same efficiency found in a normal tank.

When an identical electrically rated (with similar operating temperature) and physical size transformer replacement is desired, the addition of cooling fans to the radiators should lower the temperature of the new unit to desired specifications.

e) Thermal Stability
Silicone transformer liquid has more heat stability than either mineral oil or askarel. Thus, silicone can be subjected to very high temperatures—above the operating temperatures for transformers—without

creating excessive vapor pressure or corrosive by-products. There is no solvent in silicone as there has been in most askarels. In contrast to mineral oil, silicone fluids do not begin to oxidize until 175° C (342° F), therefore minimizing the formation of degradation products (acids or sludges).

f) Material Compatibility
Compatibility and thermal stability are closely related. Materials which are compatible at room temperature may become incompatible at elevated temperatures because of increased solvent action or chemical activity at the higher temperatures. The result of incompatibility is failure of the insulation system.

Silicone fluids are compatible with the materials used in askarel transformers, with the exception of the silicone rubber, and highly plasticized vinyl rubber. Silicone fluids are compatible with most but not all of the materials that are compatible with transformer oil. Conversely, silicone fluids are also compatible with a number of the materials that are not compatible with transformer oils.

g) Absorption of Moisture
While silicone oil absorbs a much larger content of moisture, it maintains good dielectric properties with a much higher level of water than askarel or mineral oil (Figure 9.20).

h) Lubricity
Silicone liquid is usually not recommended as a metal-to-metal lubricant. Thus, use of silicone fluid requires different design considerations for tap changers, pumps, and so forth. It is one of the best lubricants for fiber and plastic gears or bearings, for natural and synthetic rubber, polystyrene, phenolics, and most other plastics.

b. High Fire Point Hydrocarbon Liquids
Major PCB replacement fluids fall into two catagories—silicones, which we have considered in some detail, and high temperature hydrocarbons, about which relatively little is known at this time.

Each high fire point hydrocarbon has electrical and environmental properties similar to conventional transformer mineral oil, but the liquid has been modified to increase its fire resistance (high temperature hydrocarbon-HTH).

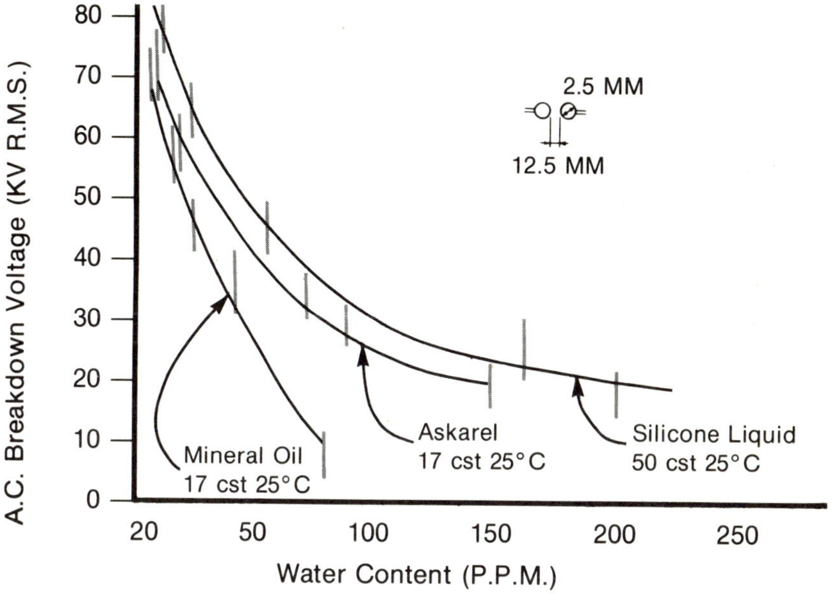

Figure 9.20 - Electrical breakdown as a function of water content. Dow-Corning ***Silicone 561 Instruction Manual*** *(1979).*

Whether the liquid is a modified or synthetic hydrocarbon is not a practical matter, since the respective characteristics are very similar (Table 9.24).

One modified hydrocarbon RTEmp® fluid is a mineral oil refined by a two-stage hydrogenation process from a high molecular weight *paraffinic* type oil.* This results in an oil that has a substantially higher fire point than is produced by the traditional one-stage process. It has been further modified to improve its pour point (Figure 9.21).

Two other high fire point modified hydrocarbons (FR or fire resistant) are WEMCO FR and Gulf FR, both included in the Factory Mutual evaluation (Table 9.25).

**Most in-service oil-filled transformers of the past 50 years have used naphthenic crudes.*

One commercially available synthetic hydrocarbon is PAO-13-C or poly-alpha-olefin (Uniroyal Incorporated, New York, NY) with similar characteristics. Failure to submit it for evaluation may account for its not being listed in Table 9.25.

A definite advantage of the high fire point compounds, in comparison to silicones, in addition to cost may be an improvement of heat transfer capability under heavily loaded conditions.

For each of these liquids, the greatest unknown is its long-term stability. With paraffinic type crudes the success of future reclamation efforts is unknown. Higher viscosities also limit the cold weather applications.

c. Vaporization Cooling

In mid-1978 a completely different liquid-filled transformer was introduced—the vaporization cooled unit (Figure 9.22). This design has the same electrical characteristics found in conventional liquid-filled units. These include impedance, BIL, sound level, and paper/liquid compatibility.

In this new cooling system, the core and coil assembly are immersed in a liquid designated "R-113" that vaporizes to dissipate the heat. This liquid has been used as a cleaning agent and refrigerant under the trade name of FREON®.* The vapor then rises through a header into a condensing unit at the top of the transformer (Figure 9.23A).

In the condenser the vapor enters slanted cooling tubes where it is cooled, changing back into a liquid. By gravity the liquid flows down the cooling tubes to the bottom of the headers and back into the transformer tank (Figure 9.23B).

Table 9.25 shows that this fluorocarbon is rated as non-flammable, but it does require a completely different cooling design, as well as more vertical clearance for installation.

This transformer provides the lowest operating temperature (45° C rise) of any type because R-113 has about twice the cooling effect of conventional transformer mineral oil. Recall the standard 65° C or older 55° C rise for other liquid-filled equipment.

**Registered trademark of E.I. Dupont De Nemours & Company.*

Figure 9.21 - An RTEmp® fluid filled, 2500 KVA unit substation transformer. (Courtesy of RTE Corporation).

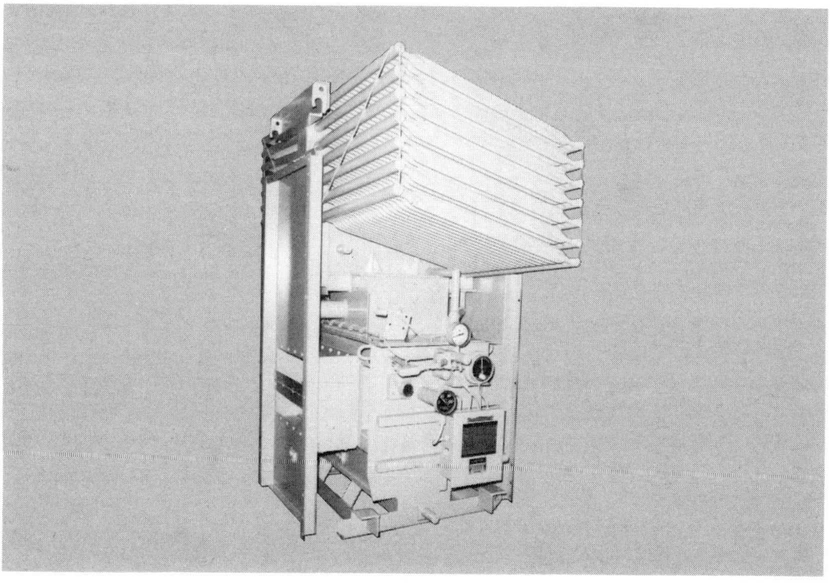

Figure 9.22 - VaporTran™ Transformer. (Courtesy of General Electric Company).

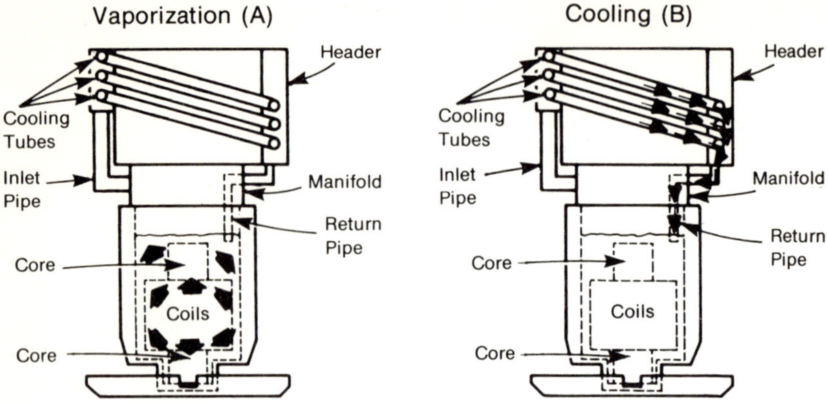

Figure 9.23 - Pictorial representation showing operation of a vaporization cooled transformer.

Figure 9.24 illustrates "transformer loading" for a vaporization-cooled unit. It is given in terms of tank pressure rather than as a function of winding temperature. Twelve pounds per square inch gauge pressure (psig) is normally considered with a 67° C fluid temperature. During operation this temperature, and thus the pressure, is determined by the condenser size. While the heat exchanger is larger, insulation temperatures are also inherently lower. In this regard, the condensing unit could be mounted outside, provided that reasonable piping distances are observed and the condenser is above the transformer tank level.

3. Latest Developments

a. Gas Vapor and Mineral Oil

Gas vapor and mineral oil combination type transformers were placed into service in late 1979 by several utilities to monitor and evaluate prototype performance.

The transformer core and coil assembly is totally immersed in a mixture of perchloroethylene (C_2Cl_4) and mineral oil. C_2Cl_4 is widely used in industry as a dry cleaning and metal cleaning liquid.* It is considered one of the most stable of

**Perchloroethylene illustrates the dilemma that exists in trying to find a "perfect" substitute for PCBs. While all agree C_2Cl_4 is not bioaccumulative like PCBs, some consider it a suspected carcinogen. This view depends upon "The eye of the beholder". Nonetheless, the overwhelming evidence supports the opinion that the liquid is not a likely carcinogen.*

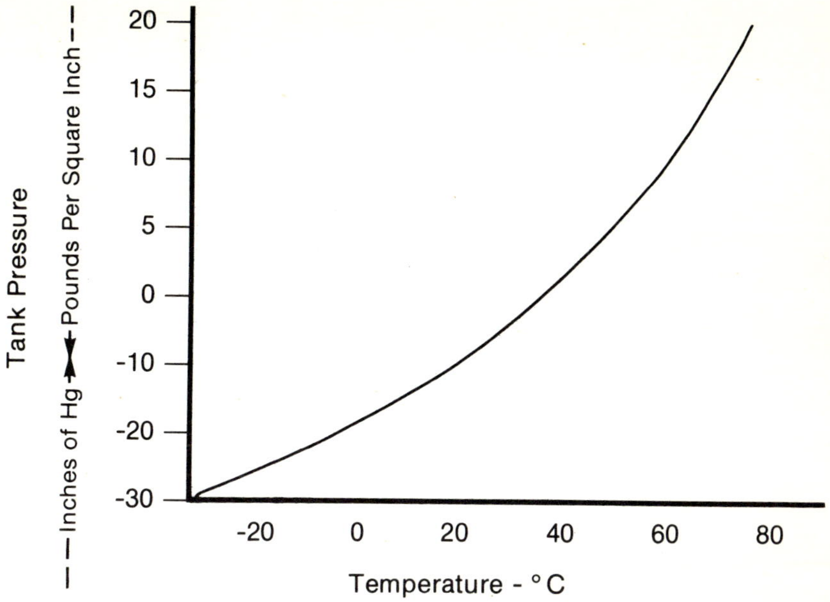

Figure 9.24 - Trichlorotrifluoroethane vapor pressure allows lower transformer operating temperature.

chlorinated solvents. In a mixture with oil, the combination coolant proves even more advantageous. The transformer oil imparts improved freezing and boiling points, reduces cellulosic deterioration, and of course the C_2Cl_4 imparts its non-flammability characteristic.

In the latest application of vapor cooling the core and coil assembly is totally immersed in a mixture of C_2Cl_4 and oil. Vapors form as bubbles at hot-spots but condense again within the liquid. This cooling system is more efficient than prior designs because no recirculatory pump is required.

Operation at lower temperatures because of the lower boiling point of the coolant should extend the life of the transformers. Commercial sales may come about by late 1980.

b. Gas and Oil Mixture

A longer range system may use the combination of sulfur hexafluoride (SF_6) or hexafluoroethylene (C_2F_6) and mineral oil. At this point, only laboratory studies have been done; no field prototypes have been built. Therefore, commercial use of either mixture may be some years in the future.

Several other approaches under investigation are listed in Table 9.29. The ultimate solution may be a combination of the best dry and liquid types of transformers. This could involve the combination of 115°C rise aramid papers (NOMEX®)* with silicone liquid (Figure 9.25). The unit should be smaller, but the great unknown is how efficient the transformer would be.

TABLE 9.29

DEVELOPMENT REPLACEMENT FLUIDS FOR PCBs

Fluids	Fire Point	Flammability of Arc Formed Gases	Electrical Characteristics	Hydrolytic Stability	Environmental Acceptability	Cost
Phosphate Esters	Good	Flammable	Fair	Poor	Good	Moderate
Fluorocarbons	None	Non-Flammable	Excellent	Excellent	Unknown	High
Water	None	Non-Flammable[1]	Poor[1]	Excellent	Excellent	Low

[1]This conclusion is subject to individual interpretation. Thus, Table 9.29 illustrates the obvious major areas of challenge which must be overcome before any one fluid could become an economically viable alternative.

The Final Decision

One thing is certain in the years ahead. A reliable energy supply demands a consistent maintenance program; thus, the bottom line may be the relative amount of in-service use, maintenance required, and desired efficiency of operation. Table 9.30 gives a capsule summary of basic preventive maintenance procedures for today's medium power transformer.

Until sufficient comparative test data is available, determination of the final choice rests on the best match between the various transformer criteria and the particular location, use, and nameplate capacity required. Therefore, it is evident that "the final chapter has yet to be written" in the development of transformers for use in maximum fire safety applications.

**Registered trademark of E.I. Dupont De Nemours & Company.*

TABLE 9.30

COMPARATIVE MAINTENANCE REQUIREMENTS — SECONDARY SUBSTATION TRANSFORMERS (AVAILABLE)

	TYPE OF TRANSFORMER	YEAR FIRST IN SERVICE	MAINTENANCE FREQUENCY	INSULATING FLUID TESTING POWER ON	ASTM D-3612 ANALYSIS OF DISSOLVED GASES IN OIL POWER ON	ASTM D-1533 MOISTURE PARTS PER MILLION POWER ON	INTERNAL UNIT CLEANING POWER OFF	PERMANENT ACCESSORY REQUIRED	FLUID RECLAMATION POWER ON	COMMENTS
TRADITIONAL	Mineral Oil Filled	1887	Medium	X	Yes	X	NA	NA	X	IFT, acid number, color; gas analysis a proven companion test (IEEE)
	Askarel	1932	Low	X	No	X	NA	NA	X	Dielectric, color, acid number, visual
	Dry Type 1. Ventilated	1884	High	NA	NA	NA	Frequent	Electric Heater	NA	Field accessibility, absorption of moisture (humidity)
	2. Totally Enclosed Non-ventilated	1930	Medium	NA	NA	NA	Less Frequent	Electric Heater	NA	Absorption of moisture (humidity)
	3. Gas-filled Sealed (Sulfur Hexafluoride)	1956	Low	NA	NA	Dew Point	Annual	Leak Detector	NA	Low frequency with high maintenance complexity; difficult to test; can't see or smell
	4. Cast Coil	1960	Low	NA	NA	NA	Annual		NA	Annua cleaning (vac. or steam)
NEW IN-SERVICE	Silicone Fluid	1972	Medium	X	Yes*	X Modified	NA	NA	X	Dielec ric, power factor, visual, color
	High Temperature Hydrocarbon	1972	Medium	X	Yes	X	NA	NA	X	Complete desludging may not be possible with paraffinics
	Vaporization Cooled	mid 1978	Low (?)	X	No	Dew Point	NA	Molecular Sieve: removal of moisture	X (?)	Dielectric, acid number

Table 9.30 continued on next page

Table 9.30 continued

TABLE 9.30

NA - Not applicable
X - Applicable
IFT - Interfacial Tension Test
**Acceptable method of gas-in-oil analysis unavailable, and therefore needed.*

Figure 9.25 - 1500 KVA silicone liquid filled power transformer 13800Δ-600Y/347 volts. (Courtesy of Ferranti-Packard).

Chapter 10

Implementing Your Own Protective Maintenance Program

Part 1—A New "Mind-Set" Mandated For Maintenance

Introduction

Eric Hoffer, the free thinking long shoreman turned philosopher and author was asked, "What is the most important word in the English language?" His one word answer to the national T.V. audience was profound! "Maintenance." How ironic that countless schooled individuals tasked with establishing transformer maintenance directives, are apparently less knowledgeable.

Typical Traditional Responses

Two maintenance perspectives have prevailed over the years. One school calls for routine, periodic tests, repairs, and scheduled down-times (preventive maintenance). The other hopes for extended operation with little or no maintenance unless malfunction is imminent or has already occurred (ordinary maintenance).

Traditionally, requests for PM budget monies will prompt a reply of, "We haven't done anything to our transformers for 20 years. Why should I budget money now?" The fact is transformers do not improve with age and use. When the pay day for neglect does come around, the next generation will pay the bill. We have a problem however. We are the next generation!

We philosophize that ordinary maintenance is counter productive. Several catch-words tell it all: replacement costs, downtime costs, customer dissatisfaction, etc.

By contrast, we believe maintenance should be protective — preventive, predictive, and corrective.*

**Chapter 6, Part 1.*

Economic Justification of a Protective Maintenance (PM) Program

In addition to the "Great Northeastern blackout" of 1965, another power outage affected New York in July, 1977. In this event, an important intertie was tripped well below its capacity because a relay had a bent contact. In 1979, a Canadian utility had a major system failure. It required 36 hours to restore the system to normal at an estimated cost of $8,000,000! These examples of unscheduled power outages will inevitably be repeated unless a proper PM program is established and periodically reviewed.

A recent Niagara Mohawk Power Corporation experience illustrates the benefits that can be expected from a good PM program. Tests indicated that a 700 MVA generator step-up transformer on a nuclear station had a particular problem. The unit was de-energized, repaired, and back on line sixteen days later. Undetected, this fault would have inevitably led to transformer failure. Thus, a $200,000 per day cost to repair or replace the unit was saved. A time factor of many months, or years would not be unusual to repair or replace respectively a transformer of this size.

Industrial consumers of electricity generally suffer costs due to power outages because materials and products are spoiled, while normal production cannot take place. In a survey of twenty principal U.S. industrial users of electricity to determine outage costs at various durations (that is, one minute to five hours), time of day, the type of industry, etc., the cost was $1-$6 per kilowatt-hour (KWH).

An outage which affects the general public — homes, schools, hospitals — precludes certain community benefits; namely, security, safety, health care, comfort, education, law enforcement, etc. Again, costs are high but a realistic estimate is difficult.

Countless other case histories demonstrate conclusively that PM is prudent. But do not forget these . . .

There are at least **four** other pressing **issues** which make it increasingly necessary to implement an adequate PM program. Each issue continues to demand your attention and refuses to go away.

Increasing Energy Costs

In the electrical world, everything has its price. Today's accelerating energy costs are causing management to take a hard look at possible cut-backs. Contaminated insulators, loose connections and overloaded transformers represent a few of the numerous causes for transformer electrical losses. And the cost of the energy to feed these losses can no longer be ignored. The cost of fuel to generate one kilowatt-hour represents approximately 85 percent of direct costs. These fuel costs increased 850 percent from 1969-1979! Where will electrical energy costs be ten years hence?

This escalating cost factor alone justifies many times over a PM budget. Substation infrared inspection, electrical insulation testing, and oil gas chromatography analysis can each detect expensive heat producing losses.

Inflation

Compare the cost, say of 10,000 KVA 13.4 KV 1955 transformer to a 1965 unit, and then to a current unit. Note the trend. Likewise, compare the cost of a gallon of insulating oil in 1955, 1965, and 1981!

At one time the per unit cost of reclaiming sludged oil was considered prohibitive. Not so now! Reclaiming oil is not only less expensive than new oil, it holds the added advantage of having the easily oxidized hydrocarbons removed.*

Because of higher replacement costs, various options available to decrease transformer aging become correspondingly more attractive. Inflation may justify the added cost of keeping the top oil temperature acceptable (60°C) and of maintaining oil in the sludge-free range.

Decreasing (Naphthenic) Oil Supplies

As we have no track record on the newer oils now being marketed, one cannot assume they will perform as oils of the past.† Regular oil tests on in-service transformers must become common practice.

Chapter 7, Part 3, pp. 525, 533. At least two transformer manufacturers now use reclaimed oil in* ***new* *equipment.*

†Chapter 2, Parts 1 and 2.

Supplementing these with periodic insulation tests should supply a running account of any changes in the oil and cellulose.* This combined data will be necessary more often as paraffinic oils become more prevalent.

Environmental Pressures

The environmental movement has made an invaluable and significant contribution to humanity. Wisdom demands, therefore, PM procedures which include compliance with existing Federal, State, and local regulations.†

All of this and more is the thinking person's fiat to revise, upgrade, and implement a **comprehensive** protective maintenance program with long range and immediate goals.

**Chapter 3 and Chapter 5, respectively.*

†Chapter 9, Part 2.

Part 2—Beginning a Protective Maintenance Program

Understandably, the natural response here may be "I agree with all this, but where do I begin?" Several areas must have prompt attention. Probably of first concern is the end product — What do I hope to accomplish?

Establish Goals

A reasonable long-range goal is that modern transformers should have a life expectancy of fifty years. How then can this be accomplished? The content of the nine foregoing chapters answer this question. Thus, at this point, we will only mention key points.

1. Schedule annual oil screening tests.
2. Insist on "sludge free" operation of all oil-filled transformers, especially, IFT, NN, and dielectric.*
3. Establish 60°C (on cost/benefit basis) as the maximum top oil operating temperature. Each 10°C rise above 60°C will halve the "sludge free life" of the oil.
4. Schedule annual dissolved gas-in-oil tests.
5. Run electrical tests periodically — five year maximum.
6. Tolerate no condition that will accelerate cellulosic deterioration from excessive moisture, oxygen, heat, and acid.
7. Regularly inspect, calibrate, and maintain all transformer protective devices.
8. Weather proof transformer case and seals.
9. Clean and service bushings/insulators.
10. Inspect connectors and cables.

These key points are the road bed for a track on which to operate — and it leads toward the goal — 50 years life expectancy. Incorporate new expertise as it becomes available, making the necessary modifications.

Chapter 6, Part 2, Table 6.5, p. 448.

Sell Management

Here again a typical problem emerges. Management personnel may not be aware of the basic requirements and unique problems in the substation. One electrical engineer lamented, "How can I convey this to management? My boss has a major in History!" But before attempting to plan and initiate a program, help make management understand your problems and solutions. Only active support will do — not lip service. Then only will the "changes" occur, as budgeted money is made available.

It will be valuable to review current PM practices and contrast this with the proposed plan. Key personnel as well as management must understand the significance and value of what is being proposed and what changes are anticipated. Without this support, no program should be activated.

Another key consideration is the use of terms used to describe the new project. Too often a program which involves an element of control is called just that — maintenance control. The worker looks upon control (and you cannot blame him) as something that is threatening his freedom and his license to be creative . . . Therefore, eliminate the word 'control' from any new program.

Who will be responsible for developing and administering the program? Most programs that fail do so at this point. Somehow, the importance of incorporating the new group into the overall company plan is missed.

Maintenance planning should not be under, but on the same level as the plant maintenance supervisor or foreman (Figure 10.1).

At the same time, though one person must be charged with the responsibility of instituting the program and follow-up, in a broader sense it must be the responsibility of all plant personnel. It must be "our" program.

A Tool of Service

Service must be the watchword of maintenance planning. It is only a tool, providing a service to aid cost improvement. To paraphrase a well known quotation, "The 'plan' was made for man and not man for the 'plan'."

Frequently, questions are asked about the effect of PM on maintenance manpower. One may anticipate that initially manpower hours will increase as indicated in Figure 10.2.

Observe that the PM program results in reduced total maintenance hours after the initial introductory time. Even more significant is the reduction in production breakdown-repair man hours. An attractive corollary to this is significantly reduced downtime costs. Nonetheless, expect a **good** PM Program to require increased manpower at its inception.

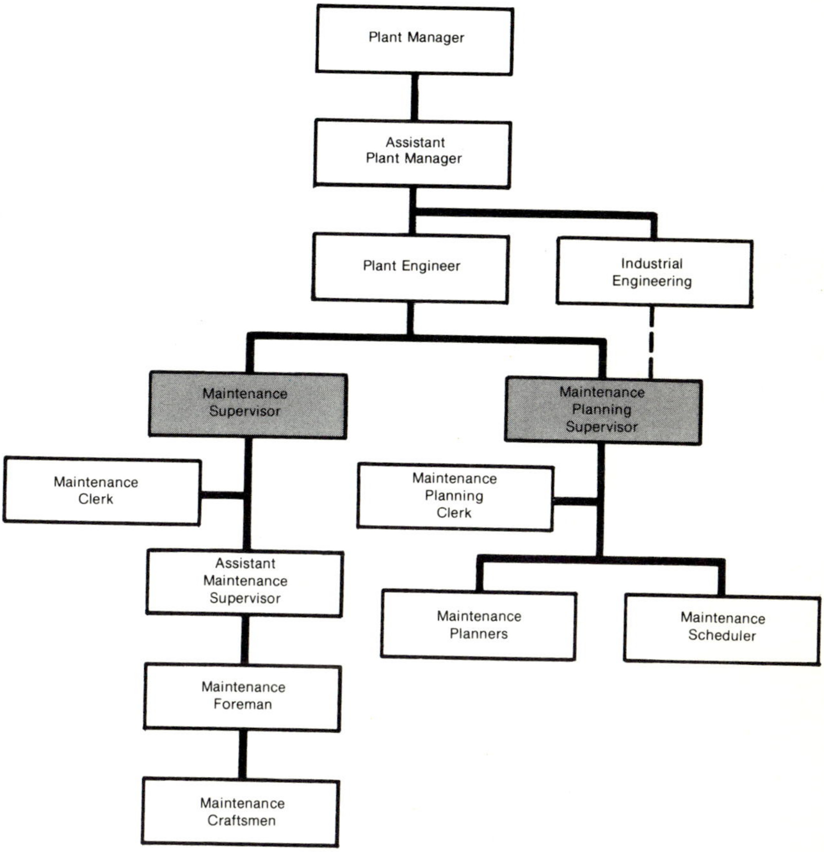

Figure 10.1 - Recommended organization has head of Maintenance Planning reporting to Plant Engineer. Industrial Engineering has staff relationship to Maintenance Planning-Scheduling function. Courtesy of ***Plant Engineering®***.

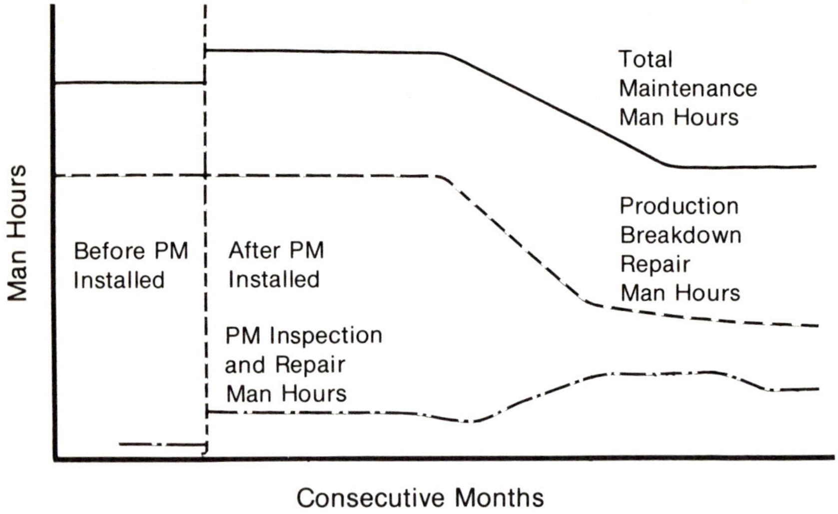

*Figure 10.2 - Projected maintenance man hours—before and after PM program instituted. Courtesy of **Plant Engineering**®.*

Education

As part of the above general guidelines, some thought should be given to personnel, training, and education. Failure here will preclude any general cooperation and support as time progresses. Others must understand both "what" and "why". One possibility is to schedule periodic training sessions. This book would make an excellent test. An up-to-date reference library should be considered mandatory.*

Attend seminars — yes; your plant will get along without you — you take vacations, don't you?

Purchase and make available relevant ANSI, IEEE, and ASTM standards. Other sources of information would include Doble Conference Minutes, manufacturer's literature, and electrical journals.

> Subscribe to five electrical magazines and read every article on a given subject for two years and you will be an expert on the subject.
>
> Stanley D. Myers

**A selected chapter and subject bibliography follows the Color Section in this book.*

Files and Records

Keep a detailed record on each transformer. Implement a topical file on data accumulated from oil tests, electrical tests, GC tests, IR inspections, etc. Maintain a daily log of unit operating temperature, pressure, load and outside temperature. Record the time and extent of any repair work done. Provide a running account of such occurences as overloading, through faults, and any adverse operating conditions. Observe Figure 10.3 as an example of an actual case history. This information will later prove invaluable as decisions become necessary. Both time, experience, and the above information are usually necessary to make an intelligent decision. At best, the evaluation that can be made from a given set of data is fallible and the entire procedure should be viewed as an art rather than a science. Like the man who has just gotten a "clean bill of health" from his doctor, transformers can fail when all data indicates the unit to be "A-OKAY." Remember Murphy's Law: "Whenever everything seems to be OK, then something else will fail." This could involve, for example, the failure of a transformer's oil pump.*

T.M.I. cannot emphasize too strongly the need for a complete preventive maintenance program!

Practical Maintenance Scheduling

This book has demonstrated that the transformer insulation system has four potent enemies — water, oxygen, excessive heat, and contamination†. Likewise, we suggest that multi-testing by various methods (oil testing, dissolved gas-in-oil analysis, electrical insulation testing) and periodic inspection represent the most comprehensive tools for protective maintenance of power transformers‡.

Transformer authorities as well as plant insurers tell us that the first ten years of a transformer are the most critical, especially the first three!§ In addition, the ANSI/IEEE Loading Guide is at best only a "rough assessment" of the effect of transformer overloading

**Chapter 6, Part 2 (Full Assurance Against Power Outages).*

†Chapter 2, Part 5, Figure 2.32.

‡Chapter 6, Figure 6.4, Chapter 7, Figure 7.1.

§Chapter 1, Part 3, Figure 1.76 (The "bathtub curve").

Customer City State

LOCATION OF TRANSFORMER

TC NUMBER

Substation Name or Number #14 Transformer

Unit Or Tag Number 62-91

TRANSFORMER CONSULTANTS ®

☐ Indoor ☒ Outdoor ☐ Roof

☒ Ground ☐ Platform Ft ☐ Pole

NAMEPLATE INFORMATION

Manufacturer West. Primary Volts 13800 Gal. 4230 ☒ Oil ☐ Askarel

Serial Number 5067533 Secondary Volts 625 Impedence 6.0

KVA Rating 6940 Phase/Cycle 3/60 Other

Radiators ☒ Yes ☐ No Conservator Tank ☐ Yes ☒ No Top Valve 2" Accessibility – Power on Power off

Fans ☒ Yes ☐ No Tap Changer Comp. ☐ Yes ☒ No Bottom Valve 2" Testing

H_2O Cooled ☐ Yes ☒ No Bushings: ☒ Top ☐ Side Other Access ReRef.

Other

ReRefining Parking ☐ Yes ☐ No Hose distance Feet Power 3 ☐ Yes ☐ No Voltage ☐ 220/240 ☐ 440/480

TRANSFORMER LIQUID TEST DATA

	DATE	Service	LEVEL	TEMP	P V	DIEL	ACID	IFT	COLOR	SPECIFIC GRAVITY	VISUAL	LEAKS	PAINT	OIL CLASS	RECOMMEND
1	7/69	MTL				26	0.56	18.4	3.5	0.872	Clr			V. Bad	
2	2/70	MTL				12	0.48	19.9	4.0	0.890	Clr			Bad	SludgPurge
3	5/70	HOC				36	0.03	37.4	2.0	0.875	Clr			Good	Retest 1 yr
4	10/71	MTL				30	0.03	39.9	0.75	0.885	Clr		Fair	Good	Retest 1 yr
5	10/72	MTL				31	0.04	36.5	1.0	0.882	Clr	None	Bad	Good	Retest 1 yr
6	11/73	MTL				30	0.04	34.9	1.0	0.878	Clr	None	Good	Good	Retest 1 yr
7	12/74	MTL				32	0.03	34.6	1.0	0.875	Clr	None	Good	Good	Retest 1 yr
8	11/75	MTL				34	0.02	34.3	0.75	0.880	Clr	None	Good	Good	Retest 1 yr
9	12/76	MTL				46	0.03	33.6	0.75	0.875	Clr	None	Good	Good	Retest 1 yr
10	3/78	MTL				43	0.03	32.9	1.0	0.880	Clr	None	Fair	Good	Retest 1 yr
11	11/78	MTL				31	0.03	33.5	1.0	0.880	Clr	None	Fair	Good	Retest 1 yr
12	11/78	PCB				PCB CONTENT TESTED BY ELECTRON CAPTURE									3 PPM
13	2/80	MTL	N	30	+1	34	0.03	32.5	1.0	0.880	Clr	Yes	Fair	Good	Retest 1 yr
14															
15															
16															
17															
18															
19															
20															
21															
22															
23															
24															
25															

transformer Div. of S. D. MYERS, INC. P.O. Box 3575, Akron, Ohio 44310 Phone (216) 929-2847 **consultants**

SDM-120 PRINTED IN U.S.A.

Figure 10.3 - Typical on going field oil test data and inspection report. This is of a medium power transformer. Ten years after the oil was hot oil cleaned (5/70), it is still rated "Good". Both acid number and interfacial tension remain acceptable.

While ANSI/IEEE has established optimum temperature criteria for the cellulosic insulation, no similar mention in the IEEE maintenance guide has been made concerning the oil condition. Yet before the transformer's solid insulation can continually perform its function, it is assumed that it operates in a clean, dry, oxygen-free environment. This means the transformer oil must run in the "good" range throughout the unit's life in order to obtain the maximum life for the paper. Since we believe that heat from loading affects the insulating oil sooner than paper, preventing severe and rapid oil deterioration is a must! Maintain the oil in its sludge free range.

Over fifteen years of field experience have become the basis for suggested maintenance schedules (Tables 10.1 and 10.2.)* Observe that the servicing interval is based not on oil test data alone, but also include the factors of unit age and average top oil temperature. Tables 6.3, 10.1, and 10.2 constitute what we call the TIMS™ program, or Transformer Insulation Maintenance Service. These schedules rigidly followed, will enable the transformer owner/operator to assure the maximum useful life from his equipment. When the time limits and oil test data as defined exceed the values in the TIMS program; then corrective action is required. This may include (1) auxiliary cooling, (2) mobile oil reclaiming, and/or (3) testing and desludging.

Maverick failures may still occur, but they will not be the result of incomplete preventive maintenance. In addition, you should periodically update these schedules based on your particular situation, thereby achieving the desired objective — reducing outages, costly repairs and unplanned shutdowns.

Transformer Troubleshooting Charts

The following series of (flow) charts (Tables 10.3 through 10.10) suggest remedial action to be taken when a given test procedure reveals any of the corresponding problems. These troubleshooting procedures are adequate for general application. Recall that the appropriate remedial action, while it will correct a given problem, will NEVER restore any loss of life already suffered in the transformer's cellulosic insulation.

**See Chapter 6, Part 1, Table 6.3, p. 443.*

TABLE 10.1

POWER TRANSFORMER AND ASSOCIATED EQUIPMENT MAINTENANCE INSPECTION AND MINIMUM TESTING INTERVALS*

Recommended Frequency of Testing for Energized Transformers†**

Interval	Visual Inspection and/or Tests
Daily (or every shift)	1. Transformer load currents and voltages. (Record findings Items number 1-5). 2. Ambient top oil, winding temperatures. 3. Oil-in and oil-out temperature (forced oil cooling). Water-in and water-out temperature (water cooling). 4. Head space gas pressure (psi); it should vary under changes in loading and ambient temperature. 5. Protective relay indicating targets. Record and reset. 6. Listen for unusual noises. 7. Observe auxiliary cooling pumps and fans. (Should they be running at these temperatures?) 8. Observe all switchgear indicating lights (all trip circuits functioning?) All heater ammeters (all heaters functioning?)
Weekly	1. Leaks (especially PCB contaminated oils). 2. Oil liquid level in tank and in oil-filled bushings (if so equipped). 3. Feel cooling tubes; note temperature change. 4. Cooler screens (water cooled or forced oil-cooled).
Monthly	1. Inspection of all gauges and bushings; and for leaks from tanks, fittings and cooling tubes. 2. General condition, automatic load tap changer; note and record number of operations. 3. Protection alarms (temperature and pressure); check for proper operation. Check pressure/vacuum gauge; compare readings with manufacturer's. 4. Dehydrating breather (open and conservator types); free from moisture restriction.
3 months (Quarterly)	1. Oil Testing (OT) - Average top oil temperature; >90°C T < 100°C.‡ 2. Pressure relief device - see if it has tripped; flag in vertical position. 3. External inspection of switchgear parts by opening doors, i.e., front of CBs, rear of meters, relays, etc. Check for cleanliness, moisture, rodents, overheating, or oil or water leakage.§ 4. Gas-in-oil analysis.‡

Table 10.1 continued on next page

Table 10.1 continued

Interval	Visual Inspection and/or Tests
6 months (Semi-Annually)	1. Oil Tests (OT) - Average top oil T: 80°C range.‡ 2. Bushings/insulators and lightning arrestors—visual check for cracks, cleanliness, contamination, chips, unscrewed hardware; leaning at bad angle. 3. Grounding system - inspect for loose, broken, or corroded connections. 4. Sudden pressure relay.§ 5. Gas-in-oil analysis.‡
12 months (Annually)	1. Oil Tests (OT) - Average top oil T: <80°C.‡ 2. Bushing/insulator powder blast cleaning in polluted environments, especially metallic oxides or chemicals; plus the extra step of coating of porcelain surfaces with silicone insulating compound, especially environments of salt cake and cement dust.*** 3. Infrared inspection - for "hot spots" - loose connections, broken porcelain, etc. 4. Inspect switchgear cable terminations for cleanliness and tracking. 5. Inspect transformer cables for deformation, closeness to grounded metal parts. 6. Ground resistance test - verify good power system ground (less than 5 ohms).§ 7. Protective relay trip check; test calibration.§ 8. Gas-in-oil analysis—all units (Table 4.26).‡
24 months (2 years)	1. Bushing/insulator powder blast cleaning in non-polluted environment, plus extra step of silicone coating. 2. Power factor test on oil-filled bushings.

*Critical units, those already suspected of internal problem, or equipment located where corrosion, humidity, vibration or dust is present, *double* the frequency of the procedure; for example, oil testing (OT), recommended annually should then be done semi-annually, etc.

†Energized tests should be performed well in advance of a scheduled shutdown. (Pinpoint possible problems).

‡All recommendations *assume* the use of auxiliary cooling (considering cost-benefit factor); otherwise, testing should be even more frequent (Chapter 7, Part 4). Refer also to Figure 6.4 (include all observations) and Table 6.3.

§Qualified personnel only - Use IEEE Guide for Field Testing Power Apparatus Insulation, IEEE Std. 62-1978.

***One time* test (PCB content in mineral oil) especially units near waterways, thunderstorm prone areas, or units nearing end of useful life.

***Primary frequency criterion - The effluence of airborne contamination (up to 230 KV) from insulated bucket trucks or 161 KV from towers.

TABLE 10.2

POWER TRANSFORMER AND ASSOCIATED EQUIPMENT
RECOMMENDED FREQUENCY OF MAINTENANCE FOR
SCHEDULED TRANSFORMER *SHUTDOWN*

Interval	Visual Inspection and/or Tests
Annual* (Optional)	1. Switchgear inspection, vacuum cleaning, lubrication of circuit breaker mechanisms. 2. Inspection thru top man hole cover - especially bushings, also for moisture, rust, sludge deposits. 3. Lightning and surge protective devices, visual inspection, cleaning and repair. 4. Auxiliary cooling fans, pumps (complete cover-haul). 5. Functional trip testing of air circuit breakers. 6. Power factor of oil and air circuit breakers and bushings. 7. Load tap changer - external inspection of accessible parts for seal leakage, corrosion, wear and looseness. 8. Minor repairs - tighten bolts, replace gaskets, repair weak weld joints, etc. 9. Basic electrical tests (ET) including power factor and megohm meter series.
3 years	1. Complete series of transformer electrical tests (ET). 2. Load tap changer - electrical tests (power factor and dc resistance); each step of changer. 3. Inspect pressure relief diaphragm for cracks, or holes; mechanical pressure relief device for proper operation. 4. Switchgear complete cleaning, inspection and lubrication of all associated devices, including CB, wiring, buses, disconnect devices and insulators; High current air CB tests - contact and insulation resistance. 5. All "annual" items not considered at one or two year intervals.
6 years (Minimum)	1. Insulated cables - dc high potential testing - fault detection. 2. Undercover transformer inspection - lowering of oil level, detailed inspection of all accessible internal mechanical and electrical parts, especially tap changer contacts, loose bracing and connections. Applicable to heavy duty equipment, such as arc furnace units. Inspection should be made under *clean* conditions.

*All nine items might be done on a biennial or triennial basis, depending on the particular plant situation. ET should be performed at least once every five years (number 9).

TABLE 10.3

TROUBLE SHOOTING CHART—OIL TESTS

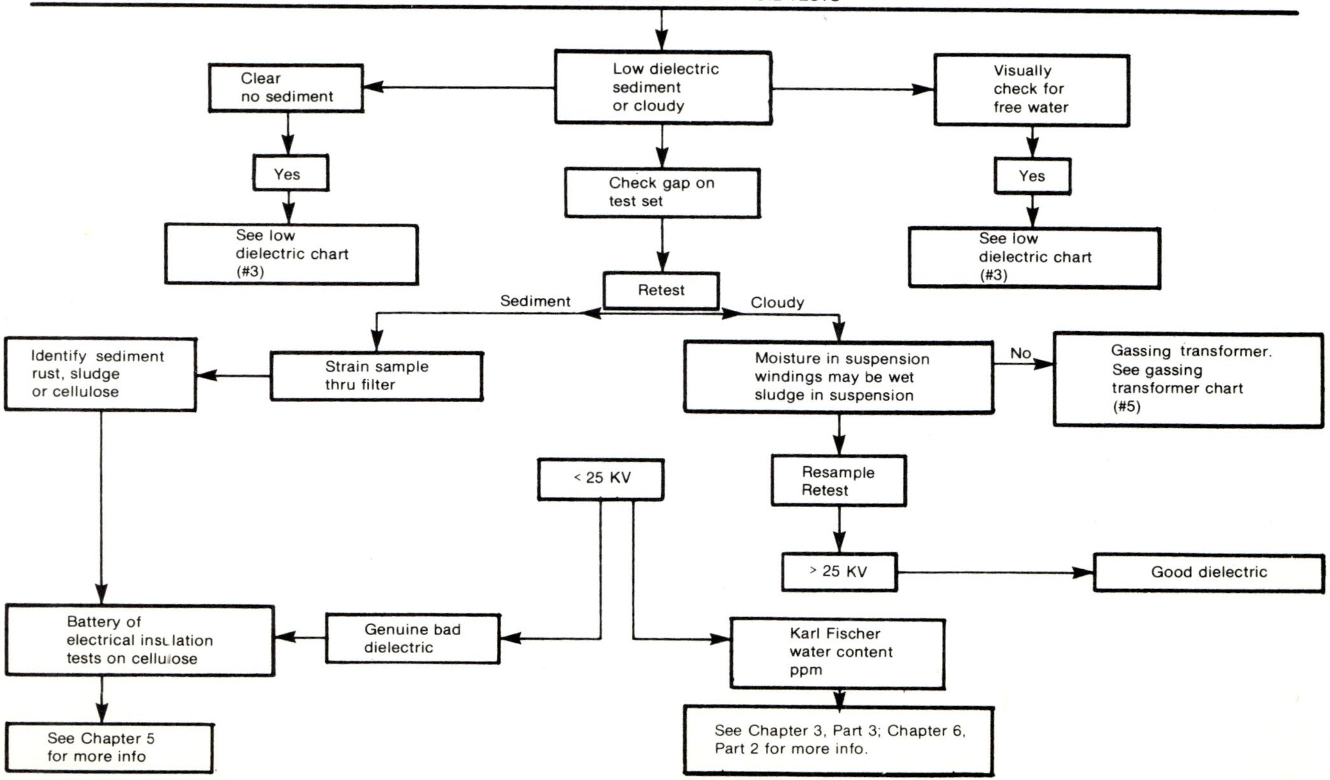

TABLE 10.4

TROUBLESHOOTING CHART—OIL CONDITION RAPID COLOR CHANGE 1½ PINTS OR MORE

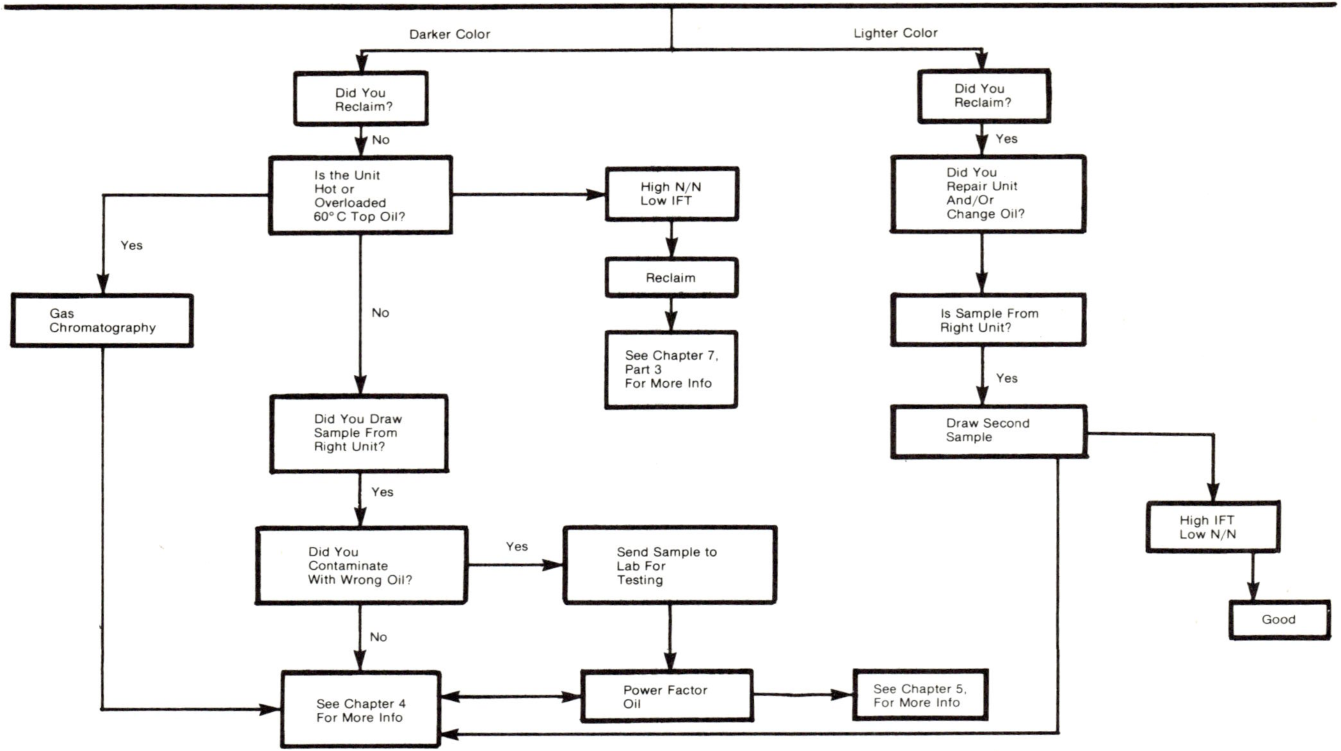

TABLE 10.5

TROUBLESHOOTING CHART—LOW DIELECTRIC

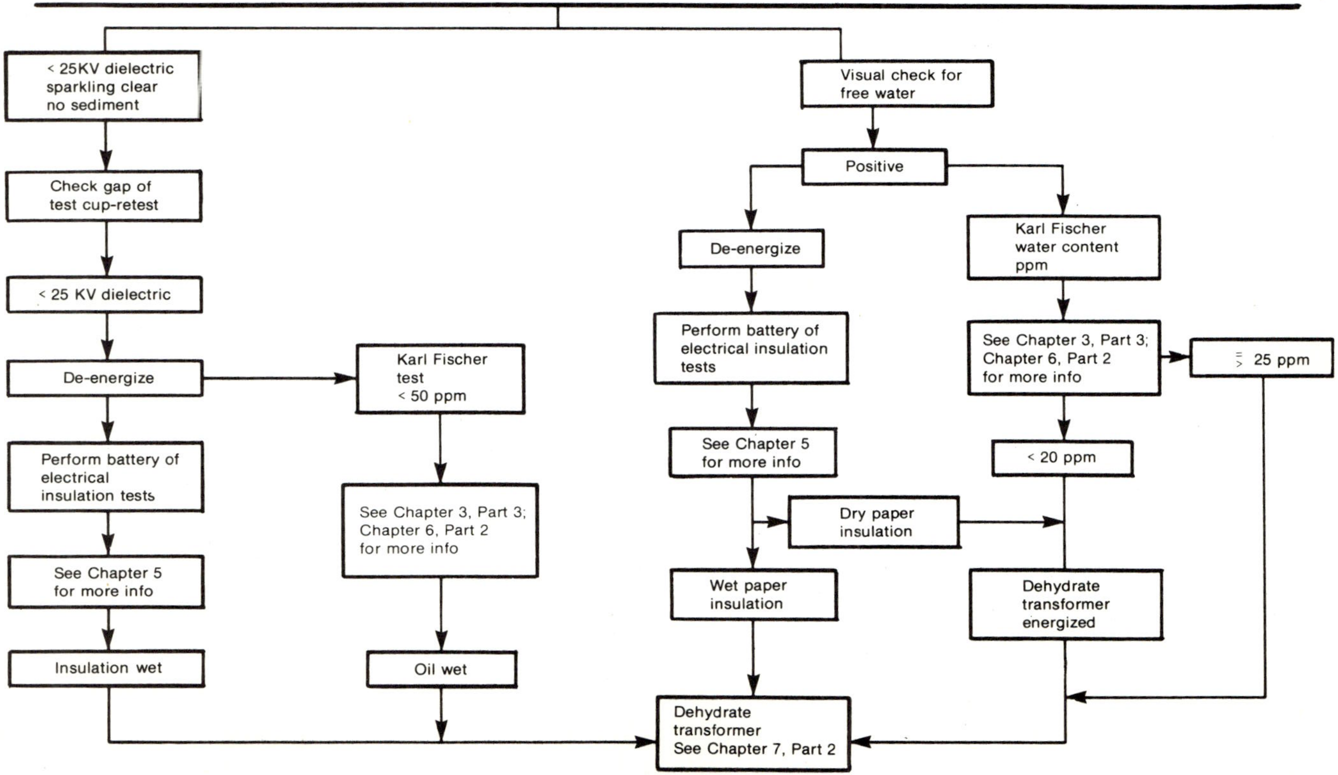

TABLE 10.6

TROUBLESHOOTING CHART—HIGH INTERFACIAL TENSION AND HIGH NEUTRALIZATION NUMBER

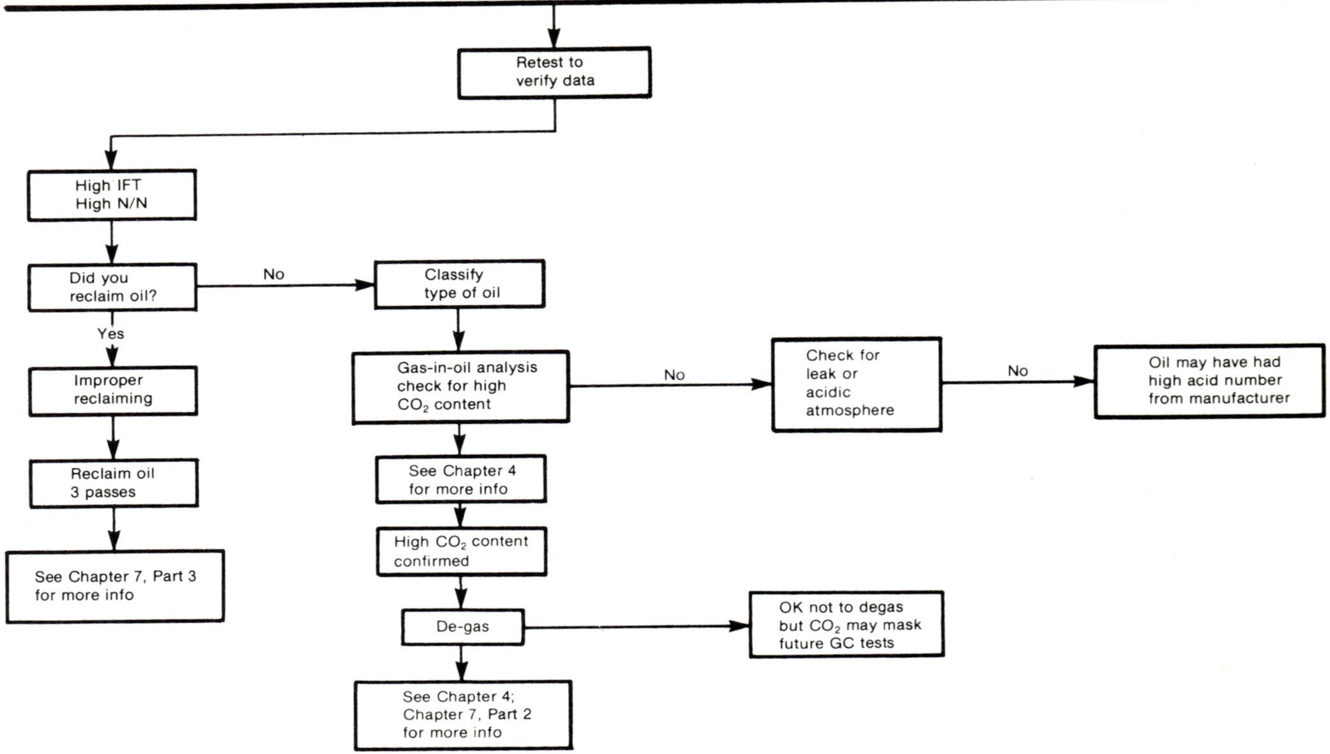

TABLE 10.7

TROUBLESHOOTING CHART—GASSING TRANSFORMERS

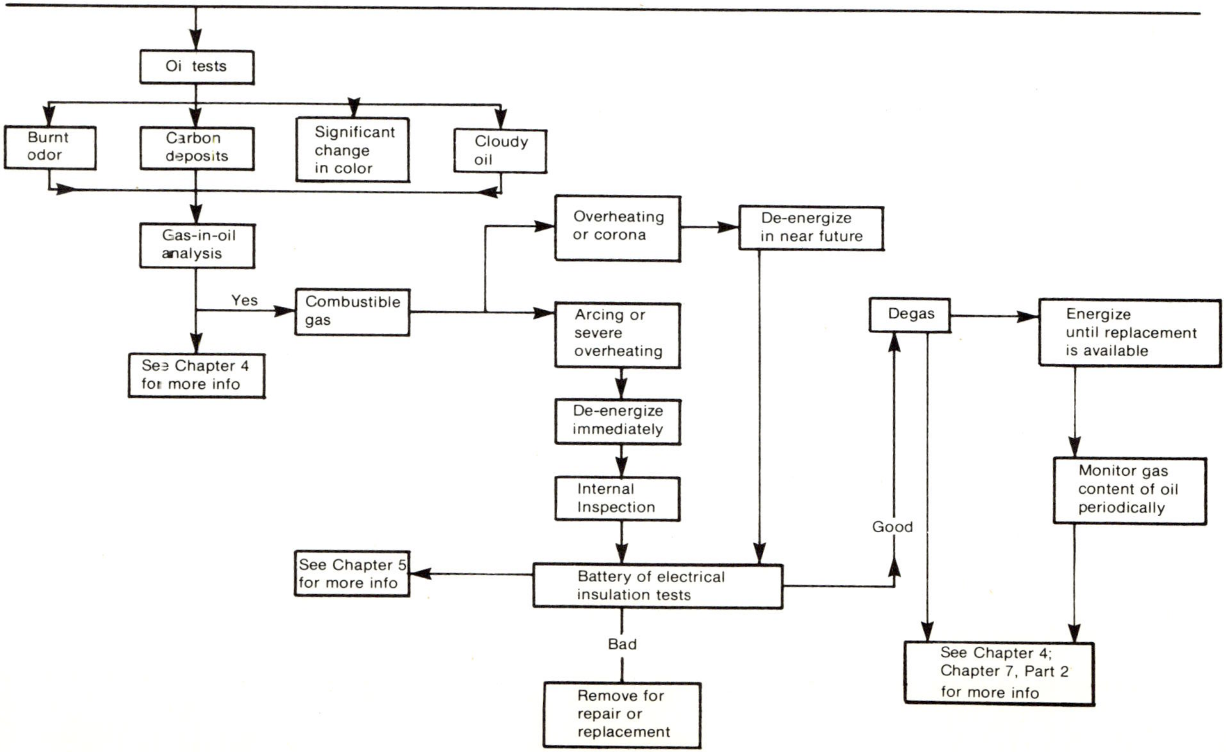

TABLE 10.8

TROUBLESHOOTING CHART—OVERLOADED AND/OR HOT TRANSFORMERS

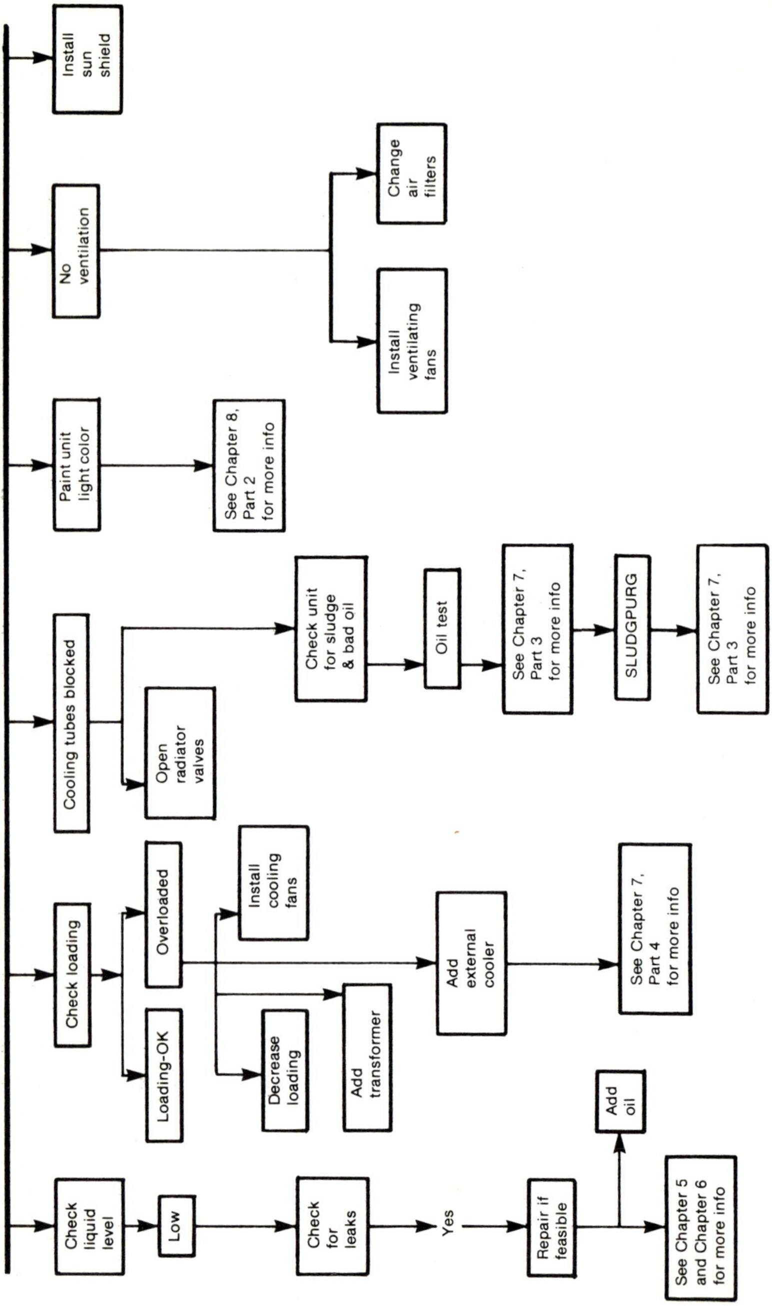

TABLE 10.9

TROUBLESHOOTING CHART—BUSHINGS AND INSULATORS

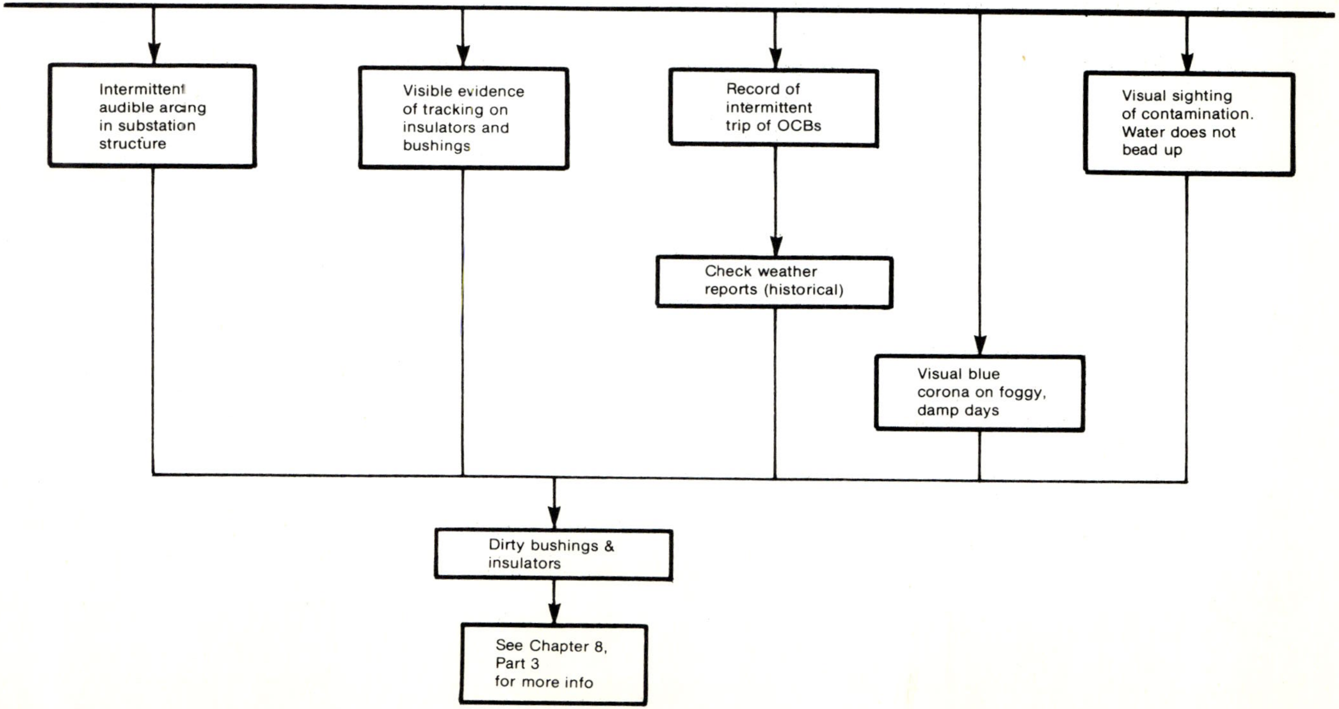

TABLE 10.10

TROUBLESHOOTING CHART—ASKAREL (PCB) UNITS

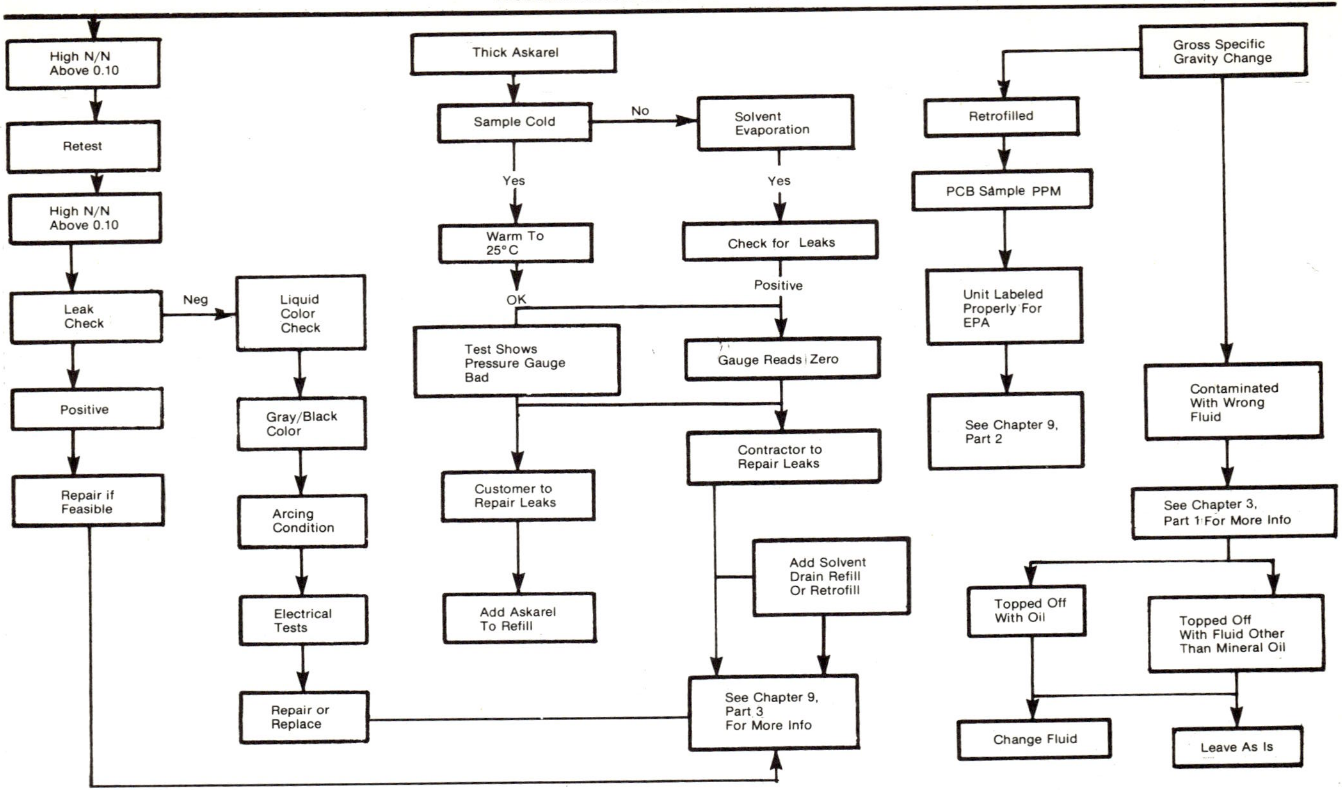

Part 3—Specific Cost Justification of a PM Program

One company has estimated a savings on replacement cost of $900,000 through the use of oil tests and Gas Chromatography. A total of 64 transformers were serviced — sludge removed, degasified, and dried — while energized for a cost of $100,000. Replacement of the 64 transformers would have been in excess of $1,000,000. No estimate was suggested for downtime costs had the transformers failed.

On the other hand, Figure 10.4 shows what happened to a medium power arc furnace transformer.* The owner chose to monitor the gas volume (dissolved combustible gas in oil) until a new unit arrived (against the advice of the maintenance contractor). The photo is self-explanatory! The obvious question concerns what total costs would have accrued if the transformer had failed **before** arrival of the new unit? A word to the wise should be sufficient!

Extending Transformer Life

Extensive experience indicates that proper transformer maintenance will preserve, protect and prolong the life of the transformer's insulating system. A modern transformer properly maintained, should operate up to 50 years. Fully loaded it will not last 50 years, unless maintained! Hence, transformers will not tolerate abuse or extended neglect. While the manufacturer may project a typical transformer's life expectancy of 20 years — a long time to him — this is never long enough to the user.†

Cost of Equipment Replacement

Further justification is apparent when one considers replacement costs. A 1,000,000 KVA transformer may cost $4-$5/KVA; a 50,000

**Figure 10.4 is found in the Color Section following Chapter 10.*

†Chapter 1, Part 3, p. 117.

KVA transformer, \$6-\$8/KVA. Distribution transformers may run \$8-\$10/KVA. A simple calculation reveals the pro-rated cost — and the advantage of extending the transformer's life.*

Field data tells us that most transformer failures can be attributed to the breakdown of the oil and paper insulating system. These so-called "weak links" can still provide almost indefinite unit life. Although no single test by itself can fully indicate the inner condition of a transformer, a complete Protective Maintenance program can detect with near certainty any unknown internal problem.

Protective Maintenance is essential to both older style transformers (before 1963) as well as the newer thermally upgraded units (since 1963).†

The Cost of Power System Maintenance

Assume that a medium sized transformer has an installed price of about \$12/KVA. Plan for a minimum per year PM expenditure of 1 percent of the installation cost. This allows 12¢/KVA/year to work with.

It is now apparent that if a good PM program can extend the life of a transformer from say, 25 years to only 50 years, the second 25 years is purchased for only half the price of a second new transformer — with no allowance for inflation of the purchase price!

Using a 1000 KVA transformer and allowing 12¢/KVA/year, what monies are available after 10 years on line when extensive maintenance work may be necessary?

12¢ x 1000 KVA x 10 years = \$1200.00

Using the same formula, a 10,000 KVA (10 MVA) unit under the same conditions would be worthy of \$12,000 of maintenance money. A reputable service company should be able to degas and dry the transformer and remove all sludge from the core, windings, and case while reclaiming the oil — for much less than this amount. This would permit a balance for other regular maintenance procedures.

Keep in mind also that a new unit is an* ***investment* *item and can only be written off for tax purposes over a number of years. Whereas a PM cost is an* ***expense*** *item and can be written off in the year incurred.*

†*Chapter 1, Part 3, p. 123.*

And even better than that, this formula will work in actual practice. Tables 10.11 and 10.12 develop this concept for transformers from 500-10,000 KVA and 1-40 years.

Downtime Costs

Downtime is the final component which must be factored into the total equation when a transformer must be repaired or replaced, several factors determine correction costs:

- Availability of a spare
- Repair feasibility and cost
- Repair time
- Replacement cost
- Replacement time factor
- Lost production costs (total)
- Increased liability claims
- High cost of purchased power

Again, this total downtime cost must be factored into any equation used to realistically evaluate the cost of PM.

The time required to replace a failed transformer when a spare is not available is a function of many variables:

- The time required to obtain repair bids and award the contract
- The time to transport the transformer to the factory or service shop
- The time for actual repair
- The time to transport the transformer back to the site
- The time needed to install it

Normally it takes about 12-15 months to repair and replace a 345 MVA/138 KV transformer if a spare is not readily available.* If a spare is available, however, replacement, transportation and installation usually take about 8-12 weeks.

For arc furnace units over 10 MVA, a spare transformer is recommended to avoid lengthy production downtime. Repairs can take several months, plus perhaps a month for transportation to and from one of the few available repair shops.

A considerably longer period of time would be required to build, transport, and install a* ***new* *transformer.*

In addition to the above there is the impact of an unannounced transformer failure on:

- Raw materials
- Lost production
- Insurance company relationships
- Fire and extended damages
- Increased cost for state unemployment benefits

Advances made during the past 15 years and as outlined in this book — predictive testing and data evaluation — have given to the industry the ability to eliminate all but maverick type transformer failures.

In our day of power crises and shortages, it behooves transformer owner/operators to recognize the benefits of sound maintenance practice — lower net cost, insurance rates and as a bonus — peace of mind.

TABLE 10.11

TRANSFORMER PREVENTIVE MAINTENANCE ECONOMIC JUSTIFICATION[1]						
(Based on Annual Maintenance Expenditure of 1% of installed Cost Per Year)						
KVA Rating	500	1000	2500	5000	7500	10,000
Installed Cost, $	6K	12K	30K	60K	90K	120K
Annual Budget for Maintenance (1% of installed costs), $	60	120	300	600	900	1200
ANNUAL COST OF TESTING PER TRANSFORMER[2]						
Oil	13 (20)[3]	13[4]	13	13	13	13
GC	55 (75)	55	55	55	55	55
Total	$68 (95)	$68	$68	$68	$68	$68
Bank Balance in Maintenance Dollars At the End of Given Year						
KVA Rating	500	1000	2500	5000	7500	10,000
Year			Bank balance, $			
1	(8)	$ 52	232	532	832	1132
5	—	260	1160	2660	4160	5660
10	—	520	2320	5320	8320	11320
15	—	780	3480	7980	12480	16980
20	—	1040	4640	10640	16640	22640
25	—	1300	5800	13300	20800	28300
30	—	1560	6960	15960	24960	33960
35	—	1820	8120	18620	29120	39620
40	—	$2080	$9320	$21280	$33280	$45280

[1]See examples below.

[2]Prices listed are subject to change.

[3]Cost per test based on non-contract price structure.

[4]Cost per test based on contract price structure (over 300 units tested).

1. Examples using Table 10.11.

For a 500 KVA unit:

Annual budgeted dollars for maintenance per unit	= $60
Annual cost of testing per unit	= $68
Bank balance in Maintenance dollars	$8

For a 1000 KVA unit:

Annual budgeted dollars for mainenance per unit	= $120
Annual cost of testing per unit	= $68
Bank balance in Maintenance dollars	$52

TABLE 10.12

REMEDIAL TYPE MAINTENANCE ECONOMIC JUSTIFICATION[1]

(5-Year Interval)

KVA Rating	Electrical Test, $	RE-3[3], $		Total	Annual Maintenance 1% Installed Cost, $	Budget $ (±)
500	325	240	=	565	—	—
1000	325	479	=	804	260[2]	(544)
2500	325	1198	=	1523	1160[2]	(325)
5000	325	2395	=	2720	2660[2]	(60)
7500	325	3593	=	3918	4160	242
10,000	325	4790	=	5115	5660	545

[1]Every 5 years:

- Run complete electrical test on all key transformers.
- Service each transformer using RE-3[3].
- Assume ½ gallon oil per KVA (1940 equipment).

[2]**Additional** cost beyond budgeted amount (Table 10.11-5 years) can be justified as follows. Advantages:

1st Eliminates failure due to neglect and other potential expenses; for example, solidified pipe if downtime in chemical plant requires replumbing.

2nd Monitors solid insulation (high percentage of failures are result of insulation failure).

3rd Transformer repairs replace rewinds.

4th Transformer life extended from 20 to 20+ years (up to 50 years).

5th Diminishes transformer failure probability.

6th Frees up capital money for other investments.

[3]RE-3 - Transformer oil reclaiming—3 passes (Chapter 7, Part 3, Figure 7.42).

Figure 1.11 - The Transformer - The Heart of Transmission and Distribution. Contrasting a large modern power transformer with the first practical transformer in 1885, designed by Blathy, Deri, and Zipernowsky (core type, approximately 1400 va). These three engineers had the "privilege" of employing the term transformer for the first time. (Courtesy of Trafo-Union, Stuttgart, West Germany).*

*Chapter 1, Part 1, p. 9.

Figure 1.19 - Transformer Design and Construction. 1020 MVA, core-form generator transformer. (Courtesy of Trafo-Union, Stuttgart, W. Germany).

Figure 1.21 - Transformer Design and Construction. Typical internal construction of a large power shell-form transformer. (Courtesy of McGraw-Edison Company).

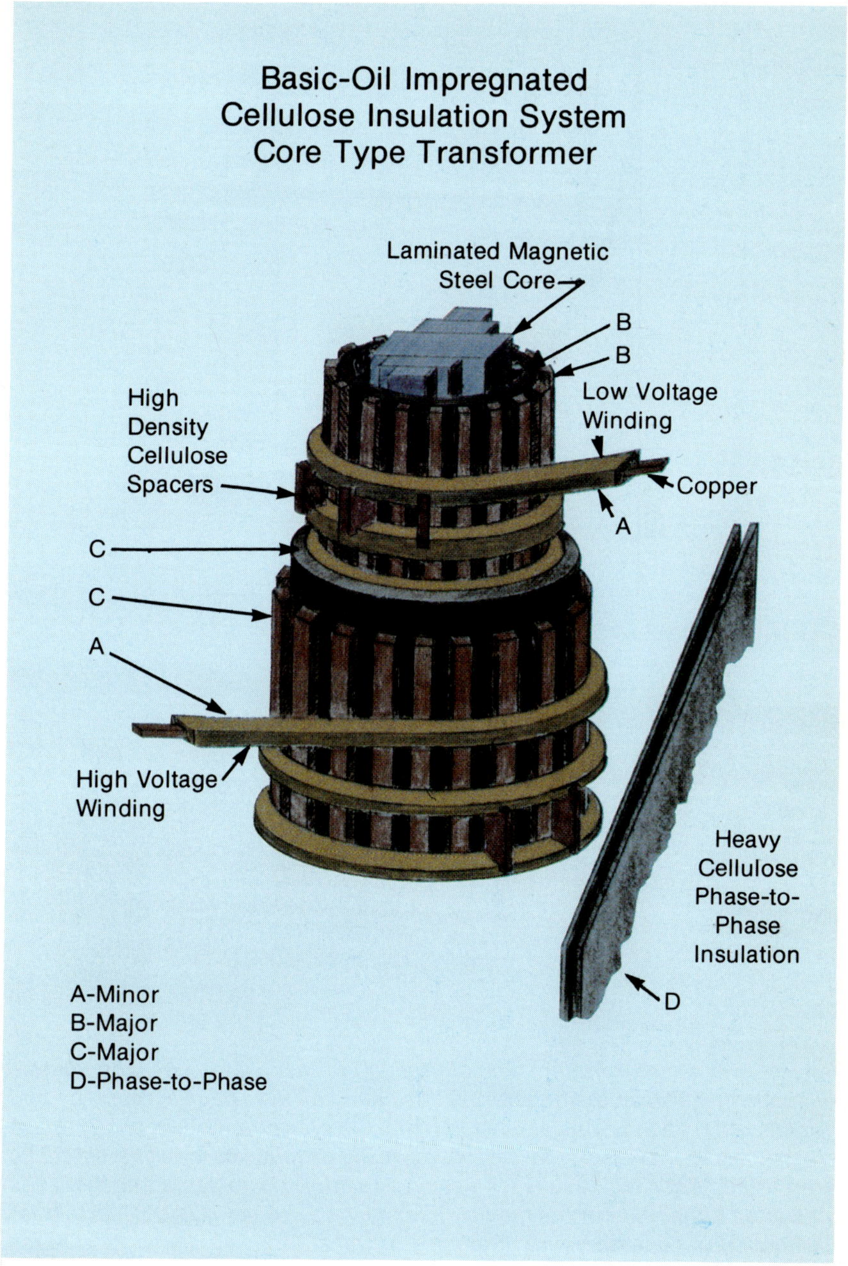

Figure 1.55 - The Lifeline of a Transformer. Basic oil-impregnated cellulose insulation system/core type. (For details, see Chapter 1, Part 2, Figure 1.55).

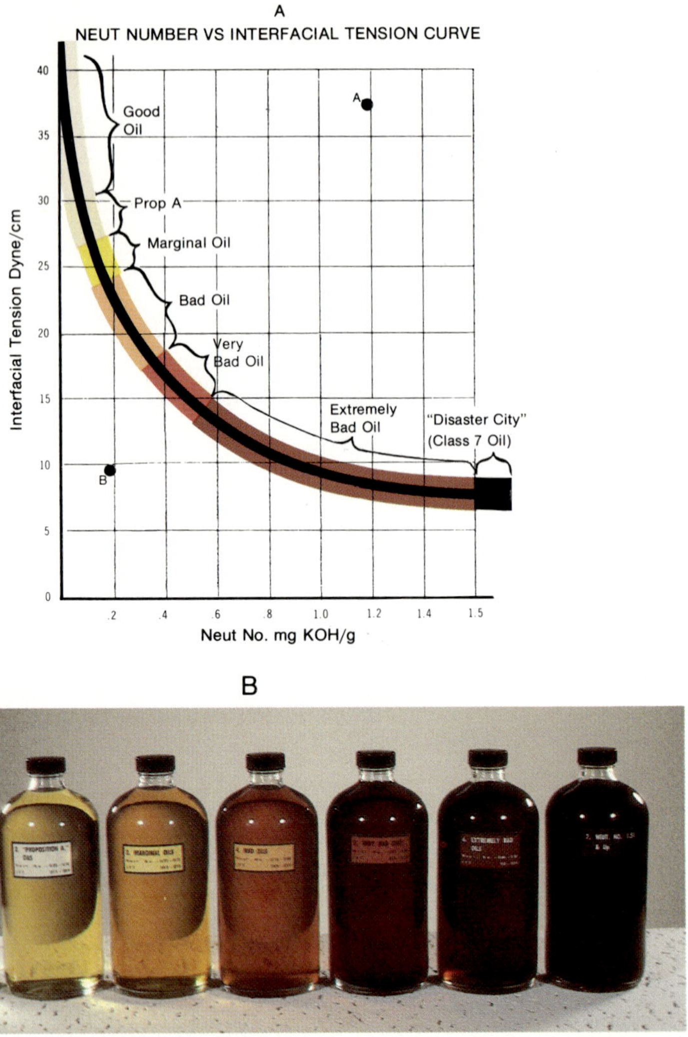

Figure 3.11 - Interpreting Transformer Oil Test Data. (A) The classic relationship between the transformer insulating oil tests neutralization or acid number (NN) and interfacial tension (IFT). Several independent studies have shown that an increase in NN should normally be followed by a characteristic drop in IFT. When test results for a given oil sample do not fall within the range shown on either side of the median line (as in the points A and B) further investigation is necessary. (B) Samples of transformer oil, illustrating seven color classifications.*

**Chapter 3, Part 2, pp. 284, 288.*

TABLE 6.10

BASIC PROTECTIVE MAINTENANCE PRACTICE FOR OIL-INSULATED TRANSFORMERS

Authoritative transformer oil color and classification chart

TRANSFORMER OIL COLOR GUIDE	TEST DATA NORMALLY RELATED WITH COLORS		EFFECT OF THE OIL ON TRANSFORMER

TRANSFORMER OIL
Color Chart

EFFECT ON TRANSFORMER

Color Classification	Acid (Neut.) No. mg KOH/g	Interfacial Tension Dynes/CM	NOTATION
GOOD	0.03 to 0.10	30-45	Providing these functions 1. Efficient Cooling 2. Preserving Insulation
PROP A	0.05 to 0.10	27-29	Polar Compounds (sludges) in solution (Products of oil oxidation) causes the drop in IFT.
MARGINAL	0.11 to 0.15	24-27	Fatty Acids coat the windings. Sludges in solution ready for initial fall-out. Sludges in insulation voids highly probable.
BAD	0.16 to 0.40	18-24	In almost 100% of the transformers in this range sludges are deposited on core and coils. Sludges are first deposited in fin areas.
VERY BAD	0.41 to 0.65	14-18	Deposited sludges continue to oxidize and harden. Insulation shrinkage is taking place. Premature failure a good possibility.
EXTREMELY BAD	0.66 to 1.50	9-14	Sludges insulate cooling fins, block vents causing higher operating temperatures.
CLASS 7 OILS	1.50 and higher	6-9	Vast quantities of sludges may require other means than SludgPurg® procedures.

*Figure 8.8 - Protective Apparatus Coatings. A faithful, rusting transformer prior to blast cleaning surface preparation and primer coat (**before** energized painting).*

*Figure 8.14 - Protective Apparatus Coatings. U.S. Navy specification marine formula alkyd-silicone enamel finish coat (**after** energized painting).*

Figure 8.20 - Energized Cleaning of Transformer Bushings and Standoff Insulators. As the conductive path became more pronounced, the final step of arcing-over or flashover took place on the porcelain surface of this pin type standoff insulator. This advanced stage of insulator aging resulted in higher voltage stresses, with porcelain fracture occurring and subsequent power outage. (Courtesy of Bonneville Power Administration).

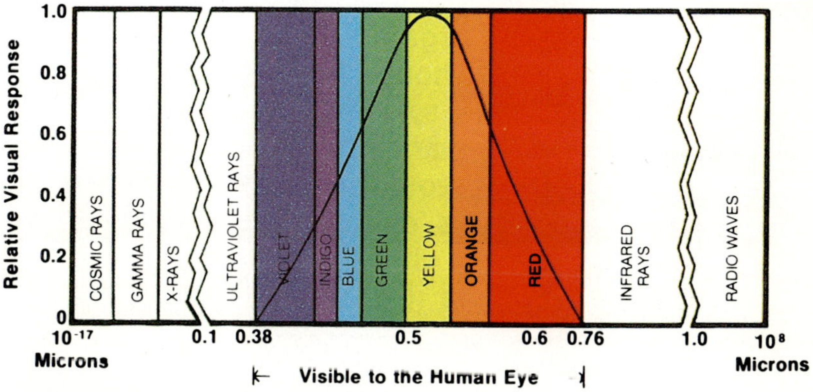

Figure 8.31 - Infrared (IR) Scanning of Electrical Systems. Sir William Herschel is credited with discovering this infrared electromagnetic spectrum about 1805. However, today's use of IR technology is based on the pioneering work of ***many*** *of the world's scientists. Development of the technology depends on specialists in the fields of physics, optics, electronics and mechanical design.*

Figure 10.4 - Implementing Your Own Protective Maintenance Program. Early detection by dissolved combustible gas-in-oil analysis (Chapter 4) could have prevented removal from service of this arc furnace transformer. The industrial facility was unable to shut down the unit due to the fact that they would have to stop all production. The client did monitor the gas volume and ***fortunately*** *a new transformer arrived* ***before*** *inevitable failure would have occurred. Upon subsequent visual examination it was found that the transformer had arced from its primary winding to ground. Upon further examination it was found that the insulating spacers were completely burned on the primary winding. (See Chapter 1, Part 2, Figure 1.55).*

Selected Chapter Bibliographies*

Acronym Identification

AIC - Annual International Conference of the Clients of Doble Engineering Co., Watertown, Mass.

AIEE - American Institute of Electrical Engineers (Superseded in 1962 by IEEE)

APPC - Annual Pulp and Paper Conference of IEEE/IAS (USA)

ASTM - American Society for Testing and Materials

CEA - Canadian Electrical Association of IEEE

EEI - Edison Electric Institute

EIC - Biennial Electrical/Electronic Insulation Conference (USA)

EPRI - Electric Power Research Institute, Palo Alto, Calif.

IAS - Industrial Applications Society of IEEE

IEC - International Electrotechnical Commission

IEE - Institute of Electrical Engineers (United Kingdom)

IEEE - Institute of Electrical and Electronics Engineers

IEEE Transactions on PA & S - IEEE Transactions on Power Apparatus & Systems (Power Engineering Society)

IEEE Transactions on Industry Applications - IEEE Industrial Applications Society (IAS)

IEEE Transactions on Electrical Insulation - IEEE Electrical Insulation Society (EIS)

NEMA - National (U.S.) Electrical Manufacturer's Association

PES - Power Engineering Society of IEEE

Only the **first reference to a particular item is normally listed in this bibliography. Many of the cited publications are also used in subsequent chapters. A complete bibliography by subject is available in the "TMI Evaluates" series.*

Chapter 1—Part 1

Brittain, J.E. **Turning Points in American Electrical History**, IEEE (1977).

Hamilton, J.W. **Men and Volts: The Story of General Electric**. J.B. Lippincott, 1941.

Hingorani, N. "The Reemergence of DC in Modern Power Systems." **EPRI Journal** (June 1978).

Holden, C. "Bern Dibner: Science Bibliophile." **Science** (May 30, 1980).

King, W.J. "Contributions from the Museum of History and Technology." **The Development of Electrical Technology in the 19th Century**, Bulletin 288 (1962).

Chapter 1—Part 2

Bailey, I.W. **Industrial Engineering Chemistry**, 1930.

Bean, R., et al. **Transformers for the Electric Power Industry**. McGraw-Hill, 1959.

Clark, F.M. **Insulating Materials for Design and Engineering Practice**. John Wiley, 1962.

Factory Mutual Insurers. Loss Prevention Data "Transformers" Sheet 5-4/14-8 (April 1978).

Feinberg, R. **Modern Power Transformer Practice**. John Wiley, 1979.

Gilman, J.J. "Metallic Glasses." **Science** (May 23, 1980).

Greenwood, A. "A Perspective on Insulation Systems in the Electric Power Industry." **IEEE Transactions on Electrical Insulation** (June 1980).

IEEE Standard Dictionary of Electrical and Electronic Terms (IEEE Std. 100-1977).

IEEE/PES Tutorial Course—Application of Distribution and Power Transformers. Text #76 CH 1159-3-PWR (1976).

Pansini, A.J. **Basic Electric Power Transformers**. Hayden Press, 1976.

Sillars, R.W. **Electrical Insulating Materials and Their Application**. Peter Peregrinus, Ltd., for the British IEE, 1973.

Skinner, C.F. "The Story of the Insulations." **The Electric Journal** (April 1920).

Smith, R.L. "The Meaning Behind Transformer Nameplates." **Power** Magazine (January 1975).

Stigant A., et al. **The J & P Transformer Book**, 10th ed. Newnes-Butterworths, 1973.

Yasafuku S., et al. "Tube Aging of Oil-Filled Transformer Electrical Insulation Systems and Aging of Oil-Filled Distribution Transformers." EIC (1979).

Chapter 1—Part 3

Factory Mutual Insurers. Loss Prevention Data—"Electrical Maintenance." Sheet 5-20/14-22 (May 1977).

Chapter 1—Part 3 continued

"Few Utilities Know Failure Rates; EEI Sees Need for Equipment Testing." **Electric Light and Power** (February 1978).

Maczka, W.J., "Playing the BIL game-or How to Juggle Savings and Security." **Electric Light and Power** (December 1978).

Ravindranath, B., et al. **Power System Protection and Switchgear**, Halsted Press, 1978.

Sahu, R. "Using Transformer Failure Data to Set Spare Equipment Inventories." **Electric Light and Power** (October 1980).

Sorrell, E.V. (Hartford Steam Boiler Inspection and Insurance Co.), "Transformers—What Price Reliability?" **The Locomotive**, Vol. 60, No. 4 (1976).

Chapter 2—Part 1

Doble, F.C. "The Reclamation of Insulating Oils." **American Society for Testing and Materials (ASTM) Special Technical Publication 152** (1952).

Laskowski, L., et al. "765 KV Gas-Insulated Substation is World's First." **Transmission & Distribution** (July 1980).

Raab, E.L., et al. "U.S. Transformer Oil Supply and Demand 1975-1985." **EPRI Report** No. EL-303 (November 1976).

Rouse, T.O., et al. "Evaluation of Alternative Insulating Oils for Use on Transformers and Other Electrical Apparatus." EPRI Research Project, Palo Alto, Calif. (February 1980).

Wilson, A.C.M. **Insulating Liquids: Their Uses, Manufacture and Properties**. Peter Peregrinus Ltd., for the British IEE, 1980.

Chapter 2—Part 2

Baker, A.E. "Doble Transformer Oil Purchase Specifications" (TOPS), Doble AIC (1979).

Doble, F.C. "A New Approach to the Insulating Oil Problem." Doble AIC (1954).

Donohue, E.P., et al. "A Comparative Analysis of Paraffin- and Naphthene-Based Oils in a 34.5 KV Circuit Breaker." **IEEE Transactions on PA & S** (May/-June 1980).

Flanagan, P.E., et al. "Utilization of Insulating Oils by North American Utilities—A Survey in 1977." EPRI Research Project, **Milestone Report No. 10** (1977).

Frey, R.M., et al. "Final Comparison of ASTM D-2440 and Minicapacitor Aging of Naphthenic and Paraffinic Transformer Oils." **IEEE/PES Summer Meeting** (1979). Paper No. A-79-405-2.

IEEE Guide for Acceptance and Maintenance of Insulating Oil on Equipment. IEEE Std. 64-1977 (1977).

Chapter 2—Part 2 continued

Lindenbaum, J.J. "An Oil is...What?" **Plant Engineering®** (October 26, 1975).

Lipshtein, R.A., et al. **Transformer Oil**. Jerusalem; Israel Program for Scientific Translations, 1970.

Pugh, D.R. "Electrical Insulating Oil." Doble AIC (1974).

St. Jean, G., et al. "Comparative Performance of Paraffinic and Naphthenic Insulating Oils Under Arc Quenching Conditions" **IEEE Transactions on PA & S** (November/December 1980).

Chapter 2—Part 3

Downey, A.M. "A Review of Testing Methods as Applied to Insulating Oils," Doble Engineering Co., (1973).

Salomon, T. "The Performance Characteristics of Used Insulating Oils." **Symposium on Insulating Oils, ASTM Special Technical Publication No. 172** (1954).

Chapter 2—Part 4

Benko, I.S., et al. "Loading of Substation Electrical Equipment With Emphasis on Thermal Capability." **IEEE Transactions on PA & S** (July/August 1979).

Brancato, E.L. "Insulation Aging: A Historical and Critical Review." **IEEE Transactions on Electrical Insulation** (August 1978).

Cassant, R.J., et al. "Designing Transformers to Withstand Short Circuits." Doble AIC (1968).

Heinrichs, F.W. "The Effect of Gassing During Overloads on the Impulse Strength of Transformer Insulation." **IEEE/PES Summer Meeting** (1979). Paper #A-79-434-2.

Lahoti, B.D. "Evaluation of Transformer Loading Above Nameplate Rating," IEEE/PES Summer Meeting (1980); Paper #80SM 708-8.

McNutt, W.J. et al. "The Combined Effects of Thermal Aging and Short Circuit Stresses in Transformer Life." **IEEE Transactions on PA & S** (July/August 1976).

McNutt, W. et al. "Short-Time Failure Mode Considerations Associated with Power Transformer Overloading." **IEEE Transactions on PA & S** (May/June 1980).

Chapter 2—Part 5

Bellaschi, P.L. "Residual Water in EHV Transformer Insulation." Doble AIC (1967).

Flanagan, P.E., et al. "The Determination of Moisture on Cellulosic Electrical Insulation." **Insulation/Circuits** (February 1972).

Flowers, A.E. "Some Effects of Moisture in Insulation Oils," Doble AIC (1939).

Chapter 2—Part 5 continued

Lampe, W., et al. "Continuous Purification and Supervision of Transformer Insulation Systems in Service." **IEEE/PES Winter Meeting** (1978). Conference Paper A78 111-7.

Moser, H.P. **Transformerboard**, St. Johnsbury, Vt. EHV-Weidmann, 1979.

Piper, J.D. "Moisture Equilibrium Between Gas Space and Fibrous Materials in Enclosed Electrical Equipment **AIEE Transactions**, Vol. 6 (1946).

Smith, L.W. "Water, Electrical Insulation's Enemy Number One." Doble AIC (1950).

Stannett, A.W. "The Problems of Water in Power Transformers." **Electrical Times** (March 25, 1965).

Summer, W.A., et al. "Life Expectancy of Oil-Immersed Insulation Structures." **AIEE Transactions** (October 1953).

Teague, W.L., et al. "Dielectric Measurements on New Power Transformer Insulation." **AIEE Transactions** (October 1952).

Chapter 3—Parts 1 and 2

Barresi, J. "An Old Approach to a New Problem." Doble AIC (1974).

Gedemer, T.J., et al. "An Improved Infrared Method for Determining Acidity in Transformer Oils." **IEEE/PES Winter Meeting** (1978). Paper A78-561-3.

Gerell, G.W. "Interfacial Tension Testing of Transformer Oil." Doble AIC (1943).

McConnell, T.A. "Evaluation of Laboratory Tests as Indicators of the Service Life of Uninhibited Electrical Insulating Oils." **ASTM Special Technical Publication #218** (1957).

Smith, L.W. "Interfacial Tension of Oil Against Water as a Diagnostic Tool." Doble AIC (1974).

Walsh, E.F. "Progress Report on Quality Rating of Insulating Oils." Doble AIC (1947).

Chapter 3—Part 3

ASTM Bulletin, "A More Sensitive Method for Dielectric Strength of Insulating Oil," (December 1960).

Burns, C.P. "New Approaches to Testing Insulating Oils." **Electrical Times** (1976).

Cornelius, H.A. "Developing a Program to Keep Electrical Equipment Insulation High in Quality." Doble AIC (1948).

Franklin, E.B. "Distribution of Water in Transformer Insulation." **Electrical Times** (May 27, 1965).

Guidi, W.W. "Mathematical Methods for Prediction of Moisture Take-Up and Removal in Large Power Transformers." **IEEE/PES Winter Meeting** (1974). Paper No. C-74-242-4.

Chapter 3—Part 3 continued

Ierusalimov, M.E., et al. "Methods for Evaluating the Moisture Content of Large Power Transformers." **Elektrotekhnika**, Vol. 49, No. 1 (1978).

Ishii, T., et al. "Effect of Moisture on AC Breakdown Voltage of Chlorinated Insulating Oil at Low Temperature." **Electrical Engineering in Japan**, Vol. 91, No. 3 (May/June 1971).

McCrae, G.G. "Transformer Insulation Drying Methods..." Doble AIC (1975).

Morin, F. "Vacuum Drying of Oil Transformers." Doble AIC (1967).

Povazan, E. "Modern Techniques in Processing Transformer Oils." **Transactions of CEA** (March 1975).

Wentz, E.C. "Transformer Insulation Power Factor." **Westinghouse Engineer**, Vol. 14, No. 6 (November 1954).

Chapter 4

Hubacher, E.J. "Analysis of Dissolved Gas in Transformer Oil to Evaluate Equipment Condition." Pacific Coast Electric Association (March 18-19, 1976).

Jakob, F. "An Analysis of Dissolved Gas in Transformer Oil to Evaluate Equipment Condition." **Proceedings of Annual Northeast Public Power Association**. Engineering and Operations Conference (April 24-27, 1979).

Luke, L.E. "Gas Detection: A Key to Transformer Health." **Transmission & Distribution** (January 1980).

McNair, H.M., and E.J. Bonnelli. **Basic Gas Chromatography**, Consolidated Printers for Varian Instruments, 1969.

Rogers, R.R. "IEEE and IEC Codes to Interpret Incipient Faults on Transformers Using Gas-In-Oil Analysis." **IEEE Transactions of Electric Insulation** (October 1978).

S.D. Myers, Inc. **TMI Evaluates: Gas-In-Oil Analysis vs. All Other Methods** (1980). A complete list of footnote references is given.

Chapter 5

Doble Engineering Co. Proposed "Transformer Test and Maintenance Guide" (July 1980).

James G. Biddle Company. "A Stitch in Time..." **Manual on Electrical Insulation Testing** (1972).

Chapter 6—Parts 1 and 2

Kemper Insurance Company. "Costly Transformer Failures Often Can be Traced to Neglect." (1976).

Marinello, R.L. "Predictive Maintenance." **Plant Engineering®** (February 2, 1978).

National Fire Protection Association. **Electrical Equipment Maintenance** (1977).

Chapter 6—Parts 1 and 2 continued

"Transformer Oil Pumps." **EPRI Journal** (October 1980).

Wolff, R.F. "Utilities Seek Improved Testing Methods." **Electrical World** (July 1, 1980).

Chapter 7—Parts 1 and 2

Baranowski, L.B. "Latest Trends in Transformer Drying and Filling Practices." Doble AIC (1971).

Casey, R.F. "Transformer Drying." Doble AIC (1977).

McCoy, J.R. "Oil Conditioning with CFC Honan-Crane Purivac." Doble AIC (1960).

Tudor, R.I. "Detection of Transformer Fault Using Exciting Current Test." Doble AIC (1969).

Chapter 7—Part 3

Eggleston, W.W. "Reclaiming Oil in Energized Transformers at Remote Locations." Doble AIC (1969).

"Guide for the Reclamation of Insulating Oil and Criteria for Its Use." **IEEE Transformer Committee Project No. 637** (Draft, April, 1980).

S.D. Myers, Inc. **TMI Evaluates: Transformer Oil Treatment vs. Transformer Desludging** (1980). A complete list of footnote references is given.

Chapter 7—Part 4*

Bellaschi, P.L. "Large Portable Spare Transformers for Present-Day HV/EHV Systems." Doble AIC (1977).

Brubaker, J.F. "Fault Protection and Indication on Substation Transformers." **IEEE Transactions on Industry Applications** (May/June 1977).

Kovacs, J.P. "Economic Considerations of Power Transformer Selection and Operation." **IEEE Transactions on Industry Applications** (September/October 1980).

Chapter 8—Part 1

Orth, H.S. "Your Plant Can Survive an Abnormal Electrical Condition If..." Paper presented at Open Pit Mining Association Meeting, June 19, 1969.

Fink, D.G., and H.W. Beaty. **Standard Handbook for Electrical Engineers**. 11th ed. McGraw-Hill, 1978.

Freund, A., et al. **Overcurrent Protection, (Electrical Construction and Maintenance**, McGraw-Hill), 1980.

Lythall, R.T. **J & P Switchgear Book**, 9th ed. John Wiley, 1972.

See also Chapter 2, Part 4.

Chapter 8—Part 2

General Electric Co. "Repainting Transformers." #E-236 (1971).

Lee, R.H. "The Effect of Color on Temperature of Electrical Enclosures Subject to Solar Radiation." **IEEE/PES Summer Meeting** (1973). Paper C73-388-6.

Munger, C.G. "Repairing Protective Coatings...Evaluating Coating Condition." **Plant Engineering**® (November 11, 1976).

Tator, K.B. "How Coatings Protect and Why They Fail." **Corrosion,** National Association of Corrosion Engineers (November 1978).

Trimber, K.A. "Systematic Approach to Coating Systems Selection." **Plant Engineering**® (April 19, 1979).

Chapter 8—Part 3

Holte, K.C., et al. "Application of Insulators in a Contaminated Environment." **IEEE Transactions on PA & S** (September/October 1979).

Lambeth, P.J. "Effects of Pollution on High Voltage Outdoor Insulators." **Proc. IEE, IEE Review** (September 1971).

Nasser, E., et al. "A Survey of the Problem of Insulator Contamination in the United States and Canada - Part 1." **IEEE/PES Winter Meeting** (1971). Paper 71 TP 133-PWR.

Orbeck, T., et al. "Development of a Long-Life Arc-Resistant Silicone Insulator Compound." Doble AIC (1975).

Chapter 8—Part 4

Brice, C.W. "Infrared Detection of Hot Spots in Energized Electrical Equipment." **IEEE Transactions on Industry Applications** (May/June 1979).

Heffner, C.W., "An Economical Approach to Infrared Inspection." **Electrical Construction and Maintenance** (September 1979).

Jorgensen, A. "Experiences, Equipment, Contract Services, Costs: A State-of-the-Art Report on Infrared Inspection as an Electrical Preventive Maintenance Technique." **Plant Engineering**® (February 8, 1979).

Chapter 9

Ames, B. "Identifying Environmental Chemicals Causing Mutations and Cancer." **Science** (January 12, 1979).

Corbett, T.H. **Cancer and Chemicals**. Nelson-Hall, 1977.

Doble Client Committee on Insulation Fluids. "Review of Status of PCB Substitutes." Doble AIC (1979).

EPA (U.S.) "Polychlorinated Biphenyls: An Alert for Food and Feed Facilities." (December 1979). p. 3. Compare with **Federal Register**, Vol. 42, No. 22 (February 2, 1977), p. 6537.

Federal Register, Vol. 43, No. 243 (December 18, 1978), pp. 58962-58963.

Federal Register, Vol. 44, No. 169 (August 29, 1979), pp. 50777-50778.

Federal Register, "Regulatory Dilemma," Vol. 45, No. 15 (January 22, 1980), p. 5008.

Federal Register (May 22, 1980), 49CFR p. 34560 ff. (Parts 171-173).

Federal Register, (February 26, 1980), pp. 12722-12754; (May 19, 1980), p. 33066 and p. 33086; and (January 12, 1981), p. 2846.

EPRI-Disposal of PCB and PCB-Contaminated Materials. Palo Alto, CA: Electric Power Research Institute, 1979. FP-1207, Vol. 1, 2, 3 (1979); Vol. 4 (1980).

Hutzinger, O., et al. **The Chemistry of PCBs**. CRC Press, 1974.

McGraw, M.G., "PCB-Control Plan Moves Into High Gear," **Electrical World** (September 18, 1980), p. 125.

"More PCB Destruction Methods Developed." **Chemical and Engineering News** (September 27, 1980).

Palko, E. "Selecting Secondary Substation Transformers." **Plant Engineering®** (June 14, 1979).

Palko, E. "The Options Available to Owners of Transformers." **Plant Engineering®** (November 26, 1980).

"PCB Inspections: What to Expect" **Electric Light and Power** (November 1980).

Polychlorinated Biphenyls, National Academy of Science, 1979.

S.D. Myers, Inc. **TMI Evaluates: What to do About Askarel (PCB) Transformers** (1980). A complete list of footnote references is given.

Sheppard, H.R. "Alternatives to PCB Transformers." APPC-IAS (1979).

Chapter 10

"Detecting Incipient Transformer Faults." **Transmission & Distribution** (November 1980).

Margolis, H.B. "Evaluating Power System Losses." **Transmission & Distribution** (June 1980).

Munasinghe, M. "A New Approach to Power System Planning." **IEEE Transactions on PA & S** (May/June 1980).

The Plant Engineering® Library - Maintenance Volume, Barrington, IL; Technical Publishing Company (1977 thorugh 1980).

"Symposium on Power System Reliability," Doble AIC (1979) Section 2.

Trytko, E., et al. "Transformer Oil Maintenance Prevents $1 Million Expenditure; Avoids Electrical Disruption and Production Loss." **Plant Services** (May 1980).

INDEX

INDEX

INDEX